PARAPSYCHOLOGY:

New Sources of Information, 1973-1989

by
RHEA A. WHITE

The Scarecrow Press, Inc.
Metuchen, N.J., & London
1990

British Library Cataloguing-in-Publication data available

Library of Congress Cataloging-in-Publication Data

White, Rhea A.
    Parapsychology : new sources of information, 1973-1989 / Rhea
A. White.
        p.   cm.
    Includes indexes.
    ISBN 0-8108-2385-3 (alk. paper)
    1. Parapsychology--Bibliography. I. Title.
Z6878.P8W47 1990
[BF1031]
016.1338--dc20   90-21327

to the

EAST MEADOW PUBLIC LIBRARY

without whose resources ——
staff, collection, and
equipment —— this work
would not have been undertaken

TABLE OF CONTENTS

Acknowledgments

I am indebted to many for providing me with the information needed to complete this book. First, I want to express my appreciation to the Reference Department of the East Meadow Public Library and to various members of the library's staff: to Rita Dworkin, Head of Adult Services, for building a strong parapsychology collection; to Irving Adelman, Head of the Reference Department, for the magnificent reference collection he has established, and for encouraging me to take full advantage of the library's services; and to Maureen Dwyer (and Caroline Rocek, before her), for interloaning many books and articles with dispatch and a smile. I also want to express my gratitude for the assistance of James G. Matlock, Librarian and Archivist of the American Society for Psychical Research Library; Wayne Norman, Librarian of the Eileen J. Garrett Library of the Parapsychology Foundation; K. Ramakrishna Rao, who has sent me important documents from the Foundation for Research on the Nature of Man and has always given me warm encouragement; and Alexander Imich, who made substantial improvements in the glossary.

I am very much indebted to the Staff of the Parapsychology Sources of Information Center: to Linda Henkel, for checking and editing an earlier draft; to Mark Feggeler especially, for checking every book in the bibliography to insure that the information cited was correct, for his editing of the manuscript and for compiling much of the author and title indexes; and to Maria Cassano, for locating book review citations and checking them and for making substantial editorial changes in the computer. Stephanie Becker initially entered the bulk of the material into the computer, and the manuscript has also benefited from her agile mind, which is just as swift as her touch. The camera—ready copy would not have become a reality without the masterful Tom Feggeler, who was responsible for the format, the fonts, and for smoothing out any and all snags and rough spots, and for entering and editing the book review citations. Frances Bonamo and Barbara Heimanson assisted in compiling the indexes. Harriet Edwards fine—tuned both the prose and the copy, in her usual astute and good—humored manner.

Finally, special thanks to Eileen J. Coly and the Parapsychology Foundation, Trammell Crow, the Fetzer Institute, and an anonymous donor, whose financial assistance helped to make it possible for all of us to work together in preparing the manuscript for publication.

Foreword

K. Ramakrishna Rao
Director, Institute for Parapsychology

Few in the field of parapsychology, past or present, can equal Rhea A. White in her versatile contributions to the development of the field. Her achievements as author, bibliographer, editor, experimenter, and leader are significant and lasting. She was elected President of the Parapsychological Association (1984) and thus received the highest professional recognition. She is the founder and the moving spirit behind the Parapsychology Sources of Information Center, and of its journal, *Parapsychology Abstracts International*, and its database, *PsiLine*, and she is also the editor of the *Journal of the American Society for Psychical Research*. Along with Margaret Anderson she conducted highly significant experimental research relating teacher—pupil attitudes to ESP scores. Her 1964 report reviewing a variety of techniques used for enhancing psi receptivity was, as John Palmer pointed out, a major inspiration for the subsequent work on ESP and altered states of consciousness. While Rhea White's contributions in all the above endeavors are quite significant, I believe she will be remembered most of all as an outstanding bibliographer of parapsychological information.

Careful and cautious by nature, meticulous and methodical by temperament and training, Rhea White is obsessively thorough and entirely dependable in whatever she does. The present volume is a remarkable testimony to her discerning scholarship and sound judgment. The book is a much needed supplement to an earlier volume she published along with Laura Dale.

Parapsychology is a difficult field for anyone seeking credible information. Like the phenomena it seeks to investigate, its publications are also notorious for their unreliability. Its literary bag is a wild mix of the credible and the outlandish, the discriminating and the disconcerting, the religiously believing and the irrationally skeptical. In fact, the sensational books seeking to exploit the popular interest in the subject outnumber the sober and the scientific. The proliferation of books providing substandard and sometimes misleading information

viii

is a serious impediment to the healthy growth of the field. Therefore, in a field like parapsychology, the need for dependable source books that guide one to materials that satisfy professional standards of reliability and scholarship cannot be overemphasized. Rhea White has done a valuable service to the field by bringing out the annotated bibliographies of books she considers to be among the "best," a worthy continuation of her earlier effort.

The annotated book list is arranged alphabetically under twenty—six heads beginning with altered states of consciousness and ending with unorthodox healing. There are separate sections for criticisms, education, and history as well as for the various kinds of psychical phenomena from mediumship to survival and reincarnation. There is also a section for books for young people. Thus the bibliography captures the various facets of psychic phenomena.

This book is more than an annotated bibliography of important books published during the last fifteen years; it is also a guide to organizations and periodicals in the field. It contains a chapter on parapsychology and U.S. government interest. Not the least valuable is a chronological list of theses on parapsychology. A more personal touch is evident in the section on "New Views of Parapsychology and Parapsychological Phenomena," which reflects the author's 35—year association with parapsychology, and "Information on Parapsychology in General Sources of Information," which reflects her experience as a reference librarian. Other useful materials in the book include addresses of publishers not readily available and lists of books containing glossaries and illustrations.

Parapsychology: New Sources of Information is a necessary and valuable addition to anyone's reference collection. It is indispensable to all libraries. I unreservedly recommend it as a must for anyone seeking credible parapsychological information. Rhea White has rendered a significant and laudable service to parasychology in bringing out this book and she deserves our appreciation and thanks.

Institute for Parapsychology
Foundation for Research on the Nature of Man
Box 6847, College Station
Durham, North Carolina 27708

Introduction

Parapsychology (the modern and more restrictive term for psychical research) is the field which uses the scientific method to investigate phenomena for which there appear to be no normal (that is, sensory) explanations. Basically this refers to phenomena subsumed under the general term psi,* which in its motor aspect is called psychokinesis and in its more familiar mental aspect, extrasensory perception (comprising telepathy, clairvoyance, and precognition). All these phenomena have been observed under laboratory conditions. In the vastly more complex and intricate world of real life, some form of psi often seems to be a probable explanation of such human experiences as dreams that come true, waking visions of events occurring at a distance, inexplicable hunches, and similar occurrences. [See the articles by Ian Stevenson on the importance of studying such experiences.**] *Psi is also a useful concept in explaining much that happens in mediumship. Since parapsychologists have established that psi is a part of living behavior,* many have hypothesized that what in the early years of psychical research was thought to be evidence of communication with the dead can better be explained in terms of the combination of some form of psi with the dramatizing propensities of the unconscious minds of the medium and the other persons involved. [For a discussion of the status of the evidence for mediumistic and other survival-related topics, see the article by Emily W. Cook.]

It is these building blocks of telepathy, clairvoyance, precognition, and psychokinesis that parapsychology uses to extend the bridge of knowledge into the unknown. But contrary to uninformed popular opinion, parapsychology does not deal with astrology, numerology, Tarot cards, theosophy, UFOs, witchcraft, or other occult systems or practices——or, if so, only insofar as they empirically demonstrate that at their base some form of psi is operating.

---

*For definitions of this and other terms, see Chapter VIII, "Glossary of Terms."

**The publications cited here are listed at the end of this Introduction.

The italicized portions of the preceding paragraphs of this Introduction were taken from the first edition of this book, *Parapsychology: Sources of Information*. At the popular level they are still accurate, but at the growing edges of parapsychology they are not, due to developments in anomalies research, quantum physics and the role of the observer, the philosophy of science, changes in parapsychology itself (principally the use of electronic data processing equipment and random generators plus the growth of new theories, such as the observational theories and the intuitive data—sorting hypothesis), and interchanges between parapsychologists and critics. At the frontier the picture grows exceedingly fuzzy. In the space of an introduction it is impossible to explain adequately what may become major waves of change in parapsychology. What it boils down to is that parapsychologists are holding much more loosely to their terminology. In the second paragraph we referred to ESP, PK, precognition, etc. as "building blocks." For everyday purposes, both in life and in the laboratory, they may still be regarded as such, but at theoretical levels what has been taken for psi is now thought of as a significant correlation between a subjective and an objective event. There is less need to label it one thing or another. Parapsychologists are more ready to admit that they don't know exactly what it is they are studying, but that does not render the research less exciting, or the data less real. Something is undoubtedly taking place, whatever the label or labels that eventually will be chosen to call these intriguing phenomena, once we reach a fuller understanding than we possess now.

To fully understand these trends it is essential to read the professional parapsychological journals. A book claiming to provide sources of information on parapsychology should in some way aid the reader in locating information on what is happening at the growing edge. I shall do so in two ways. First, as Appendix 5, I am reprinting what I consider to be a very important paper on parapsychology——what it is and where it is going——by John Palmer, a psychologist and leading parapsychologist, and a past president of the American Society for Psychical Research and of the Parapsychological Association. Second, I have written a bibliographic essay entitled "New Views of Parapsychology and Parapsychological Phenomena," which describes the changes that have taken place since I entered parapsychology in 1954, and which cites many journal articles and some books and individual chapters to guide readers who want to follow the leading edge of parapsychology as of early 1989.

Palmer's views and the studies cited in my chapter on "New Views" are based primarily on parapsychology experiments, whereas the librarians who will be the primary users of this book may be interested mainly in mediumship, dreams that come true, and "ghosts." I do not want to give the impression that parapsychologists have left these phenomena behind. Thus, for the edification of librarians and their patrons, I recommend the articles by Cook and Stevenson cited on page xiv. I have also provided a chapter on using the library to locate information in parapsychology in *non*parapsychological sources.

The main purpose of this book is to enable librarians, students, and others who are interested in parapsychology to locate sound information on all aspects of the subject. Although Chapter I, the annotated list of books, comprises half the book, books are only one source of information on parapsychology. The most important is the professional journal literature. The journals of the field are described in Chapter III. Also described are parapsychological organizations, theses on parapsychology, government publications, and terminology.

It is essential that the user realize that this is not a new edition of *Parapsychology: Sources of Information*, compiled by my colleague Laura A. Dale (now deceased) and myself in 1973 and published by Scarecrow Press. Since that time, many reputable publications dealing with parapsychology have been published, new terms have been introduced, centers have been organized, and new periodicals initiated. Too much has happened to warrant simply an updated edition of *Sources*. Instead, I have opted to produce *Parapsychology: New Sources of Information*, which is conceived as a continuation of *Sources*. The cutoff point for information supplied in *Sources* was 1972. This volume covers information available from 1973 through July 1, 1989, with one 1990 book [803a] inserted at the last minute because of its value and to fill a blank page.

The fact that many more items in each category have been covered in *New Sources* than in *Sources* is due both to the expansion of activity in parapsychology itself (especially as regards terminology and degrees) and to the fact that the overriding policy governing *Sources* was selectivity, whereas in *New Sources* it is to indicate all possible sources of information on all aspects of parapsychology that can be considered to be reasonably reliable.

To provide some conception of the amount of activity that has taken place in the 17 years since *Sources* was published, the number of books, periodicals, organizations, and degrees granted for work on a parapsychological topic in both *Sources* and *New Sources* are compared in the following table. It should be borne in mind that none of the books described in *Sources* appears in *New Sources*, but all of the periodicals, organizations, degrees, glossary terms, and several of the encyclopedias found in *Sources* are in this volume. This is because in 17 years significant changes have taken place in the periodicals and journals, and the degrees and terms were repeated for sake of completeness. However, many new periodicals, organizations, encyclopedias, thesis titles, and terms are described as well.

Table

Comparison of Number of Items in *Sources* and *New Sources*

Type of Document	*Sources*	*New Sources*	New Items in *Sources*
Books	282	483	483
Periodicals	14	46	33
Organizations	15	58	43
General Encyclopedias	19	9	3
Specialized Encyclopedias	12	13	12
Special Periodical Issues	3	16	16
Degrees	49	225	176
Glossary Terms	143	193	50

The layout of *New Sources* follows the format of *Sources* in most respects. Some new subject sections have been added to the book chapter and a new chapter has been added covering government publications. The chapter on encyclopedias has not been continued, but information on both general and specialized encyclopedias is provided in Chapter VI on general sources of information on parapsychology.

The numbering system continues that in *Sources*. The last item described in *Sources* was 342. The first item in *New Sources* in 343. Chapters I–IV (and each section of the first chapter), contain an introduction in which reference is made by number to items in the earlier book, especially to those titles still considered to be of value. *Parapsychology: Sources of Information* is *still* useful and is in print. Those who buy *New Sources* but who do not own *Sources* may find it useful to purchase it as well, although as a guide to parapsychology information sources from 1972 to date *New Sources* can stand on its own.

Item numbers also are assigned to the titles in Chapter II on periodicals, the organizations listed in Chapter III, and to the annotated items in Chapter VI on general sources of information. Item numbers have been assigned to annotated materials only. They have not been assigned to the documents cited in the reference lists at the end of some chapters, nor have they been assigned to publications listed in the bibliographies in Chapters IV–VI.

This same rule is followed in the indexes. All annotated items are listed in the author, title, and subject indexes by item number in boldface. The items for which only bibliographic citations are given are listed only in the author and subject indexes by page number, which is indicated by a p or pp preceding the number.

REFERENCES

Cook, E.W. (1986). The survival question: Impasse or crux? *Journal of the American Society for Psychical Research, 81,* 125–139.

Stevenson, I. (1971). The substantiality of spontaneous cases. *Proceedings of the Parapsychological Association,* No. 5, 1968, 91–128.

Stevenson, I. (1987a). Guest editorial: Changing fashions in the study of spontaneous cases. *Journal of the American Society for Psychical Research, 81,* 1–10.

Stevenson, I. (1987b). Guest editorial: Why *investigate* spontaneous cases? *Journal of the American Society for Psychical Research, 81,* 101–109.

Parapsychology Sources of Information Center
2 Plane Tree Lane
Dix Hills, NY   11746
July 14, 1989

I
Annotated List of Books

GUIDE TO THE ARRANGEMENT OF CHAPTER I

The books in this annotated list represent what are considered by the compiler to be the most informative of those published since *Parapsychology: Sources of Information* (hereafter referred to as *Sources*) went to press in 1972. It is not exactly a list of "best" books, for if it were, far fewer titles would be listed. Nor are all titles considered to be beyond reproach in the sense of containing no inaccuracies whatsoever. But a book has to be reasonably reliable to be included. (Book review citations have been included; in order to determine how reliable a given work is, consult the reviews, especially those in the parapsychological journals.)

Cut–Off Date

The original cut–off date for the inclusion of the books was August 1, 1988. However, an Addenda was included at the end of the book section. The cut–off date for the Addenda was December 31, 1988. The insertion of new titles since then was accomplished by adding a letter to the preceding number. Following this rule, additions were made through July 1, 1989. (One number, **448**, had to be eliminated because the same book was listed in two sections; **803a**, a 1990 book, was added at the last minute.)

General Arrangement

The books described in Chapter I are arranged alphabetically by author in 27 subject sections plus an Addenda (these headings are listed in the Table of Contents). In each section there is an introduction to the subject matter of that section, some mention of the treatment of that subject in *Sources*, and an overview of the books included; this is followed by a paragraph which lists additional books also treating that subject found in other subject sections in the chapter.

1

Books by Item Number

Each of the titles is listed under its own individual item number (which in the case of books is referred to as a book number), beginning with item 343. (*Sources* covered items 1–342.) Whenever a title is mentioned elsewhere in this compilation, often its author, followed by its number, or the number alone, rather than the author, title, or page number, is used to identify its location in all chapters and sections of the book and in the indexes. Item numbers have also been assigned to the periodicals, organizations, and the annotated titles in Chapter IV. When reference is made to items in *Sources*, the item number will also be used in order to facilitate locating titles in that volume. Item numbers are in boldface type inside brackets.

Since the manuscript was completed some organizations and journals have ceased. These have been omitted so there will be a skip in the item number when this is the case. Or, if a book or other item has been added, it will be inserted alphabetically and then will be assigned the same number as the item preceding but with an *a* added.

Bibliographic Citation of Books

For each title the following bibliographic information is given: author(s) and/or editor(s); title; subtitle; translator, if any; name of person writing the introduction, preface, or other front matter; place of publication; publisher; date of publication, number of pages; series title if important; and original publisher and date, if the in–print edition is revised or a reprint. In the case of books translated into English, the original foreign–language title is also given wherever possible. Some of the titles were published in both the United States and Great Britain. When this is the case, the American edition is listed, unless only the English edition was available for inspection in compiling this bibliography.

Additional Information

Following the bibliographic description, information is given concerning illustrations, glossaries, type of bibliographic citation, indexing, tables, graphs, etc. This information is indicated by means of several terms used consistently throughout the annotations, as follows: (a) If a book contains a list of definitions, this is shown by the word "Glossary," preceded by the number of items included.* (b) Several types of bibliogra–

---

*An index of books with glossaries is provided in Append–

phical citations are used, sometimes two or three in the same title. The terms indicating them are "Annotated bibliography," "Bibliographic footnotes" (i.e., at bottom of page), "Bibliography"; "Bibliographical notes" (i.e., at end but not by chapter) "Chapter notes" (i.e., at end of book) "Chapter references" (i.e., at end of chapter); "Citations in text"; and "Classified bibliography." When other specially titled lists are provided, the title used in the book is given in quotation marks, e.g., "Suggested readings," "Recommended books," etc. In the case of bibliographies or other lists, either the number of pages or the number of items is given, with the exception of "Bibliographic notes" and "Chapter notes," because those usually contain many lines of text in addition to the citation and so reporting the length does not provide an accurate assess‐ ment of the number of bibliographic citations. (c) The type of material indexed is indicated by "Name index," "Subject index," and "Index of cases" or other special index title that may be used. "Index" alone means that both names and subjects are included in a single index. "Illustrations" indicates the presence of pictorial matter, whether in the form of drawings, paintings, or photographs.* Also listed are figures, graphs, and tables. All of the items listed as "Additional Information" are in al‐ phabetical order.

A key to the abbreviations used in the bibliographic cita‐ tions is given in Appendix 3.

Book Annotations

The purpose, principal contents, and any special features of each book are described in an annotation which follows the bibliographic citation. Occasionally, brief biographical informa‐ tion about the author is also included if it is helpful in elucidat‐ ing the contents. Specialized parapsychological terms used in the title or annotations are defined in the glossary (Chapter VIII). In the case of controversial publications, sometimes quotations are given from, or the reader is directed to, specific book reviews.

---

dix 1, but users are urged to consult the dictionary compiled by Thalbourne [687]. Thalbourne's dictionary is now used for the definitions given in the glossary published in each volume of the *Journal of Parapsychology* [825].

*Because illustrative parapsychological material is often needed, but difficult to locate, an index of books containing il‐ lustrations is given in Appendix 2.

Book Review Citations

It was considered a useful adjunct to the annotations to give citations to reviews of the books in order to facilitate check—ing the compiler's opinion against those of others as well as to supplement it. Reviews are especially useful in pinpointing shortcomings of a given work. Reviews also are useful in that they often place a given title in a broader context of ongoing work. Reviews in all the major English—language para—psychological periodicals have been included, and also those in general periodicals such as *Time* and *Nation*, in major newspapers, in library and book—trade journals such as *Library Journal* and *Publishers Weekly*, and in professional journals from fields such as anthropology, psychology, philosophy, religion, psychiatry, and sociology. Because more of the books listed were reviewed in parapsychological periodicals than in non—parapsychological ones, and because the reviews in the para—psychological journals are generally longer, we decided to cite these reviews separately. Moreover, because the same para—psychological journals are cited often, we have used very brief abbreviations for them. The titles of both parapsychological and nonparapsychological periodicals, with the abbreviations used for them, are given in Appendix 3.

Included with nonparapsychological periodicals are a few titles such as *American Dowser* and *Spiritual Frontiers*, which could be considered parapsychological journals, but since their reviews tend to be shorter than the standard parapsychology journals and because they cover a broader subject area than parapsychology proper they were classed with the nonpara—psychological periodicals.

The citations to reviews in parapsychological periodicals are given immediately following the book annotations. The reviews in nonparapsychological periodicals are listed in the next paragraph. In citing these periodicals, it was decided to spell out in full any titles consisting of only one word. For titles composed of more than one word, we followed as closely as pos—sible the rules for periodical title abbreviation suggested by Com—mittee Z39 of the American National Standards Institute.* Words such as "the" and "of" are omitted in the abbreviations, but the ampersand is used for "and," for example,

---

*American National Standards Institute. Standards Com—mittee Z39. *NCPTWA Word—Abbreviation List*. Columbus, Ohio: National Clearinghouse for Periodical Title Word Abbrevia—tions, 1971.

JNerv&MentDis. The word—by—word method of alphabetizing is used and the titles are in alphabetical order by abbreviation rather than by title spelled out in full. The ampersand is treated as if it were spelled out. To illustrate, periodicals begin— ning with the words "science" or "scientific" are alphabetized as follows: *SciAm*; *Sci&Soc*; *SciBks*; and *Science.*

All references to reviews in parapsychological periodicals were checked, but no effort was made to check the citations in nonparapsychological periodicals, except for an occasional scan— ning of *Choice, Contemporary Psychology,* and *Library Journal.* For the remainder the information was obtained from the follow— ing indexes: *Book Review Digest*; *Book Review Index*; *Index to Book Reviews in the Humanities*; *Index to Religious Periodical Literature/Religious Index II*; *Philosopher's Index*; and *Social Sciences and Humanities Index.*

If more than one edition of a title was reviewed, we listed all available reviews regardless of edition, even if in some cases the actual title was not the same for all editions. In such cases the variant titles are listed as *see* references in the title index. Wherever possible the volume of the periodical is given, followed by the pages, with the month and the year in parentheses. However, some newspaper—type periodicals such as *Spectator* and the *Times Literary Supplement* do not use volume numbers, or they were not given in the indexes used to locate the reviews. In these cases, the pagination is given first, indicated by "p," followed by the date. Not all indexes gave the in— clusive pagination for book reviews, so if only one page number is given in a citation to a nonparapsychological journal, there is no way of determining the length of the review; that is, the page number may indicate only the first page, or that first page may be the only page of the review. Finally, where reviews are cited in periodicals that have changed titles, whichever title was in use when any given review appeared is the one used in the book review citation. For example, *A.L.A. Booklist* and *Booklist* are both cited, referring to different issues of the same peri— odical. (The one exception is the *Journal of Near—Death Studies,* which used to be *Anabiosis.* All the reviews in the latter are listed under the former.)

Altered States of Consciousness
and Psi

INTRODUCTION

In the section entitled "Altered States of Consciousness" in
Sources, 25 books were listed but only six dealt substantially
with altered states and psi: Cavanna and Servadio [6], Dingwall
[9], Green [11], Parapsychology Foundation proceedings edited
by Cavanna and Ullman [18], *Proceedings of Two Conferences
on Parapsychology and Pharmacology* [19], and Van Over [25].
Of those, Dingwall is still *the* work on hypnosis and psi and the
two proceedings volumes are still the best books available on
various drugs and psi.

Of the 22 books listed here, five are reviews of altered
states and psi. The most important of these is the one by
Kelly and Locke [350], followed by Parker [356]. Holroyd [349]
covers the subject at a popular level. The volume edited by
Shapin and Coly [358] is a collection of conference papers on
various aspects of altered states and psi. Neher [355] and Reed
[813] is recommended because of their comprehensive and critical approach to altered states, including those often associated
with parapsychological phenomena. Also, the chapter by Krippner and George in Wolman and Ullman [361] is an important
(and the most recent) review of altered states and psi. Mishlove [353] and Myers [354] are useful for historical approaches to
the subject. The psychology of altered states, with some
reference to psi phenomena, is presented by Donahoe [345],
Evans [345a], Gowan [346], LeShan [352], and Tart [359].
Several books deal with specific altered states in relation to psi:
LSD (Grof [348]), hypnagogic states (Mavromatis [812]), lucid
dreams (LaBerge [351]), possession trance (Sargent [357]), and
dreams (Ullman and Zimmerman [360]). Crabtree's [344] bibliography on animal magnetism is an invaluable resource on early
studies of hypnosis and ESP. The volume edited by Barber
[343] is an anthology on altered states, including a seven—item
section on parapsychology.

**Additional books with information on altered states of con-
sciousness and psi:** Panati's [531] introduction to parapsychology emphasizes altered states and psi. Tart [412] specu-

6

lates and reports on research into the nature of consciousness
and psi. The U.S. Army report [462] is critical of several tech—
niques thought to engender psi—conducive states, such as guided
imagery, biofeedback, split—brain effects. In a chapter on para—
psychology, it is also critical of the Ganzfeld and remote—
viewing experiments, although the former is considered promis—
ing enough to be worth monitoring. There is a chapter review—
ing imagery and psi and one on Ganzfeld research in *Advances
in Parapsychological Research 4* [523]. The altered states in—
volved in shamanism are described in detail in Doore [806], Har—
ner [565] and Nicholson [569]. Mindell [651] theorizes about a
special state called the "dream body." Several books in the sec—
tion on religion deal with altered religious and mystical states
and psi: Gowan [705], Heaney [709], Heron [710], Kelsey [713],
Rossner [719], Tart [721], and Whiteman [722]. Haynes [731]
makes many observations regarding altered states and psi. Grof
[758] and Jacobson [761] deal with altered states and psi. A
number of the books in the "Training and Development of Psi"
section deal with altered states and psi: LeShan [780], Mishlove
[782], Reed [814], and Rogo [783]. Almost all the books in the
section on "Unorthodox Healing" [792-803] deal with healing in
altered states such as shamanic healing and possession trance.
The Maimonides dream ESP experiments are the subject of a
book by Ullman and Krippner with Vaughan [819].

BIBLIOGRAPHY

343 Barber, Theodore X. (Ed.). *Advances in Altered States of
Consciousness and Human Potentialities, Volume I.* New
York: Psychological Dimensions, 1976. 698p. Chapter
references; 10 figures; 22 graphs; 6 illustrations; Name
index: 695—698; Subject index: 693; 7 tables
  Barber has compiled an anthology of some of the best
papers and selections from books dealing with human poten—
tialities and altered states of awareness published in recent
years. In the preface, the editor, who is a well—known research
psychologist in ASCs, especially hypnosis, writes: "These
volumes, which will be published every two or three years, will
cover this new emphasis in psychology. Each volume will
reprint the 40 to 50 most important papers that were published
during the preceding years. The papers will focus on modes of
experience or awareness that differ from our ordinary or
'normal' experience. The topics that will be considered include:
hypnosis, biofeedback, meditation, yoga, Zen, autogenic training,
and acupuncture; fantasy, daydreams, nocturnal dreams, and
sleep; religious and mystical experiences; parapsychology and

psychic research; psychological effects of cannabis and
psychedelic drugs; and methods for heightening human
capabilities and fulfilling human potentialities" (p. v). This initial volume contains 43 selections, of which seven are under
"Parapsychology." These are by G.R. Schmeidler, W.G. and
L.W. Braud, J. Palmer, C.T. Tart (two), R.V. Johnson II, and
E.L. Smith. Barber points out in the Preface that included are
"three 'tight' papers that are within the 'hard—headed' scientific
tradition; two somewhat 'far—out' papers that discuss topics
which are not typically considered in 'hard—headed' journals;
and two critiques of 'far—out' topics" (p. ix). The three "tight"
papers are by the Brauds, Palmer, and Schmeidler. The two
"far—out" papers are by Tart. Johnson criticizes C. Backster
and E.L. Smith criticizes Raudive. Unfortunately, as of 1988,
no further volumes in this series have been published.
    JSPR 49:614—615 (Sep 1977).

344 Crabtree, Adam. *Animal Magnetism, Early Hypnotism,
    and Psychical Research, 1766–1925: An Annotated Bibliog—
    raphy.* White Plains, NY: Kraus International, 1988.
    522p. Bibliography: xxix—xxxiii; Glossary: xxiii—xxvii;
    Name index: 437—453; Subject index: 517—522; Title in-
    dex: 455—516
    In an "Historical Introduction," Crabtree examines the his—
tory of animal magnetism, and points out its impact on
medicine, psychology, and parapsychology. He adds: "The histories of animal magnetism, hypnotism, and psychical research
are inextricably intertwined. As will be evident from the annotated entries in the bibliography, the literature of any one of
these areas cannot but include the literature of the other two"
(p. xvi). Although the bibliography is aimed at including "the
literature of animal magnetism and those streams of thought
that can be identified as flowing directly from it" (p. xvii), the
literature is too vast to include all of it. Thus it is limited to
only the most significant writings arising from animal mag—
netism and to "only those works that are immediately connected
with animal magnetism and the themes that arise directly from
it" (p. xvii). The list of areas omitted "are works that deal
exclusively with occultism, possession, or witchcraft; theosophy,
anthroposophy, Christian Science, or other spiritual philosophies;
theology or religious thought; and conjuring or stage magic.
While spiritualist writings have true importance for the history
of animal magnetism and its offshoots, only those works have
been included here that depict its development from mesmeric
influences or that play a significant role in the rise of psychical
research. This means that the bibliography does *not* include
stories about clairvoyants, seers, or prophets; books relating

communications from spirits; spiritualistic speculations about the afterlife or related matters; and collections of ghost—lore" (p. xvii). The bibliography ranges from 1766—1925. Annotations are provided for about one—third of the items listed (those considered to be the most important works). The bibliography is arranged by year with works listed alphabetically by author under each year. Each item contains a letter designating that it is about hypnotism [H], psychical research [P], or both [H & P]. There is a useful glossary that contains many parapsychological terms used in the literature of the time period covered by the book.
AmRefBksAnn 20:281—282 (1989).

345 Donahoe, James J. *Enigma: Psychology, the Paranormal and Self—Transformation.* Oakland, CA: Bench Press, 1979. 199p. Annotated—classified bibliography: 171—193; Glossary: 167—170; Index: 195—199

Psychologist Donahoe is a first—person explorer of altered states and psi. He makes important observations concerning the paradigms guiding parapsychological and psychological research. He proposes that parapsychology should be based on breaking limits rather than setting new ones, and he emphasizes the imagination as the optimal approach to psi. Outer reality (i.e., empirical results) is dependent on inner conception. The same holds for psychology as the science of consciousness. Instead of mimicking the physical sciences, psychology must first take an independent stand by examining imaginative faculties and conscious intent. He proposes that the findings of parapsychology and altered states research indicate "that the world is much more fluid than the conservators of the old paradigm would acknowledge" (p. 112). The reality with which parapsychologists must deal, whether in investigating the paranormal before or after death, cannot be determined. "How we arrange existences gets determined by experiment, as always, a combination of discovery and invention in which the new scientist and new artist are simultaneously the investigators, the experimental subjects and the experiments themselves" (p. 113).

345a Evans, Hilary. *Alternate States of Consciousness: Unself, Otherself, and Superself.* Wellingborough, Northamptonshire, England: Aquarian Press, 1989. 256p. Bibliography: 244—251; Index: 252—256.

The author, who directs the Mary Evans Picture Library [888], has written two books in which he is mainly concerned with visions seen in various altered states [725, 726], but in this book he concentrates on classifying the various forms of alternate states (his preferred term), their characteristics, their trig-

gers, their benefits, their negative aspect, and in theorizing
about alternate states of the self. Occasional reference is made
to psi—based altered states. Example cases are given in all
categories discussed. There are whole chapters or sections of
chapters devoted to characteristics that recur in ASCs, and to
the several ways in which ASCs occur (spontaneous, externally
triggered, mentally triggered, voluntary) as well as various types
of "triggers" (physiological, environmental, psychological, cul—
tural, do—it—yourself) and interactions between and combina—
tions of the foregoing. A chapter is devoted to the types of
motivation in alternate states. In a final chapter he discusses
whether or not ASCs are "real," whether they are distinct from
one's usual conscious state or from other ASCs, are ASCs
"better" than the usual state or are some ASCs "better" than
others, and whether or not ASCs can be controlled. Evans has
deliberately kept technical terminology to a minimum, so this
book should be of value to laypersons, but his scheme and ideas
can be pondered with profit by experts as well.

346 Gowan, John Curtis. *Trance, Art, and Creativity.* Buffalo,
     NY: Creative Education Foundation, 1975. 448p. Bibliog—
     raphy: 409—447; Chapter notes; 3 figures; Glossary: in—
     side front cover; 3 graphs; Index: 448; 13 tables
     Gowan, an educational psychologist, discusses and describes
the various procedures used in three methods of contact between
the conscious ego and the collective preconscious (or numinous)
level. He observes that the union of the individual mind with
that of the "General Mind" can only take place in an altered
state of consciousness. The three major modes of consciousness
discussed are trance, art, and creativity. He also examines
various psychic healing powers that occur as by—products in
people pursuing the various modes.
     JTranspersPsychol 7:198 (1975).

347 Green, Elmer, and Green, Alyce. *Beyond Biofeedback.* New
     York: Delacorte Press/Seymour Lawrence, 1977. 369p.
     Bibliography: 344—354; 6 figures; 7 graphs; 13 illustra—
     tions; Index: 355—368; 3 tables
     An account of the pioneer research in biofeedback con—
ducted by the authors at the Menninger Foundation. In addition
to examining the role of biofeedback training in treating various
diseases and conditions, they report on their study of voluntary
control of mind and body in such areas as creativity, medita—
tion, altered states of consciousness, and yoga and psychic
powers. Included are reports of their research with Swami
Rama, Rolling Thunder, and Jack Schwartz.
     InstNoeticSciNewsl 6:16 (Spr 1978).

348 Grof, Stanislav. *Realms of the Human Unconscious: Obser—
vations from LSD Research.* New York: Viking, 1975.
257p. Bibliography: 245—246; 1 figure; 49 illustrations;
Index: 247—257; 1 table

This volume in the SUNY Series in Transpersonal and
Humanistic Psychology is in two sections: "Dimensions of Con—
sciousness and New Perspectives in Psychotherapy." The former
presents the extended cartography of the psyche that Grof
has developed during his clinical work with psychedelics. "It
describes the basic types of experiences that become available to
an average person whenever he or she gets involved in serious
self—exploration with psychedelics or various powerful nonphar—
macological experiential techniques. While the model of the
human psyche used in traditional academic psychotherapy is con—
ceptually limited to the *recollective—analytical level,* this new car—
tography includes two additional levels that are transbiographi—
cal. These are the *perinatal level,* characterized by emphasis on
the twin phenomena of birth and death, and the *transpersonal
level* that can in principle mediate experiential connection with
any aspect of the phenomenal world and with various mythologi—
cal and archetypal domains. I consider the knowledge of this
cartography to be indispensable for safe and effective inner
quest" (p. xvi). In the second section, he presents for the first
time the basic principles of what he calls "holotropic therapy,"
or "holonomic integration," which "combines in a particular way
controlled breathing, music and other types of sound technology,
focused body work, and mandala drawing" (p. xiv). Grof notes
that people find this method "to be an effective and exciting
tool for self—exploration with an unusual potential for mediating
transformative and mystical experiences" (p. xiv). It contains a
special section on the effective healing and personality transfor—
mation mechanisms that operate in nonordinary states of con—
sciousness, and which have since earliest times had an impor—
tant place in healing and shamanic rituals and in rites of pas—
sage. He has attempted to rediscover these age—old practices
and reformulates them in today's scientific terminology. The
book ends "with a discussion of the potential and the goals of
experiential self—exploration utilizing the therapeutic and trans—
formative power of nonordinary states of consciousness. It
describes how, in this process, emotional and psychosomatic heal—
ing is combined with a movement toward a more fulfilling
strategy of life and a search for answers to the fundamental on—
tological and cosmological questions of existence" (pp. xvi—xvii).
Parapsychological phenomena are considered in various sections
of the book: ESP and OBEs under "Transcendence of Spatial
Boundaries," reincarnation and precognition under

"Transcendence of the Boundaries of Linear Time," survival and mediumistic experiences under "Experiential Extension Beyond Consensus Reality and Space—Time," and synchronicity, PK, and RSPK under "Transpersonal Experiences of Psychoid Nature."

   P 7:25 (Mar/Apr 1976); *PR* 7:16—18 (May/Jun 1976); *PS* 1:242—243 (1976).

   JAmAcadRelig 44:309—315 (Jun 1976); *JTranspersPsychol* 1975, 7:197—202 (1975).

349　Holroyd, Stuart. *Psi and the Consciousness Explosion.* New York: Taplinger, 1977. 235p. Chapter notes; Index: 231—235

   This book by a British author proposes that parapsychology is of central importance in establishing a radically new image of humans in the universe. Holroyd feels that a changed view of human nature is essential if humans are to survive on this planet. He presents basic parapsychological findings and then relates them to innovations in other fields such as physics, philosophy and psychology. Other books are better at documenting parapsychological data but Holroyd does an excellent job of pointing out the importance of these data and placing them in a larger perspective.

   JIndPsychol 2:150—151 (Jul 1979); *JP* 42:229—233 (Sep 1978); *JSPR* 49:551—553 (Jun 1977); *PR* 8:13—15 (Sep/Oct 1977); *T* 6:13—14 (1978).

   Brain/MindBul 2:4 (Mar 21, 1977); *ContempPsychol* 3:208—209 (Mar 1980); *SpirFron* 12:51—53 (Win 1980).

350　Kelly, Edward F., and Locke, Ralph G. *Altered States of Consciousness and Psi: An Historical Survey and Research Prospectus* (Parapsychological Monographs, No. 18). New York: Parapsychology Foundation, 1981. 91p. Bibliography: 85—91; 3 tables

   Kelly, a parapsychologist whose training is in psycholinguistics, and Locke, an anthropologist, review not only the literature of psychical research but also that of historical and comparative religions and cultural anthropology for material bearing on the question of altered states and psi. Their goal is a generation of hypotheses that can be tested experimentally. They cover early studies of trance mediumship and hypnosis, psi phenomena associated with saints, yoga, shamanism, and the induction of ecstatic and other altered states. In their "research prospectus" they conclude that "unusually strong manifestations of psi are systematically related to various kinds of strong ASCs" (p. 43) and that these altered states are capable of systemic cultivation. They propose that these techniques be inventoried, analyzed, and incorporated in psi—testing situations. An appendix presents a preliminary model of ASC induction and psi, and sug—gestions for its application.

JASPR 76:388—390 (Oct 1982); *JP* 46:276—279 (Sep 1982); *JSPR* 52:77—79 (Feb 1983); *T* 10:85—86 (Win 1982).

351 LaBerge, Stephen. *Lucid Dreaming.* Los Angeles, CA: Jeremy P. Tarcher, 1985. 277p. Chapter notes; Index: 269—277

This book has some interesting things to say about psi in a chapter entitled "Dreaming, Illusion, and Reality." LaBerge agrees with Nietzsche that belief in gods, ghosts, and survival has its origins in dreams. He proposes that "OBEs are actually variant interpretations of lucid dreams" and "that dream telepathy will provide the basis for an explanation of the oc—casional accuracy of paranormal OBE vision" (p. 206). He sug—gests that out—of—body experiences (OBEs) be called OBSs, for "out—of—body sensations." He also has a chapter on near—death experiences (NDEs) and the experience of transcendence in lucid dreams in which he discusses their significance for our views of survival of death.

PAI 5:44—45 (Dec 1987).

Choice 23:206 (Sep 1985); *ContempPsychol* 31:508 (Jul 1986); *PublWkly* 227:107 (Mar 15, 1985); *PublWkly* 229:71 (May 16, 1986).

352 LeShan, Lawrence. *Alternate Realities: The Search for the Full Human Being.* New York: M. Evans, 1976. 232p. Bibliographical notes; 2 figures

This book can be viewed as a continuation of LeShan's ear—lier books, *Toward a General Theory of the Paranormal* [**129**] and *The Medium, the Mystic, and the Physicist* [**780**]. LeShan, a psychologist and self—trained healer, says in the introduction that "the central idea of this book is that we human beings in—vent reality as much as we discover it, and that if this is com—prehended, we have a wide choice as to how we invent it and therefore, what sort of world we live in. There are . . . a num—ber of basically different, valid ways of organizing whatever is 'Out there,' and we can choose from among these the one that will satisfy our needs and advance us toward our goals at the moment" (p. x). LeShan further suggests that his thesis provides "an acceptable and useful solution to the 'impossible paradox' of ESP, and to the problems of 'survival' of the per—sonality after biological death and of the existence or nonexist—ence of spirit entities" (p. x). In addition to the modes of being that he called the IR (individual reality) and clairvoyant reality which he described in his earlier books, here he develops the concepts of the transpsychic mode and the mythic modes of being. The basic limiting principles are described for each mode.

JASPR 71:310—314 (Jul 1977); *JP* 41:226—230 (Sep 1977); *JSPR* 49:540—545 (Jun 1977); *PR* 8:14 (Jan/Feb 1977). *NewReal* 1:26 (1977).

812 Mavromatis, Andreas. *Hypnagogia: The Unique State of Consciousness Between Wakefulness and Sleep.* [See Addenda for annotation]

353 Mishlove, Jeffrey. *The Roots of Consciousness: Psychic Liberation Through History, Science and Experience.* New York: Random House & Berkeley, CA: Bookworks, 1975. 341p. Chapter notes; 56 figures; 7 graphs; 287 illustrations; Index: 339—341; 2 tables

The core ideas on which this survey of consciousness exploration is based are that (1) psychic, mystical, and occult phenomena "are not paranormal, and therefore unreal or insignificant; . . . they are rooted in the essential core of our cosmic existence, in our cultural history, in our scientific knowledge, and in our social institutions" (p. xxix); and (2) the exploration of consciousness "involves intense personal commitment as well as objective understanding" (p. xxix). Section I covers historical explorations of consciousness: prehistoric, Mesopotamian, Egyptian, Indian, Chinese, Greek, Roman, Judeo—Christian, alchemical, Arabic, medieval, Renaissance, eighteenth century, and nineteenth century to the SPR and William James. Section II surveys "Scientific Approaches to Consciousness," starting with ESP research from J.B. Rhine to PMIR; OBEs; psychic healing; firewalkers; PK; life and death; other worlds; physiological mechanisms; and rhythms. The last section reviews theories, practical applications of psi, and organizations. The material in this book was prepared in partial fulfillment of Mishlove's examination requirements for his doctoral program in parapsychology at the University of California, Berkeley.

JASPR 70:325—326 (Jul 1976); *JP* 40:248—252 (Sep 1976); *P* 6:42—44 (Jan/Feb 1976); *PS* 2:166 (Feb 1977); *T* 5:14 (1977). *AmDows* 16:30—31 (Feb 1976); *JTranspersPsychol* 8:69 (1976); *SpirFron* 10:115 (Spr 1979).

354 Myers, Frederic W.H. *The Subliminal Consciousness.* New York: Arno Press, 1976. 593p. Bibliographical footnotes

This Arno Press anthology consists of articles originally published in the *Proceedings of the Society for Psychical Research,* 1888—1895. They were selected by James Webb for the Arno series, *The Occult.* Myers was a pioneer not only in

parapsychology, but is being increasingly recognized in psychol—
ogy as well. This volume may be considered as a supplement
to Myers' classic *Human Personality and its Survival of Bodily
Death* [169]. In these papers he documents how he arrived at
the concept of the "subliminal self." Some of the areas covered
were not included in *Human Personality*, and if included, were
not as fully treated as here. The selections, by title, are as fol—
lows: "Automatic Writing"; "The Daemon of Socrates"; "General
Characteristics of Subliminal Messages"; "The Mechanism of
Suggestion"; "The Mechanism of Genius"; "Hypermnesic
Dreams"; "Sensory Automatisms and Induced Hallucinations";
"The Mechanism of Hysteria"; "Motor Automation"; "The Rela—
tion of Supernormal Phenomena to Time——Retrocognition" and
"The Relation of Supernormal Phenomena to Time——
Precognition."
    Spectator 68:330—331 (Mar 5, 1892).

355 **Neher, Andrew.** *The Psychology of Transcendence.*
    Englewood Cliffs, NJ: Prentice—Hall, 1980. 361p. Bibliog—
    raphical notes; Glossary: 341—350; 11 figures; 38 illustra—
    tions; Index: 353—361; 1 table
    The author is a Professor of Psychology at Cabrillo Col—
lege where he has taught a course in transcendence for several
years. The curriculum he developed for the course serves as the
basis of this book. He defines transcendence as those
"experiences that go beyond our usual levels of functioning and,
for that reason, often seem inexplicable" (p. xiv). This over—
view of various altered states of consciousness, several of them
associated with parapsychology, performs two important func—
tions. In a sane and scholarly fashion, it describes how to cul—
tivate and benefit from transcendent experiences (which are
broken down into experiences commonly called occult, psychic,
or mystical). Second, and most important for those interested
in parapsychology, it presents some legitimate counterhypotheses
for some experiences generally assumed to be "psychic." Yet
this is not a book by a debunker, but by one who has a
genuine interest in the psychology of unusual experiences that
transcend our everyday state of mind. The basic psychological,
physiological, and cultural processes involved in transcendental
experiences are presented in Chapters 2—4. Chapters 4—6 deal,
respectively, with specific examples of mystical, psychic, and oc—
cult experience. With cartoon illustrations, a glossary, index,
and bountiful references, this is a very helpful and useful book.
    JASPR 76:189—191 (Apr 1982); *JP* 46:69—71 (Mar 1982);
T 10:41—44 (Sum 1982).
    SkptInq 6:57—58 (Sum 1982).

356 Parker, Adrian. *States of Mind: ESP and Altered States of Consciousness.* New York: Taplinger, 1975. 198p. Author index: 195—196; Bibliography: 178—191; Glossary: 173—176; Subject index: 197—198

This is the first general survey of altered states of consciousness and ESP to be published in book form, and it is by a clinical psychologist who has engaged in parapsychological research. The range of topics covered is reflected in the chapter headings: "ESP and Consciousness"; "Hypnosis and ESP"; "Trance Mediumship"; "Dream States and ESP"; "Out—of—the—Body Experiences and Lucid Dreams"; "Pathological States"; "Psychedelic States"; "Meditation, Mysticism, and Alpha States"; and "Is there a Post—Mortem ASC?" The emphasis throughout is on empirical research. Parker shows how various interpersonal factors are related to both ESP and ASCs and offers suggestions for further research.

JASPR 70:320—324 (Jul 1976); *JSPR* 48:236—238 (Dec 1975); *P* 6:42—43 (Jan/Feb 1976); *PR* 7:18—20 (May/Jun 1976); *T* 6:17—18 (1978).

813 Reed, Graham. *The Psychology of Anomalous Experience: A Cognitive Approach* (rev. ed.) [See Addenda for annotation]

357
Sargant, William. *The Mind Possessed: A Physiology of Possession, Mysticism and Faith Healing.* Philadelphia: J.B. Lippincott, 1974. 212p. Bibliography: 200—203; Chapter references; 31 illustrations; Index: 205—212

A British psychiatrist examines the mechanism underlying ritual possession, mysticism, and faith healing which enables them to work. Pounding music, wild dancing, hell—fire preaching, and psychedelic drugs can, when purposefully applied, so strain the nervous system that normal judgment is impaired and the person involved becomes highly suggestible. In such an altered state extraordinary experiences and feats may occur. Sargant points out that such things as mystical ecstasy, faith healing, snake—and—poison—immunity, etc. should not be interpreted as evidence of supernatural factors but as the nature of human beings. He concludes: "We need faith and yet . . . we must suspect it" (p. 199).

ContempPsychol 20:244—245 (Mar 1975).

358 Shapin, Betty, and Coly, Lisette (Eds.). *Psi and States of Awareness: Proceedings of an International Conference Held in Paris, France, August 24—26, 1977.* New York: Parapsychology Foundation, 1978. 278p. Chapter

references; 6 figures; 1 graph; 6 tables
In these proceedings of a conference held in 1977, there were 14 participants who contributed papers on psi and ASCs in general, or on specific altered states such as those induced by the Ganzfeld technique, sleep—dream research, or in Biblical prophecy. The discussions following each paper are included, as well as a concluding roundtable discussion. The participants were B. Bennett, Jr., W. G. Braud, F. Danest, J. C. Dierkens, J. Ehrenwald, C. Honorton, P. Janin, L. LeShan, C.A. Meier, A. Parker, C. L. Sargent, E. Servadio, I. Strauch, and C. T. Tart.

JASPR 73:303—305 (Jul 1979); *JIndPsychol* 2:168—169 (Jul 1979); *JSPR* 50:114—117 (Jun 1979); *PN* 2:2 (Jul 1979).
ContempPsychol 3:107—108 (Jun 1979).

359 Tart, Charles T. *States of Consciousness.* New York: E.P. Dutton, 1975. 305p. Bibliography: 287—295; 19 figures; 11 graphs; Index: 297—305; 7 tables
As a psychologist, Tart has been a pioneer in studying con—sciousness in various forms. In this book he presents "a way of looking at what people tell us about and how they behave in various altered states of consciousness that I have been slowly developing in a decade of research" (p. ix). He calls this a "systems approach" to viewing consciousness (so—called because "in trying to understand human consciousness, we must get the feel of the whole system as it operates in its world, not just study isolated parts of it" [p. 14]). The book is in two sec—tions. Section I, "States," describes Tart's systems approach, its implications, and provides a summary of what we now know about states of consciousness. Section II, "Speculation," he says, "presents ideas that, while consistent with the systems approach, are not a necessary part of it and are more unorthodox" (p. 11). Although some of the points Tart makes are based on parapsychological findings, basically this book is not about parapsychology. However, it is necessary reading for para—psychologists, for they will not make significant progress until they can relate their findings to the larger field of psychology, a psychology that, as Ornstein and Tart see it, is a science of con—sciousness.
JP 40:167 (Jun 1976); *JRPP* 1:70—72 (1976); *PS* 2:147—148 (Feb 1977); *T* 5:14—15 (1977).
Brain/MindBul 1:4 (Jan 5, 1976); *ChristCent* 93:1058—1060 (Nov 24, 1976); *ContempPsychol* 22:359 (May 1977); *NewReal* 3:42 (Feb 1980).

360 Ullman, Montague, and Zimmerman, Nan. *Working with Dreams.* New York: Dell, 1979. 368p. Chapter notes; 11

figures; 1 graph; Index: 361—368; 2 tables
This is a book about dream interpretation. Ullman's
philosophy is that no one but ourselves can be an expert about
our own dreams. However, a knowledgeable person can help
others to be experts about their own dreams, and it is the aim
of this book to provide such guidance. There is one chapter on
ESP in dreams, and the authors point out that "they persuade
us to look at dreams as events occurring in a much larger and
more complex reference frame than we are accustomed to" (p.
349). The value of this book for parapsychology lies in the fact
that it sets forth a technique for working with dreams in a
group situation that might be viewed as a living laboratory for
the observation of the dynamic role ESP may play in dreams,
and vice versa.
SpirFron 12:53—54 (Win 1980).

361 Wolman, Benjamin B., and Ullman, Montague (Eds.).
Handbook of States of Consciousness. New York: Van
Nostrand Reinhold, 1986. 672p. Bibliography: 657—659;
Chapter bibliographies; 14 figures; Name index: 661—668;
Subject index: 669—672; 5 tables
This anthology of 20 articles (only one of them previously
published) describes different approaches to states of conscious—
ness. There are seven contributions to Part One: "Theory."
Three of these might be of interest to parapsychologists:
"Hypnosis, Multiple Personality, and Ego States as Altered
States of Consciousness," by J.G. Watkins and H.H. Watkins;
"Lucid Dreaming," by S. LaBerge and J. Gackenbach; and
"Personal Experience as a Conceptual Tool for Modes of
Consciousness," by E.M. Pattison and J. Kahan. Part Two:
"Manifestations," contains eight chapters, of which three seem
especially important: "Working in Isolation: States that Alter
Consensus," by F. Jeffrey; "Trance and Possession States," by
E.M. Pattison, J. Kahan, and G.S. Hurd; and "Psi Phenomena
as Related to Altered States of Consciousness," by S. Krippner
and L. George. Part Three: "Accessibility," contains five chap—
ters, all of some interest to parapsychologists. They are:
"Meditation as an Access to Altered States of Consciousness,"
by P. Carrington; "Access to Dreams," by M. Ullman;
"Biofeedback and States of Consciousness," by E.E. Green and
A.M. Green; "Becoming Lucid in Dreams and Waking Life" by
J.R. Malamud; and "The Mystical Way and Habitualization of
Mystical States," by J.H.M. Whiteman. In addition to the Krip—
pner and George chapter, parapsychology is touched on in two
other chapters. Psi experiences in relation to lucid dreaming
are dealt with in the chapter by LaBerge and Gackenbach, and
B. Wolman, in a chapter entitled "Proto—conscious and

Psychopathology," has a subsection about schizophrenia and psi.
JASPR 82:177–182 (Apr 1988); *PAI* 5:45–46 (Dec 1987).
ContempPsychol 32:818–819 (Sep 1987).

Animal Psi

INTRODUCTION

The books in the section on animal psi (or anpsi) in *Sources* are still of value as far as they go.

Since *Sources* was written, four popular books on psi in animals have appeared, the best one being that by Rogo [360]. All provide useful background information, many anecdotal accounts, and some review experimental work as well. The most authoritative overview of the experimental work and the problems involved in investigating psi in animals have been written by R.L. Morris. (See his chapter in Wolman [688] and his article, "Psi and Animal Behavior: A Survey," *JASPR* 64:242−260 (Jul 1970), reprinted in White [546]).

Additional books with information on animal psi: Anpsi is touched on in the book by Barbara Ivanova [390]. There is a chapter on anpsi by Michael Scriven in Thakur [614].

BIBLIOGRAPHY

362 **Bardens, Dennis.** *Psychic Animals: A Fascinating Investigation of Paranormal Behavior.* Foreword by David Bellamy. New York: Henry Holt, 1987. 203p. Bibliography: 195−196; 14 illustrations; Index: 197−203
   Bardens is a journalist and member of the Society for Psychical Research. Here he explores the part ESP may play in the animal world. The book opens with a general chapter that is followed by two chapters each on cats and dogs, and one chapter each on elephants, lions, bears, and nonhuman primates. There also are chapters on homing, animals and magic, animals as gods, and animal intelligence. Although some scientific experiments are briefly summarized, this book is primarily a collection of anecdotes, most of them not investigated. Nonetheless, Bardens has performed a useful service in locating this material and pulling it together. The book is a useful starting point for those who are interested in research on psi in animals.
   JSPR 55:38−39 (Jan 1988).

LibrJ 113:70 (Mar 1, 1988); *PublWkly* 233:78 (Feb 12, 1988).

363 Rogo, D. Scott. *ESP and Your Pet.* New York: Tempo Books, 1982. 171p. Chapter notes; Citations in text; 5 figures; 1 table

This book serves as a general introduction to psi in animals, but the emphasis is on testing the psychic abilities of pets. After two introductory chapters, there is a chapter each on horses and psi, "communicating" animals, homing, and psi trailing. Also included is a chapter on laboratory experiments conducted with animals as subjects. It is followed by a chapter that describes in detail how one can test one's pets for psi ability. "A Final Note" is about compassion in animals.

364 Schul, Bill. *The Psychic Power of Animals.* Greenwich, CT: Fawcett, 1977. 223p. Bibliography: 215–217; In- dex: 218–223

Schul, a popular writer on the occult, has brought together many anecdotes on various anomalous phenomena associated with animals. He begins with a chapter citing cases with various animal species evidencing unusual abilities. There are chapters on animal communication; precognition in animals; ESP experiments with animals; the psychic Terrier, Missie; animals and the supernatural; the American Indians' view of animals; the question of whether animals survive death; animal language; and unaccountable experiences with animals. He closes with a chapter on implications, which deals with the meaning of animals to human beings and the universe, and the role that animals play in the inner growth of human beings.

365 Wylder, Joseph (Pseudonym). *Psychic Pets: The Secret World of Animals.* New York: Stonehill, 1978. 163p. Selected bibliography: 163; 22 illustrations; Index: 151– 161; 1 table

This work presents many examples of what appears to be psi in animals: dogs, cats, horses, birds, dolphins, and others. It also includes descriptions of some experimental attempts to test anpsi. Wylder suggests that at one time people could communicate with animals and that it is possible to regain this lost "sixth sense." ("Joseph Wylder" is a pseudonym used by a research scientist in a pharmaceutical company.)

PR 10:13–14 (Nov/Dec 1979).

LibrJ 104:639 (Mar 1, 1979); *Observer* p39 (Mar 2, 1980); *PublWkly* 214:55 (Nov 13, 1978).

Applications of Psi

INTRODUCTION

This is a new category. There was only one reference to psi applications in *Sources* (Ryzl [178]), but since that book was published interest in the practical aspects of psi has burgeoned, as evidenced by the fact that this chapter is one of the largest in the book: 17 titles. However, it is one thing to write about applying psi in daily life and another to actually do it. The laboratory can sometimes be made to serve as a model of the real world, but experiments in such situations have not been overly successful, and reports of methods that work in actual life situations do not work well under the experimental micro-scope. It is still a large question as to whether we can learn to produce psi on demand (for the current status of this problem, see the section on "Training and Development of Psi"), and un-til we can, the evidence for practical applications of psi will remain primarily anecdotal and conjectural. General sources of information listed here are Inglis [809], the conference proceed-ings edited by Jones [373], and the bibliography on psi applica-tions compiled by White [378]. Both works deal with many specific applications of psi, whereas the other books listed are about various specialized psi applications. The Agor [366] and Dean et al. [369] volumes are concerned with business applica-tions. Allison and Jacobson [367], Hibbard and Worring [371], Tabori [377], and Yeterian [379] are about applications of psi in criminal investigation. Archaeology and psi is the subject of Schwartz's books [376, 816]. Dowsing is dealt with by Bird [368], who not only covers locating water but also various geological applications of dowsing and even its use in criminal investigation. Personal guidance is the primary topic of Kautz [374], and military applications or harmful uses of psi are covered in the books by Ebon [370], Hurley [372], McRae [375], and White [819].

Additional sources of information on applications: All of the books in the section on "Biographies and Autobiographies of Healers, Mediums, and Psychics" contain accounts of psi applica-tions. One of the books for young people, by Rudley [449], is on criminal investigation and psi. Hines [474] has chapters criti-cal of psi applications. Practical applications of psi are emphasized by Bartlett [506] and by Chauvin [510] in their general introductions. Two books in the section on mediumship, posses— sion, and automatism deal with cures of possession through psychotherapy of the so—called possessing spirits: Crapanzano [574] and Rogo [584]. The role NDEs may play in peacemaking and raising the level of planetary consciousness is set forth in Flynn [585] and Ring [592]. In a second book, Ring [593] has a chapter on "Implications and Applications of NDE Research" and Grey [586] describes some applications and implications of NDEs. Several of the books in the section on "Psychology, Psychiatry, and Parapsychology" deal with psychotherapeutic applications of psi: Dean [641], Ehrenwald [642], Eisenbud [644, 645], Mintz [652], Schwarz [655], and Taub—Bynum [658]. All of the books in the section entitled "Training and Development of Psi" [778-791] touch on practical aspects of psi because if one can learn to use psi on demand one can try to use it for practical purposes. Some specific applications covered are the use of intuition in science [Goldberg, 779], healing (LeShan [780]), and forecasting (Loye [781], Vaughan [789, 790]).

BIBLIOGRAPHY

366 **Agor, Weston H.** *The Logic of Intuitive Decision Making: A Research—Based Approach for Top Management.* Foreword by Ken Blanchard. New York: Quorum Books, 1986. 182p. Annotated bibliography: 149—177; Chapter references; Glossary: 147—148; 6 graphs; Index: 179—182; 1 questionnaire; 35 tables
       The author, who is head of a management consulting firm and Professor and Director of the Master of Public Administra— tion Program at the University of Texas, El Paso, does not use any parapsychological terms in the text, but they appear in the glossary because several of the references cited are para— psychological. He summarizes the result of his research on the use of intuition in management. The primary research tool used was the survey method, using the Myers—Briggs Type In— dicator to assess intuition and a questionnaire devised by the author to learn how and under what conditions executives make use of intuition in their work. (The questions asked are given in the appendix.) The book is in three parts: In the first, he reviews his research; in the second he provides a step—by—step method for using and developing intuition to make key deci— sions; the last part is a discussion of future possibilities for intui— tion research. The "Annotated Bibliography of Resources on Intuition" contains sources (books, articles, measurement instru— ments, audio—visual materials) on the nature of intuition, tests

used to measure intuition, and examples of the use of intuition
by executives.
   Choice 24:1257 (Apr 1987).

367 Allison, Dorothy, and Jacobson, Scott. *Dorothy Allison: A
   Psychic Story.* New York: Jove, 1980. 223p.
   Although based on extensive interviews and conversations
with the psychic, Dorothy Allison, the book was written entirely
in the third person by Scott Jacobson. He says the material
for the book was gathered not only from Allison's recollections
but is "supported by newspaper and magazine articles and
signed affidavits from many of the parties involved" (Author's
Note). Most of the cases related involve solving murders and
locating missing persons.

368 Bird, Christopher. *The Divining Hand: The 500–Year–Old
   Mystery of Dowsing.* With hand–drawn illustrations by
   Thomas Heston. New York: Dutton, 1979. 340p. Bibliog–
   raphy: 319–335; 93 figures; 15 graphs; 206 illustrations;
   Index: 336–340; 9 tables.
   This well researched and illustrated (with photographs as
well as drawings) survey of the lore of dowsing by
author/journalist Bird centers on the hypothesis that "faith may
be the true substance of dowsing"––"the ultimate mysterious
'force' that moves the dowsing rod at the appropriate moment"
(p. 15). In addition to historical material, interviews with
living dowsers are included. Scientific and engineering aspects
of dowsing––other than for water––are described, and the
book closes with a theoretical section.
   Archaeus 1:55 (Win 1983); *Kirkus* 47:830 (Jul 15, 1979);
LibrJ 104:1466 (Jul 1979);   *NYTimesBkRev* p12 (Sep 2, 1979);
PublWkly 215:87 (Jun 18, 1979).

**369 Dean, Douglas, Mihalasky, John, Ostrander, Sheila, and
   Schroeder, Lynn.** *Executive ESP.* Englewood Cliffs, NJ:
   Prentice–Hall, 1974. 290p. Bibliography: 268–284; 4
   graphs; 3 illustrations; Index: 287–290; 30 tables.
   This is an account of the work of the Psi Communications
Project at Newark College of Engineering which was formed in
1962 "to develop the techniques and system, using the ap–
proaches of science and engineering, for obtaining reliable infor–
mation via the processes associated with [psi]" (p. v). Douglas
Dean is an electrochemist and parapsychologist who collaborated
on the Psi Communications Project with Project Director John
Mihalasky. They have examined the role of ESP in its applica–
tion and success in business by means of "physiological in–
strumentation, scientific and engineering techniques, and simula–

tions of applied ESP" (p. 1). This book is a report of their research that includes ESP experiments and interviews with top executives. Their research suggests that successful executives make use of ESP in business. Factors which help or hinder ESP performance are also considered. This is one of the few empirically based books on the practical application of psi.
    P 6:34—35 (May/Jun, 1975).
    LibrJ 99:2062 (Sep 1, 1974); *LibrJ* 100:457 (Mar 1, 1975).

370 Ebon, Martin. *Psychic Warfare: Threat or Illusion?* New
        York: McGraw—Hill, 1983. 282p. Bibliography: 262—272;
        Index: 273—282.
    In this book Ebon stresses the importance of research in the Soviet Union. The three chapters on Western research are on Mitchell's Apollo 14 ESP experiment, the SRI remote—viewing experiments, and U.S. government interest in psi. Ebon discusses the pros and cons of a possible psychic arms race.
    JASPR 79:288—291 (Apr 1985); *JP* 47:257—260 (Sep 1983); *PAI* 3:39 (Dec 1985); *ZS* No.12/13:171—173 (Aug 1987).
    HumanEv 45:15 (Aug 24, 1985); *Kirkus* 51:745 (Jul 1, 1983); *LibrJ* 108:1882 (Oct 1, 1983); *NewReal* 6:61—62 (Sep/Oct 1984); *PublWkly* 224:120 (Jul 22, 1983); *Reflect* 3:21 (Spr 1984); *SciBooks&Films* 20:190 (Mar 1985); *SpirFron* 16:60—61 (Win 1984).

371 Hibbard, Whitney S., and Worring, Raymond W. *Psychic
        Criminology: An Operations Manual for Using Psychics
        in Criminal Investigations.* Springfield, IL: Charles C
        Thomas, 1982. 108p. Bibliography: 101—103; Glossary:
        97—99; Index: 105—108; 1 table.
    The authors, past co—chairpersons of the Missoula City/County Crime Attack Team, report on five years of experience in mobilizing psychics for use in criminal investigations. The purpose of this book is to provide interested parties with the necessary background and procedures to establish an effective working relationship with proven psychics. The authors give a brief history of psychic criminology and an introduction to parapsychology. The bulk of the book is a practical manual for identifying, recruiting, testing, and employing psychics as criminal investigators and examples from actual cases are given. One of the chapters is devoted to Stanford's PMIR model as providing a framework in which to view synchronistic happenings in daily life that may guide the investigator in tracking down criminals.
    JASPR 78:280—282 (Jul 1984); *JP* 46:284—285 (Sep 1982); *T* 11:92 (Win 1983).

372 Hurley, J. Finley. *Sorcery.* Boston, MA: Routledge and Kegan Paul, 1985. 232p. Bibliography: 213–225; Chapter notes; Index: 226–232.

Hurley, a forensic psychologist, writes about remote control of human behavior and delves into the cultural antecedents of the phenomenon found in the literature of anthropology, magic, and shamanism. He also reviews relevant findings from psychology and parapsychology about the ability to hex or heal at a distance. He attempts to show that sorcery is theoretically possible and provides instructions on how to practice it. In the last chapter, he speculates on the ramifications of sorcery and the possibility that it may be an unsuspected force in history. His bibliography is extensive and useful, especially for its nonparapsychological citations.

> *JASPR* 82:77–85 (Jan 1988).
> *BritBkNews* p397 (Jul 1985); *GuardianWkly* 132:22 (May 5, 1985); *Man* 21:154–155 (Mar 1986); *ReligStudRev* 12:262 (Jul 1986).

809 Inglis, Brian, with Ruth West and the Koestler Foundation. *The Unknown Guest: The Mystery of Intuition.* [See Addenda for annotation]

373 Jones, C.B. Scott (Ed.). *Proceedings of a Symposium on Applications of Anomalous Phenomena, November 30– December 1, 1983, Leesburg, Virginia.* Santa Barbara, CA: Kaman Tempo, 1983. 514p. Chapter references; 39 figures; 27 graphs; 49 illustrations; 5 tables

The purpose of this volume is to provide an opportunity for several senior parapsychologists to present their assessment of the current status of the ability to apply psi phenomena to practical ends, as well as a projection of what parapsychology might be in 1990, to a group of managers and scientists from various U.S. government departments and agencies. General theoretical and methodological approaches were presented by Harman, Kelly, Roll and Montagno, and Schwartz and an overview of American and Soviet applications was presented by Targ. The ongoing research at Princeton Engineering Anomalies Research Laboratory was described by Jahn. Research at SRI International was described in two papers: Puthoff on remote viewing and May on PK research. The status of several specific applications was reviewed: in human–equipment interactions (Morris), in dowsing (Harvalik), and in criminal investigation (Osis). There are 5 appendices, two containing bibliographies of psi applications.

> *JASPR* 81:297–299 (Jul 1987).
> *SkptInq* 8:271–277 (Spr 1984); *SpirFron* 20:123–124 (Spr

1988).

374 Kautz, William H., and Branon, Melanie. *Channeling: The Intuitive Connection.* Foreword and Forecast by Kevin Ryerson. San Francisco, CA: Harper & Row, 1987. 179p. Bibliography: 177–179; Chapter notes; 2 figures

The authors define channeling as "a mental process in which an individual (the channel), partially or totally sets aside waking consciousness" so that "knowledge that lies beyond conscious awareness can then flow freely into the mind and be conveyed through speaking or writing to others" (pp. 2–3). In this book they set forth how channels can provide practical guidance as well as give psychotherapeutic assistance and foster personal growth. Also described is what individuals seeking assistance can do to benefit most from the assistance provided by channels. One chapter is devoted to advice on how to communicate with one's own intuitive source. In a chapter entitled "Catalyst for Widespread Change," practical benefits of intuition are described and two methods of collective intuition, "intuitive consensus" and "group attunement," are set forth. The authors illustrate the team approach that is beginning to "affect problem solving and decision making in today's society" (p. 124). The book closes with a compilation of forecasts made by the channel Kevin Ryerson and "A Directory of Channels."

PAI 6:44 (Jun 1988).
Booklist 84:187 (Oct 1, 1987).

375 McRae, Ronald M. *Mind Wars: The True Story of Secret Government Research Into the Military Potential of Psychic Weapons.* New York: St. Martin's, 1984. 155p. Bibliography: 139–150; Chapter references; Index: 151–155

Although less authoritative, this book complements Martin Ebon's *Psychic Warfare* in that it emphasizes U.S. government interest in psi. McRae also deals with Soviet research and discusses the possibility of a psychic arms race. Both authors include Truzzi's bibliography of English–language literature on Soviet and U.S. government efforts in parapsychology.

Booklist 80:830 (Feb 15, 1984); *Kirkus* 52:190 (Feb 15, 1984); *SkptInq* 8:271–277 (Spr 1984).

816 Schwartz, Stephan A. *The Alexandria Project.* [See Addenda for annotation]

376 Schwartz, Stephan A. *The Secret Vaults of Time: Psychic Archaeology and the Quest for Man's Beginnings.* New York: Grosset & Dunlap, 1978. 370p. Chapter references:

353—366; 20 figures; 58 illustrations; Index: 367—370.
Writer—lecturer Schwartz describes in depth the major ex—
amples of the use of psi in locating archaeological artifacts and
information, including the work of Frederick B. Bond, S. Os—
sowiecki, Scott Elliot, J. Norman Emerson, Edgar Cayce, and
Charles Garrud. Also included is a chapter about the studies of
Weiant, Swanton, and Long and of what happened at the
Rhine—Swanton Symposium on Parapsychology and Anthropol—
ogy at the 1974 meeting of the American Anthropological As—
sociation. Schwartz also discusses the importance of psychic ar—
chaeology both within the context of the history of that subject
and within science as a whole. The final chapter is an outstand—
ing treatment of the psychology of the sensitive and of the psi
investigator. In two lengthy appendices he outlines in detail how
to conduct an archaeological investigation with a psychic as part
of the team.
    PR 10:15—16 (Mar/Apr 1979).
    LibrJ 103:372 (Feb 1, 1978); *PublWkly* 212:59 (Nov 14,
1977); *SpirFron* 10:125 (Spr 1979).

377 Tabori, Paul. *Crime and the Occult: How ESP and Para—
        psychology Help Detection.* New York: Taplinger, 1974.
        260p. Classified bibliography: 257—258; Index: 259—260
    Paul Tabori, a native of Hungary, is a crime reporter who
has written popular books on parapsychology. In this book he
relates many instances in which a psi factor may have been in—
volved in crime investigation. His main concern, in his words,
has been to include cases "in which someone has solved or
presented a crime purely by some mental process for which we
have no materialistic explanation——at the moment. The men
and women whose stories are told and examined here have ac—
tually achieved this; nor do they particularly care whether the
skeptic believes them or not. The material has been taken from
police archives or the personal memoirs of officials——most of
whom are quite unconcerned with the pros and cons of psychi—
cal research, and who have simply registered the facts" (p. 113).
Subjects covered are hypnosis and suggestion, psychometry, clair—
voyance, motor automatisms such as dowsing or the pendulum,
and precognition. Most of the cases come from the European
countries, Hungary in particular. It is by no means an exhaus—
tive survey, because many of the better known works of
Tenhaeff in Holland with Croiset and other sensitives, are not
included. However, it can be used as a starting point in survey—
ing the applications of psi to the problems involved in solving
crimes.
    TimesLitSuppl p261 (Mar 7, 1975).

820 White, John (Ed.). *Psychic Warfare: Fact or Fiction?*
[See Addenda for annotation]

378 White, Rhea A. *Parapsychology: Sources on Applications and Implications* (2nd ed.). Dix Hills, NY: Parapsychology Sources of Information Center, 1989. 118p.
This is a listing by subject of books, dissertations, articles, and items dealing with the potential applications and implications of psi. There are 1479 items, some dealing with more than one subject. The subjects covered are Psi and Anthropology, Archeology, Behavior Control, Biology, Botanical Applications, Business Applications, Conduct of Life, Creativity, Criminal Investigations, Disaster Prediction, Education, Ethical Implications, Finding Lost Objects, Gambling, Games, Geology, Handicapped, Historical Applications, Human—Machine Interaction, Invulnerability, Journalism, Martial Arts, Meteorology, Military Applications, Locating Missing Persons, Personal Growth, Philosophical Implications, Physics Implications, Political Applications, Problem Solving, Psychiatric and Psychological Applications, Religious Implications, Scientific Applications, Social Implications, Space Exploration, Sports, Unorthodox Healing, and Use for Peace. There is a general section citing publications in which psi applications are discussed in broad terms and another citing basic research that is relevant to applications, such as the repeated—guessing technique.
JP 52:242—244 (Sep 1988); *ChrP* 7:220 (Jun 1988); *PR* 19:12—13 (Sep/Oct 1988).

379 Yeterian, Dixie. *Casebook of a Psychic Detective.* New York: Stein and Day, 1982. 197p. Index: 195—197.
Autobiographical account by a psychic whose psi abilities have been used by police to solve crimes. She tells how her ability developed. Most of the cases involve murder or locating missing persons. The names of the persons involved and, when necessary, place names have been altered to preserve anonymity. In a final chapter, "About Life and Death," she relates what she has learned from her experiences, such as that persons who will be murdered, their family members, often have awareness at some level of the impending death. She has also learned, based on reliving the death experiences of murdered persons, that "in even the most terrible of murder cases . . . death seems to be a beautiful experience" (p. 188).

Biographies and Autobiographies of Healers, Mediums, and Psychics

INTRODUCTION

This is a new section necessitated by the number of autobiographies and biographies of psychics and mediums that have appeared since *Sources* was written. However, a number of autobiographies and biographies were described in *Sources* and they still are considered to be of value. There were three autobiographical works by Eileen J. Garrett [103, 104, 105], and those by Rosalind Heywood [106] and Gladys Osborne Leonard [108]. Also listed were biographies of D.D. Home by his wife [107] and E.J. Dingwall [99], a biography of Mrs. Leonard by S. Smith [114], one of Eusapia Palladino by E.J. Dingwall [100], one of Leonora Piper by her daughter [110], and a book on Hélene Smith by T. Flournoy [102]. A biography of British spirit healer George Chapman [277] was also included, as well as E.T. Smith's [113] collective biography of several psychics and mediums.

In the present section autobiographies predominate (13 of the 19 books listed). These are the books by Cohen and Skutch [383] (psychics), DiOrio [384, 385] (healer), Geller [386] (psychic), Harribance [388] (psychic), Harrison [389] (psychic), Ivanova [390] (healer), Kraft [392] (healer), Manning [393] (psychic), Robbins [395] (psychic), Swann [396] (psychic), Tanous [397] (psychic), and Twigg [398] (medium). The biographical works are by Angoff [380] (on Eileen Garrett, medium), Brian [381] (on Jeane Dixon, psychic), Cerutti [382] (on Olga Worrall, healer), Jenkins [391] (on D.D. Home, medium), and Martin [394] (George Anderson, medium).

Additional books with biographical information on healers, mediums, and psychics: For a bibliographic guide to autobiographies, biographies, and investigations of healers, mediums, and psychics, see White's *On Being Psychic* [791]. Accounts of the psychic work of Dorothy Allison [367], Kevin Ryerson [374], Stephan Ossowiecki, Scott Elliot, J. Norman Emerson, Edgar Cayce, and Charles Garrud [376], and Dixie Yeterian [379] are noted in the section on applications. In the section describing books aimed at young people, Edmonds has written biographies about D.D. Home [432], and Katie and Margaretta Fox [433]. Kettelkamp [439] presents life histories of five psychics: Ingo Swann, Rosemary Brown, the Worralls, and Matthew Manning. Gregory [576] writes of the life of Rudi Schneider. Investigations of Geller and information about him are the subject of two books in the section on psychokinesis: Ebon [624] and Panati [629]. Harvey [798], Kautz [801], Krippner and Villoldo [801], and Villoldo and Krippner [803] provide biographical information on healers.

BIBLIOGRAPHY

380 Angoff, Allan. *Eileen Garrett and the World Beyond the Senses.* New York: William Morrow, 1974. 241p. Bibliography: 232−234; Index: 235−241

Eileen Garrett was not only one of the most gifted mediums ever to be investigated scientifically, but she was also a capable administrator (she organized the Parapsychology Foundation), writer, and publisher. This biography is written by a man who has been a journalist and editor, and for many years has worked for the Parapsychology Foundation while also serving as Director of Public Relations at the Teaneck NJ Public Library. In the introduction, Angoff states: "I worked with Eileen Garrett for twenty−five years, a skeptic like herself about psychic people. But I know her psychic feats cannot be ignored even if they are an affront to reason. I have tried to tell her story objectively. I have gone back to her earliest years, presenting her own words about her psychic experiences when she was five and a 'sinful' child, and those that occurred when she was older and famous as an awesome seer" (p. xi). The first four chapters are about Eileen Garrett as psychic and parapsychologist. The remaining 12 present a roughly chronological review of her life and accomplishments.

JASPR 69:183−185 (Apr 1975); *JSPR* 48:108−109 (Jun 1975); *P* 6:10 (Nov/Dec 1974); *PR* 5:8 (Nov/Dec 1974).

Booklist 70:1119 (Jun 15, 1974); *Choice* 11:671 (Jun 1974); *Kirkus* 42:215 (Feb 15, 1974); *LibrJ* 99:1122 (Apr 15, 1974); *Light* 95:188 (Win 1975); *PublWkly* 205:109 (Feb 25, 1974).

381 Brian, Denis. *Jeane Dixon: The Witnesses.* Foreword by Jeane Dixon. Garden City, NY: Doubleday, 1976. 216p. Bibliography (compiled by Danielle Brian): 213−216

Parapsychologists have paid little attention to Jeane Dixon because she refuses to be investigated. [Brian says "She says her gift to foresee the future comes from God and she has neither

time nor inclination to be a guinea pig" (p. 18)]. Moreover,
the two previous books on Mrs. Dixon, one an autobiography
and one on her experiences in which she collaborated, made no
attempt to substantiate the experiences described. Now, Denis
Brian, an "investigative reporter," novelist, and author who has
increasingly evidenced an interest in parapsychology has rectified
this situation. For each instance of apparent spontaneous psi
he presents Mrs. Dixon's version, the testimony of other prin—
cipals involved or quotes from newspapers or other written ac—
counts, as well as the questions he put to Jeane Dixon and to
the witnesses and their replies. Several cases stand up very
well and some are quite detailed. Scattered throughout are
clues as to the psychology of the process, such as when a
chance word spoken in Jeane Dixon's presence touches off an
ESP impression. Mr. Brian has done an excellent job in track—
ing down the witnesses and in presenting the fullest description
we have of Mrs. Dixon's abilities. In a summary chapter, he
reviews the claims of the skeptics and the believers as regards
psi in general and Jeane Dixon's psychic ability in particular.
He concludes that "the accumulation of evidence in this book is
so remarkable, that the skeptics must find other explanations
than luck, faulty reporting, weak memories, inside
information——to explain it away" (p. 207).
> *JP* 40:166 (Jun 1976).
> *Kirkus* 44:99 (Jan 15, 1976); *LibrJ* 101:724 (Mar 1, 1976);
PublWkly 209:48 (Jan 12, 1976).

382 Cerutti, Edwina. *Olga Worrall: Mystic with the Healing
      Hands.* New York: Harper & Row, 1975. 169p. 1 table
      The author, who has written books and articles in the
field of medicine and is married to an M.D., had no previous
interest in spiritual healing and even considered herself a skeptic
when she and her husband consulted the Worralls in a
desperate search for a cure for an illness of Dr. Cerutti's that
orthodox medicine had failed to cure. Impressed by Olga Wor—
rall as a person——although still skeptical about paranormal
healing——Mrs. Cerutti attended the New Life Clinic conducted
by Olga Worrall every Thursday in the Mount Washington
Methodist Church in Baltimore, interviewed Mrs. Worrall on
many occasions, and became a friend. In this book, Cerutti
gives a firsthand account of Olga Worrall's healing methods and
of her philosophy of life, how her psychic and healing powers
developed, and why she believed in survival. Olga Worrall felt
she had a mission to heal and this religious orientation per—
meated her life and comes through strongly in this book. Some
investigations of Olga Worrall by scientists are described briefly,
but those who want to read further will be frustrated by the

complete lack of any bibliographic references, although some—
times a periodical title is given.
P 6:34–35 (Nov/Dec 1975).
      LibrJ 100:1933 (Oct 15, 1975); *PublWkly* 207:76 (Jun 23,
1975); *SpirFron* 15:178–180 (Fal 1983).

383 Cohen, Tamara, and Skutch, Judith. *Double Vision.*
      Berkeley, CA: Celestial Arts, 1985. 269p. Citations in
      text; 5 figures; 60 illustrations
      This clever, enjoyable, informative book is a joint
autobiography by parapsychologist/promoter Skutch and her
daughter by her first marriage. Ms. Cohen provides many
recollections of how it feels to grow up psychic and to par—
ticipate as a subject in psi experiments. Most of her spon—
taneous psi experiences had to do with members of her family
and they often illustrated aspects of the dynamics of the
mother—child relationship. Ms. Cohen's remarks are of value
especially because they often are quoted from diary notes writ—
ten near the time of the experience described. Skutch outlines
how her interest in parapsychology became kindled and how she
served as a catalyst and educator, culminating with her in—
volvement in *A Course in Miracles*, which she now publishes.
There is much humor, wisdom, and humanity in this book,
which deserves a wide readership.
      JASPR 82:388–391 (Oct 1988); *PAI* 6:43 (Jun 1988).
      LibrJ 111:76 (Feb 1, 1986); *PublWkly* 228:44 (Nov 29,
1985); *SpirFron* 19:182 (Sum 1987).

384 DiOrio, Ralph A. *Called to Heal: Releasing the Transform—*
      ing Power of God. Garden City, NY: Doubleday, 1982.
      260p. 27 illustrations
      A sequel to the following book, here Ralph DiOrio
describes his sense of vocation in regard to healing and presents
a number of case reports and testimonies from persons who
claim to have been healed through his gift. Although expressed
in religious rather than parapsychological or medical terms, his
book is worth reading by anyone with an interest in the modus
operandi of faith healing and the contexts that foster it.
      PAI 4:43 (Dec 1986).
      America 148:174 (Mar 5, 1983); *BestSell* 42:387 (Jan 1983);
Booklist 79:540 (Dec 15, 1982); *LibrJ* 107:2345 (Dec 15, 1982).

385 DiOrio, Ralph A., with Gropman, Donald. *The Man*
      Beneath the Gift: The Story of My Life. New York:
      William Morrow, 1980. 239p. 16 illustrations
      This autobiography by a charismatic Catholic priest who is
a healer describes how healings happen and also describes his

experience of God working through him as an instrument. He discusses various experiences associated with his healing ministry such as the "word" or "energy of knowledge," which enables him to see the affliction of the person to be healed, and "slaying in the spirit," all of which are instrumental in his healing. When healing occurs he is in a trance and feels the Holy Spirit take over. He describes various visions. An especially informative chapter is entitled "How I Experience a Healing Service."

    Booklist 77:654 (Jan 15, 1981); *Kirkus* 48:1126 (Aug 15 1980; *LibrJ* 106:156 (Jan 15, 1981); *PublWkly* 218:59 (Sep 5, 1980).

386 Geller, Uri. *Uri Geller: My Story.* New York: Praeger, 1975. 282p. 41 illustrations

    Wild claims have been made both by and about Uri Geller concerning his seeming paranormal powers. Unfortunately, it has been difficult to get him to perform for bona fide parapsychologists and under strictly controlled conditions, although a few reports have appeared in the professional literature [see **629**]. Equally strained and extravagant claims have been made against Geller [see **480**]. The arguments for and against the genuineness of his phenomena have been dealt with in several books which are described in the sections on "Criticisms" and "Psychokinesis." Here we have his autobiography, which is his own attempt to describe his life, his powers, how they developed, and where he thinks they are leading. There is a genuine ring of truth about the account, and whether it portrays a myth that modern madness has shaped or a glimpse of abilities that can nullify the barriers of time and space, it reads easily and is an unusual document. Geller's story is continued in the title listed next.

    JP 39:242–250 (Sep 1975); *JSPR* 48:352–353 (Jun 1968); *P* 6:46–47 (Sep/Oct 1975); *PS* 2:155–156 (1977).

    Booklist 71:1147 (Jul 15, 1975); *Esquire* 84:30 (Jul 15, 1975); *Kirkus* 43:346 (Mar 15, 1975); *LibrJ* 100:1118 (Jun 1, 1975); *PsycholToday* 10:93 (Jul 1976); *PublWkly* 207:53 (Mar 17, 1975); *PublWkly* 209:97 (Mar 1, 1976); *SciAm* 234:134 (Feb 1976); *SchLibrJ* 22:134 (Sep 1975).

387 Geller, Uri, and Playfair, Guy Lyon. *The Geller Effect.* New York: Henry Holt, 1986. 288p. Citations in text; 6 figures; 51 illustrations; Index: 281–288

    Part I of this autobiographical/biographical treatment of Uri Geller is by popular writer Guy Lyon Playfair. Part II is by Geller and Part III again by Playfair. The book as a whole is aimed at familiarizing the reader with what has happened to

Geller since he published *My Story* [386]. The arrangement is more or less chronological. Part I recapitulates Geller's psychic life through 1975. Then, in the longest part, Geller relates what he has done during the previous decade, including business applications of psi, such as aerial prospecting for minerals, PK—ing computer operations, and dabbling in Middle East politics. In the last five chapters Playfair describes the present status of parapsychology, some firsthand experiences with Geller, and confronts Geller's critics.

    JASPR 83:281 (Jul 1989).

    Booklist 83:1077 (Mar 15, 1987); *Kirkus* 55:108 (Jan 15, 1987); *Ref&ResBkN* 2:2 (Spr 1987); *SpirFron* 20:56—57 (Win 1988).

388 Harribance, Sean, as told to The Reverend H. Richard Neff. *This Man Knows You.* Foreword by J.B. Rhine. San Antonio, TX: Naylor, 1976. 138p. Bibliography: 133—135

    Unlike Geller, Sean Harribance (whose first name in most of the professional parapsychological literature is given as Lal—singh although he has since changed it to Sean), has submitted to scientific investigations of his abilities in several laboratories and by many parapsychologists. J.B. Rhine writes in the Foreword that in this research Harribance "has on the whole had unusual success. He has shown as much patience, persis—tence, and versatility . . . as anyone on record" (p. vii). This book is not so much about the laboratory research as it is an account of Harribance's work as a professional psychic. It supplements the reports published in the parapsychological journals by placing his psi abilities in context, by showing how they provide meaning for Harribance and others, and also be—cause he provides insights into the way psi operates, both in and out of the laboratory.

    JP 40:166 (Jun 1976); *P* 7:45—46 (Sep/Oct 1976); *T* 5:14 (1977).

389 Harrison, Shirley, and Franklin, Lynn. *The Psychic Search.* Portland, ME: Guy Gannett, 1981. 152p. 2 illustrations

    This is a biography of Maine psychic Shirley Harrison, written in collaboration with investigative reporter Lynn Franklin. Part One consists of six chapters, each dealing with a documented example of Harrison's psychic ability. Emphasis is given on instances in which crimes were solved or lost objects or missing persons were found. The second part presents her psychic autobiography, a detailed account of her first experiences of psi, observations about the psi process, and details of inves—tigations of her psi ability, in particular those by W.G. Roll of the Psychical Research Foundation.

**390 Ivanova, Barbara (Maria Mir and Larissa Vilenskaya,
Eds.).** *The Golden Chalice: A Collection of Writings by
the Famous Soviet Parapsychologist and Healer.* San
Francisco, CA: H.S. Dakin, 1986. 192p. Bibliography:
180—186; 9 figures; Glossary: 188—192; 13 illustrations
    A composite work written primarily to acquaint readers
with the Soviet researcher, teacher, and healer, Barbara Ivanova,
it is written by Ivanova herself with contributions by the two
editors, one an American psychologist who met Ivanova on a
visit to Moscow (Maria Mir) and the other a Soviet psi
researcher who worked with Ivanova in the USSR but who
emigrated to the U.S. in 1979 (Vilenskaya). The editors write:
"The major issue of the book is not Barbara's theoretical or
experimental work in psi research, but Barbara herself——her
spirituality, her love for all living creatures, her commitment to
actions which will make our world a better place to live in" (p.
xiii). The work begins with Mir's story of meeting Ivanova in
Moscow and being entrusted by her with her writings on
parapsychology that she hoped could be published in the United
States as a book. These writings and some others form the
bulk of this book. Some of the chapters are written by Ivanova
and Vilenskaya and some by Vilenskaya alone. There are sec—
tions on psychic healing, on phenomena which transcend current
scientific concepts, on "exotic" areas of psi (reincarnation,
electronic voice phenomena), on psi in animals, and a miscel—
laneous section. A bibliography of Ivanova's work is included.
    JASPR 83:177—180 (Apr 1989); *PAI* 6:44 (Jun 1988).
    Light 106:185—186 (Win 1986).

391 Jenkins, Elizabeth. *The Shadow and the Light: A Defence
of Daniel Dunglas Home, the Medium.* London, England:
Hamish Hamilton, 1982. (Distributed in the United
States by David & Charles, North Pomfret, VT.) 275p.
Chapter notes; Index: 268—275; 6 illustrations
    This is a biography based primarily on Home's two
autobiographical works and the two books by his widow. Con—
siderable attention is given to Robert Browning's behavior in
regard to Home and to the role of Jane Lyon in Home's life.
    PAI 3:39—40 (Dec 1985).
    Choice 20:1372 (May 1983); *Kirkus* 51:502 (Jan 15, 1983);
LibrJ 108:590 (Mar 15, 1983); *Spectator* 249:27 (Nov 13, 1982);
Spectator 249:45 (Dec 18, 1982).

392 Kraft, Dean. *Portrait of a Psychic Healer.* New York:
Putnam's, 1981. 187p. 6 tables

Autobiographical account of how a Brooklyn musician dis—
covered he was psychic.   Inexplicable physical phenomena oc—
curred in his presence,   he discovered he was able to heal,
psychometrize, and move objects without touching them.   He
tells how he came to devote himself to healing on a fulltime
basis.   Interested in understanding the nature of his psi ability,
Kraft is willing to submit to scientific investigation.   He
describes how he was tested at Dakin Laboratories, Stanford
Research Institute, Lawrence Livermore Laboratories, and
Science Unlimited Research Foundation.   He also founded the
Foundation for Psychic—Energetic Research to promote scientific
understanding of psychic healing.   Appended is a report by Dr.
John M. Kmetz describing the experiment conducted at Science
Unlimited Research Foundation in which Kraft succeeded in in—
fluencing cancer cells.
   T 11:46—47 (Sum 1983).
   Booklist 12 (Sep 1, 1981); *LibrJ* 106:1552 (Aug 1981); *Light*
95:47—48 (Spr 1975); *PublWkly* 219:37 (May 29, 1981); *SpirFron*
16:63 (Win 1984).

393 Manning, Matthew. *The Link: Matthew Manning's Own*
    Story of His Extraordinary Psychic Gifts. Introduction by
    Peter Bander. New York: Holt, Rinehart and Winston,
    1975. 200p. 1 figure; 36 illustrations; 1 table
   Matthew Manning, an Englishman, has had unusual ex—
periences, beginning with poltergeist phenomena when he was
eleven.   In this book he tells how he has learned to live with
his abilities (automatic drawing, metal bending and other
physical phenomena, automatic writing in languages unknown to
him, e.g., Japanese and Russian, and ESP). Manning has sub—
mitted to scientific investigations in England and Canada. Peter
Bander provides a survey of the scientific investigations of
Manning in a 26—page introduction, and a 21—page appendix
by A.R.G. Owen provides an overview of his research.
   JSPR 48:104—106 (Jun 1975); *P* 6:30—34 (Nov/Dec 1975);
PS 2:157 (1977).
   Choice 12:1371 (Dec 1975); *Kirkus* 43:498 (Apr 15, 1975);
LibrJ 100:1138 (Jun 1, 1975); *PublWkly* 207:53 (Apr 14, 1975);
PublWkly 209:98 (Jun 28, 1976).

394 Martin, Joel, and Romanowski, Patricia. *We Don't Die:*
    George Anderson's Conversations with the Other Side.
    New York: Putnam's, 1988. 302p. Bibliography: 299—
    302; Glossary: 274—277
   Joel Martin, host of his own Joel Martin Show on radio
and cohost, with medium George Anderson, of a weekly cable
TV talk show, *Psychic Channels*, presents Anderson's life story

plus near—verbatim transcripts of questions asked by members of the audience and by phone—in calls. Anderson's method is to write on a pad what he sees and hears when communicating with people who have "passed on" on the behalf of those still living. Several examples are given of the sometimes symbolic mode of communicating that can cause difficulty in transmitting the intended message. Also included are some informal reports of attempts to investigate George by persons who call them—selves parapsychologists but who are not members of the Parapsychological Association. Appendix I is a glossary of some of the symbols George receives psychically along with their in—terpretations. A second appendix presents a "selective history of mediumship." There is a bibliography on the history of psychic phenomena, seances, and religion as well as some general interest books to assist those who want to learn more about the subject.

    Fate 42:105, 108 (May 1989); *Kirkus* 56:43 (Jan 1, 1988); *LibrJ* 113:86 (Apr 15, 1988); *NYTimesBkRev* 93:39 (Mar 27, 1988); *VillVoice* 33:52 (Jul 19, 1988).

395 Robbins, Shawn, as told to Milton Pearce. *Ahead of Myself: Confessions of A Professional Psychic.* Englewood Cliffs, NJ: Prentice—Hall, 1980. 200p. Index: 197—200

    This autobiography is especially insightful into the voca—tional problems involved in being psychic. This not only in—volves finding a way to make a living as a psychic but, before that, discovering that one is psychic and the difficulty involved with "coming out" as a psychic. She describes some of her major predictions and also makes 59 predictions for the next decade. Many of these involve scientific discoveries or world events that could be inferred from facts already known. A sec—tion is devoted to noticing whether one is psychic and how to develop psychic ability. Robbins is not afraid to stress the fact that psychic ability is best when it rests on a foundation of normally acquired information.

    LibrJ 106:360 (Feb 1, 1981); *PublWkly* 218:48 (Nov 14, 1980).

396 Swann, Ingo. *To Kiss Earth Good—Bye.* New York: Haw—thorn Books, 1975. 217p. Bibliography: 199—207; 6 figures; 4 graphs; 16 illustrations; Index: 211—217; 2 tables

    Ingo Swann, although trained as a biologist, is an artist who has strong psychic abilities. *To Kiss Earth Good—Bye* is his personal account of his psi abilities (PK, precognition, and out—of—body experiences), including the research done with him as a subject by Targ and Puthoff at Stanford Research In—

stitute, by Karlis Osis at the ASPR, and by Gertrude Schmeid—ler at City College. Understandably, most psychics are primarily interested in their abilities primarily from a practical and personal point of view, but Swann is intellectually stimu—lated by his experiences and has a scientific interest in under—standing them. There are frequent references to the scientific literature throughout the book, which has a nine—page bibliog—raphy. For his recent ideas on ESP, see his latest book [**784**].

JP 39:347—350 (Dec 1975); *P* 6:44 (Sep/Oct 1975); *PR* 6:6—8 (May/Jun 1975); *PS* 2:156 (1977); *T* 4:15 (Sum 1976).

Kirkus 42:1246 (Nov 15, 1974); *LibrJ* 100:490 (Mar 1, 1975); *PublWkly* 206:58 (Dec 2, 1974).

397 Tanous, Alex, with Ardman, Harvey. *Beyond Coincidence: One Man's Experiences with Psychic Phenomena.* Garden City, NY: Doubleday, 1976. 195p.

Alex Tanous is a psychic who has a doctorate of divinity, formerly taught theology and philosophy, and who now makes his living as a lecturer. In this book Tanous provides an autobiography that stresses the conflict he had in younger years because his psi ability made him "different." He also gives ex—amples of the various ways in which his psi ability has been expressed: precognition, psychic diagnosis and healing, psychic photography, out—of—body experiences, and communication with the dead. Most importantly, he devotes considerable space to describing the scientific investigations in which he has served as subject, those conducted by Dr. Karlis Osis at the American Society for Psychical Research in particular. Also useful are his comments on how it feels to receive psi impressions and his suggestions for developing psi ability.

T 5:14 (1977).

Booklist 72:1220 (May 1, 1976); *Kirkus* 43:1374 (Dec 1, 1975); *LibrJ* 101:725 (Mar 1, 1976); *PublWkly* 209:62 (Jan 5, 1976).

398 Twigg, Ena, with Ruth Hagy Brod. Introduction by Rt. Rev. Mervyn Stockwood. *Ena Twigg: Medium.* New York: Hawthorn Books, 1972. 297p. Citations in text; Index: 289—297

Part I, which comprises two—thirds of the book, consists of an autobiography of the British medium, Ena Twigg. She describes childhood ESP experiences and how her mediumship developed. Considerable space is devoted to the sittings Bishop James A. Pike had with her concerning the death of his son. Part II describes various cases in which persons were helped via Twigg's psychic abilities. Five cases deal with postmortem communications, five involve healing, three involve crime and

violence, and six are cases of persons who turned to her for en—
lightenment, creative inspiration, or the pursuit of knowledge
about mediumship. All are persons of some note. The final
brief section describes some cases of information received
through Mrs. Twigg's mediumship that defy explanation simply
by use of her powers of telepathy or clairvoyance.
    P 4:19—20 (Dec 1972).
    Bks&Bkmn 18:62 (Jun 1973); *Light* 93:50 (Spr 1973); *Lis-
tener* 89:156 (Feb 1, 1973); *PublWkly* 201:50 (May 15, 1972);
PublWkly 203:261 (Jan 29, 1973); *SpirFron* 4:248—249 (Aut
1972); *TimesLitSuppl* p221 (Feb 23, 1973).

Biographies and Autobiographies
of Parapsychologists

INTRODUCTION

A sufficient number of autobiographies or biographies of parapsychologists have appeared to warrant a new section. In *Sources* there was only one autobiography (that of Sir Oliver Lodge [84]) and two semi—autobiographical works that were primarily firsthand accounts of the author's research: one by J.G. Pratt [174] and one by J.B. Rhine [53]. The only biography was Baird [80], although biographical information was included in books about the work of William James [85] and William McDougall [86]. Alan Gauld gives considerable biographical information on F.W.H. Myers, E. Gurney, and E.M. and H. Sidgwick in his *Founders of Psychical Research* [82]. The *Biographical Dictionary of Parapsychology* [200] is still a useful source, though dated.

As in the preceding section, there are more autobiographies than biographies listed here (eight, compared with four biographies). In alphabetical order, they are by Blackmore [400], Dodds [402], Krippner [405], McConnell [406], Moss [407], Rogo [411], and Tart [412]. The biographies are on Sir Oliver Lodge (Jolly [403]), Gaither Pratt (Keil [404]), and J.B. Rhine (Rao [409], L.E. Rhine [410]). Pilkington's [408] work is a collection of autobiographical remarks by 12 parapsychologists. The Berger [399] work is a welcome reference source on American parapsychologists, primarily of the modern period. (A new biographical dictionary of parapsychology is being prepared by Arthur and Joyce Berger but at this writing it has not yet been published.)

Additional books with biographical information on psychical researchers and parapsychologists are Shepard [685], which reproduces the material in the *Biographical Dictionary* [200]—— sometimes updating it and also including many new entries. McConnell's book [497] is to some extent auto—biographical. Rogo [815] presents a largely autobiographical account of his poltergeist investigations. Berthold Schwarz [655] presents an autobiographical account of how his interest in psi developed in an introduction. A biography of Louisa E. Rhine written by her

daughter is included in Rao [740].   There is a biography of
Elisabeth Kübler—Ross in the section on survival [756]
(Although Kübler—Ross is a thanatologist, her subject matter
sometimes overlaps with parapsychology.)

BIBLIOGRAPHY

399 Berger, Arthur S. *Lives and Letters in American Para-*
psychology: A Biographical History, 1850–1987. Jefferson,
NC: McFarland, 1988. 391p.   Bibliographical notes: 329–
362; 11 illustrations; Index: 363–391
   Berger, who is president of the International Institute for
the Study of Death, provides scholars, teachers, and general
readers with an overview of parapsychology in the United States
by focusing on the major figures in the field.   There are six sections, the first five of which cover chronological periods.   The
first, covering 1851–1905, consists of a biography of Richard
Hodgson.   The second covers 1906–1920 and consists of a
portrait of James H. Hyslop.   The third, for the years 1921–
1938, contains portraits of Walter Franklin Prince and William
McDougall.   The years 1939–1980 are covered in the fourth,
which is about Gardner Murphy, Gaither Pratt, and J.B. Rhine.
The fifth section is on the period from 1981–1987, with accounts about or by Laura Dale, Louisa E. Rhine, Gertrude R.
Schmeidler, and Montague Ullman.   The last section contains
summaries of the careers, interests, and work of 25 present—day
parapsychologists.   In "Respice, Adspice, Prospice," Berger discusses the future prospects of parapsychology.
   PR 20:12–15 (May/Jun 1989).
   Choice 26:1252–1253 (Mar 1989).

400 Blackmore, Susan. *The Adventures of a Parapsychologist.*
Buffalo, NY: Prometheus Books, 1986. 249p.   Bibliography: 243–249
   Susan Blackmore became interested in parapsychology
before entering Oxford University, and persisted in that interest
while at college, receiving one of England's first doctorates for
work in parapsychology.   However, as she became involved in
research and discovered that she was not what is known in
parapsychology as a psi—conducive experimenter, her early en-
thusiasm waned and was gradually replaced by skepticism.
Today she is a voluble exponent of the view that there is as
great a need as ever to study the experiences people have that
have been identified as paranormal or parapsychological——but
with no need to involve psi.   Blackmore feels that psychology is
the answer to parapsychology. This informally   frank and in-

timate autobiography tells how she came to this position. She still considers herself a parapsychologist, yet her book was published by Prometheus Books, a publisher that specializes in books debunking parapsychology and various pseudosciences.

    ChrP 7:31–32 (Mar 1987); *JASPR* 82:374–384 (Oct 1988); *JP* 52:177–180 (Jun 1988); *JSPR* 54:219–221 (Jul 1987); *PAI* 6:42–43 (Jun 1988).

    Choice 24:1292 (Apr 1987); *Listener* 17:31 (Mar 26, 1987); *Nature* 325:670 (Feb 19, 1987); *PsycholRec* 37:441 (Sum 1987); *SciBooks&Films* 23:6 (Sep 1987).

401 Brian, Denis. *The Enchanted Voyager: The Life of J.B. Rhine.* Englewood Cliffs, NJ: Prentice–Hall, 1982. 367p. Chapter notes; 21 illustrations; Index: 359–367

    The author, who has written a book on Jeane Dixon, presents the first full–length "authorized" biography of pioneer parapsychologist Joseph Banks Rhine. In reviewing Rhine's life, Brian perforce must also review the history of American parapsychology. Much of the book Brian devotes to a chronicle of Rhine's presentation of his results, the response from critics, and his response to their criticisms. The book is based largely on interviews with Rhine, his colleagues, critics, subjects, and scientists in other fields, as well as Rhine's correspondence and published writing.

    JP 46:273–275 (Sep 1982); *PR* 14:10–11 (Jul/Aug 1983); *T* 11:93 (Win 1983).

    Booklist 78:1046 (Apr 15, 1982); *Discover* 3:76 (Jul 1982); *Kirkus* 50:243 (Feb 15, 1982); *LibrJ* 107:722 (Apr 1982); *NewReal* 4:46 (Jul 1982); *SciBooks&Films* 18:87 (Nov 1982); *SciFic&FanBkRev* p9 (Sep 1982).

402 Dodds, Eric Robertson. *Missing Person: An Autobiography.* New York: Oxford University Press, 1977. 202p. 9 illustrations; Index: 197–202

    This is the autobiography of a prominent classical scholar and British parapsychologist. Chapter XI, "Universal Question Mark," provides an account of his lifelong interest in psychical research, with a special interest in survival, and a summation of an entire period in the history of parapsychology. He concludes that the future of the subject belongs not to "scientifically illiterate amateurs like myself" but to the natural scientists—— those who will apply quantitative methods of investigation to the study of psi phenomena.

    JSPR 49:754–755 (Mar 1978).

    Choice 15:858 (Sep 1978); *GuardianWkly* 118:22 (Jan 22, 1978); *Listener* 99:90 (Jan 19, 1978); *NewYorker* 54:143 (Mar 13, 1978); *NYRevBks* 25:31 (Apr 6, 1978); *NewRepub* 177:28 (Nov

26, 1977); *Observer* p28 (Nov 20, 1977); *Observer* p21 (Dec 18, 1977); *SewaneeRev* 86: pR90 (Jul 1978); *TimesLitSuppl* p1311 (Nov 11, 1977).

403 Jolly, W.P. *Sir Oliver Lodge: Psychical Researcher and Scientist.* Rutherford, NJ: Fairleigh Dickinson University Press, 1975. 256p. Chapter notes; 10 illustrations; Index: 251—256

Lodge was a pioneer in parapsychology as well as in physics. He assisted in an experiment with Malcolm Guthrie, the first well—designed successful ESP test. His deserved renown as a scientist and educator lent respectability to the new science of psychical research, which he and others were bringing to birth. This is the first full—length biography of Lodge, and Jolly, himself a physicist and a biographer of Mar—coni, describes Lodge's scientific, social, and personal life. At least half of the book, as was at least half of Lodge's life, is devoted to his work on psychical research, including his study of cross correspondences and mediumship and how he became convinced of survival.

JP 39:147—148 (Jun 1975); *JSPR* 48:101—103 (Jun 1975); *PR* 5:8—9 (Nov/Dec 1974).

Economist 252:85 (Aug 3, 1974); *LibrJ* 100:1208 (Jun 15, 1975); *TimesLitSuppl* p146 (Feb 7, 1975).

404 Keil, Jürgen (Ed.). *Gaither Pratt: A Life for Para—psychology.* Jefferson, NC: McFarland, 1987. 295p. Bibliography: 283—287; 3 figures; 28 illustrations; Index: 293—295; 12 tables

Gaither Pratt was one of the first full—time para—psychologists and his contributions to the field were many and varied. He was J.B. Rhine's chief assistant at the Para—psychology Laboratory at Duke University for five years. Then, at the Division of Parapsychology of the University of Virginia he solidified a position as a world—wide ambassador and leader of parapsychology. The purpose of this book is both to present details of his life and to document his contributions to the field. In a frank Introduction, Keil describes how Pratt's leadership abilities were diminished while he worked with Rhine even though he was a mainstay of the Duke laboratory. Keil also contributes a 40—page biography of Pratt. Seven of Pratt's key publications are reprinted with editorial notes. Martin Johnson's "Gaither Pratt as a Friend and Colleague," Ian Stevenson's "Gaither Pratt as a Scientist," and a bibliography of Pratt's work by Laura A. Dale, complete the volume.

JASPR 83:258—263 (Jul 1989); *JP* 52:237—240 (Sep 1988); *PAI* 6:44 (Jun 1988); *PR* 19:10—13 (Sep/Oct 1988).

405 Krippner, Stanley. *Song of the Siren: A Parapsychological Odyssey.* New York: Harper and Row, 1975. 311p. Bibliography: 293–307; Index: 309–311

Stanley Krippner, a psychologist as well as parapsychologist, gives an autobiographical account of the 10 years of his life which he devoted to full–time research in parapsychology. Because he has travelled to centers of parapsychological research all over the world, especially in Eastern Europe and the USSR, and because he has also engaged in many different aspects of parapsychological research, this book can serve as a readable introduction to who has done what and where in the world of modern parapsychological research. The final chapter is a distillation of his thought concerning the nature of psi: what we know and how best to discover more.

IJParaphys 9:124–125 (1975); *JP* 1976 (Jun), 40:155–156 (Jun 1976); *JASPR* 70:419–423 (Oct 1976); *JSPR* 48:345–347 (Jun 1976); *P* 6:42–44 (Jan/Feb 1976); *PR* 8:24–26 (Mar/Apr 1977); *PS* 2:162–163 (Feb 1977).

Choice 12:1503 (Jan 1976); *GiftedChildQ* 20:15–16 (Spr 1976); *LibrJ* 100:2331 (Dec 15, 1975); *RevBks&Relig* 5:9 (Mar 1976); *StLukeJTheol* 19:304–307 (Sep 1976).

406 McConnell, R.A. *Parapsychology in Retrospect: My Search for the Unicorn.* Pittsburgh, PA: Author, 1987. 228p. Chapter references; 3 figures; 15 graphs; Index: 222–228; 4 tables

Biophysicist/parapsychologist McConnell writes that this "is a collection of essays, memoranda, and research reports concerning parapsychology as a science in an irrational world" (p. ix). Most of the chapters have an autobiographical slant. The first part, called "Light," contains seven chapters, the first being the editor/publisher's autobiographical account of why he left physics for parapsychology. The second describes his attempt to distribute 4,800 copies of his privately published books to scientists world wide and contains an appendix on self–publishing in parapsychology. The third chapter describes his efforts to obtain opinions about one of his PK experiments (which appears as Chapter 4) by sending a copy to over 500 scientists and scholars, only three of whom made substantial replies (those by Luis Alvarez and J.S. Bell are reproduced in the chapter). In the following chapters he describes what he learned from counseling persons claiming psychic experiences who sought him for help. In Chapter 5, McConnell speculates that PK may be operating in the work of scientists in other fields, and presents Neal Miller's rat experiments as an example. In the next chapter he discusses the possibility of experimenter

psi effects in research in biofeedback, hypnosis, multiple per—
sonality, and Ego—State Therapy.   In Part II, entitled
"Shadow," he says he "has assembled parapsychological docu—
ments of a sociologically illuminating kind. Detailed case his—
tories [involving the author] illustrate how, in parapsychology as
in life generally, by loose symbolism and false logic, all of us
distort the real world for our individual gain and to our collec—
tive disadvantage" (p. x). The eighth chapter tells how the
author took up self—publishing because in his opinion the
editors of parapsychological journals discriminated against him.
The next chapter is aimed at demonstrating reality avoidance in
some reviews of his books. Chapter 10 tells of the manipula—
tion techniques used by administrators to mismanage the
author's parapsychology funding.   Chapter 11 relates self—
deception within parapsychology in the reports of J.W. Levy,
the assistant J.B. Rhine (rightly) reported for doctoring his
research results. The following chapter is about discrimination
within the Committee for the Scientific Investigation of the
Paranormal. In the final chapter, McConnell examines the col—
lective future of humankind, which he considers to be quite
bleak, but he closes with the hope that parapsychology might
save civilization as we know it.
     PAI 6:45—46 (Jun 1988).

407 Moss, Thelma. *The Body Electric: A Personal Journey Into
     the Mysteries of Parapsychological Research, Bioenergy,
     and Kirlian Photography.* Los Angeles, CA: J.P. Tarcher,
     1979. 223p. 53 illustrations; Index: 221—223
     This is an autobiographical account of the academic career
of a UCLA medical psychologist (and former actress—script
writer). Moss founded a laboratory at the UCLA Neuropsych—
iatric Institute to investigate psychic healing, ESP, and Kirlian
photography, and is a pioneer in this country in the latter,
having consulted with Soviet researchers.)   Although her main
interest is in unusual biological energies, whether or not they
are parapsychic, described here is her doctoral dissertation on
subliminal perception and ESP, and her well known investiga—
tions of agent—percipient relationships in GESP experiments
with emotional stimuli as targets. This book is included
primarily because of the autobiographical element. It is impor—
tant to study the motivation of psi experimenters who are able
to elicit results.   Moss has been a psi positive experimenter,
and she has aimed to capture that "something more" in human
nature, and to understand that "none of us, none of us, is
made completely of matter" (p. 220).
     LibrJ 105:110 (Jan 1, 1980); *PublWkly* 216:69 (Nov
19, 1979); *SocForces* 16:177—179 (Sum 1984); *SpirFron*

16:177—179 (Sum 1984).

408 Pilkington, Rosemarie (Ed.). *Men and Women of Para-psychology: Personal Reflections.* Foreword by Stanley Krippner. Jefferson, NC: McFarland, 1987. 173p. Bibliography: 161—167; Chapter references; Index: 169—173

The author interviewed 12 persons over 65 who had devoted most of their careers to psychical research (parapsychology). Her aim was to present "personal reflections by the 'elder statespersons' of the field . . . [as] a means of sharing at least a little of their wisdom with those not fortunate enough to know them" (p. vii). Each of the persons interviewed was asked to address five areas: how they became interested in parapsychology, what they felt were their most important contributions, what they might have done differently or how their beliefs might have changed as a result of being in the field of parapsychology, what unusual experiences might they have had that exceeded their previous expectations, and what advice they would give to newcomers to the field. Those interviewed were four psychiatrists (Jan Ehrenwald, Jule Eisenbud, Emilio Servadio, Montague Ullman), a psychologist (Gertrude Schmeidler), a biologist (Bernard Grad), a physicist (Joseph H. Rush), two individuals with literary careers (Renée Haynes and George Zorab), and three full—time parapsychologists (Hans Bender, Eileen Coly, and Karlis Osis). Bibliographies are included for each author and there is a general bibliography at the end of the book. In an interesting Foreword, Stanley Krippner describes a pattern he discerned while reading these essays. This pattern is evident in the factors that predisposed the persons included to enter the field, in the precipitating factors that led each to become identified with parapsychology, and in the maintenance factors that made possible their continued involvement in the field. He uses examples from the book to illustrate his points.

JSPR 55:287—289 (Jan 1989); *PAI* 6:46 (Jun 1988). *Choice* 25:974 (Feb 1988).

409 Rao, K. Ramakrishna (Ed.). *J.B. Rhine: On the Frontiers of Science.* Jefferson, NC: McFarland, 1982. 263p. Chapter references; 1 illustration; Index: 257—263; Suggested reading list: 233—256

This volume is a collection of essays about J. B. Rhine and the significance of his work written by parapsychologists who knew him and were intimately familiar with his work. Most of the papers were given originally at a conference held in memory of Rhine. Those by Chari, Dommeyer, and Robinson were written specifically for this volume. There is a chronology of important events in J. B. Rhine's life, a bibliography of his works, and one on works about him. The other essays are by Beloff, Child, Feather, Hall, Inglis, Johnson, Kanthamani, Koestler, Mackenzie, Mauskopf, McMahan, Palmer, Rao, L. E. Rhine, Schmeidler, Truzzi, and McVaugh. The subject of this book is not only J. B. Rhine and his work but parapsychology itself for roughly the years 1930—1960. Many of the essays place Rhine and parapsychology in the larger context of history, psychology, philosophy, and science.

JP 46:355—359 (Dec 1982); JSPR 52:207—209 (Oct 1983); PAI 4:69 (Jun 1986); PR 15:9—12 (May/Jun 1984).

ContempPsychol 29:80—81 (Jan 1984); LibrJ 108:137 (Jan 15, 1983); Light 103:133—135 (Aut 1983); SciBooks&Films 19:94 (Nov 1983).

410 Rhine, Louisa E. *Something Hidden.* Jefferson, NC: McFar—
land, 1983. 287p. 29 illustrations; Index: 283—287
    This book began as a biography of J. B. Rhine by his wife but instead became an account of their life and work together; thus, it can also be viewed as L.E. Rhine's autobiog—raphy. Based on letters, notes, and the author's diary, it con—tains many details not available elsewhere. The history of American parapsychology is here revealed by L. E. Rhine's in—timate account of her husband's participation as well as of the impact on him of the words and action of others. His reaction, in turn, influenced the entire field of that period.
    JP 47:361—365 (Dec 1983).
    Choice 21:204 (Apr 1984); LibrJ 109:82 (Jan 1984); Light 104:78—81 (Sum 1984).

411 Rogo, D. Scott. *In Search of the Unknown: The Odyssey of
a Psychical Investigator.* New York: Taplinger, 1976.
190p.
    Most young people who enter parapsychology today, as Rogo points out, take the academic route, and become trained in experimental science with the hope of eventually joining a laboratory. Rogo is an exception. He has specialized, in his terms, as a "field investigator——that person who goes out criti—cally and cautiously in search of the paranormal and tries to get to the root of a report of a psychic or haunting. This area is . . . difficult because one must be aware of how people will willfully fake psychic phenomena, the many normal and anomalous physical phenomena which often pass for psychic phenomena and psychological effects that can smack of pseudo—ESP" (p. 13). This book is Rogo's account of his ex—periences as a field investigator of psi phenomena. Its value lies not so much in the genuine phenomena he observed (he would

be the first to say that there are very few grains of wheat compared to all the chaff), but in describing the many normal means that can account for what naive persons might assume to be psi phenomena. He covers psychometry, billet reading, haunted houses, poltergeists, psychic healing (including surgery), Kirlian photography, and direct voice. Rogo has also engaged in laboratory testing for ESP, and devotes a chapter to out—of—body research with Blue [now Keith] Harary. He closes with a chapter on Attila von Szalay, whose tape—recorded direct voice phenomena Rogo is convinced were genuine, albeit ignored by establishment parapsychologists.

    JP 4094:333 (Dec 1976); *P* 7:46—47 (Sep/Oct 1976).

    Choice 13:1362 (Dec 1976); *LibrJ* 101:1539 (Jul 1976); *PsycholToday* 10:135 (Nov 1976); *PublWkly* 209:53 (Apr 26, 1976).

412 Tart, Charles T. *Psi: Scientific Studies of the Psychic Realm.* New York: Dutton, 1977. 241p. Bibliography: 223—234; 17 figures; 3 graphs; Index: 235—241; 6 tables

    Although this book is an account of the author's research and thinking about psi, it can also be viewed as a parapsy—chological autobiography of this well—known parapsychologist, psychologist, and pioneer in altered states research. There are five chapters in the first part dealing with models of psi opera—tion, the fifth of which, "Altered States of Consciousness and Psi," is an important contribution. Part II, "Studies," consists of four chapters on his research. The final part, "Speculation," concerns the nature of consciousness and of psi.

    JIndPsychol 2:156—157 (Jul 1979); *JP* 42:210—219 (Sep 1978); *JASPR* 72:274—279 (Jul 1978); *JSPR* 49:758—759 (Mar 1978); *PN* 2:2 (Jan 1979); *PR* 9:17—18 (Sep/Oct 1978); *T* 7:15 (Spr 1979).

    Kirkus 45:1084 (Oct 1 1977); *LibrJ* 102:2353 (Nov 15, 1977); *SpirFron* 12:60—61 (Win 1980).

Books on Parapsychology
for Young People

INTRODUCTION

This is a new category reflecting the increase in books on parapsychology, or aspects of it, directed at children and young adults. There was a scattering of books suited to young adults in *Sources*: Gaddis and Gaddis [26], Litvag [38], J.B. Rhine [55], J.B. Rhine and Brier [56], Sinclair [61], MacKenzie [76], Ebon [81], Garrett [105], Heywood [106], E. Smith [113], S. Smith [122], Freud [147], Gudas [159], Pierce [172], MacKenzie [233 and 234], L.E. Rhine [237, 274]. With the possible exception of Gudas and L.E. Rhine [274], none of these titles was written specifically for young people, but they were all at a level that young people could understand.

In this section, the books aimed at younger children are those by Akins [413], Aylesworth [418], Cohen [424, 427, 428], Deem, Kettelkamp [439], Lambert [444], and Maynard [445]. Most of the 38 books in this section are about historical aspects of parapsychology, such as accounts of spiritualism and mediums of the nineteenth century or the first half of the twentieth century, or else accounts of haunts, poltergeists, apparitions, and instances of spontaneous ESP. Those that deal to some extent with laboratory investigations are Aylesworth [417], Cohen [421], Curtis [429], Ebon [430], [431], Feola [435], Heaps [437], Kettelkamp [441], and Klein [442]. Of the latter, those by Cohen, Ebon, Feola, and Rhine are the most detailed.

Additional books with information for young people: Armstrong [451] includes useful annotated bibliographies on nonordinary experiences in children's literature. L.E. Rhine's nontechnical introduction to parapsychology [532] contains information aimed at students. Two nontechnical books with a question and answer format that would be of interest to young people were written by Braud [509] and by Hintze and Pratt [517]. Ebon's book [778] on the dangers of dabbling with the occult and efforts to develop psi is a readable and useful book for young people.

BIBLIOGRAPHY

413 Akins, W.R. *ESP: Your Psychic Powers and How to Test Them.* Illustrated by Terry Fehr. New York: Franklin Watts, 1980. 59p.

2 figures; Glossary: 53—54; 8 illustrations; Index: 57—59; Suggested reading list: 55; 7 tables

This is "a first book" describing some elementary tests for ESP (clairvoyance, precognition, telepathy) that young people can do themselves. It contains a glossary   and a brief list of further readings.

SchLibrJ 27:66 (Sep 1980).

414 Anderson, Jean. *The Haunting of America: Ghost Stories from Our Past.* Boston: Houghton Mifflin, 1973. 171p. 18 illustrations

This is a collection of some two dozen of the more interesting American ghost stories. Emphasis is placed on those involving apparitions of famous Americans, haunted landmark houses, and the legendary spirits that emerged during our history. The entire range of American history is spanned from the colonies to the 20th century.

Booklist 70:442 (Dec 15, 1973); *Commonweal* 99:214 (Nov 23, 1973); *Kirkus* 41:1171 (Oct 15, 1973); *LibrJ* 98:3720 (Dec 15, 1973); *PublWkly* 204:61 (Oct 15, 1973).

415 Andreae, Christine. *Seances & Spiritualists.* New York: J.B. Lippincott, 1974. 157p. 18 illustrations; Index: 155—157; Suggested reading list: 151—153

According to the author, this is "an introduction to some of the more intriguing people and phenomena connected with spiritualism" (p. 8). She describes many of the mediums of the nineteenth century such as Andrew Jackson Davis,   the Fox sisters, and Daniel Dunglas Home, but she also covers some twentieth century psychics such as Edgar Cayce, Arthur Ford, Eileen Garrett, and Robert Monroe. She deals with the pros and cons concerning the existence of such subjects as tabletipping, spirit photography, reincarnation, survival of death, and out—of—body experiences, and provides a firsthand account of a seance. There is an annotated list of suggestions for further reading.

Booklist 71:240 (Oct 15, 1974); *Kirkus* 42:1015 (Sep 15, 1974); *LibrJ* 99:3050 (Nov 15, 1974).

416 Atkinson, Linda. *Psychic Stories Strange but True.* Illustrated by Marc Cohen. New York: Franklin Watts, 1979. 95p. 21 illustrations; Index: 93—95

The author presents cases of telepathy, clairvoyance, precognition, out—of—body experiences, and psychokinesis. There is also a chapter discussing the views of scientists about psi, for and against. Several simple tests that can be done at home are described.

Booklist 76:347 (Oct 15, 1979); SchLibrJ 26:52 (Feb 1980).

417 Aylesworth, Thomas G. *ESP.* Illustrated by George Mac—Clain. New York: Franklin Watts, 1975. 64p. Bibliography: 60; Glossary: 58—59; Index: 61—63; 10 illustrations; 1 table

Parapsychology, the study of psi, is defined and a brief history of it is given. Cases and some research findings are presented for clairvoyance, telepathy, precognition, and psychokinesis. There are chapters on dreams and ESP, Russian parapsychology, and one that presents pro and con views.

ACSB 9:6 (Spr 1976); *Booklist* 72:229 (Oct 1, 1975); *Kirkus* 43:779 (Jul 15, 1975); *SchLibrJ* 22:51 (Jan 1976); *SciBooks&Films* 12:2 (May 1976).

418 Aylesworth, Thomas G. *Spoon—Bending and Other Impossible Feats.* Illustrated by Wendy Mansfield. St. Paul, MN: EMC Publishing, 1980. 64p. 5 figures; 47 illustrations

Aylesworth, author of several books on the occult, aims this work at young people. He explores levitation, fire—walking, dowsing, precognition, eyeless vision, metal—bending, and physical invulnerability. He gives well—known examples, describes how some phenomena may be explained away or duplicated by magicians, but concludes that a residue remains. "The newest frontier in science is the mind. As we further explore and understand the brain, we may find that it contains a reservoir of unused abilities. Someday, we may routinely perform mind—over—matter tricks that will put all of our magicians out of business" (p. 64).

419 Berger, Melvin. *The Supernatural: From ESP to UFOS.* New York: John Day, 1977. 117p. Index: 115—117; Suggested reading list: 108—111

This book, by the author of over 30 books on science for young readers, consists of nine chapters, six on parapsychology: ESP, parapsychology, psychokinesis, spiritualism, faith and healing, and three on the subjects of astrology, witchcraft and UFOS. It is a good introduction to young people, presenting evidence for each aspect and alternative explanations. Suggestions for experimentation are also given. A list of books and a directory suggest further sources of information.

BablBookworm 6:4 (Apr 1978); *Booklist* 74:809 (Jan 15, 1978); *CenChildBksBul* 31:154 (Jun 1978); *HornBook* 54:308 (Jun 1978); *Kirkus* 45:1200 (Nov 15, 1977); *SchLibrJ* 24:135 (Mar 1978); *SciBooks&Films* 14:142 (Dec 1978).

420 Cohen, Daniel. *The Encyclopedia of Ghosts.* New York: Dodd, Mead, 1984. 307p. Annotated bibliography: 305—307; 4 figures; 65 illustrations

Skeptical science writer Cohen's *Encyclopedia of Ghosts* is an historical casebook of haunts and apparitions, some with their origin in folklore and legend. He narrowed the selection down to English—language accounts from the last 40 years. Even within those confines, it would be impossible to be ex—haustive, so he "tried to include cases that are famous, sig—nificant, representative, or unique" (p. xii).

AmRefBksAnn 16:245 (1985); *Booklist* 81:667 (Jan 15, 1985); *EngJ* 74:80 (Oct 1985); *LATimesBkRev* p8 (Dec 30, 1984); *SchLibrJ* 31:92 (Feb 1985); *VillVoice* 30: 39 (Jan 1, 1985); *VoiceYouthAdv* 8:206 (Aug 1985).

421 Cohen, Daniel. *ESP: The New Technology.* New York: Julian Messner, 1986. 116p. Selected bibliography: 115—116; 16 illustrations

This excellent up—to—date introduction to parapsychology for young people begins with a visit to a modern parapsychol—ogy research facility, Psychophysical Research Laboratories. There are informative chapters on remote viewing, psi games, magicians and psi, animals and psi, poltergeists, and survival of death.

ASBYP 20:16 (Fal 1987); *Booklist* 82:1611 (Jul 1986); *CenChildBksBul* 39:204 (Jan 1986); *LATimesBkRev* p10 (May 10, 1987); *SciBooks&Films* 23:22 (Sep 1987).

422 Cohen, Daniel. *ESP: The Search Beyond the Senses.* New York: Harcourt Brace Jovanovich, 1973. 187p. Classified bibliography: 179—180; Glossary: 175—178; 1 figure; 13 illustrations; Index: 181—187

This is a somewhat skeptical history of parapsychology by journalist and free—lance writer Cohen. He begins with the period before ESP was scientifically studied——when it was as—sociated with prophets and visionaries. He covers the investiga—tions of the Society for Psychical Research in England, and then the heyday of ESP and tests. There are chapters on the "Current Scene" and "Psychic Spectaculars." In a chapter en—titled "Strange Abilities" he discusses dowsing, eyeless vision (or dermo—optic perception), psychic photography, healing, stigmata, poltergeists, and OBEs. In each instance he tries to offer nonpsi hypotheses to account for the results, and he concludes that the case for psi is not proven, but that "the search remains as frustrating, yet as compelling" (p. 174) as it did in SPR founder Henry Sidgwick's day.

JP 39:76–79 (Mar 1975);
Booklist 70:733 (Mar 1, 1974); *CenChildBksBul* 27:155 (Jun 1974); *Kirkus* 41:1214 (Nov 1, 1973); *Kliatt* 12:32 (Win 1978); *LibrJ* 99:1225 (Apr 15, 1974).

423 Cohen, Daniel. *Ghostly Animals.* Garden City, NY: Doubleday, 1977. 81p. Bibliography: 77–78; 1 figure; 16 illustrations; Index: 79–81
Here Cohen presents the best–evidenced accounts of animal ghosts. The author points out that although many apparently reliable people believe the stories are true, he remains skeptical. The reader must decide.
PublWkly 212:69 (Aug 8, 1977); *ReadTeacher* 32:608 (Feb 1979); *SchLibrJ* 24:68 (Nov 1977).

424 Cohen, Daniel. *How to Test Your ESP.* Photographs by Joan Menschenfreund. New York: Dutton, 1982. 52p. 16 illustrations; Index: 48–52
This introduction to ESP and PK by author Daniel Cohen is simple enough for gradeschoolers to understand. He covers clairvoyance, telepathy, precognition, dreams, and psychokinesis, devoting a chapter to each. In each instance he describes a well–known case to illustrate the phenomenon in question and then describes a simple do–it–yourself test for it, including precautions against sensory cues and mathematical evaluation of the results. The index is very complete, and information, including addresses, is given for those who want to find out more.
Booklist 78:1519 (Aug 1982); *Booklist* 80:1111 (Apr 1 1984); *HiLo* 4:4 (Nov 1982); *SchLibrJ* 29:72 (Jan 1983)

425 Cohen, Daniel. *In Search of Ghosts.* New York: Dodd, Mead, 1972. 182p. Bibliography: 174–178; 1 figure; 44 illustrations; Index: 179–182
The author says this is a "book of 'real' or 'true' ghost stories——that is, stories that people believed in at one time, and that many people still believe in today" (p. 14). He has tried to present those stories in which the evidence for the existence of ghosts seemed to be the soundest. Also included are mediumship, magicians, haunted houses, poltergeists, apparitions, and spirit photographs.
PR 3: 17–18 (Sep/Oct 1972);
Booklist 68:907 (Jun 15, 1972); *Kirkus* 40:337 (Mar 15, 1972); *LibrJ* 97:3471 (Oct 15, 1972); *NYTimesBkRev* p8 (Apr

16, 1972).

426 Cohen, Daniel. *Real Ghosts.* New York: Dodd, Mead, 1977.
125p. 31 illustrations; Index: 123—125
   Most of the accounts of ghosts presented here were taken
from the files of the Society for Psychical Research. Cohen
devotes one chapter to what he thinks is the best ghost story,
the Lt. McConnell case. Next he describes the Hornby case,
which he considers the worst case and he shows how it il—
lustrates some of the problems involved in investigating cases.
Also presented are the Cheltenham ghost, Katie King, spirit
photography, the *Exorcist*, the Seaford case, and a case of
retrocognition, *An Adventure.*
   Booklist 74:548 (Nov 15, 1977); *Kirkus* 45:729 (Jul 15,
1977); *SchLibrJ* 24:110 (Oct 1977).

427 Cohen, Daniel. *The World's Most Famous Ghosts.* New
York: Dodd, Mead, 1978. 112p. 38 illustrations; Index:
110—112
   Aiming the book specifically at young people, Cohen
describes some well—known ghosts and haunted places.
   Booklist 74:1738 (Jul 15, 1978); *NYTimesBkRev* p45 (Apr
30, 1978); *SchLibrJ* 25:132 (Sep 1978); *TopNews* 36:199 (Win
1980).

428 Cohen, Daniel. *Young Ghosts.* New York: Dutton, 1978.
1978. 90p. Bibliography: 85—86; 2 figures; 13 illustra—
tions; Index: 87—90
   This work consists of accounts of child ghosts or ghosts
whom only children have seen. The author's aim is not to
prove or disprove the existence of ghosts but to present some
enjoyable tales that will make the reader wonder.
   Booklist 75:808 (Jan 15, 1979); *CenChildBksBul* 32:132
(Apr 1979); *SchLibrJ* 25:143 (Oct 1978).

429 Curtis, Robert H. *On ESP.* Illustrated by Allen Davis
and Joan Freed Curtis. Englewood Cliffs, NJ:
Prentice—Hall, 1975. 86p. 6 figures; Glossary: 82—83; 51
illustrations; Index: 84—86
   This is a nonsensational elementary introduction to para—
psychology written by a skeptical medical doctor in consultation
with several parapsychologists.   He covers ESP, telepathy,
precognition, clairvoyance, PK, faith healing, and dreams and
psi.   Suggested tests that young people may carry out are given
in each chapter. It is suitable for grade school readers.
   CenChildBksBul 29:25 (Oct 1975); *Kirkus* 43:515 (May 1,
1975); *SchLibrJ* 22:52 (Jan 1976); *SciBooks&Films* 11:207 (Mar

4, 1976).

429a Deem, James M. *How to Find a Ghost.* Illustrated by
   True Kelley. Boston: Houghton Mifflin, 1988. 138p.
   Bibliography: 132—134; For further reading: 135; Index:
   136—138; Copious illustrations.
   This book by a communications skills professor is an out—
growth of his own interest in ghosts (apparitions and hauntings)
and his ambition to show that the media do not portray ghosts
accurately. Deem describes what ghosts look like, their origin,
where they are likely to be found, and how one can prepare
oneself to see one. Advice is also given on what to do to prove
a ghost is real (how to become a "ghost detective"), including
instructions on how to write a report of a ghost. Reputable
cases from the parapsychological literature are used as examples
throughout.
   Booklist 85:484 (Nov 1, 1988); *CenChildBksBul* 42:69 (Nov
1988); *Kirkus* 56:1148 (Aug 1, 1988); *SchoolLibrJ* 35:119 (Nov
1988); *VoiceYouthAdv* 11:250 (Dec 1988).

430 Ebon, Martin (Ed.). *The Psychic Scene.* New York: New
   American Library, 1974. 188p. Citations in text; 3
   figures
   This anthology covers many subjects of interest to young
people. Ebon has supplied introductions for each selection.
The first of five sections is "Telepathy and Clairvoyance," which
contains articles on ESP tests with drawings, dreams and ESP,
and Mitchell's Apollo 14 space flight ESP tests. The second
section, "Unusual Psychic Phenomena," has selections on plant
psi, a talking dog (Antoine), a corroboration of Edgar Cayce's
statements on Atlantis, UFOs and psi, astrology, and a case of
a Polynesian hex death. Section 3, "Mysticism in Our Time,"
has nine chapters, on: Edgar Cayce, the I Ching, biofeedback,
Transcendental Meditation, Arthur Ford's spirit control, the
Kidd bequest for survival research, Rosemary Brown, the
music—writing medium, a case involving communication from
the dead, and electronic voice phenomena. Section 5,
"Reincarnation," has three chapters on Bridey Murphy, religious
magic in Spanish Harlem, and an article on reincarnation and
karma by an Indian scholar.

431 Ebon, Martin. *What's New in ESP.* New York: Pyramid
   Books, 1976. 192p. Index: 187—192; Suggested reading
   list: 185—186
   This is a popular survey of current happenings in para—
psychology. Its style and language and subject matter make it
of interest to high school students. The combination of non—

technical language and subject matter of current interest makes it a useful book for newcomers to the field, for students, and for laypersons. After an introductory chapter on self—testing for ESP and PK, the book is divided into three sections. The first, "Leading Contemporary Psychics," consists of chapters on Uri Geller, Matthew Manning, Ingo Swann, and Ganzfeld experiments. Part II, "ESP Enters Other Areas" has three chapters on ESP and animals, archaeology, and religion, respectively. The last part is on survival and has chapters on exorcism, the experiments conducted by Karlis Osis funded by the Kidd legacy, and the instances of seeming xenoglossy and reincarnation investigated by Ian Stevenson. A concluding chapter relates ESP to other sciences. A list of the names and addresses of six "major research centers" is appended.

JIndPsychol 2:153 (Jul 1979); *JP* 41:230—232 (Sep 1977); *SF* 11:120—121 (Fal 1979).

432 Edmonds, I.G. *D.D. Home: The Man Who Talked With Ghosts.* Nashville, TN: Thomas Nelson, 1978. 192p. Bibliography: 181—182; 8 illustrations; Index: 183—192

This is a popular biography of D.D. Home by a prolific author of reliable books for young people. Edmonds reports on the feats of Home witnessed by others including levitation, fire—immunity, and spirit raps. In the last chapter he describes techniques by which a magician could repeat most of Home's phenomena. However, as he also points out, no one ever proved that Home was fraudulent.

Choice 16:516 (Jun 1979); *CurrRev* 18:349 (Oct 1979); *LibrJ* 103:1277 (Jun 15, 1978); *SchLibrJ* 25:60 (Jan 1979).

433 Edmonds, I.G. *The Girls Who Talked to Ghosts: The Story of Katie and Margaretta Fox.* New York: Holt, Rinehart and Winston, 1979. 160p. 6 illustrations; Index: 157—160; Suggested reading list: 154—156

This is a well—researched and neutral account of the Fox sisters who initiated Spiritualism in America. Although emphasis is placed on their early years, the later details of their lives are sketched as well, especially the period of Margaretta's confession. Edmonds, who has written over 30 books for young people, tries to present the arguments on both sides of the question of whether or not the "Rochester rappings" were genuine or fraudulent.

T 9:19 (Aut 1981).

Booklist 75:1625 (Jul 15, 1979); *CenChildBksBul* 33:5 (Sep 1979); *EngJ* 68:76 (Nov 1979); *SchLibrJ* 26:135 (Sep 1979).

434 Edmonds, I.G. *Second Sight: People Who Read the Fu—*

ture. Nashville, TN: Thomas Nelson, 1977. 160p. Bibli—
ography: 153—154; Glossary: 151; Index: 155—160
This is an interesting yet scholarly history of spontaneous
precognition and prophecy aimed at young people and written
by a retired Air Force combat photographer. He examines the
evidence which he deems sufficient to establish that precognition
occurs and discusses some of the ideas put forth to explain it.
The historical survey covers Biblical prophecies and ancient
methods of divination, but most of the book is devoted to ac—
counts of individual seers: Roger Bacon, St. Odile, Mother
Shipton, William Lilly, John Dee. Two chapters are devoted to
Nostradamus, and individual chapters to Cheiro (Count Louis
Hamon), Edgar Cayce, and Jeane Dixon. Of particular interest
are two chapters, "Prophecies of and by Famous People" and
"Science Fiction as Prophecy." He does not mention precogni—
tion experiments except to say that they have dealt primarily
with establishing the existence of precognition. He adds, "The
need now is for deeper research into what prophecy really is,
how it works, and whether it can be developed, amplified, and
possibly taught. Our present research methods are not producing
results. Fortunately when one generation lacks either the tools,
will, or intelligence to solve a fundamental problem, a new
generation always comes along with fresh ideas. Young people
usually see the problem from a new angle and make sense out
of what was incomprehensible to those who came before them"
(148—149).
    Booklist 74:542 (Nov 15, 1977); *EngJ* 67:91 (Apr 1978);
JRead 22:772 (May 1979); *Kirkus* 46:9 (Jan 1, 1978); *LibrJ*
102:1658 (Aug 1977); *PublWkly* 212:111 (Aug 1, 1977).

435 Feola, José. *PK: Mind Over Matter.* Minneapolis, MN: Dil—
        lon Press, 1975. 175p. Chapter notes; 4 figures; Glos—
        sary: 163—167; 1 graph; 11 illustrations; Index: 173—175;
        Suggested reading list: 169—172; 5 tables
This is a volume in the "Psychic Explorations" series
being published by Dillon Press. Each book in the series serves
as an introduction to an aspect of parapsychology written in
language understandable by high school students and laymen.
The author, whose doctorate is in environmental health, is also
a parapsychologist and currently teaches an introductory course
at the University of Minnesota. After an introductory chapter,
the next five chapters center on spontaneous and mediumistic
expressions of PK covering poltergeists, haunts, levitation,
materializations, and PK phenomena associated with death. The
last five chapters provide a history and review of experimental
PK, including a chapter on PK with animals and plants. The
emphasis is on encouraging the reader to conduct his or her

own experiments. A glossary, tables, an index, and an annotated guide to further reading are useful aids for the student who wishes to go further.

JP 40:85–87 (Mar 1976); *P* 7:22–23 (Mar/Apr 1976); *PR* 7:17–18 (Mar/Apr 1976); *T* 4:14 (Spr 1976).

SchLibrJ 23:116 (Oct 1976); *SciBooks&Films* 12:65 (Sep 1976).

436 Hall, Elizabeth. *Possible Impossibilities: A Look at Parapsychology.* Boston, MA: Houghton Mifflin, 1977. 169p. Index: 165–169

Elizabeth Hall, formerly Managing Editor of *Psychology Today* and currently Editor–in–Chief of *Human Nature*, has specialized in writing books on psychology aimed at young people. This particular book is about the pros and cons of the science of parapsychology. Although she accepts the legitimacy of the evidence for psi phenomena, her approach is critical. She does a good job of pointing out the normal explanations of what sometimes give the superficial appearance of being in-stances of ESP or PK. Regrettably, no bibliographic references are cited, although numerous investigations are described.

PR 8:15 16 (Nov/Dec 1977).

BablBookworm 6:3 (Apr 1978); *Booklist* 73:1645 (Jul 1, 1977); *CatholLibrWorld* 49:188 (Nov 1977); *CurrRev* 17:385 (Dec 1978); *EngJ* 66:77 (Oct 1977); *HornBk* 53:458 (Aug 1977); *Kirkus* 45:430 (Apr 15, 1977); *SchLibrJ* 23:69 (May 1977); *SciBooks&Films* 14:2 (May 1978).

437 Heaps, Willard A. *Psychic Phenomena.* Nashville, TN: Thomas Nelson, 1974. 192p. Bibliography: 172–183; In-dex: 184–192

This introduction to parapsychology describes psychic (psi) experiences and has chapters on ESP, clairvoyance, telepathy, precognition, and psychokinesis. The field of parapsychology is described, including "new frontiers" (unorthodox or psychic healing). There is an appendix listing sources about ESP tests and a fairly extensive bibliography of books and articles ar-ranged by topic.

JP 39:350–356 (Dec 1975).

Booklist 71:612 (Feb 15, 1975); *BestSell* 35:97 (Jul 1975); *CenChildBksBul* 29:27 (Oct 1975); *Kirkus* 43:26 (Jan 1, 1975); *PublWkly* 207:58 (Jan 6, 1975); *SciLibrJ* 21:65 (Apr 1975).

438 Kettelkamp, Larry. *Haunted Houses.* New York: William Morrow, 1969. 94p. 1 figure; 17 illustrations

Ten documented poltergeist and haunting cases are described in the book. Various scientific theories that have been

suggested as possible explanations for these phenomena are dis—
cussed.
    Booklist 65:1124 (Jun 1, 1969); *BkWorld* 3:12 (Jul 27,
1969); *Kirkus* 37:322 (May 15, 1969); *LibrJ* 94:2103 (May 15,
1969); *NYTimesBkRev* p26 (May 11, 1969); *PublWkly* 195:56
(Apr 7, 1969).

439 Kettelkamp, Larry. *Investigating Psychics: Five Life His—
    tories.* New York: William Morrow, 1977. 128p. 6
    figures; 18 illustrations; Index: 125—128.
    The first section introduces parapsychology by describing
key experiments and theories. The next four chapters are
devoted to psychics Ingo Swann, Rosemary Brown, Olga and
Ambrose Worrall, and Matthew Manning and provide
biographical information, an account of each psychic's abilities,
and the results of investigations of them. The final section dis—
cusses the subject of psi in the normal life situation and how to
discover it.
    Booklist 74:926 (Feb 1, 1978); *Kirkus* 45:1202 (Nov 15,
1977); *SchLibrJ* 24:130 (Mar 1978); *SciBooks&Films* 14:142 (Dec
1978).

440 Kettelkamp, Larry. *Mischievous Ghosts: The Poltergeist and
    PK.* New York: William Morrow, 1980. 127p. 4 figures;
    37 illustrations; Index: 125—127
    Kettlekamp reviews what is known about psychokinesis,
including poltergeists, laboratory research, and theories. Half of
the book consists of cases, including "Philip," the man—made
ghost. The second half deals with experiments and well—known
PK subjects. Some home tests are described.
    T 9:19—20 (Win 1981).
    Booklist 77:810 (Feb 15, 1981); *HornBk* 57:68 (Feb 1981);
Kirkus 48:1397 (Nov 1, 1980); *ReadTeacher* 34:967 (May 1981);
SchLibrJ 27:146 (Mar 1981).

441 Kettelkamp, Larry. *Sixth Sense.* New York: William Mor—
    row, 1970. 95p. 4 figures; 17 illustrations
    Describes and gives examples of a variety of phenomena
studied by parapsychologists: ESP (telepathy and clairvoyance),
psychometry, precognition, retrocognition, reincarnation, out—
of—body experiences, possession, and PK. There is a chapter on
magicians and fraud and one that presents a simple theory to
understand psi and describes some methods of self—testing.
    CenChildBksBul 24:138 (May 1971); *Commonweal* 93:205
(Nov 20, 1970); *LibrJ* 95:4351 (Dec 15, 1970); *NYTimesBkRev*
Pt.I, p22, (Feb 21, 1971); *SciBooks&Films* 6:276 (Mar 1971).

442 Klein, Aaron E. *Beyond Time and Matter: A Sensory Look at ESP.* Garden City, NY: Doubleday, 1973. 117p. Bibliography: 111–112; 8 illustrations; Index: 113–117

This low–key introduction to parapsychology by a biology teacher and science editor/writer is aimed mainly at young people. He introduces the reader to the basic problems, methods, and findings of parapsychological research, concluding with some suggestions for possible home experiments.

Booklist 70:484 (Jan 1, 1974); *CenChildBksBul* 27:97 (Feb 1974); *Kirkus* 41:1112 (Oct 1, 1973); *LibrJ* 99:218 (Jan 15, 1974).

443 Knight, David C. *Poltergeists: Hauntings and the Haunted.* Philadelphia, PA: Lippincott, 1972. 160p. 16 illustrations; Index: 158–159; Suggested reading list: 157

Knight presents accounts of 11 famous poltergeists of the last few centuries. The final chapter gives some theories of pol–tergeists and suggestions for what to do if you meet one.

Books&Bkmn 23:62 (Dec 1977); *Booklist* 69:649 (Mar 1, 1973); *ChildEd* 50:32 (Oct 1973); *JrBkshelf* 42:104 (Apr 1978); *Kirkus* 40:1207 (Oct 15, 1972); *SatRev* 55:78 (Nov 11, 1972); *TimesEdSuppl* p39 (Feb 3, 1978).

444 Lambert, David (Jaqui Bailey, Ed.). *Ghosts.* Illustrated by Harry Bishop. Windermere, FL: Ray Rourke, 1981. 24p. 14 illustrations; Index: 24

This book attempts to reveal a truer picture of ghosts than folklore shows them as being––shadowy shapes or transparent skeletons. The picture of ghosts presented in this book is built up from many ghosts described by hundreds of persons who believed in what they saw. There are also excellent illustrations by Harry Bishop. Topics covered are: Apparitions, ghosts of the living and of the dead, phantom armies, ghostly creatures, ghostly objects, poltergeists, haunted buildings, ghost hunting, and what ghosts may be.

445 Maynard, Christopher. *The World of the Unknown: All About Ghosts.* St. Paul, MN: EMC Corporation, 1978 (Originally published in England in 1977). 32p. 5 figures; Glossary: 30–31; 113 illustrations; Index: 32; Suggested reading list: 32

The folklore of ghosts, the different types of ghosts, and some of the more famous ghosts are described briefly. Maynard explains the techniques of ghost hunting and ways in which "ghosts" have been exposed as fakes or as caused by natural means. Recent ways of explaining ghosts are also included. The illustrations are excellent, as are the photographs and graphs.

CurrRev 17:384 (Dec 1978); *SchLibrJ* 25:60 (Apr 1979).

446 Myers, Arthur. *The Ghost Hunters.* Illustrated by Peter
Coes. New York: Julian Messner, 1980. 158p. 8 illustrations; Index: 155—158; Suggested reading list:
151—154

This is an introduction to the study of apparitions,
haunts, possession and poltergeists by a journalist. It is based
primarily on experiences described in personal interviews, which
provides a note of freshness and currency. Alternative explanations are discussed, which makes it a worthwhile introduction
for young people, toward whom the book is aimed.

Booklist 77:966 (Mar 1, 1981).

447 Myers, Arthur. *The Ghostly Register: Haunted Dwellings—
Active Spirits: A Journey to America's Strangest
Landmarks.* Chicago: Contemporary Books, 1986. 378p.
94 illustrations; Index: 369—378

One of the newest and best journalistic accounts of haunts
one can visit. Descriptions of 64 haunted places are included.
For each one, Myers provides the location, a description of the
place, the ghostly manifestations that occur there, a history of
the place, the identity of the presumed ghost, his or her personality, witnesses, best time to witness, whether or not it is
haunted today, investigations that have been made, and the
names of persons submitting the data. Illustrations are also
included. The last chapter is a directory of parapsychologists,
psychics, and organizations the reader can contact for assistance
and information.

PAI 6:46 (Jun 1988).

ABBookman 79:298 (Jan 26, 1987); *SchLibrJ* 33:18 (Mar
1987); *Trav/Hol* 167:81 (Jan 1987).

449 Rudley, Stephen. *Psychic Detectives.* New York: Franklin
Watts, 1979. 118p. Bibliography: 113—114; Index: 115—
118

Written by a professional photographer and author of
books for young people, this is a popular overview of the question of whether it is possible to help to solve crimes by psychic
means. There is a general introduction to ESP research and
there are individual chapters on Peter Hurkos, Gerard Croiset,
Irene Hughes, and Beverly Jaegers.

CenChildBksBul 33:17 (Sep 1979); *SchLLibrJ* 26:161 (Sep
1979); *SpirFron* 12:188 (Aut 1980).

450 Williams, Gurney, III. *Ghosts and Poltergeists.* Drawings
by Michael Deas. New York: Franklin Watts, 1979. 86p.

8 illustrations; Index: 83—86

The author maintains an open mind regarding ghosts and tries to encourage skepticism in believers and appreciation of the need to think twice or thrice in skeptics. He presents cases of poltergeists, apparitions, and hauntings. He tells how apparitions may be faked and what to do if you see one.

Booklist 76:282 (Oct 1, 1979).

Children and Psi

INTRODUCTION

This is a new subject category consisting of five titles. There were no books in *Sources* dealing specifically with the subject of children and psi with the possible exception of a few autobiographies or biographies of mediums that include chapters on childhood psychic experiences such as those of Eileen Garrett [103-105], Rosalind Heywood [106], Gladys Osborne Leonard [108], Leonora Piper [110], and Hélene Smith [102].

Interest in the psychic and related nonordinary experiences of children developed in the late 1970s as evidenced by the publication of the books by Tanous and Donnelly [453] and by Young [455]. The former is aimed at parents and the latter emphasizes experiences related to survival of death. The more recent books by Armstrong [451] and Peterson [452] both take developmental approaches to children who have psychic and mystical experiences. They describe cases and also attempt to provide a theoretical context for the occurrence of such ex—periences. The bibliography by White [454] is aimed at both parents and children who have psi experiences, and it guides users not only to works on psi experiences in childhood but also to any experiments that have been conducted with children as subjects and material on the parent—child relationship and psi.

Additional books with information on children and psi: Several of the books in the section on "Biographies and Autobiographies of Healers, Mediums, and Psychics" contain ac—counts of childhood psi or mediumistic experiences. Cohen [428] gives accounts of child ghosts or ghosts whom only children have seen. See the section entitled "Books for Young People" for books on parapsychology aimed at children and teenagers.

BIBLIOGRAPHY

451 **Armstrong, Thomas.** *The Radiant Child.* Wheaton, IL: Theosophical Publishing House, 1985. 203p. Annotated bibliographies: 153—174; Chapter notes; 2 figures; 2 graphs; Index: 195—203

This book explores nonordinary experiences in childhood against the backdrop of a new developmental view of children which Armstrong characterizes as being "from the spirit down" as opposed to the standard view of "from the body up." In the new view, which is complementary to the standard view, it is posited that "there is a part of the child which does *not* have its roots in the mother or earth matrix . . . . There is a dimension of the child (and therefore of all of us) that has its foundations elsewhere" (p. 6). He documents the "spirit down" potentialities of children in the book, weaving together instances of nonordinary experiences with the findings of many fields, such as anthropology, literature, mythology, philosophy, and religion, to provide a foundation for a new discipline that will explore "the child's experience of the extraordinary: visions, intuitions, ecstasies, encounters with the supernatural, confrontations with the infinite, interactions with other levels of existence, and so on" (p. 10). This discipline would explore ways of helping children to integrate these experiences when they grow up rather than simply forgetting or repressing them. In one chapter Armstrong presents a "Spectrum of Childhood Consciousness." In another he describes ways of helping children on nonordinary levels. There are three appendices consisting of useful annotated bibliographies on transpersonal children's literature, books relevant to the study of the child's higher nature, and resources in transpersonal child studies.

JTranspersPsychol 17:216−218 (1985); *SpirFron* 18:180−182 (Sum 1986); *WestCoastRevBks* 12:45 (Jul 1986).

452 Peterson, James W. *The Secret Life of Kids: An Exploration into Their Psychic Senses.* Wheaton, IL: Theosophical Publishing House, 1987. 237p. Bibliography: 233−237; 14 illustrations

This book is an expansion of the author's Master's Thesis at the University of California (Berkeley). Peterson is an educator and teaches a course on "Psychic and Spiritual Development of Children" at John F. Kennedy University. The book is founded on 17 years of exploring the psychic world of children in which he discovered that "the various forms of paranormal experience are surprisingly common, much more than in adult life" (p. 2) (occurring 7% more often, according to a survey he conducted). One chapter is devoted to the history of psi phenomena in childhood. Accounts and anecdotes of various psychic experiences, described to him by the children involved, are presented in the heart of the book. An evolutionary view of psi in children, drawing on the ideas of Joseph Chilton Pearce, is presented in Chapter 11. He discusses this developmental approach further in the next chapter, leaning on

the theories of Rudolf Steiner, Phoebe Payne and Laurence Bendit. One chapter consists of an interview with Olga Worrall regarding her childhood experiences. Another discusses how teachers, parents, and friends can relate to psychic children and help them to cope with their experiences, while yet another is critical of "new age" spiritual practices for children. In the final chapter Peterson speculates on what the future may hold for psychic children.
   AmTheos 75:401 (Nov 1987); *WestCoastRevBks* 13(3):47 (1987).

453 Tanous, Alex, and Donnelly, Katherine Fair. *Is Your Child Psychic? A Guide for Creative Parents and Teachers.* New York: MacMillan, 1979. 200p. (Reprinted under title: *Understanding and Developing Your Child's Natural Psychic Abilities: A Guide for Creative Parents and Teachers.* Simon & Schuster, 1988.) Bibliography: 189—192; Chapter notes; Glossary: 185—187; Index: 195—200; Suggested reading list: 193
   The aim of this book is to educate parents and teachers so that they will be able to understand and appropriately counsel children who are psychic and who have questions about their experiences. The book is grounded in the theory that psychic ability must be taken into account if a child is to reach his or her healthy maturity, and that "all psychic experiences should be utilized to integrate the child with his environment" (p. xxii). Both authors are psychic, and Dr. Tanous, psychologist, participates in family and group counseling and is a well—known psi subject. Part I serves as an introduction to psi in children for parents and teachers. The second section ex—plores new developments with "exceptional" children, Part 3 outlines stages of development in children in relation to psychic development, and the final section describes tests and games for developing psychic abilities.
   LibrJ 104:1707 (Sep 1, 1979).

454 White, Rhea A. *Parapsychology for Parents: A Bibliographic Guide* (2nd ed.). Dix Hills, NY: Parapsychology Sources of Information Center, 1988. 36p. Glossary: 27—35
   This is a reading guide aimed at parents of children who have had psychic or other unusual experiences. It also could be used by older children who are interested in learning more about ESP and PK phenomena in their own lives and those of their friends. It begins with a list of nontechnical general sources of information about parapsychology followed by a sec—tion on how to test for ESP and PK abilities, and how to in—vestigate spontaneous psychic phenomena, including ghosts and

haunted houses.   The third section focuses on children and parapsychology.   The first part lists references to experimental investigations with children as subjects and is followed by a sec— tion listing works dealing with children's psychic, mystical, and other nonordinary experiences, such as imaginary companions. Section IV consists of materials on the psychology of the parent—child relationship and psi and Section V lists sources specifically aimed at parents of psychic children.   The *Guide* closes with a list of books on parapsychology written for young people (Section VI), a glossary of parapsychological terms (Section VII), and a short list of parapsychological organizations (Section VIII).

    JP 52:242—244 (Sep 1988); *JSPR* 55:160—162 (Jul 1988).

    Archaeus 7:47 (Spr/Sum 1988); *SpirFron* 20:186—187 (Sum 1988).

455 Young, Samuel H. *Psychic Children.* Garden City, NY: Doubleday, 1978. 159p. Index: 153—159.

    This book, by a journalist who is now with the Associa— tion for the Understanding of Man, describes psychic experiences suggestive of survival. Chapter—length case histories are given of four children. A final chapter, "Growing Up Psychic," provides useful counsel to parents on how to handle psychic children.

    PR 8:19—20 (Jul/Aug 1977).

    Kirkus 45:213 (Feb 15, 1977); *LibrJ* 102:934 (Apr 15, 1977); *PublWkly* 211:92 (Mar 7, 1977); *SpirFron* 10:111—113 (Spr 1979).

Criticisms of Parapsychology

INTRODUCTION

The list of critical books has quadrupled since *Sources* was written. Eight books were covered in the earlier volume, and all but one are still as relevant today as they were then. The exception is Hansel's book [56], because it has been revised (see [472] below). The great increase in critical books is largely due to Prometheus Books, which is headed by Paul Kurtz, a very active member of the Committee for the Scientific Investigation of Claims of the Paranormal. Prometheus authors have been criticized for lumping parapsychology with crackpot subjects, for picking on popular cases that parapsychologists have claimed were not properly investigated, and concentrating on reports of mediums and claims of psychics that parapsychologists have not investigated. The weaker studies are criticized rather than those studies parapsychologists themselves consider to provide the best evidence. Nonetheless, these authors--many of them magicians--do perform a useful function in that they explain how fraud or deception, deliberate or unconscious, can account for many cases considered by the press, and some by parapsychologists, to be paranormal.

Many of the 29 books listed here are critical of pseudoscience, and they treat parapsychology as one of the pseudosciences. The books that are devoted just to parapsychology are Alcock [457], Booth [458], Christopher [459], [460], Couttie [461], Fuller [465] [466], Gardner [468], Gordon [470], Hall [471], Hansel [472], Harris [473], Hines [474], Keene [475], Kurtz [476], Marks and Kammann [477], Randi [479], and Taylor [483]. Most of the books in this section do not criticize hard—core parapsychological experiments. Those that do are by Alcock [457], Druckman and Swets [462], Frazier [463], [464], Hansel [472], and Kurtz [476]. Rogo's book [481] illustrates some topics toward which many parapsychologists are critical--but it is worth reading. An annotated bibliography of the skeptical literature on parapsychology by Hövelmann appears in Kurtz [476], and Stokes has a chapter on parapsychology and its critics in the same volume.

Additional books with information on criticisms of para-

psychology: John Palmer reviews some common elements of criticisms of parapsychology in Shapin and Coly [**817**]. Other reviews of the critical literature are Akers' review of methodological criticisms in *Advances: 4* [**523**], Child's review of criticisms of experimental parapsychology for the period 1975–1985 in *Advances: 5* [**524**], and Ransom's earlier review of criticisms in White [**546**]. Neher [**355**] and Reed [**813**] describe some psychological experiences and states that could account for what on the surface seem to be psi experiences. Brian [**385**] reviews the claims of skeptics and believers in regard to psi in general and the experiences of Jeane Dixon in particular. In her autobiography [**400**], Susan Blackmore explains how she turned from a believer to a skeptic in regard to psi phenomena. The criticisms leveled at early psi experiments and parapsychology's response are stressed by Brian [**401**] and L.E. Rhine [**410**]. Several of the books aimed at young people devote considerable space to criticisms of parapsychology or to pointing out various counterhypotheses, or to presenting pro and con views: [**416, 417, 418, 419, 422, 426, 432, 433, 436, 441, 445, and 450**]. Three of the books in the history section are critical of parapsychology, usually associating it with spiritualism and religion. For this approach see Brandon [**549**], Moore [**559**], and Oppeheim [**560**]. McClenon [**567**] discusses the criticisms of parapsychology offered by elite scientists. Criticisms of transcendent explanations of NDEs are found in Greyson and Flynn [**576**] and Moody [**591**]. Finucane [**726a**] argues that perceptions of ghosts have their origin *this* side of the grave. Coleman [**723a**] persuasively argues against the retrocognitive interpretation of the well–known "adventure" of Moberly and Jourdain. Two anthologies of articles on philosophy and parapsychology have several critical selections: those edited by Flew [**608**] and Ludwig [**611**]. The Leaheys [**649**] deal with the nature of pseudoscience and the pseudosciences, including parapsychology. One of the best critical accounts of parapsychological methodology is presented by Zusne and Jones [**659**]. Ian Wilson [**703**] presents a critical review of the reincarnation hypothesis and in another book he views the survival hypothesis [**772a**]. Paul Kurtz [**714**] provides a critique of religion and the paranormal. Moore [**768**] is critical of survival. Ebon [**778**] is critical of attempts by the unwary to develop psi ability and otherwise dabble in the occult (e.g., use the ouija board). Dossey [**795**] offers criticisms of holistic health.

BIBLIOGRAPHY

456 Abell, George O., and Singer, Barry (Eds.). *Science and*

the *Paranormal: Probing the Existence of the Super—
natural.* New York: Scribner's, 1981. 414p. Chapter
notes; 15 figures; 3 graphs; 31 illustrations; Index:
407—414; Suggested reading list: 392—394; 1 table

An anthology of criticial examinations of aspects of the
"paranormal" by well—known scientists and skeptics, the aim of
this work is to present the response of main line scientists to
claims of the paranormal. Of the 20 chapters, eight are on
parapsychology: "Parapsychology and Quantum Mechanics," by
Martin Gardner; "Scientists and Psychics," by Ray Hyman; "On
Double Standards" (on scientific prejudice against
parapsychology), by Barry Singer; "The Subtlest Difference," by
Isaac Asimov (on survival of death); "Life after Death," by
Ronald K. Siegel (reprint of his *Psychology Today* article on
NDES); "Psychic Healing," by William Nolen (excerpted in part
from his book, *Healing*); "Science and the Chimera," by James
Randi (on chicanery and pseudoscience); and "To Believe or Not
to Believe," a general introduction to how we know, with some
examples from parapsychology, by Barry Singer.

JASPR 76:257—281 (Jul 1982); *Psychoenergetics* 4:406—408
(1982).

Booklist 78:6 (Sep 1, 1981); *Choice* 19:899 (Mar 1982);
IntJPsychosom 32:25—26 (1985); *Kirkus* 49:1314 (Oct 15, 1981);
Kirkus 49:1351 (Nov 1, 1981); *LibrJ* 106:1430 (Jul 1981); *Light*
102:86—88 (Sum 1982); *NewStatesman* 104:20 (Aug 6, 1982);
NYTimesBkRev 87:16 (Feb 14, 1982); *SciBooks&Films* 17:244
(May 1982).

457 Alcock, James E. *Parapsychology: Science or Magic? A
Psychological Perspective.* Elmsford, NY: Pergamon
Press, 1981. 224p. Bibliography: 197—210; Name index:
213—218; Subject index: 219—224; Suggested reading list:
211; 3 tables

Canadian psychologist Alcock's stated aim in this book is
to show that parapsychology "is best described as being belief
in search of data rather than data in search of explanation" (p.
ix).   He approaches the subject from six perspectives: (1)
parapsychology is a modern instance of belief in the super—
natural; (2) by understanding that belief in psi is part of a
larger hierarchical network, it can be seen why proponents of
parapsychology resist negative information; (3) "based on our
knowledge of normal perceptual and cognitive processes, we
should *expect* to have from time to time the kinds of
'paranormal' experiences that people report" (viii); (4) humans
evidence a great tendency to impute causality in situations in—
volving simple coincidence. Alcock asserts that the most that
statistical evidence in parapsychology, which he dubs a pseudo—

science, can accomplish is to demonstrate from the model—it cannot provide explanations; (5) the public is being duped by parapsychologists who speak in the name of science; and (6) "parapsychology is indistinguishable from pseudo—science, and its ideas are essentially those of magic" (p. 196).

JASPR 76:177—186 (Apr 1982); JP 46:231—271 (Sep 1982); PAI 4:67 (Jun 1986); PR 13:24—27 (Mar/Apr 1982).

AmSpectator 14:15 (Dec 1981); Choice 19:994 (Mar 1982); ContempPsychol 27:688—689 (Sep 1982); ContempPsychol 27:689—690 (Sep 1982); SkptInq 6:54—56 (Sum 1982).

458 Booth, John. *Psychic Paradoxes.* Los Alamitos, CA: Ridgeway Press, 1984. 240p. 1 figure; 85 illustrations; Index: 233—240; Suggested reading list: 231—232

The author, who is both a mentalist/conjuror and a min—ister, views psychical research from several angles: entertain—ment, fraud and chicanery, the need to believe, and serious scientific research. The book is in five parts, and begins with a section on theoretical psi. Part 2, on mentalism, considers the many tricks of the trade. Part 3, "The Psychic Connection," consists of three chapters describing Booth's ghost—hunting ex—periences. He attributes "98 per cent of all alleged phantoms to imagination, stress, hysteria, attention—seeking, consciousness changing conditions, loneliness, drink or drugs. The two per cent of phenomena left unexplained is what keeps us ghost hunters in business" (p. 111). Part 4 consists of five chapters on mediumship in which Booth states his beliefs that the R—101 case, involving the crash of a dirigible and subsequent "message" from its deceased captain to psychic Eileen J. Gar—rett, was based on normal information sources and that D.D. Home's famous levitation was engineered by means of a double rope sling. His conclusion to a chapter on cross correspondences is that "without conscious fraud or provable collaboration, a number of insightful individuals could produce parallel or inter—locking results by normal means" (p. 177). Part 5, "Serious Psi Research," consists of a chapter on OBEs, one on motives for producing fraudulent phenomena, a chapter on precognition (which deals only briefly with experiments), and the final chap—ter, "From Hoaxes to Hopefulness," which concludes that "there may be more to [psi] than skeptics admit but less than occul—tists proclaim" (p. 229).

JASPR 80:336—338 (Jul 1986); PAI 4:67 (Jun 1986).

459 Christopher, Milbourne. *Mediums, Mystics and the Occult.* New York: Thomas Y. Crowell, 1975. 275p. Bibliogra—phy: 259—264; 50 illustrations; Index: 265—275

The subjects covered in this book are Uri Geller, psychic

surgery, Peter Hurkos, eyeless vision, Mitchell's outer space ex—
periment, Ted Serios, Arthur Ford, some "mystics from the
East," research on survival financed by the Kidd legacy,
Margery Crandon, the Scientific American investigations, and
Annie Eva Fay. In each case Christopher tries to show how the
investigators were hoodwinked by the psychics being studied.
        JP 40:161—162 (Jun 1976); *P* 6: 31—32 (Nov/Dec 1975).
        LibrJ 100:677 (Apr 1, 1975); *NYTimes* 124:19 (Aug 23,
1975); *PublWkly* 207:42 (Apr 21, 1975).

460 Christopher, Milbourne. *Search for the Soul.* New York:
        Thomas Y. Crowell, 1979. 206p. Bibliography: 189—192;
        Index: 195—206
        This is a popular but serious review of "efforts that have
been made to see, isolate, and analyze the soul" (p. 7), and to
establish that it exists after death.   Included are the
materializations of Katie King investigated by William Crookes;
a not previously reported attempt to photograph the soul at the
moment of death; attempts to weigh the soul, reveal its shape
and map its dimensions; near—death experiments of the past
and present; the Kidd trial and the research conducted with the
miner's legacy at the American Society for Psychical Research
and Psychical Research Foundation, including out—of—body ex—
periments and deathbed experiences; near—death experiences,
Mrs. Piper; Elisabeth Kübler—Ross; and reincarnation.   Chris—
topher points out why none of the attempts described suc—
ceeded, but chides modern researchers for not centering their ef—
forts on either the moment of "conception or during the period
of dying. This is when the animating power is generated or
when it flickers and is finally extinguished" (p. 187).
        JP 44:375—378 (Dec 1980); *PR* 11:14—16 (May/Jun 1980);
T 8:22 (Spr 1980); *ZS* No.6:170—171 (1980).
        America 142:107 (Feb 9, 1980); *Booklist* 76:582 (Dec 15,
1979); *Kirkus* 47:1173 (Oct 1, 1979); *LibrJ* 104:1574 (Aug 1979);
PublWkly 216:84 (Sep 3, 1979); *SkptInq* 4:65—66 (Spr 1980);
SpirFron 16:50—51 (Win 1984).

461 Couttie, Bob. *Forbidden Knowledge: The Paranormal
        Paradox.* Cambridge, England: Lutterworth Press, 1988.
        155p. Chapter notes
        This book is based on Couttie's BBC Radio 4 series of the
same name.   Couttie says he was a "believer" in parapsychology
in his teens, but became skeptical when he was not able to
repeat experiments with the same level of results, and he found,
upon looking into some reports (he says he mainly read those
in popular books available at the library), "that experiments
had been poorly designed and badly reported.   Often I dis—

covered beneath the overt evidence yet another layer of data
rarely brought out for the public eye or ear. This data is the
'forbidden knowledge' of the title of this book" (p. 2). He then
became a magician, in order to try to produce some seeming
psychic effects by normal means. He writes he became "an
open—minded skeptic, willing to look at the evidence but un—
willing to accept self—deception, fraud, outright falsehood and
downright silliness as evidence of the existence of the paranor—
mal" (p. 2). The book is in four sections. There are six
chapters in the first, "The Miracle Workers:" two chapters on
Uri Geller; one on various mediums such as D.D. Home and
Arthur Ford; one on seance magic reporting the investigations
of magician S.J. Davey; one on mental medium Doris Stokes;
and one on "misdetection," or cases in which psychics did *not*
aid the police by paranormal means. Part 2, "Hidden Forces,"
has a chapter on fraudulent children, two on astrology, one on
lunar influence, and two on dowsing. Part 3, "Towards an
Anthropology of the Paranormal," has a chapter on scientists
and the paranormal, two on the psychology of the paranormal,
and one on the role of chance. Part 4 consists of two chapters
on "How to be Psychic," in order to provide the readers "with
sufficient information so that next time [they] are faced with
someone claiming psychic powers [they] will be in a better posi—
tion to judge the truth of the claim" (p. 135). Not one ex—
perimental paper in the two major parapsychology journals, the
Journal of Parapsychology and the *Journal of the American
Society for Psychical Research*, are cited in the chapter notes.
Thus, Couttie offers many insights into mediumistic and spon—
taneous phenomena and into some experiments early in this
century or the last, but he has nothing to say on modern ex—
perimental parapsychology.
    JSPR 55:234—236 (Oct 1988).
    BritBkNews p589 (Sep 1987); *BritBkNews* p822 (Dec 1987);
ContempPsychol 32:806 (Sep 1987); *Encounter* 70:60 (May 1988);
Light 108:129—131 (Win 1988); *SciAm* 259:108 (Aug 1988).

462 Druckman, Daniel, and Swets, John A. (Eds.). *Enhancing
    Human Performance: Issues, Theories, and Techniques.*
    Washington, DC: National Academy Press, 1988. 299p.
    Bibliography: 209—231; 1 Figure; Glossary: 252—261;
    Index: 289—299; 5 tables
    The report of the Committee on Techniques for the
Enhancement of Human Performance of the Commission on
Behavioral and Social Sciences and Education of the National
Research Council assesses the "potential value of certain tech—
niques that had been proposed to enhance human performance"
and was funded by the Army Research Institute. The report

describes the Committee's research, findings, and conclusions. The techniques investigated were: accelerated learning, sleep learning; guided imagery, split—brain effects, stress management, superior performance engendered by altered mental states, biofeedback, influence strategies, and psi phenomena. The first chapter provides the context in which the research was carried out. Chapter 2 relates the findings of the Committee in relation to the techniques studied and its conclusions about evaluation procedures. The third chapter treats the relevant evaluation issues more systematically and presents the Committee's philosophy of evaluation. Chapters 4—8 are about specific psychological techniques and the ninth chapter is on parapsychological techniques. Further details concerning the Committee and its procedures are given in six appendices. The Committee found little evidence to support any of the areas investigated. As regards parapsychology, they nevertheless recommended "that research in certain areas be monitored, including work by the Soviets and the best work in the United States. The latter includes that being done at Princeton University by Robert Jahn; Maimonides Medical Center in Brooklyn by Charles Honorton, now in Princeton; San Antonio by Helmut Schmidt; and the Stanford Research Institute by Edwin May. Monitoring could be enhanced by site visits and by expert advice from both proponents and skeptics. The research areas included would be psychokinesis with random event generators and Ganzfeld effects" (p. 22). It should be noted that for its chief investigator of parapsychology the Committee chose a well—known critic of the field, Ray Hyman.

Choice 25:1478 (May 1988); *SciAm* 259:108,110 (Aug 1988).

463 Frazier, Kendrick (Ed.). *Paranormal Borderlands of Science.* Buffalo, NY: Prometheus Books, 1981. 469p. Chapter references; 14 figures; 5 graphs; 24 illustrations; 34 tables

This is an anthology of 47 articles that originally appeared in the *Skeptical Inquirer*, the journal of the Committee for the Scientific Investigation of Claims of the Paranormal, which is edited by science writer/editor Kendrick Frazier. There are 43 authors who are primarily psychologists, astronomers, philosophers, investigative journalists, and magicians. They examine various pseudoscientific claims reported in the media in such areas as astrology, ESP, PK, precognition, biorhythms, and UFOs. Parapsychology is the subject of most of the chapters in Book 1: "Skeptical Inquiries into the Paranormal." Book II, "Inquiries into Fringe Science," covers primarily astrophysical phenomena. The emphasis is on distinguishing between "true" science and "pseudo—science."

JP 46:280−283 (Sep 1982); *JSPR* 51:173−174 (Oct 1981); *PAI* 5:43 (Dec 1987).
Choice 19:394 (Nov 1981); *Humanist* 41:47 (Sep 1981); *LibrJ* 106: 1552 (Aug 1981); *Sky&Telescope* 62:362 (Oct 1981).

464 Frazier, Kendrick (Ed.). *Science Confronts the Paranormal.* Buffalo, NY: Prometheus Books, 1986. 367p.
Chapter references; 7 figures; 2 graphs; 17 illustrations; Suggested reading list: 359−360; 5 tables
  A continuation of *Paranormal Borderlands of Science,* also edited by Frazier [**463**], this work consists of articles published in the *Skeptical Inquirer* from 1981−1985. It is separated into two parts: "Assessing Claims of Paranormal Phenomena," and "Evaluating Fringe Science." The first is almost entirely devoted to parapsychology, whereas the second covers astrology, UFOs, fringe archaeology, creationism and shroud science, and cryptozoology. Thus, in Part II, parapsychology only figures peripherally in some articles that debunk psychic archaeology, a subject that is not part of hard−core parapsychology. There are six sections in Part I. The first, "Parapsychology and Belief," has four chapters on the demarcation problem, the philosophy of science and parapsychology, and skepticism regarding psi. Section 2, "Expectation and Misperception," has five articles which demonstrate how people can be duped by claims concerning psi. Section 3, "Claims of Mind and Distance," consists of four articles criticizing *The Mind Race* by Targ and Harary [**785**], remote−viewing experiments, and the Dutch psychic Gerard Croiset and his chief proponent, W.H.C. Tenhaeff. Section 4, "Claims of Mind and Matter," contains five articles on Randi's Project Alpha, in which he planted two young magicians in the guise of psychics at the McDonnell Laboratory for Psychic Research; the Columbus poltergeist; and Targ and Puthoff's research with Uri Geller. There are two papers in Section 5, "Claims of Mind and Body," one on iridology and one on palmistry−−neither of them being parapsychological. The last section contains one article on "Psychic Flim−Flam," about the Tamara Rand hoax.
  PAI 6:43 (Jun 1988).
  BookWorld p13 (Mar 23, 1986); *Choice* 23:1558 (Jun 1986); *ContempPsychol* 32:806 (Sep 1987); *LATimesBkRev* p6 (May 4, 1986); *Nature* 322:505 (Aug 7, 1986); *PsycholRec* 37:139−140 (Win 1987); *SciBooks&Films* 22:229−230 (Mar/Apr 1987); *SkptInq* 11:93−96 (Fal 1986); *TimesLitSuppl* p899 (Aug 15, 1986).

465 Fuller, Uriah (Pseudonym). *Confessions of a Psychic: The Secret Notebooks of Uriah Fuller.* Teaneck, NJ: Karl

Fulves, 1975. 41p.
In a publisher's note, Karl Fulves says "The present
manuscript has been made available to protect the reader from
the tempting bait of the fraudulent psychic. While the methods
used by hustlers in this field vary in details, the concepts explained in the following pages outline basic approaches used to
capitalize on human frailty and weakness. ESP may or may
not exist, but fraudulent psychics do, and this manuscript may
make it more difficult for them to ply their trade in the future"
(p. 8). Uri Geller was the impetus for this book, and after a
chapter of general observations, the author describes how to
fake Geller—type phenomena such as bending forks, spoons, and
keys, starting broken watches, tin—can divination, duplicating
drawings, and several others as well.

466 Fuller, Uriah (Pseudonym). *Further Confessions of a*
Psychic: The Secret Notebooks of Uriah Fuller. Teaneck,
NJ: Karl Fulves, 1980. 70p. 11 figures
In the introduction, the publisher, Karl Fulves, attempts
to give insight into the mentality of the psychic, which he
characterizes as criminal. He holds that "the true professional
will resort to blackmail, extortion, tapped telephones, tampering
with the mails, and so on if he knows he will see you on a
regular basis and intends to extort money from you . . . The
psychic is a street—wise criminal" (p. 18). "Fuller" describes
new secret methods of bending keys, duplicating drawings,
macro—PK with nylon thread, fixing broken watches, correctly
guessing ESP cards and miscellaneous tricks, such as eyeless vision, stopping an escalator, and moving compass needles.
SciAm 245:41 (Oct 1981).

467 Gardner, Martin. *The New Age: Notes of a Fringe*
Watcher. Buffalo, NY: Prometheus Books, 1988. 273p.
19 figures; 25 illustrations; Name index: 265—273
This is the latest collection of Martin Gardner's writings
reprinted from publications such as *Discover, New York Review*
of Books, New York Times, Skeptical Inquirer, and others published from 1984 through 1987. Gardner has added forewords
and/or afterwords to most of the chapters in order to update
the information, provide background, or to include responses
from his critics. Only a fraction of the book is on parapsychology, and only a fraction of that is on what parapsychologists call parapsychology, but Gardner is always fun to
read, whatever the topic. Some of the relevant pieces are
Randi's "Project Alpha," Margaret Mead and parapsychology,
the Koestler Chair, Russell Targ, SRI research with Uri Geller,
psychic surgery, D.D. Home, PK (he calls it "Psycho—krap"),

and the "channeling mania."

Booklist 84:1292 (Apr 1, 1988); *ChristCent* 105:746 (Aug 17, 1988); *ChristSciMonit* 80:17 (Jul 13, 1988); *Humanist* 48:42 (Sep 1988); *Nature* 334:114 (Jul 14, 1988); *NYTimesBkRev* 93:38 (Mar 27, 1988); *PublWkly* 233:91 (Mar 11, 1988); *TimesLitSuppl* p827 (Jul 29, 1988); *VillVoice* 33:52 (Jul 19, 1988).

468 Gardner, Martin. *Science: Good, Bad, and Bogus.* Buffalo, NY: Prometheus Books, 1981. 408p. Chapter references; 7 figures; 17 illustrations; 2 tables
This is a collection of 38 of Gardner's articles and book reviews published from 1950 to 1980. Also included are letters from his readers and his responses to them, which make for lively reading. Each piece is updated and includes fresh information. The book is billed as a sequel to his *Fads and Fallacies in the Name of Science*, and it crusades for better understanding of the difference between bad and good science on the part of the public. He deliberately uses ridicule and humor because he considers it a waste of time to give rational arguments in dealing with the "extremes of unorthodoxy in science." Approximately three quarters of the book is on aspects of parapsychology.
JASPR 78:87−95 (Jan 1984); *JSPR* 51:318 (Jun 1982); *PR* 13:24−26 (May/Jun 1982).
Booklist 78:416 (Nov 15, 1981); *ChristSciMonit* 73:17 (Oct 28, 1981); *LibrJ* 106:2035 (Oct 15, 1981); *PsycholRec* 32:452−453 (Sum 1982); *SciAm* 245:41 (Oct 1981); *SkptInq* 7:68−70 (Spr 1983); *SocSciQ* 63: 799−800 (Dec 1982); *Time* 118:71 (Aug 10, 1981).

469 Gardner, Martin (Ed.). *The Wreck of the Titanic Foretold?* Buffalo, NY: Prometheus Books, 1986. 157p. Citations in text; 8 illustrations
Skeptic Martin Gardner has reprinted Morgan Robertson's novel, *The Wreck of the Titan*, which Ian Stevenson and others have cited as an example of precognition of the sinking of the *Titanic*. He also reprints other selections (part of a novel, a short story, several poems) that also seemed to foretell the fate of the *Titanic*. In his general introduction, he argues that chance can account for any coincidence, no matter how improbable, and criticizes Stevenson's two surveys of cases foretelling the wreck of the *Titanic*. In individual introductions to all the pieces anthologized, he argues further that they could be explained by normal means and that precognition was not involved.
NYTimes 135:23 (Mar 14, 1986).

470 Gordon, Henry. *Extrasensory Deception: ESP, Psychics, Shirley MacLaine, Ghosts, UFOs* . . . Buffalo, NY: Prometheus Books, 1987. 227p.
The book is an anthology of commentaries by magician—journalist—debunker Gordon in his column in the *Toronto Sunday Star*. There are eight sections: "The Paranormal," "Prophecy," "Spirits," "Superstition," "Science," "Higher Life," "Healing," and "Truth." Throughout he shows how mentalism, or "sleight of mind," which practices the methods of deception, is the means used by psychics to accomplish their feats. The difference between a mentalist and a psychic is that the former admits he uses deception. Individual psychics discussed are Uri Geller, Kreskin, Peter Hurkos, Tommy Roberts, and Edgar Cayce. Because it lacks an index it is a difficult book to use, especially considering the many individual subjects covered.
QueensQRev 54:27 (Aug 1988); *SkptInq* 13:315—316 (Spr 1989).

471 Hall, Trevor H. *The Enigma of Daniel Home: Medium or Fraud?* Buffalo, NY: Prometheus Books, 1984. 148p. 1 figure; 1 illustration; Index: 145—148; 1 table
This book is not intended to be a biography. It is primarily a critical examination of Home's most famous levitation at Ashley House, and of some other puzzles about the medium that had not previously been explained to Hall's satisfaction. He provides an excellent bibliographic essay on Home in the Introduction.
JASPR 79:283—288 (Apr 1984); *JP* 49:103—105 (Mar 1985); *PAI* 4:43 (Dec 1986); *ZS* No.12/13:154—159 (Aug 1987). *VictorianStud* 29:613 (Sum 1986).

472 Hansel, C.E.M. *ESP and Parapsychology: A Critical Reevaluation.* Buffalo, NY: Prometheus Books, 1980. 325p. Chapter references; 11 figures; 1 graph; Index: 319—325; Suggested reading list: 317—318; 18 tables
Parts I and II of this revision of British psychologist Hansel's critical review of parapsychology are substantially the same as his earlier *ESP: A Scientific Investigation* [56], with some updating. The same can be said for Part 4, "Conclusions," although this version contains an outline of a definitive experiment involving a sorting machine. Part 3, however, consists of primarily new material. The chapter on testing with a machine has been greatly expanded by a critical review of the work of Helmut Schmidt and that of Targ and Puthoff with machines. Also covered are the Maimonides dream experiments, the Caxton Hall experiment, "the miracle men" (Croiset, Stepanek, Serios, and SRI work with Geller, who has a chapter to

himself), and remote—viewing experiments.

JASPR 75:155—166 (Apr 1981); *JP* 44:358—361 (Dec 1980); *JSPR* 51:27—28 (F 1981); *PR* 11:12—14 (May/Jun 1980) *T* 9:22 (Sum 1981).

Choice 17:456 (May 1980); *Kirkus* 47:1462 (Dec 15, 1979); *Kirkus* 48:13 (Jan 1, 1980); *LibrJ* 105:622 (Mar 1, 1980); *NewStatesman* 100:19 (Sep 19, 1980); *SciBooks&Films* 16:61 (Nov/Dec 1980); *SkptInq* 4:60—62 (Spr 1980); *TimesLitSuppl* p854 (Jul 25, 1980); *VaQRev* 57:68 (Spr 1981).

473 Harris, Melvin. *Investigating the Unexplained.* Buffalo, NY: Prometheus Books, 1986. 222p. Chapter notes; 2 figures; 14 illustrations

The author, a broadcaster and professional researcher for the BBC, in the course of searching for program material investigated several cases involving various anomalies, including psychic ones. In this book he reports on his findings. The cases covered are the Amityville Horror; the materialization mediums, Isa Northage and Paul McElhoney; the voice mediums, Leslie Flint and Keith Rhinehart; Mackenzie King's medium, William Roy; the crime—solving psychic, Frances Dymond; Doris Stokes, Nella Jones, and Robert Lees, all of whom tried to solve the Yorkshire Ripper or Jack the Ripper cases; the Empress Eugenie; the disappearance of the Fifth Norfolk Regiment; the Angels of Mons; the ghost of Sir George Tryon; the apparition seen by Lord Dufferin; Bridey Murphy; Lulu Hearst; Eileen Garrett and the R—101; the Gordon Davis case reported by S.G. Soal, and others. Harris seeks to provide normal explanations, such as fraud or self—deception. There are a few incomplete citations in his notes, but documentation is for the most part not given. Although this is not a first—rate debunking book, it is included because it discusses a number of well—known cases or mediums and because it provides insight into the many ways that we can be fooled or even delude ourselves when it comes to psi phenomena. Perhaps the British title of this book sums it up best: *Sorry——You've Been Duped.*

JASPR 82:391—396 (Oct 1988); *JSPR* 54: 221—225 (Jul 1987).

474 Hines, Terence. *Pseudoscience and the Paranormal: A Critical Examination of the Evidence.* Buffalo, NY: Prometheus Books, 1988. 372p. Author index: 353—361; Bibliography: 319—351; 8 figures; 1 graph; 13 illustrations; Index: 363—372; 2 tables

The author, a psychologist, has taught a course on "Parapsychology and the Occult" at Pace College since 1964. He found that there was no textbook that covered all of the

subjects he wanted to discuss in his class, and so he wrote this
book for his own use and for others who are interested in a
critical evaluation of paranormal claims and who also wish to
believe such claims. Although it is aimed primarily at students,
it does not require a background in psychology. The first
chapter discusses the nature of pseudoscience. A variety of psi
applications are discussed in the next chapter, "Psychics and
Psychic Phenomena," including spiritualism, psychic readings,
psychic predictions, criminal investigation and psi, the psychic
photography of Ted Serios, precognitive experiences, and
demonic possession. Three aspects of survival are considered in
the third chapter: ghosts and poltergeists, NDEs and OBEs, and
reincarnation. Laboratory parapsychology is the subject of
Chapter 4——a mere 31 pages. He discusses Uri Geller and the
early experiments of Rhine and Soal, and then briefly deals with
remote viewing, RNG experiments, and Ganzfeld research. A
chapter is devoted to faith healing and its dangers. There are
many critical (debunking) books, but Hines also provides a
"positive" contribution in his discussion of the issue of "why
people continue to believe in paranormal and occult claims in
the face of clear empirical evidence that these claims are in—
valid. Platitudes about a 'will to believe' are inadequate ex—
planations. Rather, explanations for continued belief must be
based on knowledge about how humans process information and
about the nature of human memory and perception" (p. xi).
Other subject areas covered are psychoanalysis, astrology, UFOs,
ancient astronauts, and health quackery. The final chapter
covers "Current Trends in Pseudoscience."

SciTechBkN 12:1 (Apr 1988); *SkptInq* 13:86−87 (Fal 1988);
TribBks p5 (Mar 27, 1988).

475 Keene, M. Lamar, as told to Allen Spraggett. *The Psychic
Mafia.* Foreword by William V. Rauscher. New York:
St. Martin's Press, 1976. 177p. Annotated bibliography:
165−177; 31 illustrations
This book is by a former fraudulent medium who has
since reformed and no longer engages in "mediumship" ac—
tivities. He tells how he became a fake medium and describes
all the tricks mediums use to make a living by duping people
into believing they are in communication with deceased loved
ones. He explains how to produce ectoplasm and apports, do
billet reading, make predictions, and how to heal and levitate.
Considerable space is devoted to the fraudulent practices at
Camp Chesterfield, the famed spiritualist camp in Ohio. One
chapter is devoted to sex in the séance room. There is a
chapter on the history of mediumistic fraud and a useful 13—
page bibliography of critical works in general, and of works on

specific methods of mentalism as used by fraudulent mediums. Many of the titles are annotated.
JP 40:255—256 (Sep 1976); *P* 7:46 (Nov/Dec 1976); *T* 4:14 (Sum 1976).
Humanist 37:33 (May/Jun 1977); *Kliatt* 12:54 (Win 1978); *LibrJ* 101:1539 (Jul 1976); *PublWkly* 209:78 (May 10, 1976); *PublWkly* 211:108 (Jun 27, 1977); *SpirFron* 9:48—49 (Fal 1977).

476 Kurtz, Paul (Ed.). *A Skeptic's Handbook of Parapsychology.* Buffalo, NY: Prometheus Books, 1985. 727p. Chapter references; Index: 703—727; 4 tables
This is an anthology of 30 contributions that primarily represent the views of skeptics toward the major facets of parapsychology and psi phenomena. Nineteen chapters are published here for the first time and 11 are reprinted from earlier sources. The general aim of the book is to present evaluations of "the entire history of the field of parapsychology . . . and to examine the results" (p. xi) by competent skeptics. The evidence in each case and its reliability are stressed. Five chapters present historical overviews, ten present the argument from fraud, and four deal with the question of whether parapsychology is a science or a pseudoscience. Four are critical of some methodological and theoretical issues. Three are critiques of specific aspects of psi research. One section (four chapters) is by persons associated with parapsychology. Two of these (by John Beloff and Douglas M. Stokes) present the parapsychologists' point of view. The other two (by Blackmore and Hövelmann) are written from skeptical viewpoints.
JASPR 83:263—270 (Jul 1989); *JP* 52:225—236 (Sep 1988); *JSPR* 55:90—98 (Apr 1988); *PAI* 5:41 (Jun 1987).
Choice 23:1144 (Mar 1986); *Humanist* 46:41 (Jul 1986); *NewAge* 2:68 (Jun 1986); *RefBkRev* 7(2):23 (1986); *SciTechBkN* 10:3 (Mar 1986); *SkptInq* 11:97—101 (Fal 1986); *WestCoastRevBks* 12:28 (1986).

477 Marks, David, and Kammann, Richard. *The Psychology of the Psychic.* Foreword by Martin Gardner. Buffalo, NY: Prometheus Books, 1980. 232p. Bibliography: 227—232; 21 figures; 14 illustrations; 16 tables
The professed aim of this book by two New Zealand psychologists is to report on their investigation of several of the best known psychic claims of the 1970s: remote viewing, superpsychics Kreskin and Geller, an experiment conducted by Alister Hardy, and the case history approach of Arthur Koestler. The final three chapters contain an attempt to explain why so many people believe in nonexistent phenomena, and a discussion of how people can be deceived. Although some constructive

criticisms of the research at Stanford Research Institute are of—fered, for the most part the book does not deal with the parapsychology of the parapsychologists but with that of the popular press.
    JASPR 74:425—443 (Oct 1980); *JSPR* 51:174—176 (Oct 1981); *PR* 12:25—28 (May/Jun 1981); *T* 9:21—22 (Sum 1981).
    Choice 18:166 (Sep 1980); *JApplPsychol* 93:748 (Dec 1980); *LibrJ* 105:1172 (May 15, 1980); *NewStatesman* 100:19 (Sep 19, 1980); *SciAm* 245:39 (Jul 1981); *SciBooks&Films* 16:180 (Mar 1981); *SkptInq* 5:60—63 (Win 1980/81); *TimesLitSuppl* p12 (Jan 2, 1981).

478 Randi, James. *The Faith Healers.* Foreword by Carl Sagan. Buffalo, NY: Prometheus Books, 1987. 314p. Bibliography: 307; 6 figures; 33 illustrations
    Magician James Randi herein exposes the deception and exploitation practices by the major televangelists, including Peter Popoff, W.V. Grant, Pat Robertson, and Oral Roberts. He also investigated A.A. Allen, Leroy Jenkins, Willard Fuller (the "psychic dentist"), and Ralph DiOrio. Chapter 14, "The Lesser Lights," such as Danny Davis, Kathryn Kuhlman, Daniel Atwood, David Epley, Al Warick ("Brother Al"), David Paul Ernest Angley, and Frances and Charles Hunter ("The Happy Hunters"). Randi exposes the tricks used by some of the healers, supported by evidence he and some colleagues gathered over a four—year period. He also emphasizes the gullibility and misplaced faith of the victims, which plays a part in their vic—timization and exploitation. Randi demands that legal action be taken against the exploiters. There is a chapter on "Practical Limitations of Medical Science," in which he points out that people expect too much of traditional medicine, and when disappointed, they turn to unorthodox methods of healing. A chapter is devoted to his unsuccessful efforts to obtain evidence of successful cures. This lack of evidence leads him to question whether any cures actually took place. A brief chapter reviews legal issues involved in aspects of unorthodox healing.
    PAI 6:46 (Jun 1988).
    Booklist 84:198 (Oct 1987); *ChristCent* 104:1068 (Nov 25, 1987); *Encounter* 70:58 (Feb 1988); *Humanist* 48:40 (May 1988); *LATimesBkRev* p8 (May 1, 1988); *NewAge* 4:65 (Jan 1988); *RefResBkNews* 3:2 (Apr 1988); *SchLibrJ* 34:227 (Mar 1988).

479 Randi, James. *Flim—Flam! Psychics, ESP, Unicorns, and Other Delusions* (rev. ed.). Buffalo, NY: Prometheus Books, 1982. 342p. 34 figures; 117 illustrations; Index: 335—342; Selected bibliography: 331—342; 5 tables
    Magician Randi attempts "to explain to the reader the

major events in the realm of flim—flam" perpetrated in the last 20 years. The essence of his aim is this: "I cannot prove that these powers do *not* exist; I can only show that the evidence for them does not hold up under examination. Furthermore, I in— sist that the burden of proof be placed not on me but on those who assert that such phenomena exist. Unusual claims require unusual proof" (pp. 2—3). Subjects Randi attempts to explain away are: the Cottingly fairies; Bermuda triangle; UFOs; TM levitation claims; ancient astronauts; SRI research with Uri Geller and with remote—viewing studies; biorhythms; psychic surgery and paranormal diagnosis; N Rays; spoon—bending; Schmidt machines; spiritualism; and various psychics such as Suzie Cottrell, Olof Jonsson, Geraldine Smith, Peter Hurkos, Rosemary Dewitt, Linda Anderson, Jean—Paul Girard, and other psychics he tested. He concludes: "Parapsychology is a farce and a delusion, along with other claims of wonders and powers that assail us every day of our lives" (p. 326). The annual "Uri Award" in parapsychology is described in an appendix.
 T 9:17—18 (Win 1981); *ZS* No.8:131—136 (Jul 1981).
 Booklist 77:424 (Nov 15, 1980); *ContempPsychol* 26:966 (Dec 1981); *Kirkus* 48:1343 (Oct 1, 1980); *Kirkus* 48:1472 (Nov 15, 1980); *LibrJ* 105:2578 (Dec 15, 1980); *NYTimesBkRev* 85:16—17 (Nov 23, 1980); *PsycholRec* 33:44 (Win 1983); *SkptInq* 5:58—60 (Sum 1981).

480 Randi, James. *The Truth About Uri Geller.* Buffalo, NY: Prometheus Books, 1982. 235p. Bibliography: 235; 27 figures; Glossary: 233—234; 26 illustrations
 This is a revised and expanded edition of *The Magic of Uri Geller* (1975). About three—quarters is written by Randi and the remainder consists of selections from the writings of others that Randi discusses in depth. The main purpose of the book is to show that Geller is an excellent magician, with whom Randi would have been happy to associate, if only he had admitted that fact, instead of choosing to "become a semi—religious figure with divine pretensions" (p. 21). In a chapter entitled "Objections to This Book," he responds to comments and criticisms of *The Magic of Uri Geller.*
 NewStatesman 105:25 (Apr 8, 1983).

481 Rogo, D. Scott. *The Haunted Universe: A Psychic Look at Mysteries, UFOs, and Miracles of Nature.* New York: New American Library, 1977. 168p. Chapter notes
 At first glance one could dismiss this book as ingenuous, but in fact it is both well—considered and brave. It is a ra— tional discussion of phenomena which, if true, would be con— sidered parapsychological, but which even parapsychologists tend

to reject out of hand, even as their critics dismiss phenomena parapsychologists consider legitimate.   Rogo treats such taboo topics as teleportation, religious miracles (weeping and bleeding statues, materializations, mysterious lights and apparitions as— sociated with religious figures), psychic phenomena associated with UFOs, the reality of evil, and types of "Forteana" likely to have a psi explanation.

   JP 41:368—369 (Dec 1977); *JSPR* 49:662—665 (Dec 1977); *JIndPsychol* 2:151 (Jul 1979).

482 Rothman, Milton A.   *A Physicist's Guide to Skepticism: Applying Laws of Physics to Faster—than—Light Travel, Psychic Phenomena, Telepathy, Time Travel, UFOs, and Other Pseudoscientific Claims.* Buffalo, NY: Prometheus Books, 1988.   247p.   Chapter references; 16 figures; 3 graphs; Index: 243—247; 2 tables

   Rothman, a physicist, writes: "This book is philosophy of science as understood by an experimental physicist, written for the nonspecialist" (p. 9).   He adds: "The underlying theme of this book is: How do we trace the boundary between fantasy and reality?   This question is not merely hypothetical; successful existence in the real world requires a good ability to define this boundary with some accuracy.   Because illusions and hallucina— tions abound, rational man has been forced to create a system of elaborate mechanisms and methods which aid in recognizing reality and separating it from the world of fantasy.   These methods make up the system of knowledge called science" (p. 9).   He divides the laws of nature into two major groups: laws of permission and laws of denial.   The latter "are rules that tell us what *cannot* happen to a system of objects" (p. 11).   He adds: "Understanding the laws of denial gives a logical basis for skepticism.   Pseudosciences make claims of extrasensory perception, psychic energy, poltergeists, unidentified flying ob— jects, and other exotic phenomena.   It requires the knowledge of only one basic principle of science, namely, the law of conserva— tion of energy,   to justify a position of extreme skepticism toward these claims.   Conservation of energy, as we shall see in Chapter 4, is experimentally verified to a degree of precision that makes it one of the most firmly grounded and solid pieces of knowledge in history.   With such a permanent bit of knowledge in hand, we can justify our skepticism toward claims for phenomena that purport to do what in reality cannot be done" (p. 12).   The book has nine sections: "Beliefs and Disputations," "Models of Reality: Particles," "Models of Reality: Quantum Theory," "Interactions," "Laws of Permission and Laws of Denial," "Reductionism, Prediction, and the Laws of Denial," Verification and Falsification," "Living Skeptically,"

and "Epilogue." The Epilogue concludes: "As appealing as the mysteries of the occult and the supernatural may be, solving the mysteries of the real universe is ultimately more rewarding" (p. 220). There is an appendix that explains "Why Things Can't Go Faster than Light."
Humanist 48:43 (Sep 1988); *LibrJnl* 113:165 (Aug 1988).

483 Taylor, John. *Science and the Supernatural: An Investigation of Paranormal Phenomena Including Psychic Healing, Clairvoyance, Telepathy and Precognition by a Distinguished Physicist and Mathematician.* New York: Dutton, 1980. 180p. Index: 175−180; Suggested reading list: 171−174

His curiosity piqued by Uri Geller, John Taylor, a physicist and professor of mathematics at Kings College, undertook the investigation of spoon−bending, hoping it would provide a clue to an explanation of psi in general. However, he was not able to obtain significant results under strict experimental conditions. He reviews the evidence for psychic healing, clairvoyance, telepathy, precognition, psychokinesis, and survival and concludes that certain results have normal explanations (e.g., an electrical explanation of some seeming PK phenomena), or could be explained by "mischief, fraud, credulity, fantasy, memory, cues and fear of death. With this evidence and the fact that all of these phenomena disagreed completely with scientific results, we have to conclude that the paranormal has 'disappeared'" (p. 164). He admits that a few psychic phenomena are hard to explain, namely, psychic burns, faith healing, and spontaneous human combustion, but avers "their character is almost certainly not paranormal" (p. 164).
JP 44:269−273 (S 1980); *PR* 11:16−19 (Sep/Oct 1980).
BritBkNews p590 (Oct 1980); *Kirkus* 48:572 (Apr 15, 1980); *Kirkus* 48: 594 (May 1, 1980); *LibrJ* 105:1315 (Jun 1, 1980); *NewStatesman* 100:19 (Sep 19, 1980); *Observer* p28 (Sep 7, 1980); *PublWkly* 217:65 (Apr 11, 1980); *SecondLook* 2:39 (May/Jun 1980); *TimesLitSuppl* p12 (Jan 2, 1981).

484 Vogt, Evon Z., and Hyman, Ray. *Water Witching U.S.A.* (2nd ed.). Chicago: University of Chicago Press, 1979. 260p. Bibliography: 253−260; 4 figures; 23 illustrations; 1 questionnaire; 10 tables

This is a revised edition of a classic critical work on dowsing first published in 1959 [41]. It contains a 12−page postscript that describes the more important happenings in the dowsing world occurring since 1959, especially in regard to the rise of the "urban dowser." However, the authors state that nothing has happened to change any of the basic conclusions reached about dowsing in 1959. If anything, those conclusions are more timely today. Their main conclusion, in 1959 and in 1978, is "that water divining is not an empirically reliable technique for locating shallow underground water, but rather a form of magical divination" (p. 221).

JSPR 51:28–29 (F 1981).

EarthSci 32:139 (Sum 1979).

Education in Parapsychology

INTRODUCTION

This is another new category——education in parapsychol—
ogy was not even mentioned in the books in *Sources*, although
one of the books listed, by Rhine and Pratt [58], was written as
a textbook of parapsychology.

Of the eight books listed here, five are textbooks: Edge et
al. [485], Irwin [810], the two books by Nash [487, 488], and
Shadowitz and Walsh [489]. Nash's textbooks are geared more
to undergraduates, whereas Edge et al. is suitable for graduate
students. The volume edited by Shapin and Coly [490] is a con—
ference proceedings that contains essays on various aspects of
education in parapsychology. The directory of educational op—
portunities compiled by McCormick [486] is a very useful and
popular publication. The bibliography compiled by White [491]
is strong on citing sources of information on parapsychology as
well as writings aimed specifically at teachers or students.

Additional books on education in parapsychology: Some
books cited elsewhere were also written as textbooks. See
Couttie [461], the book by the Leaheys [649], and Zusne and
Jones [659]. Neher [355] has written a textbook on transcendent
experiences and normal explanations for unusual phenomena, in—
cluding psi. Flew's anthology on philosophical problems of
parapsychology was developed for use in a course on the
"philosophy of parapsychology." Two books that should appeal
to young people using a question and answer format were edited
by Braud [509] and Hintze and Pratt [517]. The *Ashby
Guidebook* [804] serves as an introduction to psychical research
geared to new comers. Loye [781] presents a guide to educa—
tional possibilities concerned with the forecasting (i.e., precogni—
tive) mind. L.E. Rhine [448] has a section on the occult in
school, and how to get into parapsychology. Although dated,
Psi Search [508] still has value as a nontechnical yet authorita—
tive introduction to all facets of parapsychology. Many of the
bibliographies, encyclopedias, and indexes listed in the section
on "Reference Works" should prove useful to both teachers and
students. Of special note are the *Ashby Guidebook* [804],
Clarie's bibliographies [663, 664], and White and Dale [824].
White's bibliography [378] has a section on educational applica—

tions. Thalbourne's glossary [687], and Wolman's *Handbook* [688]are both useful volumes for teachers and students. The journal article bibliographies by Nester and O'Keefe [669, 670] were compiled with teachers and students in mind.

BIBLIOGRAPHY

485 **Edge, Hoyt L., Morris, Robert L., Palmer, J., and Rush, Joseph H.** *Foundations of Parapsychology: Exploring the Boundaries of Human Capability.* Foreword by T.X. Barber. Boston, MA: Routledge & Kegan Paul, 1986. 432p. Bibliography: 385—414; 3 figures; 4 graphs; 11 illustrations; Name index: 415—422; Subject index: 423—432; 6 tables

Offered as a textbook, this is "primarily a summary of the methods, research findings and theories" (p. xv) of parapsychology. The authors have also tried to present a fair summary of the skeptical viewpoint. Part I, by J.H. Rush, presents an overview of parapsychology. Part II, on "Research Methods and Findings," contains nine chapters by R.L. Morris, J. Palmer, and J.H. Rush on the methodology of psi research and the findings based on that research. Part III, "Implications: Scientific, Philosophical and Social," contains three chapters by H.L. Edge, one co—authored with R.L. Morris. Part IV, by R.L. Morris, consists of a brief summary in which he states: "This book is a combination of text, progress report, personal and intellectual statement by four colleagues, and case study" (p. 382). There are two statistical appendices and name and subject indexes. Each chapter ends with a list of "thought questions," and most also contain a summary (i.e., list) of basic terms indicative of the main subject matter of the chapter. Each chapter has its own contents listing of subsections. There is an extensive bibliography of references cited in the book.

ChrP 7:28—30 (Mar 1987); *JP* 51:353—357 (Dec 1987); *JRPR* 9:235—236 (Oct 1986); *PAI* 5:40 (Jun 1987); *PR* 17:9—12 (Nov/Dec 1986).

Light 106:132—133 (Aut 1986); *SpirFron* 19:114—115 (Spr 1987).

810 **Irwin, H.J.** *An Introduction to Parapsychology.* [See Addenda for annotation]

486 **McCormick, Donna L. (Ed.).** *Courses and Other Study Opportunities in Parapsychology* (10th ed.). New York: American Society for Psychical Research, 1987. 19p. Published every two or three years, this publication lists

courses and other study opportunities in parapsychology throughout the world. The school, department, name of course and instructor, type of course and level, course description, and name and address of person to contact for further information is provided. The schools are listed alphabetically by state for the U.S. and by country outside the U.S.

487 Nash, Carroll B. *Parapsychology: The Science of Psiology.* Springfield, IL: Charles C Thomas, 1986. 336p. Bibliography: 277–313; Glossary: 261–273; Author index: 315–323; Subject index: 325–336; Suggested reading list: 275–276; 1 table

Nash, biologist and, at the time, Director of the Parapsychology Laboratory at St. Joseph's College, says "the purpose of this work is to make a systematic and orderly presentation of the chaotic accumulation of experiences, experiments, and hypotheses pertaining to psi" (p. v) with the aim of furthering the establishment of psiology as a science. It is a revision of his earlier book, *Science of Psi: ESP and PK* [488]. The book is also designed for use as a text or reference book. Finally, it seeks to be a source of information concerning the *first* reported study of a variety of psi phenomena. Each chapter ends with a paragraph of conclusions and a list of review questions about the chapter's subject. There are 16 chapters, and a lengthy glossary and bibliography.

JASPR 82:291–295 (Jul 1988); *JRPR* 10:61–62 (Jan 1987); *JSPR* 54:155–157 (Apr 1987); *PAI* 5:44 (Dec 1987).

Choice 24:826 (Jan 1987); *ContempPsychol* 32:830 (Sep 1987).

488 Nash, Carroll B. *Science of Psi: ESP and PK.* Springfield, IL: Charles C Thomas, 1978. 299p. Bibliography: 250–277; Name index: 281–282; Subject index: 289–299; Suggested reading list: 279–280; 14 tables

Nash intends this book to serve both as a textbook and as a reference guide for researchers. Its purpose "is to make a systematic and orderly presentation of the chaotic accumulation of experiences, experiments, and hypotheses pertaining to the paranormal, with the desire of furthering the establishment of the study of psi as a science" (p.v). Review questions are given at the end of the 14 chapters and of the author's introduction. The chapters cover the history, phenomena, and methodology of parapsychology, as well as the characteristics of psi, criticisms, and theories. The descriptions of findings and experiments are overly brief and the discussions abbreviated. For the revised edition see the work above.

JASPR 73:409–412 (Oct 1979); *JIndPsychol* 2:155–156 (Jul

1979); *JP* 42:321–323 (Dec 1978); *JSPR* 50:180–181 (Sep 1979);
PN 3:3 (Oct 1980); *PR* 10:9–11 (Jul/Aug 1979); *T* 8:18–20
(Sum 1980).
ContempPsychol 24:876–877 (Oct 1979).

489 Shadowitz, Albert, and Walsh, Peter. *The Dark Side of
Knowledge: Exploring the Occult.* Reading, MA:
Addison–Wesley, 1976. 305p. Chapter notes; 15 figures;
2 graphs; 13 illustrations; Index: 297–305; 9 tables
This book is based on a course on the occult taught by
two physicists at Fairleigh Dickinson University who show how
the scientific method may be applied to all kinds of phenomena,
"internal" and "external," and that while many strange happen–
ings can be explained away, others may have a basis in fact.
Included in the latter category are several parapsychological
phenomena. This is a useful introduction to the scientific
method as applied in parapsychology and to "unexplained"
phenomena in general.
    JP 41:137–139 (Jun 1977).
    Booklist 73:862 (Feb 15, 1977); *Choice* 14:518 (Jun 1977);
HumBehav 6:75 (Mar 1977); *NYTimesBkRev* p20 (Jan 9, 1977);
SciBooks&Films 13:126 (Dec 1977); *SkptInq* 1:127–131 (Nov
1978); *WestCoastRevBks* 3:38 (Mar 1977).

490 Shapin, Betty, and Coly, Lisette (Eds.). *Education in
Parapsychology: Proceedings of an International Con–
ference held in San Francisco, California August 14–16,
1975.* New York: Parapsychology Foundation, 1976.
313p. Chapter references
    This volume consists of the 16 papers presented at the
Parapsychology Foundations's 1975 conference on education in
parapsychology. The discussion which followed each paper is
also provided. It is a must volume for teachers of parapsychol–
ogy as well as for students, and for anyone aspiring to enter
the field. In addition to the goals of parapsychology, much prac–
tical information is provided. The conference participants were
John Beloff, Irvin Child, Frederick Dommeyer, Wilbur Franklin,
Martin Johnson, Stanley Krippner, Robert Morris, James Mor–
riss, Enrique Novillo Pauli, Brenio Onetto, K. Ramakrishna Rao,
J.B. Rhine, D. Scott Rogo, Gertrude Schmeidler, Rex Stanford,
and Rhea White.
    JASPR 71:319–324 (Jul 1977); *JIndPsychol* 2:167–168 (Jul
1979); *JP* 40:317–321 (Dec 1976); *T* 6:36–37 (1978).
    Choice 14:123 (Mar 1977); *LibrJ* 102:209 (Jan 15, 1977);
Light 97:183–185 (Win 1977).

491 White, Rhea A. *Parapsychology for Teachers and Students:*

A Bibliographic Guide (2nd ed.). Dix Hills, NY: Para-psychology Sources of Information Center, 1989. 45p. Author index: 37—39; Glossary: 27—36

This is a listing of 388 books, dissertations, and journal, magazine, and newsletter articles. The guide is aimed at three audiences: teachers of parapsychology, enrolled students, and amateurs who want to try some parapsychological research on their own or who simply want to become better informed on what parapsychology is and the methodology of psi research. The bibliographic guide begins with information sources on education in parapsychology. There are six subcategories: (1) general works on education in parapsychology, (2) historical works, (3) sources aimed at teachers, (4) sources aimed at students, (5) courses and training programs in parapsychology, and (6) scholarships. Next are listed general sources of information on parapsychology broken down into eight subcategories: (1) textbooks, (2) books of readings, (3) literature surveys, (4) bibliographies, (5) other reference works, (6) criticisms of parapsychology, (7) parapsychology journals, and (8) organizations. The third section provides information sources on the methodology of parapsychology in three subcategories: (1) testing ESP and PK, (2) investigating spontaneous cases, and (3) investigating the survival problem. Section IV is on locating parapsychology information sources——how to get the most out of the library, computerized searching, abstracts, indexes, nonbook materials, etc. Section V lists information on the possibilities and pitfalls of parapsychology as a career, and the final section consists of a glossary of 157 parapsychological and related terms.

JP 52:242—244 (Sep 1988); *JSPR* 55:160—162 (Jul 1988).

Archaeus 7:46 (Spr/Sum 1988); *SpirFron* 20:186—187 (Sum 1988).

Experimental Parapsychology

INTRODUCTION

There were 20 titles listed in this category in *Sources*, and there are 16 listed here. This does not necessarily indicate that fewer psi experiments are being conducted. Rather, it is prob—ably due to the fact that psi research is becoming too complex and sophisticated to be suitable for publication in book format. (Some notion of this may be obtained by reading Houtkooper [822], the volume edited by Rao [498], Sargent's monograph [501], and the two conference proceedings edited by Shapin and Coly [502, 503].) Many of the books listed in this category in *Sources* are now chiefly of importance for historical reasons. However, Schmeidler [59], the two books by Thouless [63, 64], and Ullman and Krippner [66] are still of value. For systematic reviews of the experimental literature prior to 1940, the best book is still *Extrasensory Perception After Sixty Years* [51], and Rao's survey [52] still provides good coverage for the years 1940—1965.

For experiments after 1960, a number of sources can be consulted: the series *Advances in Parapsychological Research* edited by Krippner [520-524], Wolman's *Handbook of Parapsy—chology* [688], Nash's textbooks [487, 488], and the bibliographies in Rao's *Basic Experiments* [498] provide good coverage. The *Advances* series is also excellent for methodology, especially the first three volumes. The textbook by Edge et al. [485] provides in—depth discussions of all forms of methodology. Several good discussions of methodology are found in the two books edited by Shapin and Coly [502, 503], and the Eysenck and Sargent [495] book is a guide to psi experimentation, and Ryzl [500] emphasizes methods of handling subjects. Specific experiments are described and discussed by Collins [492], Duplessis [494], Hardy et al. [496], Houtkooper [822], McConnell [497], Roll [499], Sargent [501], Targ and Puthoff [504], Tart et al. [505], and Ullman and Krippner with Vaughan [819]. Collins [492], and Collins and Pinch [493] provide a perspective of psi research from the viewpoint of the sociology of science.

Additional books with information on experimental para-psychology: The books in the section on animal psi [362-365]

describe psi experiments with animals as subjects. Dean et al.
[369] report on their experiments with executives, primarily
using precognition tests. Ebon [370] describes Mitchell's Apollo
14 ESP experiment and remote—viewing experiments at SRI.
Several experiments are reported in the book edited by Jones
[373] which also contains methodological discussions (see especially the contributions by Jahn, Kelly, May, Morris, Osis,
Puthoff, and Schwartz). Schwartz [376, 816] describes several
experiments in psychic archaeology, and includes a chapter on
the psychology of the sensitive and the investigator. Under the
heading of "Basic Research," White's bibliography [378] cites psi
research that is relevant to investigating applied psi, such as the
repeated—guessing technique, and another of her bibliographies
[454] cites research with children and still another lists research
bearing on the voluntary control of psi [791]. A Parapsychology
Foundation proceedings [817] describes several experimental
research approaches of the 1980s. W.G. Roll's experiments with
the psychic Shirley Harrison are described in Harrison and
Franklin [389]. John Kmetz describes his experiments with
Dean Kraft in the latter's book [392]. Peter Bander describes
experiments with Matthew Manning [393]. Tanous [397] devotes
considerable space to descriptions of experiments in which he
has been the subject, particularly those conducted by Osis at
the ASPR. Swann [396] describes research conducted with him
by Osis, Schmeidler, and Targ and Puthoff. A history of the
Rhinean school of experimental parapsychology is provided by
the books about J.B. Rhine [401, 409, 410]. The book on J.G.
Pratt [404] covers the same years but his work took him
beyond FRNM to the University of Virginia, Czechoslovakia, the
USSR, and the ASPR. McConnell's autobiography [406] also
deals largely with aspects of experimental parapsychology.
Charles Tart [412] describes many aspects of his research
program. Several books in the chapter on "Books on Psi for
Young People" describe ESP and PK tests: [413, 416, 421,
429, 431, 440, 441, 442, and 448]. One of them, by Daniel
Cohen [424], is a handbook of ESP tests for young people.
Several of the books provide nontechnical introductions to psi
experiments: [417, 411, 429, 430, 436, 437, 442, 448]. The
most notable is the one by L.E. Rhine [448]. Feola [435] concentrates on PK experiments and Cohen [421] provides the most
up—to—date introduction, including a chapter on a visit to the
Psychophysical Research Laboratories. The Report of the Committee on Techniques for the Enhancement of Human Performance [462] is one of the few critical works that deals with
mainline psi experiments. ESP experiments are also criticized
by Hansel [472]. Mauskopf and McVaugh [558] provide a
detailed history of experiments in parapsychology from 1920—

1940. The chapters in Beloff [507] emphasize the experimental approach to psi. Experimentation is also emphasized in all of the annual volumes of *Research in Parapsychology*, listed in the section on "General Parapsychology," as do Eysenck and Sargent [514]. The PK and precognitive remote viewing research conducted at the Princeton Engineering Anomalies Research Laboratory is described in Jahn and Dunne [617]. Review articles on parapsychology experiments, covering both ESP and PK, are found in the *Advances in Parapsychological Research* series [520-524]. Schmeidler [654] presents a theory of psi and backs it up with reference to many of her experiments and those of others. She also questions the criticism that parapsychology lacks a repeatable experiment, and she provides an excellent experimentally oriented glossary. There are two long sections on experimental ESP and PK methodology in Zusne and Jones [659]. For Tart's experiments on the application of learning theory to ESP performance see [786] and [787]. Ingo Swann [784] presents a method for developing ESP in the experimental situation. Targ and Harary [785] describe remote-viewing experiments in general terms. Hearne [731a] reviews precognition experiments and offers research proposals.

BIBLIOGRAPHY

492 Collins, H.M. *Changing Order: Replication and Induction in Scientific Practice.* Beverly Hills, CA: Sage Publications, 1985. 187p. Bibliography: 175−181; 10 figures; 2 graphs; Name index: 183−184; Subject index: 185−187

Collins, a British sociologist, views parapsychology in the larger picture provided by the history, philosophy, and sociology of science. In connection with the repeatability of scientific findings, three cases from varied fields are studied and compared: laser building, the detection of gravitational radiation, and two areas of parapsychology: so−called plant psi and mind−over−matter (PK). He points out that "the parapsychology and the gravity wave debates . . . look very like each other in terms of the structure of argument which surrounds the claims of replicability; and they both look quite unlike the TEA−laser case. Thus . . . it is the TEA−laser that is the odd one out! This marks a difference between the perspective of this work and more orthodox ways of looking at science. The important dimension turns out not to be the scientific subject matter but the phase of science that is represented" (p. 4). He is concerned with how change occurs in science and how conceptual order is brought about. Through the cases examined he shows how individuals create the potential for change, how other in−

dividuals can help or hinder this process, how these efforts are embedded in the wider society, how the wider society is the locus of conceptual order, and how facts are made fact—like in spite of their human creators.
  JASPR 81:187—192 (Apr 1987); *JSPR* 53:397—398 (Jul 1986); *PAI* 5:42—43 (Dec 1987); *PR* 18:10—12 (Jul/Aug 1987).
  ContempSociol 15:619 (Jul 1986); *Nature* 319:274 (Jan 23, 1986); *Science* 230:1267 (Dec 13, 1985).

493 Collins, H.M., and Pinch, T.J. *Frames of Meaning: The Social Construction of Extraordinary Science.* Boston, MA: Routledge and Kegan Paul, 1982. 210p. Bibliography: 195—204. Chapter notes; 3 figures; Name index: 205—207; Subject index: 208—210; 3 tables
  The authors, then sociologists at the University of Bath, were invited to participate in some experiments with children who claimed to be metal—benders. Participation in the experiments made evident the need to question the nature of the scientific method, particularly the role of rationality in science and the relation between different cultures within science itself. The authors discuss the difficulties involved in communication between groups with differing rational categories, e.g., between psychology and parapsychology. They propose that participation is needed to promote communication. They present an excellent and detailed survey of metal bending research and the conflicts aroused by it between believers and skeptics. Next they examine the question of parapsychology's "compatibility with science" (physics and psychology). They devote a chapter to quantum theory, concluding that this area is compatible to parapsy—chological ideas, although it is itself a low—status area, and ob—serving that parapsychology is not necessarily incompatible with science. They next describe the metal—bending experiments at Bath, and note the problems involved in observing the data and the role of preconceived notions in the interpretation of evidence. Alternative explanations are discussed, including the experimenter effect. They close with a discussion of the conse—quences of their findings for the idea of scientific revolution.
  JSPR 51:309—314 (Jun 1982); *Light* 102:86—88 (Sum 1982); *PR* 14:20—21 (Mar/Apr 1983); *Psychoenergetics* 5:83—84 (1983); *T* 11:69—71 (Aut 1983).
  ExplorKnowl 3:82—85 (1986); *NewStatesman* 104:20—21 (Aug 6, 1982); *SociolRev* n.s.31:169—170 (Feb 1983); *Sociology* 16:620—621 (Nov 1982).

494 Duplessis, Yvonne. *The Paranormal Perception of Color* (Parapsychological Monographs No. 16). New York: Parapsychology Foundation, 1975. 117p. Bibliography:

103—107; 5 figures; 1 graph; Index: 109—117

Dr. Duplessis is a French parapsychologist who has special—
ized in research with blind or partially blind subjects with the
emphasis on extra—retinal vision and dermo—optic sen—
sitivity. Some of this research is described here. The first
chapter deals with normal perception of a color when an organ
other than the eye is involved. In the next chapter she deals
with the paranormal perception of color, or, color perceived at a
distance. The third section is on extra—retinal vision in special
subjects. Duplessis reviews historical cases as well as her own
research with the blind. The final chapter is devoted to
theories and applications of paranormal color perception.

JSPR 48:341—342 (Jun 1976).
Light 96:139—140 (Aut 1976).

495 Eysenck, Hans J., and Sargent, Carl. *Know Your Own
Psi—Q: Probe Your ESP Powers.* New York: World Al—
manac Publications, 1983. 192p. 29 figures; 22 illustra—
tions; 12 tables

This is a guide to 30—odd ways of testing oneself or
others for psi (mainly ESP). Both quantitative and qualitative
tests are described. Some tests are in the form of games. There
is a chapter on computer tests written in BBC Basic, with Ap—
pleSoft conversions. Psychological factors in psi testing are
described throughout.

ASPRN 10:3 (Apr 1984); JASPR 79:558—561 (Oct 1985).
SkptInq 8:180—182 (Win 1983—84).

496 Hardy, Alister, Harvie, Robert, and Koestler, Arthur. *The
Challenge of Chance: A Mass Experiment in Telepathy
and Its Unexpected Outcome.* New York: Random House,
1975. 308p. Chapter references; 55 figures; 18 illustra—
tions; Index: 305—308; 10 tables

British biologist/parapsychologist Sir Alister Hardy
describes his Caxton Hall experiment, which is notable for the
fact that nearly 200 agents were used in a GESP testing situa—
tion with 20 subjects simultaneously trying to guess target pic—
tures. The fact that the main significance was the high cor—
respondence among the subjects' responses themselves rather
than in their relation to the targets led Hardy to enlist the aid
of psychologist/statistician Robert Harvie. He reports on some
controlled runs he did in which similar results were obtained.
To explain these results he discusses the nature of probability
and chance. In the third section, Arthur Koestler describes
many spontaneous cases involving unusual coincidences in
everyday life. In the fourth part, Koestler suggests that such
experiences can only be accounted for by some principle that is

beyond psi or at least an extension of it. Three appendices
provide further details concerning the Caxton experiment and
the later control experiments.

JASPR 69:174—177 (Apr 1975); *JP* 38:423—427 (Dec 1974);
JSPR 47:319—326 (Mar 1974); *P* 6:10—11,52 (Nov/Dec 1974);
PR 5:16—19 (Jan/Feb 1974); *PS* 1:239—240 (1976).
Books&Bkmn 19:23 (Feb 1974); *ContempPsychol* 21:890
(Dec 1976); *Choice* 11:1386 (Nov 1974); *Economist* 249:115 (Dec
15, 1973); *GuardianWkly* 109:23 (Nov 17, 1973); *Kirkus* 42:562
(May 15, 1974); *LibrJ* 99:1965 (Aug 1974); *Listener* 91:22 (Jan
3, 1974); *Observer* p39 (Nov 4, 1973); *Parabola* 2(3):90—91
(1977); *PsycholToday* 8:28 (Nov 1974).

822 Houtkooper, Joop M. *Observational Theory: A Research
Programme for Paranormal Phenomena.* [See addenda
for annotation]

497 McConnell, R.A. (Ed.). *Parapsychology and Self—Deception
in Science.* Pittsburgh, PA: Author, 1983. 150p. Chapter
references; 8 figures; 6 graphs; Index: 147—150; 7 tables
    The editor observes that parapsychological phenomena have
been seriously and competently investigated for four generations
(100 years), yet they are still ignored by scientists. He suggests
that the answer as to why this is so lies not in parapsychology
but in science as a whole. The chapters in this book are offered
as a contribution to the study of truth—avoidance in science as
well as a contribution to the experimental parapsychology litera—
ture. Three chapters are included as examples of self—deception
among parapsychologists, because they consist of papers rejected
for publication in parapsychological journals. One is physicist
C.K. Jen's account of current parapsychological investigations in
China, one is an historical review and reanalysis of the editor's
1948—1950 PK experiment reported under the title "Wishing
With Dice Revisited," and the third chapter presents a further
analysis of the earlier paper and a theory to explain the scoring
patterns revealed. In it the operation of an "organizing prin—
ciple" is postulated as well as a "principle of ambivalence"
which it is hoped will help to provide a unified explanation of
"target missing" and the decline effect.   Another chapter dis—
cusses the rejection of these papers and what it might reveal
about parapsychological journal editors (he suggests they are
modeling themselves after psychology journal editors). There is
a chapter by McConnell describing the unusual difficulties his
assistant, T.K. Clark, experienced during 12 years of graduate
study working on a topic that was not parapsychological. It is
implied that the resistance she encountered was due to her as—
sociation with him. In the last paper, McConnell discusses what

is likely to happen on planet Earth before the year 2000. He is
of the opinion that if parapsychology has not become dominant
by then, the field will disappear. What occurs in those two
decades will have momentous consequences for humankind. The
planet's physical and biological resources are rapidly being used
up, and from the viewpoint of human resources he thinks the
prospect for the year 2000 is also dim. He holds that, as a
species, *Homo sapiens* is doomed unless we experience a moral
rebirth. Parapsychology could provide the solution to our
problems because it can change our philosophic outlook, and
with it our ethical standards.

JASPR 78:70–75 (Jan 1984); *JP* 48:236–237 (S 1984);
PAI 3:40 (Dec 1985).

498 Rao, K. Ramakrishna (Ed.). *The Basic Experiments in
Parapsychology.* Jefferson, NC: McFarland, 1984. 264p.
Bibliography: 243–259; Index: 261–264

This book consists of a selection of experimental ESP and
PK reports from the last 50 years. The editor, who is Director
of the Foundation for Research on the Nature of Man, points
out that his "concern is not to present the best from the point
of evidence and safeguards but to collect a few basic experiments that illustrate diverse procedures and broadly reflect the
major trends of psi research. In making this selection, I have
had in mind, among others, the student who wishes to carry
out an experiment and is looking for useful experimental procedures" (p. 3). The Introduction to the book itself in effect is a
bibliographic guide to modern experimental parapsychology.
Rao briefly discusses several key topics and cites relevant literature. The topics are card–guessing tests, early PK experiments,
ESP and personality variables, animal psi, EEG and ESP
studies, memory and ESP, subliminal perception and ESP, psi-
missing and the differential effect, and replication. Fifteen experiments are included, nine of which were originally published in
the *Journal of Parapsychology*, three in the *Journal of the
American Society for Psychical Research*, and one each in
Biological Psychiatry, the *Indian Journal of Psychology*, and the
International Journal of Parapsychology. Each experiment is
preceded by an introduction by the editor, which provides the
context of the article and suggestions for additional advanced
reading. The experiments included are: "A Review of the
Pearce–Pratt Distance Series of ESP Tests," by J.B. Rhine and
J.G. Pratt; "A Further Investigation of Teacher–Pupil Attitudes
and Clairvoyance Test Results," by Margaret Anderson and
Rhea White; "ESP Card Tests of College Students With and
Without Hypnosis," by Jarl Fahler and Remi J. Cadoret; "A
Telekinetic Effect on Plant Growth. II. Experiments Involving

Treatment of Saline in Stoppered Bottles," by Bernard Grad; "A Laboratory Approach to the Nocturnal Dimension of Paranormal Experience: Report of a Confirmatory Study Using the REM Monitoring Technique," by Montague Ullman and Stanley Krippner; "A Second Confirmatory ESP Experiment with Pavel Stepanek as a 'Borrowed' Subject," by J.G. Blom and J.G. Pratt; "Personality Characteristics of ESP Subjects: III. Extraversion and ESP," by B.K. Kanthamani and K. Ramakrishna Rao; "An ESP Drawing Experiment with a High–Scoring Subject," by J. Ricardo Musso and Mirta Granero; "PK Tests with a High–Speed Random Number Generator," by Helmut Schmidt; "Further Studies of Relaxation as a Psi–Conducive State," by Lendell W. Braud and William G. Braud; "Awareness of Success in an Exceptional Subject," by H. Kanthamani and E.F. Kelly; "Psi Information Retrieval in the Ganzfeld: Two Confirmatory Studies," by James C. Terry and Charles Honorton; "Yogic Meditation and Psi Scoring in Forced–Choice and Free-Response Tests," by K. Ramakrishna Rao, Hamlyn Dukhan, and P.V. Krishna Rao; "Transcontinental Remote Viewing," by Marilyn Schlitz and Elmar Gruber; and "Transcontinental Remote Viewing: A Rejudging," by Marilyn Schlitz and Elmar Gruber. The bibliography contains all references cited in the book––those originally listed in the experimental reports and those cited by Rao.
    JSPR 53:461–462 (Oct 1986); *PAI* 3:41–42 (Jun 1985); *PR* 16:10–12 (May/Jun 1985).
    Choice 22:1077 (Mar 1985); *ContempPsychol* 31:180 (Mar 1986).

499 Roll, William G. *Theory and Experiment in Psychical Research.* New York: Arno Press, 1975. 510p. Bibliography: 468–495; 1 figure; 8 graphs; 7 illustrations; Name index: 496–500; 1 questionnaire; Subject index: 501–510; 30 tables
    This was originally submitted in 1959 as a thesis at Oxford University and was later revised and expanded. Roll describes his research on hypnosis and ESP, personality variables and ESP, and habit–patterns in ESP testing. He also provides a review of psychometry. Several models of psi are described and discussed, and hypotheses capable of being tested empirically are presented. The book reflects an extensive acquaintance with the literature of parapsychology, and there is a 30-page bibliography.
    T 45:8 (Sum 1975).

500 Ryzl, Milan. *ESP Experiments Which Succeed.* San Jose, CA: Author, 1976. 212p. Chapter references; 6 figures;

6 graphs; 7 tables
This is a guide to ESP testing methodology by a success—
ful parapsychology experimenter. There are chapters of general
orientation to the scientific method, hints on observation, infor—
mation on handling subjects, experimental design, analyzing, and
reporting results. A number of different types of experiments are
suggested. In four appendices he presents experiments he has
conducted that exemplify topics discussed in the text. The book
closes with a reprint of his article in *Psychic* entitled "ESP in
Eastern Europe and Russia."

501 Sargent, Carl L. *Exploring Psi in the Ganzfeld*
(Parapsychological Monographs, No. 17). New York:
Parapsychology Foundation, 1980. 124p. Bibliography:
119—124; 29 tables
Sargent received his Ph.D. for research in parapsychology
from Cambridge University, where he is now a member of the
Psychology Department. Here he reports on the first six
Ganzfeld experiments he carried out at Cambridge. He makes
suggestions for further research in this area. He emphasizes
process—oriented research and discusses the question of why the
Ganzfeld test situation is psi—conducive.
JSPR 51:165—167 (Oct 1981); *T* 9:18—19 (Win 1981).
LibrJ 106:1302 (Jun 15, 1981).

502 Shapin, Betty, and Coly, Lisette (Eds.). *Parapsychology and
the Experimental Method: Proceedings of an International
Conference Held in New York, New York, November 14,
1981.* New York: Parapsychology Foundation, 1982.
120p. Chapter references
Presented are the proceedings of the 30th international con—
ference held by the Parapsychology Foundation in New York
City, November 14, 1981. There were a total of six contribu—
tions dealing with various aspects of methodology. Two papers,
those by Richard Broughton and Donald J. McCarthy, were on
the use of computers in psi research. Anita Gregory spoke
about the investigation of macro—PK. Rex Stanford's paper was
entitled "On Matching the Method to the Problems: Word As—
sociation and Signal—detection Methods for the Study of Cogni—
tive Factors in ESP Tasks." Hoyt Edge presented "Some Sug—
gestions for Methodology Derived from an Activity
Metaphysics," and K. Ramakrishna Rao dealt with "Science and
the Legitimacy of Psi." Each paper is followed by the inter—
change between discussants from the floor and the author. In
addition, the transcripts of a "morning general discussion" and
an "afternoon general discussion" are presented.
JASPR 77:239—242 (Jul 1983); *JP* 47:68—74 (Mar 1983);

PsiR 2:124–126 (S 1983).

503 Shapin, Betty, and Coly, Lisette (Eds.). *The Repeatability Problem in Parapsychology: Proceedings of an International Conference Held in San Antonio, Texas, October 28–29, 1983.* New York: Parapsychology Foundation, 1985. 264p. Chapter references; 7 figures; 1 table
This proceedings of the 32nd Parapsychology Foundation conference contains nine papers, contributed by Beloff, Black—more, Braud, Edge, Honorton, Rao, Schlitz, Stanford, and Walker. All the papers dealt with aspects of repeatability——or nonrepeatability——some methodological, some theoretical, and some experimental.
JP 50:274–279 (Sep 1986); *JRPR* 9:112–113 (Apr 1986); *JSPR* 53:458–460 (Oct 1986).
ContempPsychol 32:390 (Apr 1987); *SpirFron* 28:126–127 (Spr 1986).

504 Targ, Russell, and Puthoff, Harold. *Mind–Reach: Scientists Look at Psychic Ability.* Introduction by Margaret Mead. Foreword by Richard Bach. New York: Delacorte Press/Eleanor Friede, 1977. 230p. Chapter notes; 25 figures; 55 illustrations; Index: 223–230; 12 tables
Two physicists offer a popular account of the "remote view—ing" experiments they have conducted at Stanford Research In—stitute with subjects Phyllis Cole, Duane Elgin, Uri Geller, Hella Hammid, Marshall Pease, Pat Price, and Ingo Swann. They also present a method of developing psi ability and describe practical applications of psi phenomena.
PAI 3:40 (Dec 1985).
Books&Bkmn 23:60 (Mar 1978); *BestSell* 37:47 (May 1977); *Choice* 14:449 (May 1977); *ContempPsychol* 27:5–18 (Jul 1982); *ContempRev* 231:279 (Nov 1977); *Kirkus* 44:1338 (Dec 15, 1976); *LibrJ* 102:716 (Mar 15, 1977); *Listener* 98:540 (Oct 27, 1977); *NYRevBks* 24:18 (Mar 17, 1977); *NYTimesBkRev* p6 (Mar 13, 1977); *NoeticNews* 5:5 (Spr 1977); *TimesEdSuppl* p24 (Jun 16, 1978); *TimesLitSuppl* p1360 (Nov 18, 1977).

505 Tart, Charles T., Puthoff, Harold E., and Targ, Russell (Eds.). *Mind at Large: Institute of Electrical and Electronic Engineers Symposia on the Nature of Extrasen—sory Perception.* New York: Praeger, 1979. 267p. Chapter references; 19 figures; 15 graphs; 42 illustrations; Index: 261–267; 21 tables
This is a collection of papers originally presented at two symposia of the Institute of Electrical and Electronic Engineers. The contributors are: Edgar Mitchell, Harold Puthoff, Russell

Targ, Edwin May, John Bisaha, B.J. Dunne, Charles Tart, O. Costa de Beauregard, Michael Persinger, Helmut Schmidt, Ingo Swann, Edward Wortz, A.S. Bauer, R.F. Blackwelder, J.W. Eerkens, and A.J. Sayr. The papers are both experimental reports and theoretical contributions on physics and psi. Of special note are the last two papers: one by psychic Ingo Swann, and the other a review of Soviet psychical research by five California engineers.

JASPR 74:445–454 (Oct 1980); *PR* 11:13–14 (Jul/Aug 1980).

Brain/MindBul 5:4 (Dec 17, 1979); *ContempPsychol* 27:548 (Jul 1982); *InstNoeticSciNewsl* 8:24,26 (Win 1980); *LibrJ* 104:2470 (Nov 15, 1979); *Light* 100:32–34 (Spr 1980); *SpirFron* 12:120 (Spr 1980).

819 **Ullman, Montague, and Krippner, Stanley, with Alan Vaughan.** *Dream Telepathy: Experiments in Nocturnal ESP* (2nd ed.). [See addenda for annotation]

General Parapsychology

INTRODUCTION

This category contains books that attempt to survey the entire field of parapsychology. It was the largest category in *Sources*: 36 titles. Several of the titles listed there are still cited by reputable authors: Broad [154], Heywood [160], Koestler [165], Murphy with Dale [168], Myers [169], Pratt [174], a Parapsychology Foundation proceedings [175], and West [185].

Here, too, this is the category with the largest number of books: 44. Three of them are general nontechnical introductions to modern parapsychology: Bowles and Hynds [508], Eysenck and Sargent [514], and the book by L.E. Rhine [532]. In the same class, but with a question and answer format, are the books by Braud [509] and by Hintze and Pratt [517]. Several are anthologies: Beloff [507], which emphasizes research methodology; the Parapsychology Foundation proceedings [817] presents a number of accounts of current research approaches. Ebon [512], which covers all aspects, including the relation of psi to other disciplines; Grattan—Guinness [516], which covers psychical research and parapsychology from a British perspective; and the books compiled by Mitchell [526], and White [546], which bring together state—of—the—art reviews. One Parapsychology Foundation proceedings [542] contains papers on the direction of parapsychology after a century of research and another emphasizes concepts and theories [541]. Perhaps the most important volumes in this section are those comprising the series entitled *Advances in Parapsychological Research*, edited by Krippner [520-524]. Each one includes review chapters by specialists who assess the past and current findings and methodology of the major facets of parapsychology. For the best overview of what is happening in parapsychology in any given year starting with 1972, there is *Research in Parapsychology* [527-529, 533-540, 543-545, 547, 548], which provides lengthy abstracts (1—3 pages) of the papers given at the annual convention of the Parapsychological Association and the full text of the presidential and sometimes other invited addresses. This series was just beginning when *Sources* went to press, and has since proven its worth. The remaining titles emphasize certain aspects of the subject: Bartlett [506], Chauvin [510], and Eisenberg [513] all cover the entire field, but they emphasize applications and im—

plications. Dubrov and Pushkin [511] present a Soviet view of
parapsychology. Fairley and Welfare [515] provide a skeptical
view of the phenomena; Owen [530] surveys parapsychology in
Canada; and Panati [531] emphasizes altered states research.
William James [519] surveys psychical research as it was in the
first part of this century. McConnell [525] presents parapsychology through the words of many prominent persons who have
been interested in it. Inglis [518] reviews the evidence and at—
tempts to refute the views of skeptics.

Additional books with information on general parapsychology: White's bibliography [378] lists 1,479 items that suggest
possible applications and implications of psi. Much general information on parapsychology can be gleaned from the biographical
sketches found in Berger [399] and Pilkington [408] and the
books by Ashby [804], Brian [401], Keil [404], Krippner [405],
and Rhine [410]. Ashby [804] also doubles as a popular introduction to psychical research. Several good introductions to parapsychology are listed in the section "Books on Parapsychology for
Young People:" [416, 417, 421, 422, 429, 436, 437, 441, and
448]. Several of the books listed in the section "Histories of
Parapsychology" can double as general introductions——especially
to the subject as it was prior to 1940. See Cerullo [550],
Douglas [551], Evans [552], Haynes [554], Inglis [555, 556], Maus—
kopf and McVaugh [558], and Oppenheim [560]. The chapters in
Rao's anthology [498] provide a general overview of current psi
research. Braude's [605] book on ESP and PK can be viewed
as a scholarly introduction to parapsychology that emphasizes
theoretical aspects. Krippner [648] provides a general overview
of Soviet parapsychology. Several textbooks in the section on
education also double as introductions to parapsychology: Edge
et al. [485], Irwin [810], and Nash [487, 488]. Many of the bibli—
ographies, encyclopedias, and indexes in the section on reference
books cover the entire field of parapsychology. Of special note
are the bibliographies by Clarie [663, 664] and, although dated,
the journal literature bibliographies of Nester and O'Keefe [669,
670], White's guide to books in print [679], the Soviet, Chinese,
and East European bibliography by White and Vilenskaya [680],
Thalbourne's glossary [687], and Wolman's *Handbook* [688].

BIBLIOGRAPHY

506 **Bartlett, Laile E.** *Psi Trek: A World—Wide Investigation
     into the Lives of Psychic People and the Researchers
     Who Test Such Phenomena as: Healing, Prophecy,
     Dowsing, Ghosts, and Life After Death.* New York:

McGraw—Hill, 1981. 337p. Bibliography: 325—332; Chap—
ter notes; Index: 333—337

This is a popular assessment of the current parapsy—
chological scene worldwide, based primarily on interviews. The
author, whose background is in sociology, emphasizes the practi—
cal applications of psi. Part One is devoted to seven chapters
on "uses" in business, dowsing, healing, detection, protection,
science, and general utility. The second part consists of six
chapters on "The Quest." They deal with the implications of
psi and parapsychology and society.

JP 46:72—76 (Mar 1982); *PAI* 1:30 (Aug 1983); *PR* 12:24—
25 (Jul/Aug 1981); *PsiR* 1:106—107 (Jun 1982); *T* 10:21—22
(Spr 1982).

AmDows 21:17 (Aug 1981); *Booklist* 77:1064 (Apr 1981);
Kirkus 49:396 (Mar 15, 1981); *LibrJ* 106:667 (Mar 15, 1981):
SpirFron 15:43—44 (Win 1983).

507 Beloff, John (Ed.). *New Directions in Parapsychology.*
Postscript by Arthur Koestler. Metuchen, NJ: Scarecrow
Press, 1975. 174p. Chapter references; 8 figures; Glos—
sary: xiv—xxvi; 2 graphs; 19 illustrations; Index: 171—
174; 7 tables

This collection of articles written especially for this book is
oriented toward empirical research in parapsychology.
Psychologist/parapsychologist John Beloff of the University of
Edinburgh, who is a past president of both the Society for
Psychical Research and the Parapsychological Association,
provides an overview of parapsychology, its findings, methodol—
ogy, and theoretical implications. Various key aspects of para—
psychology are treated in depth by experts in each area: Hel—
mut Schmidt on "Instrumentation in the Parapsychological
Laboratory," Charles Honorton on "ESP and Altered States of
Consciousness," K.R. Rao on "Psi and Personality," John Ran—
dall on "Biological Aspects of Psi," J.G. Pratt's "In Search of
the Consistent Scorer," Hans Bender on "Modern Poltergeist
Research," and W.G. Roll takes "A New Look at the Survival
Problem." In a postscript, Arthur Koestler discusses parapsy—
chology in relation to the sciences, especially physics. Each chap—
ter has its own synopsis and bibliography, and Beloff provides a
glossary of terms.

JASPR 70:95—101 (Jan 1976); *JP* 39:64—67 (Mar 1975);
JSPR 47:510—512 (Dec 1974); *P* 7:45—46 (Jul/Aug 1976); *PR*
7:10—12 (Jul/Aug 1976); *T* 4:14—15 (Win 1976)

BritJPsychol 66:111—112 (Feb 1975); *Choice* 13:589 (Jun
1976).

508 Bowles, Norma, and Hynds, Fran, with Joan Maxwell. *Psi*

Search. New York: Harper & Row, 1978. 168p. Chapter
notes; 10 figures; Glossary: 153–155; 104 illustrations;
Index: 165–168; 1 map; Suggested reading list: 145–147
    Basic compendium on modern parapsychology and an out—
growth of the museum exhibit "Psi Search," organized by the
authors. A reliable, comprehensive introduction to the basic
facts, major investigations and investigators, research areas, and
possible applications of psi. This is also an excellent book for
students, libraries, and schools. It is very good on what para—
psychology is and is not, its aims and methodology, and where
to find more information.
    JASPR 73:204–206 (Apr 1979); *JP* 42:323–325 (Dec 1978).
    Booklist 74:1666 (Jul 1, 1978); *Brain/MindBul* 3:4 (Feb 20,
1978); *Choice* 15:1280 (Nov 1978); *Kirkus* 46:29 (Jan 1, 1978);
Kliatt 12:38 (Fal 1978); *LibrJ* 103:177 (Jan 15, 1978); *NewReal*
2:34 (Aug 1978).

509 Braud, William G. *Psi Notes: Answers to Frequently Asked
        Questions About Parapsychology and Psychic Phenomena.*
        San Antonio, TX: Mind Science Foundation, 1984. 172p.
        Subject index: 169–172
    The purpose of this book is both to answer the demand of
the general public for information on psychical research, and to
reduce misunderstanding and apprehension engendered by inac—
curate information. A question—and—answer format is used and
much of the material is taken from Braud's 1978 weekly radio
program. The material is presented under six headings: "The
Field of Parapsychology"; "Varieties of Psychic Phenomena";
"Factors Influencing Psychic Functioning in Psi Practitioners";
"Practical Applications of Psi"; and "Some Survival Issues."

510 Chauvin, Remy. *Parapsychology: When the Irrational
        Rejoins Science.* Translated by Katherine M. Banham.
        Jefferson, NC: McFarland, 1985. 164p. Bibliography:
        157–160; 11 figures; Index: 161–164; 1 map
    French biologist Chauvin reviews what the Rhinean school
has accomplished, but he feels it is in error. He seeks new ap—
proaches to psi, emphasizing practical aspects. He proposes that
we study psi and try to understand its characteristics first, and
then experiment——but not experiment initially, as is now the
case. He advocates an ethological approach, in which one could
look for the most propitious conditions for psi operation by ex—
amining the age—old empirical practices and prophetic tech—
niques. Some of these practices are discussed in the book. Also
included are chapters on psi and physics and on parapsychology
in the Soviet Union.
    JASPR 82:165–169 (Apr 1988); *JP* 52:175–177 (Jun 1988);

PAI 4:43 (Dec 1986).
Choice 23:667 (Dec 1985).

511 Dubrov, A.P., and Puskin, V.N. *Parapsychology and Contemporary Science.* NY: Consultants Bureau, 1982. 221p.
Bibliography: 198−214; 18 figures; 6 illustrations; Subject index: 215−221; 4 tables
This is a review of parapsychology through the eyes of two Soviet scientists that has not yet been published in the Russian language. There are three introductions, by Christopher Bird, Joseph Brozek, and William Tiller. The authors emphasize the importance of facts, holding that they are not in conflict with materialistic philosophy. They develop a new classification of psi phenomena based on space and time, materials and energetics, and field and force. Parapsychology is presented in six chapters in Part I as an aspect of psychology. Part II consists of five chapters dealing with the experimental psychology of paranormal phenomena, and the third consists of 11 chapters on the fundamentals of natural science and parapsychology. It describes the general future of psi and offers a theory to explain the nature of psi and the place of psi in contemporary science.
IJParaphys 16:79−86 (1982); *JASPR* 78:166−169 (Apr 1984); *JSPR* 52:149−150 (Jun 1983); *PR* 14:12−13 (Nov/Dec 1983); *T* 11:90−91 (Win 1983).

512 Ebon, Martin (Ed.). *The Signet Handbook of Parapsychology.* New York: New American Library, 1978. 519p.
Chapter references; 10 figures; Glossary: 509−512; Suggested reading list: 513−518
Ebon's is a fairly technical anthology of 25 articles by experts on various aspects of parapsychology, originally published in scientific journals and books. There are eight major subject areas: history, survival of death, and parapsychology in relation to psychotherapy, religion, biology, consciousness, dreams, and science. The majority of the selections are review articles, but a few research reports are also included. This handbook should be of value to serious students of parapsychology and to scientists in other fields who want to be exposed to a sampling of the parapsychological literature.
JP 44:179−180 (Jun 1980).

513 Eisenberg, Howard. *Inner Spaces: Parapsychological Explorations of the Mind.* Don Mills, Ontario: Musson Book, 1977. 184p. Bibliography: 163−177; 2 figures; 2 figures; Glossary: 133−139; Index: 178−184
Eisenberg, a Canadian parapsychologist and psychotherapist, provides a conceptual framework for parapsychology

consisting of brief summaries of all aspects of the field: spon—
taneous and experimental, theoretical and practical, and fringe
areas and core subjects. Of special value are two sections: the
commercial guide, which deals with exploitations of psi such as
mind—development courses and fortune—telling, and a section
reviewing practical applications of psi.

JASPR 72:380—383 (Oct 1978); JIndPsychol 2:155 (Jul
1979); JP 42:79—82 (Mar 1978); JSPR 49:890—892 (Sep 1978);
PR 9:11—13 (Nov/Dec 1978); T 7:13—14 (Spr 1979).

514 Eysenck, Hans J., and Sargent, Carl. *Explaining the Un—
explained: Mysteries of the Paranormal.* London: Weiden—
feld and Nicolson, 1982. 192p. 23 figures; 7 graphs; 134
illustrations; Index: 190—192; Suggested reading list:
188—189; 3 tables

Two British psychologists summarize the evidence for psi
and attempt to show how parapsychological findings are of use
in creating a general scientific world view. They begin with an
examination of the evidence for psi, concluding that it is strong
enough to warrant the framing of theories as to how and why
psi functions. They also examine psychokinesis, psychic healing,
and survival. Two chapters are devoted to altered states and
psi, one to Stanford's PMIR theory, and another to physics and
psi, with emphasis on Walker's theory. The final chapter
provides a response to critics of parapsychology and calls for a
multi—disciplinary laboratory approach to the investigation of
psi. Readers are invited to take part in a mass ESP experiment
called "The Psi Factor Experiment" conducted by Sargent.

JP 48:227—231 (S 1984).

BritBkNews p. 536 (Sep 1982); ContempRev 241:109 (Aug
1982); LondRevBks 4:10 (Jul 1, 1982); NewStatesman 104:20
(Aug 6, 1982); Observer p. 30 (May 16, 1982); TimesEdSuppl p.
29 (Apr 30, 1982).

515 Fairley, John, and Welfare, Simon. *Arthur C. Clarke's
World of Strange Powers.* New York: Putnam's, 1984.
248p. 10 figures; 242 illustrations; Index: 245—248; 1
map

In the Foreword to this compendium of accounts of
anomalous phenomena, Arthur C. Clarke says: "At a generous
assessment, approximately half this book is nonsense. Unfor—
tunately, I don't know which half" (p. 4). The editors did the
research for the book (and for a television series presenting
some of the material entitled *Arthur C. Clarke's Mysterious
World*). The phenomena covered are hexing, poltergeists, clair—
voyance, PK, hauntings, telepathy, stigmata, fire—immunity,
dowsing, survival, and reincarnation. The editors tried to

present the best cases and they are well–illustrated. In an Epilogue, Clarke attempts to offer explanations for some of the phenomena and gives his personal ratings as to the validity (or lack thereof) of the subjects covered in the book. *PAI* 5:40 (Jun 1987).

Analog 106:173 (Apr 1986); *BestSell* 45:302 (Nov 1985); *FantRev* 8:21 (Sep 1985); *Kirkus* 53:460 (May 15, 1985); *LibrJ* 110:103 (Aug 1985); *PublWkly* 227:66 (Jun 28, 1985); *SciFicChron* 7:34 (Jan 1986); *SciFicRev* 14:5 (Nov 1985); *SciTechBkN* 10:2 (Apr 1986); *SkptInq* 5:71–73 (Sum 1981); *TimesLitSuppl* p1259 (Nov 2, 1984).

516 Grattan-Guinness, Ivor (Ed.). *Psychical Research: A Guide to Its History, Principles and Practices.* Wellingborough, Northamptonshire, England: Aquarian Press, 1982. 424p. Chapter references; 2 figures; Glossary: 387–399; 10 illustrations; Name Index: 413–418; Subject index: 419–424; 1 table

This is an introductory guide to parapsychology consisting of 30 original contributions by specialists. History is covered in five sections. Under "The Range of Psychical Phenomena" there are ten chapters, one each devoted to mediumship, OBES, apparitions, clairvoyance and telepathy, survival, poltergeists, psychic healing, precognition, psychokinesis, and photography (psychic and Kirlian). There are ten chapters on "Aspects of Psychical Research," such as techniques for investigating spontaneous cases, hypnosis, statistics, computers, ethics, and teaching. The fourth section consists of eight chapters relating psychical research to psychology, psychiatry, brain research, physics, the established sciences, other unorthodox disciplines (Ufology, astrology, dowsing), religion, and the media. This volume was compiled by mathematician and SPR Council Member Ivor Grattan–Guinness to celebrate the centenary of the founding of the Society for Psychical Research in 1882.

ChrP 4:266–267 (Dec 1982); *JASPR* 77:242–249 (Jul 1983); *JP* 47:163–166 (Jun 1983); *JSPR* 52:135–137 (Jun 1983); *PR* 15:10–12 (Jul/Aug 1984); *T* 11:87–89 (Win 1983). *ContempRev* 242:51 (Jan 1983); *Light* 102:177–178 (Win 1982); *Nature* 301:353 (Jan 27, 1983).

517 Hintze, Naomi A., and Pratt, J. Gaither. *The Psychic Realm: What Can You Believe?* New York: Random House, 1975. 269p. Chapter notes; Index: 265–269

Naomi Hintze is a novelist and mystery writer and Gaither Pratt was one of the first persons in the world to make a living fulltime as a parapsychologist. They cooperated to provide a comprehensive introduction to parapsychology using fresh material and language the average intelligent reader could easily comprehend. Their method of collaboration is unusual: Hintze contributes the first half of each chapter by vividly describing the subject of that chapter. Pratt, in the second half, evaluates the scientific evidence for the phenomenon under discussion. The subjects covered are ESP, clairvoyance, precognition, self-testing, anpsi, altered states of consciousness and psi, hauntings, poltergeists, Kulagina's PK, unorthodox healing, automatism of Patience Worth, possession, mediumship, survival, and reincarnation. This method of presentation makes it a useful introduction to parapsychology for students lacking a technical background. Its usefulness is enhanced by a seven—page bibliography.

EJP 1:83 (Nov 1976); *JASPR* 70:313—316 (Jul 1976); *JP* 40:156—158 (Jun 1976); *JSPR* 49:316—317 (Sep 1976); *P* 7:25—26 (Mar/Apr 1976); *PR* 7:12—13 (Jul/Aug 1976); *T* 6(1):15—16 (1978).

Booklist 72:1033 (Mar 15, 1976); *Choice* 12:806 (Sep 1975); *Kirkus* 43:1155 (Oct 1, 1975); *LibrJ* 100:2254 (Dec 1, 1975); *PsycholToday* 9:39 (Apr 1976); *PublWkly* 208:105 (Oct 13, 1975).

518 Inglis, Brian. *The Hidden Power.* London, England: Jonathan Cape, 1986. 312p. Bibliography: 292—301; Chapter notes; 23 figures; 22 illustrations; Index: 302—312

In Part 1, Inglis presents a history of the evidence for psi and asks "Why, given that there has been so much evidence for psi in all ages, . . . has it not been accepted that there is a powerful case on its behalf, which ought not to be dismissed out of hand?" (p. 7). Next, he discusses and refutes the four most sensible counterhypotheses put forth by the skeptics. In a chapter on "The Significance of Psi Effects," he examines the question of How, when "the evidence against psi is so inadequate . . . can the pretense be maintained that the evidence for psi does not fulfill science's requirements?" (p. 143). In Part 3, he presents the case against scientism, or the creed scientists have adopted to legitimatize how they operate. He shows how unscientific scientism really is, and he closes with a chapter in which he points out how useful to science the study of psi could be.

JSPR 54:151—155 (Apr 1987); *PAI* 6:43—44 (Jun 1988).

Listener 116:23 (Jul 24, 1986); *Observer* p27 (Jul 13, 1986).

519 James, William. *Essays in Psychical Research.* Cambridge, MA: Harvard University Press, 1986. 684p. Bibliographical footnotes; 10 figures; 17 illustrations; Index: 669—684; 1 questionnaire; 23 refs; 1 table

This volume is a reprint of all James' writings on psychical research from 1885 until 1910. They are presented in chronological order (he wrote something nearly every year of that time span). The writings consist of addresses, reports of experiments or sittings, book reviews, and letters. Robert A. McDermott, in a 24-page Introduction, provides an excellent summary of James's interest and work in psychical research against the backdrop of his other writings. There is a name and subject index to James's writings which enhances the value of the book. Every amendation to the text is listed in an appendix.

PAI 5:44 (Dec 1987).

520 Krippner, Stanley (Ed.). *Advances in Parapsychological Research 1: Psychokinesis.* New York: Plenum Press, 1977. 235p. Chapter references; Name index: 229-232; Subject index: 233-235

Volume I of this series, edited by psychologist Stanley Krippner, contains six chapters by parapsychologists. One half is devoted to two survey articles on psychokinesis: one dealing with methodology, by physicist Joseph Rush, and one dealing with research findings, by psychologist Gertrude Schmeidler. There are additional chapters on therapeutic applications, by psychiatrist Jan Ehrenwald; on implications for philosophy, by philosopher J.M.O. Wheatley; and on religious implications, by psychologist Robert Thouless. The last chapter is "A Select Bibliography of Books of Parapsychology, 1974-1976," by Rhea A. White.

JASPR 72:279-284 (Jul 1978); *JP* 42:76-79 (Mar 1978); *JSPR* 50:297-299 (Mar 1980); *PN* 2:3 (Apr 1979); *PR* 9:10-12 (Jul/Aug 1978); *T* 7:13 (Win 1979).

Brain/MindBul 3:4 (Apr 3, 1978); *JMentImag* 7:169-172 (Spr 1983).

521 Krippner, Stanley (Ed.). *Advances in Parapsychological Research 2: Extrasensory Perception.* New York: Plenum Press, 1978. 308p. Chapter references; Name index: 297-302; Subject index: 233-235

The volume consists of three descriptive and evaluative essays: "A Survey of Methods and Issues in ESP Research," by Robert L. Morris; a long review (194 pp.) of research findings in ESP, by psychologist John Palmer; and "Theories of Psi," by the Director of the Institute of Parapsychology, K. Ramakrishna Rao. These contributions, especially that of Palmer, make this a basic compendium of information on ESP research.

JSPR 50:297-299 (Mar 1980); *PN* 2:3 (Apr 1979); *PR* 10:23-27 (May/Jun 1979); *T* 8:19-20 (Fal 1980).

Brain/MindBul 4:4 (Aug 6, 1979); *ContempPsychol* 24:766–767 (Oct 1979); *JMentImag* 7:169–172 (Spr 1983).

522 Krippner, Stanley (Ed.). *Advances in Parapsychological Research 3.* New York: Plenum Press, 1982. 338p. Chapter references; 1 figure; Name index: 325–331; Subject index: 333–338

With this third volume of *Advances*, the original concept of the series becomes clear. Whereas Volume 1 emphasized psychokinesis, and Volume 2 ESP, Volume 3 provides surveys updating the chapters in the first two volumes. There are six chapters: "An Updated Survey of Methods and Issues in ESP Research," by R.L. Morris; "ESP Research Findings: 1976–1978," by J. Palmer; "Problems and Methods in Psychokinesis Research," by J.H. Rush; and "PK Research: Findings and Theories," by G.R. Schmeidler are all continuations of basic review chapters found in Volumes 1 or 2. This volume is dominated, however, by Chapter 5: "The Changing Perspective on Life after Death," by W.G. Roll, a 144–page survey of findings and theories. The book closes with R.A. White's annotated "Select Bibliography of Books on Parapsychology, 1976–1979," which extends the bibliography in Volume 1.

JASPR 78:363–369 (Oct 1984); *JP* 47:243–251 (Sep 1983); *T* 13/14:18–20 (Spr 1985/86).

ReligStud 20:697–699 (Dec 1984).

523 Krippner, Stanley (Ed.). *Advances in Parapsychological Research 4.* Jefferson, NC: McFarland, 1984. 254p. Bibliography: 215–239; Name index: 241–247; Subject index: 249–254; 1 table

This book contains reviews of several subject areas: PK (Schmeidler), mental healing (Solfvin), imagery and psi (George and Krippner), Ganzfeld research (Stanford), methodological criticisms (Akers), and implications of parapsychology for psychology (Child). Like its predecessors, it provides some of the best in–depth views of parapsychological topics available.

JP 51:170–175 (Jun 1987).

524 Krippner, Stanley (Ed.). *Advances in Parapsychological Research 5.* Jefferson, NC: McFarland, 1987. 301p. Bibliography: 259–285; 9 graphs; Name index: 289–294; Subject index: 295–301

This volume is dedicated to J.A. Greenwood and T.N.E. Greville, and contains a tribute outlining their contributions to parapsychology by Stanley Krippner and Patrick Scott. In his Introduction, Marcello Truzzi documents evidence for a transformation taking place in the relationship between parapsychology and the broader scientific community as regards changes in definitions and roles, theoretical changes, changes in evidence, and in the emergence of peer dialogue. The reviews included in this volume consist of "Psychokinesis: Recent Studies and a Possible Paradigm Shift," by Gertrude R. Schmeidler; "Ganzfeld and Hypnotic–Induction Procedures in ESP Research: Toward Understanding Their Success," by Rex G. Stanford; "Theoretical Parapsychology," by Douglas M. Stokes; "Criticism in Experimental Parapsychology, 1975–1985," by Irvin L. Child; and "A Select Bibliography of Books on Parapsychology, 1982–1985," by Rhea A. White. The references for all chapters are placed in the Reference List at the end of the book.

JASPR 83:251–258 (Jul 1989); *JP* 52:67–72 (Mar 1988); *JSPR* 55:157–159 (Jul 1988); *PAI* 6:45 (Jun 1988). *ContempPsychol* 34:649–650 (Jul 1989).

525 McConnell, R.A. (Ed.). *Encounters with Parapsychology.* Pittsburgh, PA: Author, 1981. 235p. Annotated bibliography: 223–228; Chapter references; 5 graphs; Name index: 229–232; Subject index: 233–235; 6 tables

This nontechnical anthology is aimed at educated laypersons and busy scientists. It is offered as a guide for those who wish to come to grips with the scientific nature of parapsychology without examining the laboratory evidence. McConnell operates from the premise that the best initial approach to parapsychology may not be directly to the phenomena but through the persons who have investigated those phenomena. The 16 selections included represent "perspectives from persons in the last hundred years who have thought at length about parapsychology and who, for one reason or another, have a claim to our attention. The result can be most simply described as a collection of personal encounters with parapsychology by outstanding people" (p. 2). Most of the selections have been shortened, modified, and each is introduced by a brief explanatory comment by the editor. The authors included are W.F. Barrett, W. James, W.F. Prince, W. McDougall, U. Sinclair, L.L. Vasiliev, R. Heywood, G. Murphy, L.E. Rhine, J.B. Rhine, H. Margenau, J.G. Pratt, C. Honorton, R. Hyman, R.G. Jahn, and R.A. McConnell. In a postscript, McConnell expresses his belief that the "seed facts for a scientific explanation of consciousness have been found in the little known field of parapsychology" (p. 198). "Training, Belief, and Mental Conflict Within the Parapsychological Association" by McConnell and T.K. Clark is reprinted in an appendix. There is an "Annotated Sociohistorical Bibliography of Parapsychology" by McConnell which includes papers dealing with the separation of folly and truth in parapsychology and with the scientific integration of

this field.
    JASPR 77:185—189 (Apr 1983); *JP* 47:65—68 (Mar 1983);
PAI 3:41 (Jun 1985).

526 Mitchell, Edgar D. (John White, Ed.). *Psychic Explora-*
    tion: A Challenge for Science. New York: G.P.
    Putnam's, 1974. 708p. Chapter references; 35 figures;
    Glossary: 687—696; 5 graphs; 55 illustrations; Index:
    697—708; 2 tables

    Anthology of chapters by 29 contributors written specifi—
cally for this volume at the instigation of former astronaut,
Edgar D. Mitchell. The first of the four main sections consists
of basic information on parapsychology with chapters on: a his—
tory of the field (M. Ebon), famous Western sensitives (A.
Vaughan), psychic personality (G. Schmeidler), telepathy (S.
Krippner), clairvoyance (R.G. Stanford), precognition and
retrocognition (E.D. Dean), psychokinesis (H. Schmidt), and
parapsychology today (R.A. White). The second section is en—
titled "The Expanding Range of Psychic Research," and con—
tains the following nine chapters: "Psychobiology of Psi" (R.L.
Morris), "Psi and Psychiatry" (M. Ullman), "Anthropology and
Psychic Research" (R.L. Van de Castle), "Man—Plant Com—
munication" (M. Vogel), "Psychic Photography" (J. Eisenbud),
"Unorthodox Healing" (H.K. Puharich), "OBEs" (C.T. Tart),
"Apparitions, Hauntings, Poltergeists" (D.S. Rogo), and
"Survival Research" (W.G. Roll). The third section is on the
"Emergence of a New Natural Science" and consists of six chap—
ters, on "Paraphysics" (J.B. Beal), "Theoretical Paraphysics" (B.
O'Regan), "Soviet Parapsychology" (T. Moss), "Devices for
Monitoring Nonphysical Energies" (W.A. Tiller), "Modern
Physics and Parapsychology" (H.E. Puthoff and R. Targ), and
"Consciousness and Quantum Theory" (E.H. Walker). The final
section is on the convergence of parapsychology and transper—
sonal psychology. It consists of five chapters on: "Mystical Ex—
perience and Psi" (L. LeShan), "Myth, Consciousness and Psi
Research" (J. Houston), "Consciousness and Extraordinary
Phenomena" (R. Masters), "Psi—Conducive States" (C.
Honorton), and "The Social Implications of Psi" (W.W.
Harman). Mitchell contributed an "Introduction: From Outer
Space to Inner Space . . .," in which he explains how he be—
came interested in parapsychology, and a "Conclusion: . . . And
Back Again," in which he provides an overview of why "psychic
research is clearly a significant part of the contemporary scene"
(p. 670), based on the chapters of the book. In an appendix
Mitchell describes his experiments with Uri Geller. The writing
is somewhat uneven, some chapters being of a higher quality
than others. Although no longer useful as "state of the art"

reviews, many of the chapters still serve well in providing basic information on parapsychology.

P 6:45−46 (Jul/Aug 1975).

AmJPsychiat 132:777 (Jul 1975); *Choice* 12:806 (Sep 1975); *Kirkus* 42:726 (Jul 1, 1974); *LibrJ* 100:452 (Mar 1, 1975); *NYTimes* 124:19 (Aug 23, 1975); *PublWkly* 206:55 (Sep 16, 1974).

527 **Morris, J.D., Roll, W.G., and Morris, R.L. (Eds.).** *Research in Parapsychology 1974.* Metuchen, NJ: Scarecrow Press, 1975. 265p. Chapter references; Glossary: 229−231; Name index: 233−236; Subject index: 237−265

The proceedings of the seventeenth convention of the Parapsychological Association contains comments on the Levy affair (the case of a fraudulent experimenter) by James Davis, and 31 research briefs and 17 papers under the following session headings: "Altered States of Consciousness," "Animal Studies," "Theory and Methods," "Psychokinesis with Human Subjects," "Out−of−Body Experiences," "Spontaneous Case Trends," and "ESP and 'Normal' Cognition." There were two symposia: "Energy Focusing and Lingering Effects in Poltergeist Cases and Experimental Studies" and "Ethical Issues Confronting Parapsychologists." The full text is given of the presidential address, "Tacit Communication and Experimental Theology," by R. L. Morris, and the invited address, "Biofeedback and Pattern Self−Regulation in Biological Perspective: A Critical Analysis of Extrasensitive Perception," by Gary E. Schwarz.

JP 40:71−73 (Mar 1976); PS 1:238−239 (1976).

528 **Morris, J.D., Roll, W.G., and Morris, R.L. (Eds.).** *Research in Parapsychology 1975.* Metuchen, NJ: Scarecrow Press, 1976. 277p. Chapter references; Glossary: 243−245; Name index: 247−250; Subject index: 251−277; 3 tables

The eighteenth convention of the Parapsychological Association as presented in this volume consists of 41 research briefs under the following categories: "Levy Replications," "Instrumentation," "Psychophysiological Studies," "Attitudinal Variables," "Variations on Forced−Choice Testing Procedures," "Utilization of Natural Targets," "Psychokinesis on Stable Systems," "Commercial Claims," "Application of Conventional Learning Theory," "Psychokinesis on Dynamic Systems," "Mental Retardation," "Subliminal Stimulation," "Field Studies," "Cognitive Variables," "Internally−Deployed Attention States: Relaxation," and "Internally−Deployed Attention States: Perceptual Isolation." There were 15 papers given in five sessions devoted to the following topics: "Unconscious Psi Processing,"

"Internal States and ESP," "Psychokinesis: Experiments and Field Investigations," "Memory and ESP," and "Imagery and ESP." The texts of Charles Honorton's presidential address, "Has Science Developed the Competence to Confront Claims of the Paranormal?" and Willis W. Harman's invited address, "The Societal Implications and Social Impact of Psi Phenomena," are included.
  JP 41:53—57 (Mar 1977); *PR* 8:15 (Jan/Feb 1977).

529 **Morris, Joanna D., Roll, William G., and Morris, Robert (Eds.).** *Research In Parapsychology 1976.* Metuchen, NJ: Scarecrow Press, 1977. 285p. Chapter references; Glos—sary: 263—265; Name index: 267—269; Subject index: 270—285
  The nineteenth convention of the Parapsychological Associa—tion as presented in this volume consists of 47 research briefs describing ongoing or recently—completed experiments or other short studies. There are condensed versions of 25 full papers and three symposia. The latter are entitled "Psi and Physics," "Geller—Type Phenomena," and "New Concepts in RSPK Research." Complete texts are given of Martin Johnson's presidential address, "Problems, Challenges, and Promises," and Sir John Eccles' invited dinner address, "The Human Person in Its Two—Way Relationship to the Brain."
  JP 42:70—76 (Mar 1978); *JSPR* 49:881—883 (Sep 1978).

530 **Owen, A.R.G.** *Psychic Mysteries of the North: Discoveries from the Maritime Provinces and Beyond.* New York: Harper & Row, 1975. 243p. Chapter notes; Index: 239—243
  A former Fellow of Trinity College, Cambridge, where he lectured on mathematics and biology, Dr. Owen now lives in Canada where he directs the New Horizons Research Founda—tion, and the Toronto Society for Parapsychological Research, which conducts parapsychological investigations. This book may serve as a good introduction to parapsychology for newcomers because throughout it the author distinguishes between the scien—tific and pseudoscientific approaches to psi phenomena and be—cause many of the subjects dealt with are of popular interest such as Uri Geller, brainwaves and ESP, and "pyramid power." (In regard to the latter, several members of the Toronto group "while following the same agreed—upon procedures were unable to discover any significant differences between material placed in a pyramid and material placed in control containers" [p. 115].) The book is also notable because it is a history and survey of parapsychology in Canada, beginning with psi phenomena among the Indians. Emphasis is placed on current research, in

much of which Owen himself has participated. Much of the Canadian material will be new to parapsychologists as well as to the general reader. It is the best source of information there is on individual Canadian psychics, parapsychologists, and investigations.

  JP 39:343—347 (Dec 1975); *P* 6:42—43 (Sep/Oct 1975); *PS* 2:157—158 (F 1977).

  Choice 12:1075 (Oct 1975); *LibrJ* 100:590 (Mar 15, 1975); *PublWkly* 207:74 (Apr 7, 1975).

531 Panati, Charles. *Supersenses: Our Potential for Parasensory Experience.* New York: Quadrangle/New York Times Book Co., 1974. 274p. Author index: 273—274; Chapter notes; 3 figures; 19 illustrations; Subject index: 269—271; Suggested reading list: 266—268

  In his introduction entitled "Impossible Things," the author, a physicist and a *Newsweek* science writer, gives a brief history of parapsychology and documents how, in the 1970s, "paranormal phenomena are no longer being viewed as impossible things. They have finally moved off the magician's stage and come full swing into the scientific laboratory. What is happening there, and its implications for dramatically reshaping our image of ourselves and our place in the universe, are the subjects of this book" (p. 12). Although this book can serve as a general introduction to parapsychology, the emphasis is on altered states and psi. There are chapters on the dream studies at Maimonides, on hypnosis and psi, on unorthodox healing, on telepathy, "the twilight mind" (automatic writing, creativity, OBES), clairvoyance, mysticism and psi, precognition, "Psychokinesis and Life Fields," and an epilogue in which he concludes: "We are more than the mere composite of our wakeful perceptions, for the purely 'wakeful person' is only half human. Perhaps as research into the paranormal accelerates, the day will come when the prefix *para* will be dropped . . . . After all, it is only an admission of our present inability to grasp the full spectrum of things that are humanly possible" (p. 254).

  JP 39:150—152 (Jun 1975); *JSPR* 48:164—168 (Sep 1975); *P* 6:44—45 (Jul/Aug 1975); *PR* 6:4—5 (May/Jun 1975); *T* 4:15 (Sum 1976).

  BkWorld p.3 (Jan 12, 1975); *Choice* 11:1700 (Jan 1975); *ContempPsychol* 21:890 (Dec 1976); *Economist* 257:133 (Oct 18, 1975); *Kirkus* 42:859 (Aug 1, 1974); *LibrJ* 99:2487 (Oct 1, 1974); *Light* 95:88—91 (Sum 1975); *Newsweek* 84:102—103 (Oct 28, 1974); *PublWkly* 206:64 (Sep 9, 1974); *TimesLitSuppl* p172 (Feb 13, 1976).

532 Rhine, Louisa E. *Psi: What Is It? The Story of ESP and*

PK. New York: Harper & Row, 1975. 247p. Chapter
notes; Index: 243–247; 4 tables

Rhine, a pioneer parapsychologist, presents a nontechnical
"account of parapsychology today, written for inquiring minds of
any age. In this book are outlined a sufficient number of the
more significant aspects of the field to give a general concept of
it [and to] help those with personal problems better to under–
stand them" (p. 9). It is primarily the story of how para–
psychology developed at the Duke University Parapsychology
Laboratory and its successor, the Foundation for Research on
the Nature of Man. There are chapters on the basic types of
psi, on various recent advances in psi research, on spontaneous
psi, survival and the meaning of psi. Useful to students is a
section on the occult in school; how to get into parapsychology,
and how to conduct psi tests.

JASPR 70:239–241 (Apr 1976); *JP* 39:254–257 (Sep 1975);
JSPR 48:238–239 (Dec 1975); *P* 6:42 (Sep/Oct 1975); *PR* 6:13
(Sep/Oct 1975); *PS* 2:298–299 (S 1977).

Choice 12:807 (Sep 1975); *Choice* 12:1076 (Oct 1975);
Kirkus 43:168 (Feb 1, 1975); *LibrJ* 100:678 (Apr 1, 1975);
PublWkly 207:50 (Feb 10, 1975).

533 Roll, William G. (Ed.). *Research in Parapsychology 1977.*
Metuchen, NJ: Scarecrow Press, 1978. 271p. Chapter
references; 8 figures; Glossary: 251–253; 12 graphs;
Name index: 255–257; Subject index: 258–271

This is the proceedings of the 20th annual convention of
the Parapsychological Association held at American University,
Washington DC. Twenty–five research briefs are presented in
the form of roundtable discussions on the following topics: the
twentieth anniversary of the Association, out–of–body research,
psi–conducive states, research methodology, and survival
research. Also included are the texts of 18 papers and the
presidential address by Charles T. Tart, "Space, Time, and
Mind."

JIndPsychol 2:165–166 (Jul 1979); *JP* 42:313–319 (Dec
1978).

534 Roll, William G. (Ed). *Research in Parapsychology 1978.*
Metuchen, NJ: Scarecrow Press, 1979. 211p. Chapter
references; Glossary: 185–186; Name index: 189–192;
Subject index: 193–211

This is the proceedings of the 21st annual convention of
the Parapsychological Association held at Washington University
in St. Louis. Included are three symposia, two roundtables, 36
research briefs, 19 papers, and the text of the presidential ad–
dress by K. Ramakrishna Rao, "Psi: Its Place in Nature."

Several areas were emphasized in the the contributions at this convention: psychophysiological studies, experimental PK, the role of the experimenter/observer, feedback in psi testing, al—tered states, and novel testing situations and subjects. Also rep—resented were such topics as spontaneous psi, psychic healing, near—death experiences, statistical issues, European parapsychol—ogy, and women in parapsychology.

JIndPsychol 2:166—167 (Jul 1979); *JSPR* 50:299—301 (Mar 1980); *JP* 43:232—238 (S 1979); *PN* 3:3 (Apr 1980); *T* 8:18—19 (Spr 1980).

535 Roll, William G. (Ed.). *Research in Parapsychology 1979.*
Metuchen, NJ: Scarecrow Press, 1980. 232p. Chapter references; 6 figures; Name index: 217—220; Subject in—dex: 221—232; 1 table
The 22nd annual convention of the Parapsychological As—sociation was held in Moraga, California. This proceedings of that convention includes abstracts of 17 research papers, 22 research briefs, 20 poster sessions, four symposia, and three roundtables. The full text of John Palmer's presidential address, "Parapsychology as a Probabilistic Science: Facing the Implications," is included. The major emphasis throughout was on methodology, from computer analysis to the need for a more creative application of the scientific method on the study of psi.
T 9:18—20 (Sum 1981).
Phoenix 5:121—126 (1981).

536 Roll, William G., and Beloff, John (Eds.), and McAllister, John (Asst. Ed.). *Research in Parapsychology 1980.*
Metuchen, NJ: Scarecrow Press, 1981. 167p. Name in—dex: 153—155; Subject index: 157—167
The 23rd annual convention of the Parapsychological As—sociation was held in Reykjavik, Iceland. Its proceedings includes abstracts of a symposium on apparitions and three roundtables entitled "Social and Ethical Issues in Parapsychology," "Reliability and Other Ignored Issues in Parapsychology," and "The Distribution of Psi." Abstracts of papers and research briefs are presented in ten subject categories: "Remote Viewing," "Studies with Children," "Metal—Bending," "Problems of Randomness," "Philosophy of Parapsychology," "Poltergeists," "Personality Variables," "Psi—Conducive States," "Ganzfeld Techniques," and "Miscellaneous Topics." Included are the full texts of the presidential address by Ian Stevenson, "Can We Describe the Mind?" and the J.B. Rhine Lecture (invited ad—dress) by Brian Inglis, "Power Corrupts: Skepticism Corrodes."
JASPR 76:379—385 (Oct 1982); *JP* 45:337—348 (Dec 1981); *JSPR* 52:79—82 (F 1983); *T* 10:65—66 (Aut 1982).

Phoenix 6:173 (1982).

537   Roll, William G., Beloff, John, and White, Rhea A. (Eds.).
*Research in Parapsychology 1982: Jubilee Centenary
Issue.* Metuchen, NJ: Scarecrow Press, 1983. 366p. Chap—
ter references; Name index: 337—343; Subject index:
344—366
This book consists of the proceedings not only of the 25th
annual convention of the Parapsychological Association but of
the Centenary Conference of the Society for Psychical Research.
The joint conference was held at Trinity College, Cambridge
University. This is the largest *RIP* to date, containing four sym—
posia, four roundtables, 72 papers and research briefs, five in—
vited addresses (including speakers from the People's Republic
of China), the J. B. Rhine Lecture, "Parapsychology: Status and
Prospects," by Hans Eysenck, and Beloff's presidential address,
"Three Open Questions."
JASPR 80:322—332 (Jul 1986); *JP* 48:149—160 (Jun 1984);
JSPR 53:399—401 (Jul 1986).
Light 104:135—136 (Aut 1984).

**538   Roll, William G., Morris, Robert L., and Morris, Joanna
D. (Eds.).** *Research in Parapsychology 1972.* Metuchen,
NJ: Scarecrow Press, 1973. 249p. Chapter references;
Glossary: 217—219; Name index: 220—224; Subject index:
225—249
This is the first in a series of yearly volumes recording the
proceedings of the annual convention of the Parapsychological
Association, the international professional society of parapsy—
chology. (It succeeds the Proceedings of the Parapsychological
Association.) *RIP 1972* covers the 15th annual convention
which was held at the University of Edinburgh. There are
lengthy abstracts (averaging two pages each) of research briefs,
papers, and symposium papers. The sessions were on PK, psi
in dreams and altered states, ESP tests with gifted subjects,
case studies, precognition, subject variables, cognitive variables,
animal psi, group ESP tests, historical studies, and mood,
motivation, meditation, and ESP. There were three symposia
entitled "Potential Use of Space Vehicles in Parapsychological
Research," "Psi, Science, and Society," and "What Evidence, If
You Had It, Would Convince You of Survival?" The full texts
are included of Arthur Koestler's invited dinner address, "Out
on a Tightrope: Parapsychology and Physics," and John Beloff's
presidential address, "Belief and Doubt."
JP 37:346—348 (Dec 1973); *JSPR* 48:49 (Mar 1975).

539   Roll, William G., Morris, Robert L., and Morris, Joanna

D. (Eds.). *Research in Parapsychology 1973.* Metuchen, NJ: Scarecrow Press, 1974. 249p. Glossary: 219—221; Name index: 223—225; Subject index: 226—249

This proceeding of the 16th annual Parapsychology Association convention held at the University of Virginia contains lengthy abstracts of papers on "Work with Special Subjects" (3), "Psychokinesis" (2), "Animal Psi" (3), "Altered States of Consciousness" (2), "Methodology" (2), and "Nonintentional Psi Events: Experimental Approaches" (2). There were two symposia, "Research on Out—of—Body Experiences" and "Psychokinesis on Stable Systems." There also were 20 research briefs. The full text is given of Rex G. Stanford's presidential address, "Concept and Psi" and C.T. Targ's invited address, "On the Nature of Altered States of Consciousness with Special Reference to Parapsychological Phenomena."

JP 38:427—432 (Dec 1974).
Light 95:46—47 (Spr 1975).

540 **Roll, William G., Morris, Robert L., and White, Rhea A. (Eds.).** *Research in Parapsychology 1981.* Metuchen, NJ: Scarecrow Press, 1982. 245p. Chapter references; Name index: 223—225; Subject index: 227—245

This volume of papers from the 24th annual Parapsychological Association convention held at Syracuse University includes a symposium on precognition, five roundtables on social, ethical, and methodological issues (with emphasis on anthropology), and 56 papers and briefs on experimental PK, theoretical approaches, psychological variables and ESP testing, personality variables, replications, efforts to improve psi performance, and observer variables. It contains the presidential address of Irvin Child, "Parapsychology and Psychology."

JASPR 77:349—356 (Oct 1983); *JP* 46:359—373 (Dec 1982); *JSPR* 52:280—283 (Feb 1984); *PsiR* 1:121—122 (Dec 1982); *T* 12(2):36—40 (Sum 1984).
Phoenix 7:171—174 (1983).

541 **Shapin, Betty, and Coly, Lisette (Eds.).** *Concepts and Theories of Parapsychology: Proceedings of an International Conference Held in New York, New York. December 6, 1980.* New York: Parapsychology Foundation, 1981. 112p. Chapter references; 1 table

This is the proceedings of the 29th annual Parapsychology Foundation Conference held in the Foundation's new quarters in New York City. Six papers were given, each followed by a discussion: "Lability and Inertia in Psychic Functioning," by William Braud; "Taxonomy and Theory in Psychokinesis," by Stephen E. Braude; "Psi, Internal Attention States and the

Yoga Sutras of Patanjali," by Charles Honorton; "Physical Models of Psychic Process," by Robert G. Jahn and Brenda Dunne; "Developing 'Extreme Case' Causal Models for Synchronistic Phenomena," by R.L. Morris; and "Cognitive Restraints and ESP Performance: On Testing Some Implications of a Model," by R. Stanford. The volume closes with a general discussion period.

   JASPR 76:287—288 (Jul 1982); *JP* 46:157—161 (Jun 1982); *JSPR* 52:71—73 (F 1983); *T* 11:17—18 (Spr 1983).
   Choice 19:1326 (May 1982).

817 **Shapin, Betty, and Coly, Lisette (Eds.).** *Current Trends in Psi Research: Proceedings of an International Conference Held in New Orleans, Louisiana August 13—14, 1984.* [See Addenda for annotation]

542 **Shapin, Betty, and Coly, Lisette (Eds.).** *Parapsychology's Second Century: Proceedings of an International Conference Held in London, England, August 13—14, 1982.* New York: Parapsychology Foundation, 1983. 156p. Chapter references; 2 figures
   This volume contains eight papers, some highly provocative, that examine the future of parapsychology. The contributors are Bierman, Blackmore, Gruber, Roney—Dougal, Sargent, Servadio, Spinelli, and Wilson. Because the future is yet to be, the papers are more subjective than usual, not only because they represent individual viewpoints, but also because the need to include subjective factors in the research program of parapsychology is stressed in several papers.
   JASPR 78:370—372 (Oct 1984); *JP* 48:160—165 (Jun 1984); *JRPR* 8:123—124 (Apr 1985).

543 **Weiner, Debra H., and Morris, Robert L. (Eds.).** *Research in Parapsychology 1987.* Metuchen, NJ: Scarecrow Press, 1988. 215p. Chapter references; Name Index: 203—206; Subject Index: 207—215
   This volume is based on the presentations at the 30th annual convention of the Parapsychological Association held at Edinburgh University. Lengthy abstracts are provided of 48 papers. (This is the first convention of the Parapsychological Association in which no symposia were held. Nor were there any round tables.) The convention consisted of 48 papers, and the presidential and an invited address. Lengthy abstracts are presented for each of the papers which are presented in 16 categories as follows: "PK Studies: PK Versus IDS" (3), "PK Studies: Observational Theories and Silent Feedback" (3), "The Remote Action Project" (5), "Belief and Affect in ESP" (2),

"ESP: Target Qualities and Personal Relationships" (3), "Psychological Correlates of Psi" (3), "Brief Papers" (4), "Spontaneous Cases and Field Studies" (4), "Concerning Fraud and Error" (3), "Statistical Issues and Methods" (3), "New Interpretations of ESP Literature" (3), "New Approaches to Traditional Issues" (3), "Philosophical Issues" (3), "The Psi Controversy" (2), "Historical Papers" (3), "Invited Address" (1), and "Presidential Address" (1). The full texts are provided of John Beloff's invited address, "Extreme Phenomena and the Problem of Credibility" and Richard S. Broughton's Presidential Address, "If You Want to Know How It Works, First Find Out What It's For." JoMarie Haight Sarling compiled the name and subject indexes.

JP 52:351–361 (Dec 1988).

544 Weiner, Debra H., and Nelson, Roger D. (Eds.). *Research in Parapsychology 1986.* Metuchen, NJ: Scarecrow Press, 1987. 240p. Chapter references; Name index: 225–228; Subject index: 229–240; 4 tables

Proceedings of the 29th convention of the Parapsychological Association which was held at Sonoma State University in California. With this volume the addresses of authors are included so that they may be contacted to send full copies of papers that are presented in abridged form in *RIP*. The subject areas of papers include PK (8), free–response ESP studies (3), forced–choice ESP and timing studies (5), displacement (4), psi and geomagnetic factors (6), methodology (4), and theories and tests of theories (3). There are two symposia: "Fifty Years of the *Journal of Parapsychology*" and "Psi and Systems Theory." There are six roundtables on: Intuitive data sorting vs. PK, what research should we be doing?, channeling, psi and mental health, clinical ethics in psi research, do we discover or create reality?, and one on stage magic and parapsychology. The full texts of Janet F. Quinn's invited address entitled "Therapeutic Touch: Report of Research in Progress," and Debra H. Weiner's presidential address, entitled "Thoughts on the Role of Meaning in Psi Research," are included.

JASPR 83:159–164 (Apr 1989); *JP* 52:157–171 (Jun 1988).

545 Weiner, Debra H., and Radin, Dean I. (Eds.). *Research in Parapsychology 1985.* Metuchen, NJ: Scarecrow Press, 1986. 242p. Chapter references; Name index: 231–234; Subject index: 235–242; 1 table

RIP 1985 covers the proceedings of the 28th annual convention of the Parapsychological Association, held at Tufts University, Medford, MA. There were nine papers on PK, six papers on Free–response and/or spontaneous–case ESP studies, six forced—choice ESP studies, seven on new test techniques, seven on statistics, methodology, and data management, and seven theoretical papers. There were two symposia: "Three Questions from 1885: Do We Have the Answers?" and "Terminological Issues in Parapsychology." There were four roundtables, on fear of psi, states of mind and psychic healing, psi and consciousness, and clinical parapsychology. There was a special session commemorating the 100th anniversary of the American Society for Psychical Research. David R. Sanders gave the invited address, entitled "Predicting Psi Performance with the Personality Assessment System." Included is the presidential address of Robert L. Morris, "The Social Context of Psi."

JASPR 82:365—374 (Oct 1988); *JP* 51:157—170 (Jun 1987). *ContempPsychol* 32:1043 (Dec 1987).

546 White, Rhea A. (Compiler). *Surveys in Parapsychology: Reviews of the Literature with Updated Bibliographies.* Metuchen, NJ: Scarecrow Press, 1976. 484p. Chapter references; 4 graphs; Name index: 475—480; Subject index: 481—484; 8 tables

An anthology of 19 state—of—the—art reviews originally published in parapsychology journals and aimed at providing in—depth coverage of key aspects of parapsychology. There are five sections: I. "Some Basic Areas of Parapsychological Study," with chapters by W.G. Roll on Precognition, W.H.W. Sabine on retrocognition, and J.G. Pratt on PK; II. "Psi in Special Subject Populations," with chapters by R.L. Morris on anpsi, R.L. Van de Castle on primitives and psi, and C.B. Nash on medical parapsychology; III. "Insights into How Psi Operates," with a chapter by J.B. Rhine on psi—missing and one by him on position effects, K. Osis on distance and ESP, C. Honorton and S. Krippner on hypnosis and ESP, C. Honorton on state of awareness factors and psi, and R.G. Stanford's "Scientific, Ethical, and Clinical Problems in the 'Training' of Psi Ability"; IV. "Theories of Psi Phenomena," with K.R. Rao on theories and psi, W.G. Roll on memory and ESP, and C.T.K. Chari on the challenge of psi for scientific research; and V. "Criticisms of Parapsychology," with chapters by C.C. Ransom, J.C. Crumbaugh, and J.B. Rhine. The already—lengthy bibliographies for each chapter have been updated by the compiler, thus making the book a bibliographic guide to further information as well as a basic source book.

P 7:38—40 (Jan/Feb 1977).

547 White, Rhea A., and Broughton, Richard S. (Eds.). *Research in Parapsychology 1983.* Metuchen, NJ:

Scarecrow Press, 1984. 184p. Chapter references; 1 figure; 1 illustration; Name index: 169—172; Subject index: 173—184

This volume differentiates between papers that describe completed work and those that describe studies that are still exploratory or at an early developmental stage. Papers in the first category are listed as "Papers, Research Briefs, and Poster Sessions," whereas papers, briefs, and poster sessions in the second category are listed under "Interim Reports." In addition, the volume includes one symposium (on psi and the nervous system), three roundtables, two invited addresses ("On the Representation of Psi Research to the Community of Established Science" by R. Jahn, and "Dream, Metaphor, and Psi" by M. Ullman), and S. Krippner's presidential address, "A Systems Approach to Psi Research Based on Jungian Typology."

JASPR 81:49—64 (Jan 1987); *JP* 49:93—103 (Mar 1985); *PAI* 3:40 (Dec 1985).

Choice 21:392 (Nov 1983).

548 White, Rhea A., and Solfvin, Jerry (Eds.). *Research in Parapsychology 1984.* Metuchen, NJ: Scarecrow Press, 1985. 205p. Chapter references; 1 figure; Name index: 191—194; Subject index: 195—205; 1 table

This volume reverted to *RIP*'s former format, with one difference. The category of "Research Briefs" was dropped from the Parapsychological Association's convention format in 1984, reducing the number of contributions. Included are 33 papers on "Free—Response Ganzfeld Studies" (2), "Novel Approaches to PK" (3), "Some Tests of Models and Theories" (4), "Brain Physiology and Psi" (2), "Enhancing and Suppressing Psi Functioning" (4), "Theoretical and Philosophical Issues" (4), "Computer—Oriented Psi Tests" (4), "Psychological Factors in Psi Research" (3), and "Criticisms and Reevaluations" (3). There were five roundtables ("Resistance to Psi Among Parapsychologists," "PK Parties," "Meta—Analysis," "Applied Psi," and "Paranormal Education"). There were four poster—session papers and two invited addresses ("Parapsychology's Four Cultures: Can the Schism be Mended?," by Steven M. Rosen and "What is Problematic in Parapsychology for Traditional Christianity?," by Mark D. Jordan), and Rhea A. White's presidential address, "The Spontaneous, the Imaginal, and Psi Foundations for a Depth Parapsychology."

JASPR 82:71—77 (Jan 1988); *JP* 50:49—64 (Mar 1986); *PAI* 5:41—42 (Jun 1987).

Historical Aspects of Parapsychology

INTRODUCTION

In *Sources* the section on history was mainly composed of books on specific historical figures in psychical research such as Richard Hodgson [80], William James [83, 85], and Sir Oliver Lodge [84], with the signal exception of Alan Gauld's history of the founders of psychical research and the early years of the Society for Psychical Research [82]. The latter, especially, is still cited.

This section contains 12 books of historical interest, but the emphasis is on spiritualism, which was a precursor of psychical research (Brandon [549], Goldfarb [553], Oppenheim [560]), and studies of mediumship conducted by the SPR or others (Cerullo [550], Douglas [551], Haynes [554], Inglis [555], [556]). A thorough history of experimental parapsychology has yet to be written, although Douglas, Mauskopf and McVaugh [558], and Moore [559] deal with it. Mauskopf and McVaugh, who are historians by profession, provide an excellent history, but of a very limited time period. The Lowe and Blacker [557] book covers the earliest works on what we now call psi but what was then known as divination. Evans [552] and Moore both emphasize cultural and social aspects of the history of psychical research.

For a history of experiments until 1940 see *Extrasensory Perception After 60 Years* [57], and until 1960 see Rao [52]. J.B. Rhine's chapter on the history of experiments in Wolman [688] carries the record through 1975. Although not organized chronologically, the chapters on PK and ESP in *Advances in Parapsychological Research* [520-524] provide the best histories of modern psi research.

Additional books with historical information on parapsychology: Crabtree's bibliography covers the history of animal magnetism and psi [344]. Mishlove [353] provides a large amount of historical information on the exploration of consciousness and also parapsychological phenomena. White's bibliography [378] has a section on historical applications of psi. Schwartz [376] describes some attempts to investigate historical artifacts by using psi. Hurley [372] suggests that sorcery (i.e., psi) may be an unsuspected force in history. Berger [399] provides a history of the past century of American parapsychology, primarily as reflected in the lives, work, and words of 39 parapsychologists who contributed to that history. A history of the Duke Parapsychology Laboratory and the early years of its successor, FRNM, through 1980, is provided by the books on J.B. Rhine [401, 409, 410]. In the section on books for young people, an historical casebook of haunts and apparitions [416] is described by Cohen. He presents key historical cases and experiments in another [422]. Finucane [726a] shows how historically, views of apparitions and haunts have altered in line with cultural changes. Both Edmonds [434] and Hearne [731a] present histories of prophecy and precognition and Feola [435] a history of PK. Coleman [723a] provides a history of the famous so-called retrocognitive "adventure" of Moberly and Jourdain. L.E. Rhine [448] gives the background of parapsychology at Duke University. Some critical historical overviews are presented in Kurtz [476]. The papers of William James [519] on psychical research have recently been made available and are of historical interest, as is the recently reprinted classic by Lombroso [811]. Several books in the section on mediumship, etc. cover historical aspects: Gregory [576], Kenny [578], Klimo [579], and Rogo [584]; multiple personality: Baldwin [572], Crabtree [573], and Kenny [578]; possession and exorcism: Ebon [575], and the ouija board: Gruss with Hotchkiss [577]. Zaleski [595] examines accounts of NDEs in medieval Christendom. Poortman [612] provides "a history of the development of the doctrine of the subtle body." Section I of Murphy and Leeds [650] presents an historical review of parapsychology. Vaughan [790] surveys the major types of prophecy throughout history.

BIBLIOGRAPHY

549 Brandon, Ruth. *The Spiritualists: The Passion for the Occult in the Nineteenth and Twentieth Centuries.* New York: Alfred A. Knopf, 1983. 315p. Bibliography: 297–303; Chapter notes; 12 figures; 40 illustrations; Index: 307–315

Social historian Ruth Brandon provides an entertaining account of spiritualism from its beginnings in 1848 through the 1930s. Her approach is that of the skeptic, and the thread on which she hinges her history is that spiritualism (and later, psychical research and parapsychology) is an attempt to restore by science the religious faith undermined by science. She accuses parapsychologists of refusing to be aware of all the necessary

128 *Parapsychology: New Sources*

facts. She says that in the works of parapsychologists, "all the facts are never mentioned——or, if they are, only in the most slanted way, so that those undermining the desired evidence will be dismissed" (p. 251). Although the pot may be calling the kettle black here, her book is entertaining and informative. She has uncovered a wealth of anecdotes, many of which will be new to most readers.

JSPR 52:209–212 (Oct 1983).

AntR 42:382 (Sum 1984); *Booklist* 80:209 (Oct 1983); *Books&Bkmn* p9 (Aug 1983); *BritBkNews* p441 (Jul 1987); *Economist* 287:95 (Jun 25, 1983); *Kirkus* 51:738 (Jul 1, 1983); *LibrJ* 108:1801 (Sep 15, 1983); *NewStatesman* 107:24 (Mar 23, 1984); *NewRepub* 189:38 (Oct 10, 1983); *Observer* p30 (Jun 26, 1983); *Observer* p25 (Sep 6, 1987); *PublWkly* 224:44 (Jul 15, 1983); *SkptInq* 8:165–167 (Win 1983/84); *Spectator* 251: 22 (Jul 2, 1983); *VictorianStud* 29:613 (Sum 1986).

550 Cerullo, John J. *The Secularization of the Soul: Psychical Research in Modern Britain.* Philadelphia, PA: Institute for the Study of Human Issues, 1982. 194p. Bibliography: 175–181; 1 illustration; Index: 183–194

Although the backdrop of this book is the history of psychical research in Great Britain, it is actually about ways of viewing the nature of human personality, typified by F.W.H. Myers on the one hand and Sigmund Freud on the other. Cerullo, who is a professor of humanities, holds that the Myersian view of human personality was rejected by psychology with the advent of Freud. Thereafter psychical research lost its impetus as Myers' concept of the subliminal self, or the secular soul, became untenable outside of psychical research. A new beginning was made by J.B. Rhine in the 1930s, but in effect he dismissed the subliminal self. "Where there had once been a whole subliminal personality under investigation, an entire secular soul to unveil, there were now only card–guessing experiments and attempts to mentally influence the fall of dice. The central paradigm of the discipline was dissipated" (p. 167). Both by returning to the origins of psychical research and by viewing the field at base as being about the nature of the self, Cerullo may aid persons currently interested in parapsychology to initiate another revolution in the basic approach to the subject. It also is an example of a new approach to history called the "sociology of knowledge," in which ideas are viewed in terms of the social climate.

JASPR 72:371–374 (Oct 1978); *T* 12:42–44 (Sum 1984).
Choice 20:137 (May 1983).

551 Douglas, Alfred. *Extra–Sensory Powers: A Century of*

Psychical Research. Woodstock, NY: Overlook Press, 1977 (c1976). 392p. Bibliography: 379–385; Chapter notes; 2 figures; 14 illustrations; Index: 387–392

This history of parapsychology deals rather sketchily with modern laboratory research, but the first two thirds give thorough coverage to early work with mediums. Douglas provides background information on many of the great mediums of the past, includes pro and con views of their phenomena, and quotes extensively from the original research reports (which are not otherwise readily available).

JASPR 72:371–374 (Oct 1978); *JIndPsychol* 2:149 (Jul 1979); *JSPR* 49:453–455 (Mar 1977); *T* (1):9 (1979).

Books&Bkmn 22:19 (Jan 1977); *Choice* 14:602 (Jun 1977); *Kirkus* 46:75 (Jan 15, 1978); *Kirkus* 46:115 (Feb 1, 1978); *LibrJ* 103:468 (Feb 15, 1978); *Observer* p21 (Aug 29, 1976); *TimesLit–Suppl* p1591 (Dec 17, 1976).

552 Evans, Hilary. *Intrusions: Society and the Paranormal.* Boston, MA: Routledge & Kegan Paul, 1982. 206p. Bibliography: 196–201; 54 illustrations; Index: 203–206

Hilary Evans, a British free–lance writer and co–founder of the Mary Evans Picture Library, provides a history of the attitudes of lay people to the paranormal. He observes that on several occasions laypersons have not been in agreement with official positions, whether religious or scientific, regarding anomalous phenomena. He deals with primitive magic, witchcraft, apparitions, spiritualism, levitation, table–tipping, mediumship and UFOs. Throughout he discusses psi as a social phenomenon. Evans is interested not so much in whether psi is an established fact but rather in the fact that people believe it. He concludes that his study highlights the persistence of the paranormal in the public mind, and he takes science to task for not coming to terms with it. Science has failed not only society but itself——not for not finding an explanation, but for not looking for one.

Atl 249:89 (Mar 1982); *BritBkN* p.3+3 (Jun 1982); *BestSell* 42:286 (Aug 1982); *CAY* 6:12 (Fal 1985).

553 Goldfarb, Russell M., and Goldfarb, Clare R. *Spiritualism and Nineteenth–Century Letters.* Rutherford, NJ: Fairleigh Dickinson University Press, 1978. 208p. Bibliography: 194–203; Chapter notes; Index: 204–208

This book provides a history of spiritualism from a special angle: that of demonstrating the widespread involvement of literary figures in the spiritualist movement and of the ways in which their interests were reflected in their work. The authors explicate many topical references and figures of speech that, unbeknownst to many of today's readers, are actually related to spiritualism.

AmLit 50:679 (Jan 1979); *Choice* 15:1356 (Dec 1978); *JAmHist* 66:139 (Jun 1979); *NewEngQ* 51:452 (Sep 1978).

554 Haynes, Renée. *The Society for Psychical Research, 1882–1982: A History.* London, England: Macdonald, 1982. 240p. Index: 235–240

This is an account of the work of the Society for Psychical Research, the founding of which in 1882 is generally cited as the beginning of a serious scientific approach to the study of psi. It is the fourth book on the subject by Renée Haynes, who edited the Society's *Proceedings* and *Journal* from 1970–1981. The first three chapters are on the SPR and its beginnings. There follow two chapters on apparitions, and one each on cross correspondences, mental mediums, haunts and poltergeists, physical mediums, controversy and fraud, and a summary of achievements and future prospects. There is a very useful chronological list with biographical sketches of the presidents of the SPR, and an appendix describing the Perrott–Warrick Studentship, and one on John Cutten's "ghost detector."

ChrP 4:267–268 (Dec 1982); *JP* 47:145–161 (Jun 1983); *JRPR* 6:314–316 (Oct 1983); *JSPR* 52:139–141 (Jun 1983).

BritBkNews p80 (Feb 1983); *Choice* 20:847 (Feb 1983); *Economist* 285:102 (Oct 9, 1982); *Light* 102:178–179 (Win 1982); *Nature* 302:181 (Mar 10, 1983).

555 Inglis, Brian. *Natural and Supernatural: A History of the Paranormal from Earliest Times to 1914.* London, England: Hodder and Stoughton, 1977. 490p. Bibliography: 455–476; Index: 479–490

British author/journalist Inglis presents a history of psychical research from earliest times through 1914. There are two chapters on primitives and psi, four on early civilizations, three on Christianity, five for the period 1200–1800, six on Mesmerism, five on spiritualism, four on psychical research, six on the Society for Psychical Research, and four on the first part of the 20th century. Throughout the book he emphasizes the ways in which psi phenomena have been received and interpreted throughout the ages. The evidence is not so much evaluated rationally as it is colored by emotions and prejudices, both for and against.

ChrP 3:16–17 (Dec 1978); *JP* 43:140–152 (Jun 1979); *JSPR* 49:826–828 (Jun 1978); *PR* 18:15–16 (Jul/Aug 1980).

Books&Bkmn 23:61 (Aug 1978); *Economist* 266:118 (Feb 4, 1978); *Encounter* 54:88 (Jan 1980); *GuardianWkly* 118:22 (Jan 22, 1978); *GuardianWkly* 118:22 (Feb 5, 1978); *JAnalPsychol*

24:269 (Jul 1979); *Light* 98:27–30 (Spr 1978); *NewScientist* 77:167 (Jan 19, 1978); *NewStatesman* 95:117 (Jan 27, 1978); *Observer* p24 (Jan 22, 1978); *Punch* 274:244 (Feb 8, 1978); *Spectator* 240:20 (Jan 28, 1978); *TimesLitSuppl* p306 (Mar 17, 1978).

556 **Inglis, Brian.** *Science and Parascience: A History of the Paranormal, 1914–1939.* London: Hodder and Stoughton, 1984. 382p. Bibliography: 345–355; Chapter notes; 8 figures; 33 illustrations; Index: 371–382

This history of parapsychology (psychical research) is a sequel to the author's *Natural and Supernatural,* which covered the period up to 1914. This book brings the history up to 1939. Inglis says he "tried to present the evidence about, rather than for, paranormal phenomena, as it accumulated between 1914 and 1939" (p. 338). He finds the evidence overwhelming, and in a Postscript deals with its lack of acceptance. In presenting the historical data, Inglis covers the phenomena of mediumship, spontaneous cases, and experimental investigations. British, Continental, and American research are all covered. Along with Sir Oliver Lodge, Inglis concludes that "psychical research had taken a wrong turning, in seeking academic recognition, if it meant losing contact with the general public; an understandable but disastrous error of strategy which vitiated much of the valuable research undertaken between the wars, and unfairly destroyed the reputations of some of the most dedicated researchers. If I have done nothing else, I hope I have done something to rehabilitate them, at least in the eyes of their successors" (p. 341).

JASPR 80:94–98 (Jan 1986); *PAI* 5:40 (Jun 1987); *T* 15:91–94 (1987).

BritBkNews p271 (May 1984); *Economist* 290:89 (Mar 10, 1984); *Light* 104:86–87 (Sum 1984); *NewStatesman* 107:24 (Mar 23, 1984); *Observer* p24 (Mar 18, 1984), *Spectator* 252:20 (Apr 21, 1984); *TimesLitSuppl* p1259 (Nov 2, 1984).

557 **Loewe, Michael, and Blacker, Carmen (Eds.).** *Oracles and Divination.* Boulder, CO: Shambhala, 1981. 244p. Chapter references; 15 figures; 12 illustrations; Index: 233–244

Loewe is a Lecturer in Chinese Studies at Cambridge University where Blacker is a Lecturer in Japanese. Eight of the nine chapters in this book were originally delivered as a series of lectures at Cambridge in 1979. The editors define divination as "the attempt to elicit from some higher power or supernatural being the answers to questions beyond the range of ordinary human understanding. Questions about future events, about past disasters whose causes cannot be explained, about

things hidden from sight or removed in space . . ." (p. 1).
Each chapter is aimed at showing how such questions were put
in various civilizations, and how the oracles (answers) were
transmitted and interpreted. The areas covered are Tibet,
China, Japan, the Classical World, the Germanic World, the
Babylonians and Hittites, Ancient Egypt, Ancient Israel, and Islam. The editors observe: "Where no higher source of wisdom
is recognized as existing outside the human world, the practices
become meaningless and 'superstitious'" (p. 2).
    BritBkNews p589 (Oct 1981); *Choice* 19:1101 (Apr 1982);
ContempRev 239:168 (Sep 1981).

558 Mauskopf, Seymour H., and McVaugh, Michael R. *The*
    Elusive Science: Origins of Experimental Psychical
    Research. Afterword by J.B. and L.E. Rhine. Baltimore,
    MD: Johns Hopkins University Press, 1980. 368p. Chap-
    ter references; Glossary: 311; 13 illustrations; Index:
    359–368

    By two science historians, this is a detailed account of the
history of parapsychology primarily in the years 1920–1940
when parapsychology was attempting to become a profession in
its own right and to gain acceptance in the scientific community. Although they begin the history with Richet and the
SPR, the bulk is devoted to work in the United States, with
emphasis on that of the Rhines. The authors treat parapsychology as an example of how new disciplines emerge in science and
how they try to become integrated with the mainstream. Thus,
the book is as much about science and the attitudes of scientists (especially psychologists) toward parapsychology as it is
about parapsychology itself.
    JASPR 75:325–343 (Oct 1981); *JP* 45:239–242 (Sep 1981);
JSPR 51:254–256 (F 1982).
    AmHistRev 87:173 (Oct 1982); *AmJPsychol* 94:549 (Sep
1981); *Choice* 18:1436 (Jun 1981); *HistRevsNBks* 9:203 (Jul
1981); *LibrJ* 106:62 (Jan 1 1981); *SciBooks&Films* 17:63 (Nov
1981).

559 Moore, R. Laurence. *In Search of White Crows:*
    Spiritualism, Parapsychology, and American Culture. New
    York: Oxford University Press, 1977. 310p. Chapter
    notes; Index: 297–300

    The author, a professor of American history at Cornell,
endeavors to show that spiritualism, and then psychical
research, were not aberrant activities but expressed a congruence
with cultural values in America. Their appeal lay in the fact
that they provided a middle ground that would accommodate
both religious and scientific interests. This is, then, a cultural

history of spiritualism and parapsychology. It is not always accurate, but it does deal with a wealth of information in an interesting and provocative way.

ChrP 3:19–20 (Dec 1978); *JASPR* 72:257–266 (Jul 1978); *JIndPsychol* 2:79–80 (Jan 1979); *JP* 42:194–209 (Sep 1978); *JSPR* 49:883–885 (Sep 1978); *PR* 9:14–16 (Mar/Apr 1978); *T* 7:12 (Sum 1979); *ZS* No.1:55+ (May 1978).

AmerRecGd p1104 (Oct 1978); *AntR* 36:131 (Win 1978); *Booklist* 74:440 (Nov 1, 1977); *Choice* 14:1705 (Feb 1978); *ChristCent* 95:86 (Jan 25, 1978); *ChronHighEd* 15:17 (Feb 6, 1978); *ContempRev* 23:471 (Jun 1978); *JAmHist* 65:464 (Sep 1978); *Kirkus* 45:911 (Aug 15, 1977); *LibrJ* 102:1847 (Sep 15, 1977); *Light* 98:140–141 (Aut 1978); *PublWkly* 212:63 (Sep 5, 1977); *RevsAmHist* 6:306 (Sep 1978); *TheolToday* 36:149 (Apr 1979).

560 Oppenheim, Janet. *The Other World: Spiritualism and Psychical Research in England, 1850–1914.* New York: Cambridge University Press, 1985. 503p. Chapter notes; 26 illustrations; Index: 487–503

The author examines the widespread public interest in spiritualism in Victorian and Edwardian England. The book is primarily concerned with motivation——why spiritualism was seen as a surrogate for religion and psychical research as the science of the future. At base, the spiritualists and psychical researchers "addressed . . . the most critical issues of science, philosophy, and religion" (p. 397). Part I, "The Setting," has two chapters, the first on the mediums and the second on the personnel of spiritualism. Part II, "A Surrogate Faith," has a chapter on spiritualism and Christianity, one on theosophy and the occult, and one on psychical research and agnosticism, which deals with the work of the SPR. The third and final part is "A Pseudoscience," and has chapters on concepts of mind, in which psychology and psychical research and the work of Gurney and Myers are dealt with; a chapter on evolution, which deals with Romanes, Chambers, and Wallace; and a final chapter on physics and psychic phenomena, which has sections on Barrett, Crookes, and Lodge.

ChrP 6:188–190 (Mar 1986); *JASPR* 81:183–187 (Apr 1987); *JP* 50:155–162 (Jun 1986); *PAI* 5:41 (Jun 1987).

AmHistRev 91:398 (Apr 1986); *BritBkNews* p397 (Jul 1985); *Choice* 23:668 (Dec 1985); *GuardianWkly* 132:22 (May 5, 1985); *LondRevBks* 8:18 (Mar 6, 1986); *Nature* 316:25 (Jul 4, 1985); *QueensQ* 94:500 (Sum 1987); *ReligStudRev* 13:81 (Jan 1987); *TimesLitSuppl* p353 (Apr 4, 1986); *VictorianStud* 29:613 (Sum 1986).

Interdisciplinary Approaches to Psi

INTRODUCTION

In *Sources* most of the interdisciplinary approaches to parapsychology were represented by various Parapsychology Foundation conference proceedings, plus three anthologies. All of these earlier works are still valuable as sources of ideas and research suggestions.

The number of books in this section has grown from 7 in *Sources* to 11. Various aspects of science and parapsychology or parapsychology and several sciences are treated in the book edited by Angoff and Shapin [562], and in McClenon [567], McConnell [568], and Schmeidler [571]. In the books listed in this section, the single field most often linked with para—psychology is anthropology (see the volume edited by Angoff and Barth [561], and the books by Harner [565], Long [566], and Nicholson [569]). A sociological approach is taken by Emmons [563] and by Greeley [564]. Randall [570] is on biology and parapsychology.

Additional books with information on interdisciplinary approaches to psi: Kelly and Locke [350] review the findings of cultural anthropology regarding psi. Hurley [372] reviews the literature of anthropology in regard to remote control of human behavior. Schwartz [376] reviews the findings of psychic archeology, including a chapter on anthropology and psi. White's bibliography [378] includes sections on the relationship of psi to a number of disciplines: anthropology, archeology, biology, botany, business, education, geology, history, journalism, meteorology, philosophy, physics, psychiatry, psychology, space exploration, and sports. Ebon [431] has a chapter relating ESP to other sciences. Several of the books in the history section emphasize social/sociological aspects of psi: Brandon [549, Cerullo [550], Evans [552], and Moore [559]. The Goldfarbs [553] write about literature and spiritualism. The books by Collins [493] and Collins and Pinch [493] approach parapsychology from the viewpoint of the sociology of science. The books edited by Ebon [512] and by Grattan—Guinness [516] have several chapters relating parapsychology to other disciplines. *Advances in Parapsychological Research 1* [520] has chapters relating para—

psychology to other disciplines. The books edited by John White [526] and Wolman [688] each contain several review chapters relating parapsychology to other disciplines. In the section on mediumship two authors deal with cultural and social aspects of mediumship, possession, and multiple personality: Ebon [575] and Kenny [578]. All the books in the section "Physics and Psi" are interdisciplinary in nature, as are those in the section entitled "Psychology, Psychiatry and Parapsychology." All of the books in the section "Philosophy and Parapsychology" [605-615] could be said to represent interdisciplinary approaches to parapsychology. The one edited by Thakur [614] is especially relevant because it contains chapters on religion and parapsychology and biological aspects of psi as well as philosophical treatments. Hardy writes about biology and psi [707]. Ways in which parapsychology can assist religion and vice versa are discussed by Perry [717] and by the participants at a Parapsychology Foundation conference [720]. A folklorist approach to spontaneous psi phenomena is taken by Davidson [224], Goss [728], and Hufford [732]. Hartmann [730] relates spontaneous psi experiences to the biology of the nightmare. The anthologies of papers on the survival problem edited by Spong [774] and Toynbee [777] approach the subject from many different disciplines. Anthropological approaches to healing are taken by Achterberg [792], Finkler [797], Katz [800], Krippner and Villoldo [801], and Villoldo and Krippner [803].

BIBLIOGRAPHY

561 **Angoff, Allan, and Barth, Diana (Eds.).** *Parapsychology and Anthropology: Proceedings of an International Conference Held in London, England, August 29-31, 1973.* New York: Parapsychology Foundation, 1974. 328p. Chapter references; 13 figures; 7 maps; 8 tables
Half of the 18 conference participants in the Parapsychology Foundation's 22nd international conference on parapsychology were anthropologists and half were identified with parapsychology. Some of the papers by parapsychologists are oriented toward empirical psi research, particularly those of Brier, the Kreitlers, and Van de Castle. The anthropological contributions tend to stress the nature of primitive beliefs about psi and the supernatural rather than empirically investigating the truth of claims. Several papers describe possible psi practices in primitive societies.
JASPR 69:285-292 (Jul 1975); *JSPR* 48:161-164 (Sep 1975); *T* 5(4):15 (1977).

562 Angoff, Allan, and Shapin, Betty (Eds.). *Parapsychology and the Sciences: Proceedings of an International Conference Held in Amsterdam, The Netherlands, August 23–25, 1972.* New York: Parapsychology Foundation, 1974. 289p. Chapter references; 2 figures; 1 graph; 1 table

This volume brings together 20 papers delivered at a Parapsychology Foundation conference on parapsychology and the sciences. The participants apparently interpreted the topic differently so their papers may be classified roughly in three major ways of relating parapsychology and the sciences: two papers treat the question of whether parapsychology can be viewed as a science (Dingwall, Mitchell); six deal with scientific methodology in parapsychology (Brier, Cahn, Cutten, Flew, Orme, Schmidt); and the remaining 11 relate parapsychology to other disciplines: psychology (Beloff, Rogo, West), psychoanalysis (Servadio), meteorology (Tromp), psychiatry (Alberti, Smythies), physics (Koestler, Tromp), genetics (Zorab), chemistry (Ellis), and biology (Poynton).
JP 39:251–253 (Sep 1975).

563 Emmons, Charles F. *Chinese Ghosts and ESP: A Study of Paranormal Beliefs and Experiences.* Metuchen, NJ: Scarecrow Press, 1982. 297p. Bibliography: 277–285; Chapter notes; 1 figure; Glossary: 286–288; 13 illustrations; Index: 289–297; 3 tables

This book by a sociologist at Gettysburg College is based on 3,600 interviews, questionnaires, and observations on paranormal beliefs and experiences made in Hong Kong from June 1980 through January 1981. The largest batch of data presented was gathered by means of a telephone survey. The questions asked were designed to provide a cross–cultural comparison by replicating a 1978 Gallup poll of paranormal beliefs in the U.S. The data have been analyzed from a sociological and anthropological point of view. The author was also interested in ascertaining if he could shed light on the question of whether or not there is a universal experience of ghosts that is unaffected by cultural differences and that fits the theories of parapsychologists. The experiences are described in several categories, e.g., hauntings and poltergeists, spirit possession and exorcism, spirit mediumship, ESP, fortune–telling, and others. No attempt was made to corroborate or verify the accounts. In a concluding chapter Emmons contrasts the results of the Chinese sample with those of the American survey.
JASPR 76:385–387 (Oct 1982); *JP* 46:284 (Sep 1982); *T* 12:20 (Spr 1984).
Choice 20:163 (Sep 1982).

564 Greeley, Andrew M. *The Sociology of the Paranormal: A Reconnaissance.* Beverly Hills, CA: Sage Publications, 1975 (A Sage research paper in the social sciences). 88p. Bibliography: 87—88; 16 figures; 19 tables

The author, a Catholic priest, is also a sociologist and Director of the Center for the Study of American Pluralism, National Opinion Research Center. As part of the Center's research program, 1,460 people responded to a poll pertaining to their views on the paranormal. This monograph reports the results of this poll and Greeley's interpretation of them. The questions were aimed at finding out what percentage of the population has ESP, survival experiences, and mystical experiences. He investigated the types of situations which appear to trigger mystical experiences, and also reports on his efforts to discover if there were significant interactions among various psychological, attitudinal and demographic variables.

JP 40:81—85 (Mar 1976).

ContempSociol 7:229 (Mar 1978; *JSciStudRelig* 16:435—436 (Dec 1977); *SpirFron* 10:77—84 (Spr 1979); *SocForces* 55:570 (Dec 1976).

565 Harner, Michael. *The Way of the Shaman: A Guide to Power and Healing.* San Francisco, CA: Harper & Row, 1980. 167p. Bibliography: 154—161; Chapter notes; 10 figures; 8 illustrations; Index: 163—166

The author, both an anthropologist and a shaman, having been initiated by the Jivaro in South America, has drawn on the worldwide literature of shamanism to supplement his firsthand experience and attempts to introduce Westerners to the practice. He defines a shaman as "a man or woman who enters an altered state of consciousness——at will——to contact and utilize an ordinarily hidden reality in order to acquire knowledge, power, and to help other persons. The shaman has at least one, and usually more, 'spirits' in his personal service" (p. 20). The shamanic state of consciousness is ecstatic, and according to Mircea Eliade is characterized by being a "trance during which his [or her] soul is believed to leave his [or her] body and ascend to the sky or descend to the underworld" (p. 21). Harner points out that shamans are healers, able to see into the past, present, and future on behalf of others, and may engage in clairvoyance. He describes guardian spirit journeys, and notes that they often involve synchronistic happenings. A number of healing techniques are described including the healing of persons at a distance. He stresses the compatibility between holistic health principles and shamanic healing methods. Perhaps the shaman is uniquely qualified among the prac—

titioners of the healing arts to "give the doctor inside a chance
to go to work" (p. 135).
    JASPR 76:78—80 (Jan 1982); *T* 11:44—45 (Sum 1983).
    AmAnthropol 83:714—717 (Sep 1981); *AmEthnol* 8:197—198
(Feb 1981); *Choice* 8:697 (Jan 1981); *ImagCogPers* 6:283—286
(1986/87); *Kirkus* 48:1130 (Aug 15, 1980); *ReligStudRev* 7:136
(Apr 1981).

566 Long, Joseph K. (Ed.). *Extrasensory Ecology: Para—
    psychology and Anthropology.* Metuchen, NJ: Scarecrow
    Press, 1977. 427p. Chapter references; 24 figures; Glos—
    sary: 397—400; Index: 407—427; 5 tables
    Anthropologist Long organized the well received interdis—
ciplinary symposium on parapsychology and anthropology at the
1974 annual meeting of the American Anthropological Associa—
tion.   In this work he extends this interdisciplinary approach,
bringing together expanded papers from that symposium plus
additional contributions by anthropologists, parapsychologists,
and other scientists.   Long provides introductory and coordinat—
ing remarks for each chapter which help to tie the book
together.   The book is aimed at anthropologists as well as any
"who wish to better understand psi in relation to culture and
social life" (p. 390). Part I, "Parapsychology and Anthropology,"
has four papers, by Long, Agehananda Buharati, Jule Eisenbud,
and Margaret Mead.   Part II, "The Physical Bases for
Paranormal Events," consists of papers by Evan Harris Walker
and James B. Beal. Part III, "Psi, Cognition, and
Communication," contains papers by Walter Griest, Alexander
Randall, and Henry Reed.   Part V, "The Origins of Psi," con—
sists of three chapters, by Berthold E. Schwarz, William S.
Lyon, and Long. The fifth part is entitled "Paranormal Dimen—
sions in Primitive Medicine," and consists of three contributions,
two by Long and one by Robert W. and Linda K. Sussman.
Part VI, "Some Practical Uses of ESP: Psi and Archeological
Reconstructions," has three papers, by J.N. Emerson, Zbigniew
W. Wolkowski, and Jeffrey D. Goodman.   The final part is
"Parapsychology and Anthropology: An Overview," and contains
three chapters by Roger W. Wescott, Charles Musés, and Long.
    JSPR 49:887—890 (Sep 1978); *JP* 43:61—66 (Mar 1979);
PR 9:14—17 (Sep/Oct 1978).
    LibrJ 103:372 (Feb 1, 1978); *SkptInq* 2:108—112 (Spr/Sum
1978): Erratum 3:69 (Fal 1978).

567 McClenon, James. *Deviant Science: The Case of Para—
    psychology.* Philadelphia, PA: University of Pennsylvania
    Press, 1984. 282p. Bibliography: 253—273; 4 figures; In—
    dex: 275—282; 1 questionnaire; 17 tables

McClenon, both a sociologist and a parapsychologist, offers a sociological approach to the problem of parapsychology's lack of acceptance as a legitimate science. Attempts to scientifically investigate claims of phenomena labeled paranormal have been underway for over a century, yet no consensus regarding the ontological status of the phenomena has been achieved. However, stable social patterns have evolved surrounding the small number of scientists investigating the paranormal. According to the major theoretical orientations within the sociology of science, this shouldn't happen! Usually such groups last about 10 or 15 years, and then either become absorbed into mainstream science or die out. Neither has happened to parapsychology. The study on which this book is based involves the social processes by which legitimacy is granted or denied new forms of scientific inquiry, using parapsychology as an example. In conducting the study McClenon surveyed the attitudes of elite scientists towards ESP, parapsychology, and anomalous experience and interviewed parapsychologists.

ASPRN 11:6 (Oct 1985); *JASPR* 80:221−226 (Apr 1986); *JP* 50:64−72 (Mar 1986); *PAI* 4:68 (Jun 1986); *PR* 18:10−15 (May/Jun 1987); *PsiR* 3:202−208 (Sep/Dec 1984).
Choice 22:1410 (May 1985)

568 McConnell, R.A. *An Introduction to Parapsychology in the Context of Science.* Pittsburgh, PA: Author, 1983. 337p Bibliography: 311−323; 25 figures; Name index: 324−328; Subject index: 329−337; 2 tables

Some steps towards a scientific answer to the question of the relation of human consciousness to the physical world are attempted by the author. Primarily, this book is based on lectures prepared by the author for a course he has given at the University of Pittsburgh since 1973. The lectures were prepared with the premise in mind that in order for psi phenomena to achieve scientific acceptability, a reassessment is required of the widely scattered areas of belief upon which they impinge. There are three parts, the first consisting of nine chapters that present the psychosocial substrata of parapsychology. The chapters are titled "Definitions and Questions," "Trance Mediumship," "The Occurrence of Multiple Personality," "An Interpretation of Multiple Personality," "Consciousness and Society," "Altered States of Consciousness," "Factors Favoring Psi," "Postmortem Survival," and "History of Parapsychology." The second part is on the observation of psi phenomena and consists of the following seven chapters: "Mechanized Methods," "Spontaneous Psychokinesis," "Experimental Psychokinesis," "Unconscious Characteristics of Psi," "Progressive Organization and Ambivalence Principles in Psychokinetic Data," "Hypnosis as Psychokinesis," and "Hypnosis, Peer Reviewed." The last part is on the sociology of parapsychology and consists of the following ten chapters: "Scientific Theory," "Progress in Science," "Fraud in Parapsychology," "With Friends Like These . . . ," "Of People and Prejudice," "Guardians of Orthodoxy," "The Closing Mind of Adolescence," "Recapitulation," "Time Present," and "The Long Tomorrow." There are also eight ap— pendices: "A Background for the Study of Parapsychology," "On the Malpractice of Medicine," "Some Persons of Prominence in the History of Parapsychology Whose Work was Published in English," "Chronology of Major, Professionally Guided Centers of Parapsychological Research in the U.S.A.," "Governmental Assistance to a Pretheoretical Science," "Analysis of the Honorton—*Science* Controversy, 1972—1975," "Listening Guide to the Controversy About the Relation of Intelligence to Inherited Biological Structure," and "Two Instances of Spontaneous, Probable ESP Known First—Hand to the Author."

   JP 48:59—64 (Mar 1984); *JASPR* 78:372—378 (Oct 1984); *PAI* 6:68—69 (Jun 1986).
   Choice 21:887 (Feb 1984); *ContempPsychol* 30:25 (Jan 1985).

569 Nicholson, Shirley. *Shamanism: An Expanded View of Reality.* Foreword by Jean Houston. Wheaton, IL: Theosophical Publishing House, 1987. 295p. Chapter references
   This collection of previously published articles on shamanism is in five sections. The first consists of five articles on the alternate reality of the shaman. The second section, "Shamanic States of Consciousness," may be the one most germane to parapsychology. It contains an article by Jeanne Achterberg on shamanic healing, Stanley Krippner on shamanic dreams, D. Scott Rogo on "Shamanism, ESP, and the Paranormal" (the most relevant chapter to parapsychology in the book), and Jim Swan on the shamanic healer, Rolling Thunder. The chapters in the third section are on shamanism as expressed in different cultures. Section IV contains four chapters about shamanism and aspects of the perennial philosophy. The last section, "Shamanism in a Changing World," consists of three articles on shamanism today.
   ChrP 7:263—264 (Sep 1988); *PAI* 5:45 (Dec 1987).
   AmTheos 75:195 (May 1987); *Booklist* 83:1232 (Apr 15, 1987); *WestCoastRevBks* 13(1):41 (1987).

570 Randall, John L. *Parapsychology and the Nature of Life.* London, England: Souvenir Press, 1975. 256p. Bibliogra— phy: 244—252; 2 figures; 3 graphs; 13 illustrations; In—

dex: 253—256; 2 tables
The author of this book is a British biologist who has been actively engaged in psi research. He attempts to assess the role of parapsychology as regards the nature of life in general and human personality in particular. He describes the books as follows in the Preface: "Part I outlines the development of those sciences which have contributed most directly to the mechanistic theory of life, and attempts to evaluate their overall effect on human life and thought. Part II——the major section of the book——deals with the problems, failures, and successes of parapsychology" (pp. 6—7). In the third part, which he grants is the most speculative section, he offers his assessment of the implications of parapsychology for our conception of nature and man. A nine—page bibliography is provided for those who wish to delve further.

ChrP 2:13—17 (Mar 1978); *JASPR* 71:109—111 (Jan 1977); *JP* 39:328—330 (Dec 1975); *JSPR* 48:275—278 (Mar 1976); *P* 7:38—40 (Jan/Feb 1978); *PR* 7:9—12 (Sep/Oct 1976); *PS* 2:161—162 (F 1972).

Encounter 54:86 (Jan 1980); *Kirkus* 44:452 (Apr 1, 1976); *LibrJ* 101:725 (Mar 1, 1976); *Observer* p19 (Dec 14, 1975); *PublWkly* 209:53 (Apr 26, 1976).

571 Schmeidler, Gertrude R. (Ed.). *Parapsychology: Its Relation to Physics, Biology, Psychology, and Psychiatry.* Metuchen, NJ: Scarecrow Press, 1976. 278p. Chapter references; 5 figures; Index: 265—278
The papers in this book are the result of a symposium sponsored by the American Society for Psychical Research held in New York City in 1974. There are two papers on para—psychology and physics by J.H. Rush and R.B. Roberts; there are two on biology and parapsychology by R.L. Morris and B.R. Grad; there are four contributions on psychology and parapsychology by I.L. Child, G.R. Schmeidler, R.L. Van de Castle, and L. LeShan; and psychiatry and parapsychology is represented in four papers by M. Ullman, S.D. Kahn, B.E. Schwarz, and J. Ehrenwald. In each subject area the sym—posiasts dealt with the following questions as stated in the Foreword: "What are the significant experimental findings? What are their implications? What are their potential applica—tions? What further research is needed?" (p. xi).

JP 41:58—60 (Mar 1977); *JSPR* 48:400—402 (Dec 1976); *P* 7:38—41 (Jan/Feb 1977); *PR* 7:13—15 (Nov/Dec 1976); *T* 7:16—17 (Spr 1979).

Choice 13:1665 (Feb 1977); *ContempPsychol* 22:429 (Jun 1977); *LibrJ* 102:1284 (Jun 1, 1977); *SpirFron* 9:61—62 (Win 1977).

Mediumship, Possession, and Automatism

INTRODUCTION

Because *Sources* included older titles, the section on mediumship was quite large (22 books), and most of the titles were about mediums or mediumship.

Of the 13 books in this section, only three are primarily on mediumship: Gregory [576], Kenny [578], and Klimo [579]. The subject of multiple personality and possession is a topic of current interest, which is reflected in several titles here: Baldwin [572], Crabtree [573], Crapanzano and Garrison [574], Ebon [575], McKellar [581], Pettiward [582], and Rogo [584]. Interest in channeling, another "hot" topic, has led to the books by Klimo and Ridall [583]. (Channeling is mediumship with a do−it−yourself twist.) Gruss and Hotchkiss [577] write about one of the channels of mediumistic communication, the ouija board, and Levine [580] writes about consulting psychics and mediums.

Additional books with information on mediumship, possession, and automatism: The classic by Lombroso [811] describes his pioneer research and theories on mediumistic phenomena. Kelly and Locke [330] review trance mediumship as does Parker [356], and Sargent [357] examines the mechanism underlying pos−session. Possession and multiple personality are discussed in chapters in Wolman and Ullman [361]. Hunt [807] deals with automatisms, the ouija board in particular. In the section on "Applications" there are books describing the use of mediums and psychics in criminal investigations and for other practical purposes: Allison and Jacobson [367], Hibbard and Worring [371], Schwartz [376], Tabori [377], and Yeterian [379]. Kautz and Branon's book [374] is an introduction to channeling. The psychology of the sensitive is discussed in Schwartz [376]. White's bibliography [378] lists hundreds of citations to sources describing practical applications of the mediumistic and psychic ability, e.g., finding lost objects, locating missing persons, and disaster prediction. The section "Biographies and Auto−biographies of Healers, Mediums, and Psychics" contains autobiographies by Geller [386, 387], Harribance [388], Harrison

[389], Ivanova [390], Manning [393], Robbins [395], Swann 396],
and Tanous [397]. It also lists a biography of D.D. Home [391].
For a bibliographic guide to autobiographies, biographies, and
investigations of healers, mediums, and psychics, as well as infor—
mation on developing mediumistic abilities, see White [791].
Several of the books in the section aimed at young people
offer nontechnical accounts of mediums and mediumship: [415,
425, 430, 432, 433, and 440]. One includes a firsthand account
of a seance [415]. Kettelkamp [440] describes the "Philip"
ghost. Booth [458] has five chapters on mediumship which
stress fraudulent practices of mediums. Several mediums are
criticized by Christopher [459], Booth [458], Couttie [461], and
Harris [473]. Gordon [470] points out how mentalism can ac—
count for many of the feats of mediums and psychics. Keene
[475], a former medium, sets forth the tricks of the trade, espe—
cially the fraudulent practices at Camp Chesterfield. Randi
[480] is critical of Geller. Several of the books in the history
section are on spiritualism and mediumship as investigated by
the early psychical researchers. See Brandon [549], Douglas
[551], Evans [552], Goldfarb and Goldfarb [553], Haynes [554],
Inglis [555, 556], and Oppenheim [560]. Braude's book on PK
[606] is largely concerned with the physical phenomena of
mediumship, or macro—PK effects. In another book, Braude
[606] deals with physical mediumship. In the section on PK,
there are two books on home circles or sitter groups: the
"Philip" group [628] and the SORRAT group [631]. Thurston
[788] writes about developing ESP based on the Edgar Cayce
readings. Spiritualistic (or mediumistic) healing is described by
Finkler [797], Harvey [798], Katz [800], Krippner and Villoldo
[801], and Villoldo and Krippner [803]. Fuller [753] provides a
book—length account of the R—101 case involving communica—
tions received through Eileen J. Garrett. Gauld [755] reviews
the phenomena of mediumship. Stevenson has written two
books on xenoglossy [775, 776].

BIBLIOGRAPHY

572 **Baldwin, Louis.** *Oneselves: Multiple Personalities, 1811—*
1981. Jefferson, NC: McFarland, 1984. 163p. Bibliogra—
phy: 157—160; Index: 161—163
There are 28 chapters, averaging 6 pages each, each con—
sisting of a sketch of a case, often well known, of multiple per—
sonality. Some, such as Ansel Bourne, Hlene Smith, Doris Fis—
cher, and Patience Worth appear in the parapsychological litera—
ture. The cases date from 1811—1981 and are listed in alphabeti—
cal order. In the Foreword, the author points out that only in

1980 did the American Psychiatric Association define multiple personality as a distinct disorder, but diagnosed cases of multiple personality were already on the rise in the 1970s (50—plus cases in that decade as opposed to 90 in the preceding century and a half). Baldwin observes that persistent time loss, chronic sleepwalking, automatic writing, and hallucinations are usually closely associated with multiple personality, and sometimes possession and mediumship abilities as well.
PAI 3:40 (Dec 1985).

573　Crabtree, Adam. *Multiple Man: Explorations in Possession and Multiple Personality.* New York: Praeger, 1985. 278p.
Bibliography: 262—270; Index: 273—278
Crabtree, a Catholic clergyman and psychotherapist, has written about multiple personality. The first of the four parts of this book is about divided consciousness and multiple personality, beginning with mesmerism. The second part is about possession and its relation to multiple personality. The third presents case histories from the author's practice, including 50 cases of apparent possession, and the final part is theoretical. Although Crabtree's speculations go beyond psychology, he admits that purely psychological theories could be considered sufficient to explain the phenomena.
ChrP 6:198—199 (Mar 1986); *JSPR* 54:76—78 (Jan 1987); *PAI* 5:44 (Dec 1987).
BksinCan p25 (Mar 1986); *Choice* 23:519 (Nov 1985).

574　Crapanzano, Vincent, and Garrison, Vivian (Eds.). *Case Studies in Spirit Possession.* New York: Wiley, 1977. 457p. Chapter references; 9 figures; Index: 451—457; 3 tables
Two anthropologists compiled this anthology of case histories of spirit possession, defined by Crapanzano as "any altered state of consciousness indigenously interpreted in terms of the influence of an alien spirit" (p. 7). There are 10 case studies by various authors, eight of which have not been previously published. Although dealing with different cultures, each case is characterized by the fact that the possessing spirits are not explained away but are empathized with and taken at face value for purposes of understanding the *Lebenswelt* of the possessed. Raymond Prince, in the Foreword, notes that for him the most interesting aspect of these studies is the therapeutic value of possession for individuals suffering from psychiatric disorders. American culture is represented by June Macklin's study of a New England trance medium, Rita M.
JSPR 49:660—661 (Dec 1977); *PAI* 1:85 (Dec 1983).
Choice 14:910 (Sep 1977); *RevsAnthropol* 5:399 (Sum 1978).

575 Ebon, Martin. *The Devil's Bride. Exorcism: Past and Present.* New York: Harper & Row, 1974. 245p. Bibliog— raphy: 243—245

Martin Ebon is one of the most prolific and one of the most reliable of the popular writers on parapsychology. This his— tory of the phenomena of possession and exorcism is a very readable survey of how possession has been expressed throughout history in various countries and cultures and of at— tempts to exorcise the so—called possessing entities. In addition, Ebon offers his own conclusions concerning the nature of exor— cism: a psychology and a sociology of exorcism, one could say.
JASPR 69:292—294 (Jul 1975); *P* 5:38 (Sep/Oct 1974).
Booklist 70:1216 (Jul 15, 1974); *BestSell* 34:131 (Jun 15, 1974); *BkWorld* p4 (May 12, 1974); *Choice* 11:1118 (Oct 1974); *ChristCent* 91:596 (May 29, 1974); *ChristCent* 91:995 (Oct 23, 1974); *ChristToday* 19:51—53 (Jul 4, 1975); *LibrJ* 99:1965 (Aug 1974); *PublWkly* 205:53 (Apr 1, 1974).

576 Gregory, Anita. *The Strange Case of Rudi Schneider.* Metuchen, NJ: Scarecrow Press, 1985. 444p. Chapter references; 30 figures; 1 graph; 43 illustrations; Name index: 427—433; Subject index: 434—444; 1 table

This book is the result of years of painstaking scholarship. Gregory reviews all of the investigations of Schneider and dis— cusses arguments for and against the authenticity of his physical phenomena. A detailed biography is presented as well. Gregory uses Schneider as a prototypical example of the difficulties in— volved in investigating star PK subjects.
JASPR 80:427—444 (Oct 1986); *JP* 59:168—176 (Jun 1986); *PAI* 4:43 (Dec 1986).
Light 105:128—129 (Aut 1985).

577 Gruss, Edmond C., with Hotchkiss, John G. *The Ouija Board: Doorway to the Occult.* Chicago, IL: Moody Press, 1975. 191p. Chapter notes; Suggested reading list: 190—191

Although the ouija board has reportedly replaced *Monopoly* as the favorite board game in this country, it is exceedingly difficult to find information on the ouija board in the literature. The author, who is Chairman of the History Department at Los Angeles Baptist College, has written a book based on Biblical answers to questions about the occult, but it is also an excellent book about the ouija board and automatism. He provides a his— tory of the ouija board, a chapter on what makes it work, and one on the use of the board in psychic development and the dangers involved in so doing. Because of the great interest in

the occult engendered by *The Exorcist,* in which Regan's posses—
sion was brought about through her use of the ouija board,
Gruss also has three chapters on possession and demonology.

578 Kenny, Michael G. *The Passion of Ansel Bourne: Multiple
Personality in American Culture.* Washington, DC:
Smithsonian Institution Press, 1986. 250p. Bibliography:
217—233; Chapter notes; 8 illustrations; Index: 235—250
   A social anthropologist familiar with the phenomenon of
possession in primitive cultures writes about multiple per—
sonality, which he views as a cultural variant of possession.
The first chapter tells the story of a pioneer case of multiple
personality, Mary Babcock.   The case of Ansel Bourne is
described in the second chapter.   In the next chapter Kenny
examines spiritism (mediumship), which he views as an impor—
tant aspect of multiple personality as it was expressed in the
1890s.   He discusses the impact Spiritualism had on
nineteenth—century psychology, and deals with a specific case——
that of Mrs. Piper and her communicating spirit, G.P.   In
Chapter 4 he deals with the work of Morton Prince with "Miss
Beauchamp" and "B.C.A."   In a concluding chapter he writes
about the current wave of interest in multiple personality, in
which many multiple selves are manifested.   Throughout the
book Kenny attempts to show how the expression of the mul—
tiple personality syndrome was influenced by and served as an
expression of cultural factors.   Just as in non—Western societies,
spirit possession occurs in individuals in difficult, contradictory,
or traditional situations, so he observed that in Western society
multiple personality appears to be "a complex metaphorical
response" to cultural and personal situations of stress and transi—
tion.
   PAI 6:44 (Dec 1988).
   AmHistRev 93:492 (Apr 1988); *BkRpt* 6:43 (Nov 1987);
Choice 24:1437 (May 1987); *JAmHist* 74:1338 (Mar 1988);
Psychiatry 50:295—297 (Aug 1987); *TimesLitSuppl* p537 (May 22,
1987).

579 Klimo, Jon. *Channeling: Investigations on Receiving Infor—
mation from Paranormal Sources.* Foreword by Charles
T. Tart. Los Angeles, CA: Jeremy P. Tarcher, 1987.
384p. Chapter notes; Glossary: 344—350; Index: 370—384;
Selected bibliography: 365—369
   Psychologist/educator Jon Klimo claims he wrote this book
for three reasons: (1) to present a balanced and objective view
of channeling, which he defines as "the communication of infor—
mation to or through a physically embodied human being from
a source that is said to exist on some other level or dimension

of reality than the physical as we know it, and that is not from the normal mind (or self) of the channel" (p. 2); (2) to ad—dress the needs of the increasingly large number of persons who are experiencing some form of channeling; and (3) to expand the reader's view of reality and to awaken him or her "to the greater possibilities of being human" (p. 3). He writes: "This book will attempt to demonstrate that the realms of the psyche and spirit may be more real and run more deeply than most of us believe" (p. 4). Chapter 1, "Channeling as a Modern Phenomenon," describes some modern channels such as Jane Roberts, Meredith Lady Young, Helen Cohn Schucman of *A Course in Miracles*, and Eileen Caddy and others of Findhorn. Chapter 2, "Channeling as a Historical Phenomenon," goes back to the prehistoric roots of channeling through the centuries up to the modern era. The third chapter is on the psychology of channeling, the development of channeling, and it presents portraits of five channels. The next three chapters are on the material that is channeled, the nature of the beings or entities that are channeled, and how——in what states of consciousness——channeling occurs. The next two chapters present a psychological and biological explanation of channeling. Chapter 9 is on "open channeling," which may include intuition, inspiration and aspects of the creative process. Chapter 10 provides instructions for becoming a channel. In his concluding remarks, Klimo writes: "Channeling holds out to us the pos—sibility of an incredibly complex, multileveled universe, filled with fellow consciousness, a universe with which we can *all* inter—act in new and meaningful ways, each of us taking ultimate responsibility for our respective growth" (p. 342).

ChrP 8:25−27 (Mar 1989).

ContempPsychol 34:381−382 (Apr 1989); *JTranspersPsychol* 19:197−199 (1987); *NewReal* 9:65 (Mar/Apr 1989); *Parabola* 13:88 (Aug 1988).

580 Levine, Frederick G. *The Psychic Sourcebook: How to Choose and Use a Psychic.* New York: Warner Books, 1988. 406p. Bibliography: 397−402; Chapter notes; Glos—sary: 379−392; Index: 403−406

This book, by a freelance writer specializing in New Age topics, is intended as a guide to those who seek information from psychics, mediums, fortune tellers, etc., with a view to making the most of the experience. Levine sees the visit to a psychic as a means of self change. The book also attempts to clear up misconceptions about psychics and what they do. Chapter 1, "What Is a Psychic?" attempts to cover too much in too small a space. Chapter 2, "How to Use a Psychic," provides a good overview of many psychic applications. The

third chapter, "How to Choose a Psychic," explains what to
look for, where to look, and what to watch out for. Chapter 4,
"The Psychic Reading," covers various psychic modes by which
psychics function: psychometry, ESP, Tarot, *I Ching*, channeling,
astrology, psychic healing, etc. The fifth chapter tells how to
get ready for a visit to a psychic, while the sixth explains how
to interpret the psychic's reading. The brief conclusion proposes
that "whatever psychic you choose will have something to give
you--if you are willing to take it" (pp. 346-347). Appendix A
provides a brief--and quite incomplete--history of psychic
phenomena. Appendix B is a directory of resources, psychics
organizations, services, and publications. Emphasis is on
popular rather than scholarly sources.
     Booklist 84:1758 (Jul 1988).

581 McKellar, Peter. *Mindsplit: The Psychology of Multiple Per-*
        sonality and the Dissociated Self. London, England: J.M.
        Dent, 1979. 188p. Bibliography: 180-184; 3 figures; Glos-
        sary: 173-179; 4 graphs; Index: 185-188
     McKellar heads the Psychology Department at the Univer-
sity of Otago in New Zealand. He reviews his research and that
of others in dissociation, particularly in its extreme forms, as in
dual and multiple personality. He views phenomena interpreted
by some as demonic possession, spirit controls, mediumistic
trance, reincarnation, clairvoyance, and OBEs as explainable in
terms of normal psychology by dissociation, multiple personality,
co-consciousness, and mental images. He calls for renewed interest in the neglected content of dissociation as casting "a fresh
slant on a wide range of psychological phenomena." He suggests
that the hypothesis of personality subsystems that are linked,
but which have amnesic barriers between them, may be useful
in making a more sophisticated classification of mental behavior
than we have had. He feels neglect of this approach by
mainstream psychology has resulted in an exaggeration of the
distinction between normal and abnormal.
     PAI 2:40 (Dec 1984).
     JMentImag 6:187-189 (Fal 1982); *LibrJ* 105:1643 (Aug
1980).

582 Pettiward, Cynthia. *The Case for Possession.* Foreword by
        G.S. Whitby. Gerrards Cross, Bucks, England: Colin
        Smythe, 1975. 129p. Bibliography: 127-129; Notes
     British teacher and healer Cynthia Pettiward proposes that
the causes of mental and physical ill-health sometimes may be
found outside the person involved and his or her environment,
epilepsy in particular. She proposes that some illnesses are due
to possession by deceased persons. In this essay she summarizes

the evidence for possession by discarnate entities and also the evidence for survival as it bears on the possession hypothesis. The characteristics and types of possessions are described and analyzed. The book is intended to serve as a stimulus to further research by psychiatrists and other medical professionals. *Light* 96:88 (Sum 1976).

583 Ridall, Kathryn. *Channeling: How to Reach Out to Your Spirit Guides.* New York: Bantam Books, 1988. 150p.

Ridall is a psychologist who has had firsthand experiences as a channel. Channeling is viewed as a capacity available to all——one that can be used to explore consciousness. She tells how she became involved in channeling and what it has meant to her. The bulk of the book is about how to channel and on the relationship with guides. There also is a chapter on "Encountering Doubts and Resistances" and one on learning special channeling techniques. She closes with a chapter on "Channeling and the Evolution of Consciousness." Dr. Ridall makes plain why channeling has become so popular——because it is a technique that can be learned and because it can be used to promote personal growth. As presented in this book it differs from trance mediumship in that one can do it oneself and need not depend on someone else.

Booklist 84:956 (Feb 15, 1988); *LibrJnl* 113:62 (Mar 15, 1988); *PublWkly* 233:88 (Feb 5, 1988).

584 Rogo, D. Scott. *The Infinite Boundary: A Psychic Look at Spirit Possession, Madness, and Multiple Personality.* New York: Dodd, Mead, 1987. 317p. Chapter notes; 15 illustrations; Index: 311—317; 2 tables

Rogo presents a history via accounts of the lives and work of those who researched the spiritistic theory of mental illness as well as of those who investigate it today. According to the spiritistic theory, some cases of psychological disorder and multiple personality may be the result of obsession or even possession by discarnate spirits. Part 1 reviews the following cases researched by James H. Hyslop: the Thompson—Gifford case, Etta de Camp, and the Doris Fischer case. In Part 2 there is a chapter which reviews the research work of Titus Bull and Carl Wickland in which, working through a medium or a spiritualist rescue circle, obsessing or possessing spirits were contacted and attempts were made to convince them to progress spiritually and to let go their hold on the living. The next chapter describes the work of Walter Franklin Prince's efforts to exorcise the possessed. Another chapter concerns the ministry of Elwood Worcester. The third part deals with research work today, and has chapters on Kardecismo groups (Brazilian spiritis-

tic followers of Allan Kardec) and the modern channel Matthew
Bronson who is involved with a rescue circle in Northern Califor—
nia.   There is a chapter on M. Scott Peck, who holds that
some multiple personality cases are a form of demonic posses—
sion.   Peck has developed a method of exorcism to deal with
them.   Also touched on is the work of Ralph Allison.   In the
next chapter Rogo considers whether or not cases of apparent
possession are literally true: that is, with the questions "Do we
ultimately survive death," and "Can we subsequently reach back
and contact the living?"   He considers the possibilities of fraud
and the super—ESP hypothesis, and concludes that at least
some cases of genuine spirit possession probably exist.
      Booklist 84:94 (Sep 15, 1987); *Kirkus* 55:1142 (Aug 1,
1987).

Near−Death Experiences

INTRODUCTION

When *Sources* was compiled the term "near−death experience" had not even been coined! However, the related subject of deathbed visions was treated in three books, by Vyvyan [184], Baird [242], and Osis [258].

Thanks to Raymond Moody and the many researchers who have followed him, there are 11 titles in this section. The three books by Moody [589-591] provide a good general introduction to the subject. Surveys of NDEs are described by Grey [586], by Moody [589], by Ring [593], Sabom [594], and Zaleski [595]. Greyson and Flynn [587] and Lundhahl [588] have compiled collections of research writings on the NDE. Flynn [585] and also Ring [592] deal with the transformative aftereffects of NDEs. It should be pointed out that NDEs as such are not relevant to parapsychology except as regards any light they may shed on the mind−body problem. However, to the extent that any form of psi is associated with NDEs or that they can be shown to be relevant to the survival problem, they become relevant.

Additional books with information on near-death experiences: LaBerge [351] has a chapter relating NDEs to lucid dreams. An article by R.K. Siegel on NDEs is reprinted in Frazier [456]. Hines [474] has a chapter critical of NDEs. Gallup [754] reports on surveys of NDErs. Significant treatments of NDEs are found in Grosso [759], Irwin [810], Kastenbaum [762], Osis and Haraldsson [769], and Wilson [777a], who devotes four chapters to NDEs and thinks they provide the best evidence for survival.

BIBLIOGRAPHY

585 **Flynn, Charles P.** *After the Beyond: Human Transformation and the Near−Death Experience.* Englewood Cliffs, NJ: Prentice−Hall, 1986. 190p. Bibliography: 182−183; Index: 185−190; 4 tables
This book is about the spiritual impact of near−death experiences (NDEs). It looks at what many NDErs have had to say about life, death, and love. In the first chapter Flynn

provides an overview of the book.   Chapter 2 examines the
NDEs of a number of persons.   Chapter 3 deals with the dif—
ficulties experienced by NDErs in getting people even to listen
to their stories, let alone accept them as valid.   Chapter 4 is
concerned with the difficulties NDErs have in incorporating their
experiences into their lives.   Chapter 5 examines some of the
value transformations that NDErs undergo and presents two in—
depth, extensive case studies of major changes in the lives of
two NDErs.   The sixth chapter covers research aimed at dis—
covering whether a brush with death without an NDE leads to
transformation.   Various religious aspects of NDEs are examined
in Chapter 7, which emphasizes how NDErs adopt a theocentric
attitude centered in a God of unconditional love.   The implica—
tions of NDEs for Christianity is the subject of Chapter 8,
which looks at the positive effects NDErs exercise on the lives
of nonexperiencers, to whom they direct the love of the Light
they were imbued with by their experience.   The 10th chapter
deals with the meaning of NDEs for those who have not had
any sort of transcendent or death—related experience and have
not been influenced by anyone who has.   Flynn presents some
of the more significant findings of an educational experiment——
the Love Project——which he conducted with several classes.   In
this project he attempted to provide his students with an oppor—
tunity to become more loving and caring after the manner of
NDErs but in a context not directly related to the NDE.   It is
his hope that "the NDE and efforts like the Love Project . . .
might provide a seedbed for large—scale transformation in a
world close to annihilation through forces opposite to those of
the love of the Light" (p. 8).   There is a Foreword by
Raymond Moody and an Afterword by Kenneth Ring.   The lat—
ter points out that Flynn goes beyond Ring's *Heading Toward
Omega* by saying we should not simply wait passively for the
coming of Omega but should "participate actively in its emer—
gence" (p. 163).   Materials for the Love Project are published
in the Appendices.

    JASPR 83:165—168 (Apr 1989); *PAI* 5:43 (Dec 1987).
    SciFicRev 15:59 (Aug 1986); *SkptInq* 12:201—202 (Win
1987/1988).

586 Grey, Margot. *Return from Death: An Exploration of the
      Near—Death Experience.* Foreword by Kenneth Ring.
      Boston, MA: Arkana, 1985.   206p.   Bibliograpy: 200—
      202; Index: 203—206; 7 tables
    Having had a near—death experience (NDE) while in India,
Margot Grey, upon her return to England, determined to learn
more about such experiences.   She began by conducting the first
British NDE survey, the results of which are reported in Part 1

of this book together with a summary of what has been learned thus far about the NDE. In the second part, she explores the after—effects of NDEs and their larger meaning. According to Kenneth Ring's Foreword, the findings reported in Grey's Part 2 independently duplicate the findings he reported in his book *Heading Toward Omega* [592]——findings that had not been revealed when she wrote her book. After a general chapter on after—effects, Grey follows with four chapters on the development of paranormal abilities following NDEs, healing manifestations, evolutionary aspects, explanations and interpretations, and reflections on some possible implications. There are two appendices, the first being her interview questions and the second providing information about the British Branch of the International Association for Near—Death Studies, which she founded.

JASPR 81:392—396 (Oct 1987); *JP* 51:176—180 (Jun 1987); *JSPR*; *PAI* 7:47 (Jun 1989).

587 Greyson, Bruce, and Flynn, Charles P. (Eds.). *The Near— Death Experience: Problems, Prospects, Perspectives.* Foreword by Michael B. Sabom. Springfield, IL: Charles C Thomas, 1984. 289p. Chapter references; 1 graph; 7 tables

This is "a sampling of the more significant scholarly attempts to understand the nature, possible explanations, and speculations regarding the meaning of the NDE" (p. x). There are five parts, each with a brief introduction by the editors. The first consists of an overview of near—death studies by Kenneth Ring. The second is on dimensions of the near—death experience and contains four selections: "The Subjective Response to Life—Threatening Danger," by Russell Noyes, Jr. and Donald J. Slymen; "Further Studies of the Near—Death Experience" and "Measuring the Near—Death Experience," both by Kenneth Ring; and "The Near—Death Experience Scale: Construction, Reliability, and Validity" by Bruce Greyson. Part III, on theories of NDEs, contains nine selections: "The Reality of Death Experiences: A Personal Perspective," by Ernst A. Rodin; "Commentary on 'The Reality of Death Experiences' by Ernst A. Rodin," by Michael B. Sabom; "The Psychology of Life After Death," by Ronald K. Siegel; "The Near—Death Experience: Balancing Siegel's View," by John C. Gibbs; "Pathophysiology of Stress—Induced Limbic Lobe Dysfunction: A Hypothesis Relevant to Near—Death Experiences," by Daniel B. Carr; "The Amniotic Universe," by Carl Sagan; "Why Birth Models Cannot Explain Near—Death Phenomena," by Carl B. Becker; "The Psychodynamics of Near—Death Experiences," by Bruce Greyson; and "Jung, Parapsychology, and the Near— Death Experience: Toward a Transpersonal Paradigm," by

Michael Grosso. Part IV is on clinical aspects of the NDE and contains four selections: "Near–Death Experiences: Dilemma for the Clinician," by Raymond A. Moody, Jr.; "Near–Death Events and Critical Care Nursing," by Annalee R. Oakes; "The Lazarus Syndrome: A Care Plan for the Unique Needs of Those Who've 'Died'," by Anthony Lee; and "Clinical Interventions with Near–Death Experiencers," by Kimberly Clark. Consequences of the NDE is the subject of the final part, and it is represented by "Near–Death Experiences and Attempted Suicide," by Bruce Greyson; "The Human Experience of Death or, What Can We Learn from Near–Death Experiences?," by Russell Noyes, Jr.; and "Meanings and Implications of Near–Death Experiencer Transformations," by Charles P. Flynn. The selections are taken from a variety of sources: *Anabiosis, Journal of Nervous and Mental Disease,* and a surprising variety of other journals such as *American Psychologist, Atlantic, Omega, RN, Suicide and Life–Threatening Behavior, Theta, Topics in Clinical Nursing,* and *Virginia Medical Journal.*

   JASPR 80:444–450 (Oct 1986); *JP* 49:265–270 (Sep 1985); *PAI* 3:40–41 (Jun 1985); *PsiR* 3:199–201 (Sep/Dec 1984).

   ContempPsychol 30:896 (Nov 1985); *ContempSociol* 14:452 (Jul 1985); *JPsychol&Theol* 13:224 (Fal 1985); *VitalSigns* 4:12–13 (Win 1984/85).

588 Lundahl, Craig R. *A Collection of Near–Death Research Readings.* Chicago, IL: Nelson–Hall, 1982. 240p. Chapter references; 1 graph; 11 tables

   This is a collection of essays dealing with scientific explorations of the subjective experiences of persons near to physical death. Lundahl, a sociologist, calls the scientific study of NDES "circumthanatology." The purpose of the book is to disseminate information on near–death research, to encourage further teaching and research in this area, and "to contribute to the reevaluation of our society's orientation toward death and the dying" (p. xii). Three criteria were used to select material for inclusion: that they deal with NDES or deathbed observations or have particular relevance for the scientific study of NDES; that they be recent; and that they be scientifically oriented. It is in five parts: (1) Science and Near–Death Experiences (Widdison); (2) The History of Near–Death Experiences (Audette); (3) Recent Research on Near–Death Experiences (Macmillan and Brown; Noyes and Kletti; Osis and Haraldsson; Moody; Ring; Sabom and Kreutziger; Garfield; Lundahl; Ring and Franklin); (4) Explanations for NDEs (Grosso); and (5) Directions in Near–Death Research (Lundahl). Five of the 13 selections were written specifically for this volume (Widdison, Audette, Ring, Sabom and Kreutziger, and Grosso).

JASPR 77:356–361 (Oct 1983); *JNDS* 3:107–111 (Jun 1983); *JP* 46:378 (Dec 1982); *JSPR* 52:215–217 (Oct 1983). *Choice* 20:1063 (Mar 1983); *SpirFron* 16:60 (Win 1984); *VitalSigns* 2:16 (Dec 1982).

589 Moody, Raymond A., Jr. *Life After Life? The Investigation of a Phenomenon: Survival of Bodily Death.* Introduction by E. Kübler–Ross. Harrisburg, PA: Stackpole Books, 1976 (Originally published by Mockingbird Books in 1975). 126p. Bibliography: 126

This is a study of near–death experiences which, as described by Moody, have the same characteristics as out–of–body experiences. The author, who holds a doctorate in philosophy and who has taught philosophy, is now training for an M.D. degree, in hopes of becoming a psychiatrist. Although the phenomena which this book treats have been investigated by parapsychologists for a century, Moody is apparently not aware of this vast literature or chose not to cite it. Yet his book may well serve two useful purposes. First, because he is not as–sociated with parapsychology he may attract readers who would never read a book on parapsychology, thus acquainting them with topics of parapsychological interest. And second, in a sense he has made independent observations on the nature of out–of–body experiences and deathbed experiences similar to those of several parapsychologists. Although he does not offer the cases described as evidence of survival, he suggests they form an important facet of human experience, and points out: "If experiences of the type . . . I have discussed are real, they have very profound implications for what every one of us is doing with his life. For, then it would be true that we cannot fully understand this life until we catch a glimpse of what lies beyond it" (p. 125).

ChrP 3:76–77 (Mar 1979); *JASPR* 70:316–320 (Jul 1976); *JP* 40:166–167 (Jun 1976); *PR* 7:18–20 (Jul/Aug 1976); *T* 4:14–15 (Fal 1976).

AngTheolRev 60:259–277 (Jul 1978); *Brain/MindBul* 1:4 (Apr 5, 1976); *ChristCent* 94:487 (May 18, 1977); *Choice* 16:188 (Apr 1979); *Commonweal* 104:473 (Jul 22, 1977); *ContempPsychol* 22:213 (Mar 1977); *Interpretation* 32:93 (Jan 1978); *Kliatt* 11:22 (Win 1977); *Light* 99:45–46 (Spr 1979); *NatlRev* 29:1004 (Sep 2, 1977).

590 Moody, Raymond A., Jr. *Reflections on Life After Life.* New York: Bantam/Mockingbird, 1977. 148p. Bibliography: 147–148; Chapter references

This book is an extension of the author's *Life After Life* and is intended to be read in conjunction with it. In the earlier book, Moody described 15 common elements found in near—death experiences; four new ones are added in this book. In ad—dition he gives historical examples, deals with religious aspects of the phenomenon, and provides answers to questions asked about NDEs. There is an important appendix on the methodol—ogy involved in studying such cases.

BestSell 37:285 (Dec 1977); *Choice* 15:1034 (Oct 1978); *LibrJ* 102: 1658 (Aug 1977); *PublWkly* 211:77 (Apr 11, 1977).

591 **Moody, Raymond A., Jr., with Perry, Paul.** *The Light Beyond.* Foreword by Andrew Greeley. New York: Ban—tam Books, 1988. 161p. Annotated bibliography: 157—161

Raymond Moody, who as a philosophy student, coined the term "near—death experience" (NDE), ten years later and as an M.D. provides a highly readable popular introduction to the find—ings of NDE researchers. He describes the characteristics of NDEs, devotes a chapter to the transformative power of NDEs, and provides another chapter on the NDEs of children. Chap—ter 4 is about "Why Near—Death Experiences Intrigue Us." The fifth chapter is "Why the NDE Isn't Mental Illness," and the sixth introduces the major NDE researchers to the reader, followed by first—person accounts of each person's interest in NDEs. The researchers included are: Melvin Morse, M.D.; Michael Sabom, M.D.; Michael Grosso, Ph.D.; Kenneth Ring, Ph.D.; and Robert Sullivan. In the seventh chapter, "Explanations," Moody discusses why he considers NDEs to be spiritual experiences and then describes the many attempts to explain NDEs, including the birth experience hypothesis, carbon dioxide overload, hallucination, fantasy, wish fulfillment, and the collective unconscious. In the "Conclusion," he expresses his conviction that NDEs provide evidence of life after life. The bibliography, with paragraph—length annotations, consists of research reports that helped him to form his "knowledge and opinion on the subject of near—death experience" (p. 157).

Booklist 84:1758 (Jul 1988); *Fate* 42:102—103 (Jan 1989); *Kirkus* 56:810 (Jun 1, 1988); *PublWkly* 233:100 (Jun 24, 1988).

592 **Ring, Kenneth.** *Heading Toward Omega: In Search of the Meaning of the Near—Death Experience.* New York: Wil—liam Morrow, 1984. 348p. Bibliography: 335—338; Chap—ter notes; 23 graphs; Index: 339—348; 7 questionnaires; 6 tables

Ring's *Heading Toward Omega* is a provocative work. In it he explores the transformative effects of NDEs and discusses their implications. Based on three years of research using ques—tionnaires and interviews, Ring suggests that those who undergo these transformations may play an important role in taking the next step in human evolution, with planet—wide implications. He hopes that the deeper spirituality of the transformed NDErs will promote peace on earth, especially if their increasing num— ber approaches the critical mass required for significant species— wide change. (An estimated 8 million Americans have thus far reported having had a near—death experience.) The chapter titles are as follows: (1) "NDEs on the Road to Omega," (2) "NDEs: What We Know," (3) "Core NDEs and Spiritual Awakening," (4) "NDEs and Personal Transformation," (5) "Value Changes Induced by NDEs," (6) "Religious and Spiritual Orientations Following NDEs," (7) "NDEs and Psychic Development," (8) "Planetary Visions of Near—Death Experiencers," (9) "Biological Basis of NDEs," (10) "NDEs and Human Evolution."

JASPR 81:67—73 (Jan 1987); *PAI* 4:44 (Dec 1986).

Booklist 81:8 (Sep 1, 1984); *Choice* 22:1078 (Mar 1985); *ContempPsychol* 31:645 (Sep 1986); *JTranspersPsychol* 16:245— 246 (1984); *Kirkus* 52:574 (Jun 15, 1984); *LATimesBkRev* p11 (Dec 9, 1984); *LibrJ* 109:1455 (Aug 1984); *Omega* 16(2): 177— 180 (1985/86); *PublWkly* 225:92 (Jun 22, 1984); *SpirFron* 17:114—116 (Spr 1985).

593 Ring, Kenneth. *Life at Death: A Scientific Investigation of the Near—Death Experience.* New York: Coward, McCann & Geoghegan, 1980. Introduction by Raymond A. Moody, Jr. 310p. Bibliography: 297—300; Chapter notes; 2 graphs; Index: 302—310; 32 tables

Psychologist Kenneth Ring describes the characteristics common to the near—death experiences (NDES) reported to him in interviews with 102 people using a structured interview schedule. He describes five stages of the core near—death ex— perience, the qualitative aspects of NDES, and the effect of the manner of death on the NDE. Two chapters are devoted to the aftereffects of NDES. The last three are entitled "Some Possible Interpretations of the Near—Death Experience," "Beyond the Body: A Parapsychological—Holographic Explanation of the Near—Death Experience," and "Implications and Applications of NDE Research."

JASPR 75:172—176 (Apr 1981); *JNDS* 2:9 (Nov 1980); *JP* 45:350—352 (Dec 1981); *JRPR* 3:235—236 (Jul 1980); *T* 8:22—24 (Fal 1980).

BestSell 40:253 (Oct 1980); *Choice* 18:588 (Dec 1980); *Con— tempPsychol* 26:225 (1981); *JTranspersPsychol* 13:70—72 (1981); *Kirkus* 48:897 (Jul 1, 1980); *LibrJ* 105:1870 (Sep 15, 1980); *NYTimesBkRev* p16 (Sep 28, 1980); *NYTimesBkRev* 87:31 (Aug 22, 1982); *Omega* 12:79—83 (1981—82); *PublWkly* 217:77 (Jun

20, 1980); *Quadrant* 14:120 (Fal 1981); *ReligStudRev* 7:139 (Apr 1981); *ReVision* 3:114—115 (Fal 1980); *SatRev* 7:61 (Jul 1980); *SciBooks&Films* 17:3 (Sep 1981); *SchLibrJ* 27:172 (Oct 1980); *TheolToday* 39:112 (Apr 1982).

594 Sabom, Michael B. *Recollections of Death: A Medical Investigation.* New York: Harper & Row, 1982. 224p. Chapter notes; 3 illustrations; Index: 219—224; 15 tables

This volume describes Michael Sabom's pioneer attempt to investigate the near—death experiences of persons suffering from cardiac arrest or other life—threatening crises. He describes the background of his research, which was an attempt to examine more carefully and repeat in as controlled an environment as possible the findings reported by Raymond Moody [589]. He hoped to answer with his research several questions raised by reading Moody. He analyzes the results of his research and describes it in the book. He distinguishes between the autoscopic NDE and the transcendental NDE. There are chapters on the "after experiences" of NDE survivors, implications of the NDE, explanations, and a final chapter in which he presents his thoughts on the meaning of the NDE. Fifteen tables setting forth his results are presented in an Appendix.

JNDS 1:172—176 (Dec 1981); *T* 11:67—68 (Aut 1983).

BestSell 41:475 (Mar 1982); *Booklist* 78:625 (Jan 15, 1982); *Brain/MindBul* 7:4 (Jul 12, 1982); *Choice* 20: 186 (Sep 1982); *LibrJ* 107:264 (Feb 1, 1982); *NatlRev* 35:1568 (Dec 9, 1983); *NewReal* 4:41—42 (Jul 1982); *Omega* 16(2): 177—180 (1985/86); *Parabola* 12:107—111 (May 1987); *ReVision* 5:108 (Spr 1982); *SpirFron* 15:48—49 (Win 1983); *TimesLitSuppl* p. 360 (Mar 26, 1982); *VitalSigns* 1:8 (Dec 1981).

595 Zaleski, Carol. *Otherworld Journeys: Accounts of Near—Death Experience in Medieval and Modern Times.* New York: Oxford University Press, 1987. 275p. Bibliography: 248—266; Chapter notes; Index: 267—275

This work began as a doctoral dissertation in the study of religion at Harvard University. It is concerned with near—death experiences, which Zaleski views as a form of otherworld journey tales. Her purpose is to examine accounts of NDEs in two widely separated settings: medieval Christendom and modern society, in the hope that "comparative study will highlight features that are not otherwise obvious, putting into sharper relief the elements that are culturally specific and at the same time drawing attention to perennial aspects of otherworld journey narration. It will disclose some of the ways in which the other—world journey narrative is shaped by the social and historical situation in which it occurs" (p. 6). The book itself is in four parts, the first of which provides an overview of the other—world journey as it appears in the world's religious traditions. Emphasis is placed in Part 2 on the medieval return—from—death story, which it considers within its cultural context. The third part discusses the modern near—death narrative and compares it with the medieval vision. Part 4 offers an interpretation of NDEs, including verification, explanations, counterexplanations, and an evaluation of near—death testimony. Zaleski concludes that "within the limits here discussed we are able to grant the validity of near—death testimony as one way in which the religious imagination mediates the search for ultimate truth" (p. 205). A "Chronological Table for Medieval Christian Other—world Journey Literature" is given in an appendix.

ChrP 7:215—216 (Jun 1988); *JASPR* 83:168—173 (Apr 1989); *JNDS* 6:258—263 (Sum 1988); *PAI* 6:47—48 (Jun 1988).

AtlMon 260:96 (Jul 1987); *Boston* 79:122—125 (Oct 1987); *Choice* 25:332 (Oct 1987); *Humanist* (U.S.) 48:42 (Mar 1988); *JSciStudRelig* 26:566—567 (Dec 1987); *JSciStudRelig* 27:293—294 (Jun 1988); *LibrJ* 112:84 (Mar 15, 1987); *Light* 107:125—128 (Win 1987); *NYTimes* 136:25 (Apr 7, 1987); *NYTimesBkRev* 136:25 (Apr 7, 1987); *Parabola* 12:107 (May 1987); *SanFran—RevBks* 12(4):29 (1988); *TheolToday* 44:525 (Jan 1988); *VaQRev* 63:140 (Aut 1987); *Vogue* 177:312—314 (May 1987).

Out−of−Body Experiences

INTRODUCTION

Since *Sources* was written the accepted term for what in that volume were out−of−the−body experiences (OOBEs) has become out−of−body experiences (OBEs). The number of books in this category in *Sources* was seven; this chapter covers nine newer books. As popular overviews and collections of cases, the older books are still of interest.

This section has some popular overviews: those by Black [596], Greenhouse [599], Mitchell [601], and Rogo [604]. Another book by Rogo [603] is a practical do−it−yourself guide; Monroe [602] also deals to some extent with technique, but primarily it consists of the "teachings" he received while out of his body. The remaining books−−by Blackmore [597], Gabbard and Twem−low [598], and Irwin [600] concern psychological approaches to the OBE. Blackmore offers a "psychological explanation," Gab−bard and Twemlow view OBEs as a specific altered state, and Irwin sees the OBE as an imagined experience.

Additional books with information on out-of-body ex-periences: LaBerge [351] theorizes that OBEs are lucid dreams. Parker [356] views OBEs as altered states. Grof [348] discusses OBEs. Alvarado reviews research on OBEs 1980−1984 in Shapin and Coly [817]. Two books aimed at young people have chapters on OBEs [415, 416] Cohen has a chapter on PK in two books [422, 441]. Booth [458] has a chapter on OBEs, as does Hines [474]. All of the books in the section on NDEs [585-595] are relevant to the subject of OBEs, because the ex−perience of being out of the body is integral to most NDEs. The book by Poortman [612] provides a thorough review of the history of the idea of a "subtle body" and is a monumental peice of scholarship. Irwin brings together the results of several surveys of OBEs, and he emphasizes the psychology of OBEs in his discussion.

BIBLIOGRAPHY

596 Black, David. *Ekstasy: Out—of—the—Body Experiences.* New York: Bobbs—Merrill, 1975. 243p. Bibliography: 215—236; Index: 237—243

An out—of—body experience is one in which a person's "I" views his/her body as if it were separate from and outside it. Black, a free—lance writer, though skeptical, became intrigued by accounts of such experiences which he kept running into and which he finally decided to investigate. This book is the result of his excursions into the literature of the subject as well as interviews with persons who feel they have experienced out—of—body experiences (OBEs), and with sensitives, such as Blue (now Keith) Harary and Robert Monroe, who feel they can control their OBEs. Black also conducted interviews with the parapsychologists and others chiefly engaged in OBE research (Lilly, Morris, Noyes, Osis, Roll, Tart). He also attempts to deal with the theories that have been offered to account for OBEs. A valuable 22—page bibliography lists nearly all of the books and articles of relevance to the study of OBEs, both technical and popular.

P 6:55 (Nov/Dec 1975); *PR* 6:12—14 (Nov/Dec 1975); *T* 4:10 (Spr 1976).
LibrJ 100:2331 (Dec 15, 1975).

597 Blackmore, Susan J. *Beyond the Body: An Investigation of Out—of—the—Body Experiences.* London, England: Heinemann, 1982. 271p. Bibliography: 253—264; 28 illustrations; Index: 265—271; 2 tables

This is a detailed and probing survey of out—of—body experiences and related questions. Psychologist Blackmore devotes a chapter to defining the OBE. Several chapters are about OBE explorers such as S. Muldoon, O. Fox, Yram, J.H.M. Whiteman, R. Monroe, etc. She also deals with collections of OBEs, the OBE in other cultures, and surveys of OBEs. Methods of inducing OBEs are described, as are subjects related to OBEs such as lucid dreams, deathbed visions, near—death experiences, the double in psychopathology, imagery and hallucinations, extrasensory perception in OBEs, OB experiments, and attempts to detect the soul. In conclusion, Blackmore reassesses theories of OBEs and offers her own "psychological approach to the OBE." This book is especially important because of the questions it asks and the way in which it sets about answering them.

ChrP 4:262—263 (Dec 1982); *IJParaphys* 16:137 (1982); *JNDS* 4:97—104 (Spr 1984); *JASPR* 78:75—81 (Jan 1984); *JP* 47:260—263 (S 1983); *PR* 15:13—14 (Nov/Dec 1984).

BritBkNews p601 (Oct 1982); *Choice* 21:1380 (May 1984);
ComGrnd 6:21 (1982); *LibrJ* 109:1135 (Jun 1, 1984); *Listener*
117:670 (Mar 26, 1987); *LondRevBks* 4:10 (Jul 1982); *NewStates—
man* 104:20 (Aug 6, 1982); *SciBooks&Films* 23:6 (Sep 1987);
SkptInq 8:74—77 (Fal 1983).

598 Gabbard, Glen O., and Twemlow, Stuart W. *With the
Eyes of the Mind: An Empirical Analysis of Out—of—body
States.* Foreword by Stephen Appelbaum. New York:
Praeger, 1984. 272p. Chapter notes; 6 figures; Name in-
dex: 263—266; Subject index: 267—272; 23 tables

The authors of this book attempt to "delineate both
descriptive and psychological aspects of the out—of—body experience" and to "differentiate the prototype from other forms
of altered mind/body perception" (p. x). They view these states
as being on "a continuum ranging from integrating and noetic
experiences to highly pathological disturbances of body boundaries" (p. x). Portions of seven of the eleven chapters are taken
from previously published articles of the authors (sometimes
joined by F. C. Jones). The book is in four parts: "The Outof—Body Experience" (two chapters), "Differentiation of the
Out—of—Body States" (four chapters), "The Near—Death Experience" (three chapters), and "Understanding the Out—of—
Body Experience" (four chapters).

   JASPR 81:299—303 (Jul 1987); *JNDS* 6:185—198 (Spr
1988).
   Choice 22:1410 (May 1985); *ContempPsychol* 31:266 (Apr
1986); *Omega* 16:177—180 (1985/86); *SpirFron* 28:111—114 (Spr
1986).

599 Greenhouse, Herbert B. *The Astral Journey.* Garden City,
NY: Doubleday, 1975. 359p. Bibliography: 341—347; Index: 348—359

Although somewhat loosely organized, this book is the
most complete survey of out—of—body experiences published to
date. Herbert Greenhouse is a popular writer who has authored
other books on parapsychology. In this book he presents an historical survey of OBEs, a review of the kinds of evidence for
OBEs, including recent laboratory experiments, and the psychology of the out—of—body experience. Its strength is not so much
in its depth as in its breadth of coverage. Each example—and
they are numerous—was selected to illustrate a particular
aspect of the OBE experience that *in toto* provide a taxonomy
of the myriad forms OBEs can take. Although most of the
cases have been published previously, Greenhouse includes some
new material that he uncovered as well as two experiences of

his own.

> *JSPR* 48:175—177 (Sep 1975); *P* 6:46 (Sep/Oct 1975); *PR*
> 6:12—14 (Nov/Dec 1975).
> *LibrJ* 100:590 (Mar 15, 1975); *PublWkly* 207:113 (Feb 24,
> 1975).

600 Irwin, H.J. *Flight of Mind: A Psychological Study of the
Out—of—Body Experience.* Metuchen, NJ: Scarecrow Press,
1985. 374p. Bibl: 325—354; Author index: 355—361; Subject index: 362—374; 2 tables

The author, a psychologist, writes that the main objective
of this book is to encourage and facilitate research by providing
a comprehensive survey of the empirical literature and by examining the aspects of OBEs amenable to scientific research.
Irwin systematically reviews the phenomenology of OBEs, the circumstances under which OBEs occur, the characteristics of persons who have OBEs, methods of OBE research, and the principal theories offered to explain OBEs. The book also serves as
a report of Irwin's empirical studies of the OBE, several published here for the first time. Irwin interprets the OBE as essentially an imaginal experience. Thus, the theory presented is
primarily psychological. However, it provides for the integral
operation of psi processes in the OBE as well. The book is
aimed at psychologists, as well as parapsychologists, in line with
the author's view that "parapsychological phenomena form a
legitimate and fruitful subject of study within their natural context of behavioral science" (p. viii).

> *JASPR* 81:303—308 (Jul 1987); *JNDS* 6:61—66 (Fal 1987);
> *JSPR* 54:69—72 (Jan 1987); *PAI* 4:67—68 (Jun 1986); *PR*
> 17:12—15 (Mar/Apr 1986); *PsiR* 5:235 (Mar/Jun 1986); *T*
> 15:90—91 (1987).
> *Choice* 23:667 (Dec 1985); *ContempPsychol* 31:645 (Sep
> 1986); *LucidLtr* 5:50—52 (Dec 1986).

601 Mitchell, Janet Lee. *Out—of—Body Experiences: A Handbook.*
Foreword by Gertrude Schmeidler. Jefferson, NC: McFarland, 1981. 128p. Bibliography: 115—124; Chapter
references; Index: 125—128

This introduction to out—of—body experiences is of value
to newcomers and seasoned researchers alike. It not only summarizes what we know about OBEs, it considers the various interpretations of them and their implications. There is a chapter
about star subject Ingo Swann——his experiences and experiments with him. She reviews the circumstances under which
OBEs occur and provides a brief history. Laboratory evidence
is summarized. Next she considers dreams and OBEs and
phenomena similar to OBEs such as various depersonalization

and possession states.   She closes with a chapter on theories
and one on ethical and social factors.   A useful feature is a
20−page section, "Questions and Comments," in which answers
are provided to the most commonly asked questions about
OBEs.

  ChrP 6:152−153 (Dec 1985); *JASPR* 76:186−188 (Apr
1982); *JP* 46:65−69 (Mar 1982); *JSPR* 51:387−389 (Oct 1982);
PR 13:22−24 (Sep/Oct 1982); *T* 11:68−69 (Aut 1983).

  Choice 19:995 (Mar 1982); *ContempPsychol* 4:232 (S 1982);
LibrJ 107:99 (Jan 1, 1982).

602 Monroe, Robert A. *Far Journeys.* Garden City, NY:
  Doubleday, 1985. 296p. 4 tables

  Monroe has written a sequel to his oft−cited *Journeys Out
of the Body* [**119**].   It is a popular account of his out−of−body
explorations and of his attempts to develop a technique for in−
ducing relaxation and sleep using audio pulses and a similar
method to synchronize the brain hemispheres ("Hem−sync").
This became known as the Gateway program because it
provided the person with access to another state of conscious−
ness.   This led to the Explorer program, in which participants,
when in the OBE state, explored beyond the solar system, some−
times meeting and communicating with intelligent entities.   The
major part of the book consists of transcripts of these explora−
tions.   The result is a modern analogue of what once was known
as "spirit teachings."   Its primary interest lies in that it is writ−
ten by Monroe.

  JP 50: 166−168 (Jun 1986);
  Kirkus 53:1068 (Oct 1, 1985); *LucidLtr* 5:53−54 (Dec
1986); *SpirFron* 18:243−244 (Fal 1986).

603 Rogo, D. Scott. *Leaving the Body: A Practical Guide to
  Astral Projection.* Englewood Cliffs, NJ: Prentice−Hall,
  1983. 190p. Bibliography: 185; Chapter references; 1
  figure; Index: 186−190

  The purpose of Rogo's *Leaving the Body* is to provide "a
complete yet inexpensive guide to the *genuine* procedures" (p.
xii) for inducing astral projection or OBEs.   It is based on his
extensive reading, contact with psychics who were familiar with
projection (such as Keith Harary and Ingo Swann), and his own
personal experiences.   Several techniques are presented: dynamic
concentration; progressive muscular relaxation; dietary control;
breathing, yoga, and mantra; the Monroe techniques; visualiza−
tion; dream control; and guided imagery.   The final chapter
briefly summarizes theories of the OBE and discusses the role
the OBE plays in self−transformation and related after−effects.

  JSPR 52:316−318 (Jun 1984); *PAI* 4:44 (Dec 1986); *PsiR*

2:122—124 (S 1983); *T* 15:88—89 (1987).
VitalSigns 3:19 (S 1983).

604 Rogo, D. Scott (Ed.). *Mind Beyond the Body: The Mystery
of ESP Projection.* New York: Penguin Books, 1978.
365p. Chapter references; 3 figures; 1 illustration; 1 table
Anthology of papers on OBEs that the editor feels "offer
the best experimental approaches . . . or that constitute insight—
ful or provocative analyses of the OBE or theories about it" (p.
11). Fifteen selections are presented in four sections: "On being
Out—of—Body"; "Laboratory Investigations"; "Reports from
Gifted Subjects"; and "Can We Explain the Out—of—Body
Experience?" Rogo provides introductory essays for each section
and a useful concluding essay in which he presents his own
reflections and ideas.
    JIndPsychol 2:152—153 (Jul 1979); *JP* 43:243—252 (S
1979).
    Choice 15:953 (Sep 1978); *Kirkus* 46:228 (Feb 15, 1978);
LibrJ 103:881 (Apr 15, 1978); *SpirFron* 12:188 (Aut 1980).

Philosophy and Parapsychology

INTRODUCTION

There were nine books on philosophy and psi in *Sources*. All are still of some interest, but especially Aristotelean Society proceedings [123], Broad [125], and Smythies [131].

There are 11 books in this chapter. All of the authors or editors, except for the Shapin and Coly [613] book, are professional philosophers (as are most of the contributors in Shapin and Coly). Most of the volumes are anthologies of (primarily) previously published articles: Flew [608], French [609], Grim [610], Ludwig [611], Thakur [614], and Wheatley and Edge [615]. Shapin and Coly are editors of a Parapsychology Foundation conference proceedings. The exceptions are Braude [605, 606], Brier [607], and Poortman [612].

Additional books with information on philosophy and parapsychology: Holroyd [349] deals with philosophy and psi. Citations to philosophical implications of psi are listed in White [378]. Several books in the history section are about spiritualism, psychical research, and/or parapsychology against the backdrop of the history of ideas. See Brandon [549], Cerullo [550], Evans [552], Inglis [555, 556], Moore [559], and Oppenheim [560]. In the section on survival a number of philosophers deal with philosophical problems of survival: Grosso [759], Hick [760], Lewis [763], Lorimer [764], Lund [765], Moore [768], and several of the authors in Spong [774].

BIBLIOGRAPHY

605 **Braude, Stephen E.** *ESP and Psychokinesis: A Philosophical Examination.* Philadelphia, PA: Temple University Press, 1979. 283p. Bibliography: 265–277; Index: 279–283

Philosopher Braude reviews the best evidence for psi phenomena as indicated by experiments conducted since 1965. He also reviews the theoretical foundations of parapsychology. He intends the book to serve as a source book on experimental parapsychology for philosophers——a sequel to C.D. Broad's *Lectures on Psychical Research* [154]——and as an exploration of "the conceptual foundations and some philosophical implications of parapsychological research" (p. xii). Part I consists of two sections on "Conceptual Foundations" and on the data. Part II is in three sections: "Psi and the Philosophy of Mind," "The Theory of Synchronicity," and "The Meaning of 'Paranormal.'"

JASPR 75:167—171 (Apr 1981); *JP* 45:242—248 (Sep 1981); *JSPR* 50:532—535 (Dec 1980); *PN* 3:2 (Oct 1980); *PR* 12:12—18 (Jan/Feb 1981); *T* 9:20—21 (Sum 1981).

Choice 18:262 (Oct 1980); *LibrJ* 105:1864 (Sep 15, 1980); *PhilosRev* 91:288—290 (Apr 1982).

606 Braude, Stephen E. *The Limits of Influence: Psychokinesis and the Philosophy of Science.* New York: Routledge & Kegan Paul, 1986. 311p. Bibliography: 290—304; Chapter notes; 7 figures; Index: 305—311

The first chapter of this courageous and exciting book about large—scale psychokinesis deals with the "importance of non—experimental evidence." In it Braude points out the limitations of experimental parapsychology. Also considered is the alleged improbability of psi. He deals at some length with the reliability of human testimony. In Chapter 2 he considers the evidence for physical mediumship, including in—depth treatments of D.D. Home, Eusapia Palladino, and, to a lesser extent, Rudi Schneider and Joseph of Copertino. Braude concludes that physical mediumship provides the best evidence for large—scale PK. Chapter 3 is about apparitions——collective apparitions in particular. He proposes that PK may offer the most plausible explanation for collective and reiterative cases. Another chapter is devoted to precognition as a form of PK. The book closes with a chapter on theories of PK in which it is suggested that efforts to theorize about PK have failed because they do not take into account the full range of PK phenomena. It is also proposed that PK may be occurring covertly in many areas of our lives unrecognized, both for ill and for good. The fact that PK is not often obvious may be due to resistance to assuming the responsibility required for its conscious use.

JASPR 82:153—160 (Apr 1988); *JP* 51:357—361 (Dec 1987); *JSPR* 54:72—75 (Jan 1987); *PAI* 5:42 (Dec 1987); *PR* 18:9—11 (Nov/Dec 1987).

Choice 24:900 (Feb 1987); *Light* 107:33—35 (Spr 1987); *Listener* 116:33 (Jul 24, 1986); *SciBooks&Films* 22:210 (Mar 1987); *SciTechBkn* 7:2 (Feb 1987).

607 Brier, Bob. *Precognition and the Philosophy of Science: An Essay on Backward Causation.* New York: Humanities Press, 1974. 105p. Bibliography: 102—103; Name index: 105; Subject index: 104

Brier, a professional philosopher as well as a parapsy—
chologist, points out in the Introduction that the data of
parapsychology raise important philosophical questions: "They
suggest that the way we have been conceiving our (physical)
universe may be wrong. Since there is a clash between the find—
ings of parapsychologists and the conceptions and theories of
philosophers, something has to give . . . Philosophers should see
what suggestions parapsychology offers as to what concepts
might be altered and in what way" (p. ix). In this book Brier
treats one such concept——that of cause——and how it "might be
altered in order to explain the phenomenon of precognition" (p.
ix). Brier attempts to show that it *is* possible for a cause to
occur after its effect. He analyzes philosophers who have dealt
with the possibility of backward causation, pro and con: Black,
Broad, Chisholm, Dray, Ducasse, Dummett, Flew, Gale, Mackie,
Mundle, Schlesinger, Scriven, Swinburne, Taylor. Brier's
treatment of this topic is advanced (it is based on his doctoral
dissertation), and considerable background in philosophy and
also physics is required to follow the complete argument.
    JP 39:74−76 (Mar 1975); *JASPR* 69:281−285 (Jul 1975);
JSPR 48:45−46 (Mar 1975).
    Choice 11:959 (Sep 1974); *PhilosRev* 86:124−125 (Jan
1977).

608 Flew, Antony (Ed.). *Readings in the Philosophical Problems
of Parapsychology.* Buffalo, NY: Prometheus Books, 1987.
376p. Bibliography: 363−369; Chapter references; Name
index: 371−376
    Philosopher Antony Flew, who earlier wrote a critical in—
troduction to parapsychology [43], has compiled an anthology of
materials basic to a philosophical discussion of parapsychology.
As he explains in his Introduction: "Whereas scientists in their
working hours always ask what is the case and why do things
happen as they do, the philosopher's question is, rather, grant—
ing that that was or is or will be so, 'So what?'" (p. 11). The
selections included were determined while developing readings for
a "philosophy of parapsychology" course at York University in
Canada. Part One, "What is Parapsychology?," contains six
selections, by J.B. Rhine, Irving Thalberg, C.D. Broad, Richard
Robinson, C.W.K. Mundle, and John W. Godbey, Jr. Part
Two, "Describing and Explaining," consists of another six pieces,
by Alan Gauld; Flew; Sidney Gendin, Clyde Hardin, and Robert
Morris; J.M.O. Wheatley; Galen K. Pletcher; and Patrick Grim.
There are four selections in "Part Three: Paranormal Precogni—
tion: Its Meaning; Its Implications." They are by C.D. Broad,
C.W.K. Mundle, Bob Brier, and Flew. Part Four, "Evidences
of the Impossible," consists of eight selections by David Hume;

F.H. Bradley; George Price; S.G. Soal; Paul Meehl and Michael Scriven; R.A. McConnell; J.B. Rhine; and Gardner Murphy. The fifth and final part, "The Question of Survival," consists of 12 pieces, by Richard Robinson and C.D. Broad; Plato; René Descartes; Thomas Hobbes; John Locke; Joseph Butler; C.J. Ducasse; C.D. Broad; E.R. Dodds; Peter Geach; Terence Penelhum; and Flew.

JSPR 54:219–221 (Jul 1987); PAI 6:43 (Jun 1988). Encounter 70:58 (Feb 1988); Fate 40:109–110 (Jul 1987); Humanist 47:42 (Jul 1987); SciTechBkn 11:1 (Apr 1987).

609 French, Peter A. (Ed.). *Philosophers in Wonderland: Philosophy and Psychical Research.* St. Paul, MN: Llewellyn Publications, 1975. 376p. Chapter references; 16 figures; 9 illustrations; 2 tables

In this anthology, Professor French, who teaches philosophy at the University of Minnesota at Morris, attempts "to set forth a series of critical examinations into the propositions and beliefs of psychical research insofar as they are related to serious philosophical questions . . . A primary consideration which affected the choice of selections . . . was each author's employment of the tools and techniques of philosophical analysis as applied to the beliefs, theories and episodic accounts which characterize the Wonderland world of psychical research . . . the prevailing attitude throughout is critical and more than a bit skeptical . . . This book's central theme is that understanding this world of the psychical researchers necessitates an understanding of the concepts used to describe such a world. This understanding is at base philosophical" (p. viii). The book consists of an introductory chapter by French and 25 selections arranged as eight symposia on the following subjects: dreams, space, time, mind and body, personal identity, psi phenomena and knowledge, survival, and the scientific acceptability of psi phenomena. French contributes an introduction to each symposium that relates the symposium topic to psychical research and points out its philosophical implications. Of the 25 selections, 12 are directly on parapsychology, three are tangentially relevant, and the remaining nine are solely on philosophy.

JP 39:341–343 (Dec 1975); PR 6:18–19 (Nov/Dec 1975).

610 Grim, Patrick (Ed.). *Philosophy of Science and the Occult.* Albany, NY: State University of New York Press, 1982. 336p. Chapter references; 8 figures; Index: 325–336; 4 tables

Philosopher Grim tries to bring together contemporary philosophy of science and traditional occult wisdom. He says this collection of 22 selections (divided equally between those

previously published and those written for this volume) can be
viewed in either of two ways: "as an introduction to philosophy
of science through an examination of the occult, or as a serious
examination of the occult rigorous enough to raise central issues
in philosophy of science" (p. 1). For pedagogical purposes, Grim
uses examples taken from astrology, parapsychology, UFOs, and
ancient astronauts "to make immediately obvious the lively ex—
citement and intellectual fascination of philosophy of science" (p.
2). He also examines these four subjects with the aim of allow—
ing proponents and critics to battle "with equally informed and
sophisticated efforts on each side . . . and without an editorially
imposed verdict" (p. 3). There are five sections dealing with,
respectively, astrology; science or pseudoscience; parapsychology;
UFOS and ancient astronauts; and other approaches to the oc—
cult. Each section has an introduction and a list of suggested
readings. The section on parapsychology consists of six chapters,
those by A. Flew and J.B. Rhine being reprints. However,
"Parapsychology: An Empirical Science," by R. Reinsel;
"Coincidence and Explanation," by G.K. Pletcher; "Philosophical
Difficulties with Paranormal Knowledge Claims," by J. Duran;
and "Precognition and the Paradoxes of Causality," by B. Brier
and M. Schmidt—Rughaven were written especially for this
volume.
　　　JP 47:77 (Mar 1983); *PAI* 4:71 (Jun 1986).
　　　Choice 20:1110 (Apr 1983); *Explorer* 1:6 (Jan 1983);
ReligStudRev 10:153 (Apr 1984).

611 Ludwig, Jan (Ed.). *Philosophy and Parapsychology.* Buffalo,
　　　NY: Prometheus Books, 1978. 454p. Bibliography:
　　　423—448; Chapter references; 2 figures
　　　The impetus for this anthology by Union College
philosopher Jan Ludwig is the fact that although most para—
psychologists believe that the existence of psi phenomena has
been scientifically verified, many philosophers and psychologists
are skeptical of these claims. This book brings together the
writings of 22 parapsychologists, psychologists, and philosophers
in debate concerning the existence of psi and various concep—
tions of its nature. All but one selection was previously pub—
lished. Ludwig provides an introductory essay, "Philosophy and
Parapsychology." Section I, "Parapsychology and Philosophy,"
has five chapters, by C.D. Broad (2); M. Kneale, R. Robinson,
and C.W.K. Mundle; J.B. Rhine; and C.J. Ducasse. Section II,
"The Argument From the Possibility of Fraud," consists of
seven selections, by G.R. Price, S.G. Soal, J.B. Rhine, P.E.
Meehl and M. Scriven, and P.W. Bridgman on the controversy
in *Science* over Price's article, "Science and the Supernatural."
The third section, "Conceptual Issues in Parapsychology," has

chapters by A. Flew, S.E. Braude, J.M.O. Wheatley, H.W. Baldwin, and G. Murphy. Section IV, "Precognition and Its Problems," has four chapters, by C.D. Broad, L. Foster, C.W.K. Mundle, and B. Brier. The fifth section, "Parapsychology and the Philosophy of Mind," has chapters by J. Beloff, H.H. Price, M. Scriven, and J.W. Godbey, Jr.. The last section, "Historical Postscript," consists of William James's "Final Impressions of a Psychical Researcher."

ChrP 3:142−143 (S 1979); *JASPR* 73:195−203 (Apr 1979); *JIndPsychol* 2:161−162 (Jul 1979); *JP* 42:319−321 (Dec 1978); *JSPR* 50:117−119 (Jun 1979); *PR* 9:13−15 (Nov/Dec 1978); *T* 8:11−12 (Win 1980).

Choice 15:1441 (Dec 1978); *PsycholRec* 31:112−113 (Win 1981); *PublWkly* 213:223 (May 22, 1978); *SkptInq* 3:63−65 (Sum 1979).

612 Poortman, J.J. *Vehicles of Consciousness: The Concept of Hylic Pluralism (Ochema).* Translated by N.D. Smith. Foreword by W.H.E. Tenhaeff. Utrecht, the Netherlands: The Theosophical Society of the Netherlands in association with the Theosophical Publishing House, 1978. 4 volumes. (Originally published in Dutch under the title *Ochema* in 1954.) Bibliographical footnotes (all volumes); Bibliography: 233−240 (Volume 4); 1 figure (Volume 1); 8 illustrations (Volume 1); Name index: 241−261 (Volume 4); Subject index: 262−308 (Volume 4)

This is considered to be the life work of Dutch philosopher and theosophist J.J. Poortman, but he did not live to complete it. He intended to write "a history of the development of the doctrine of the subtle body" (I, p. v). Ochema is Greek for "vehicle," and this work treats " a number of forms of subtler matter which serve as vehicles of an immaterial, ensouling con−sciousness" (I, p. vii). Poortman accomplishes two things in this work: in the first two volumes he reviews the literature of the subtle body in all times and places (even those portions that he did not live to complete are outlined in detail); and in the last two volumes he presents his theory of "Hylic Pluralism" and applies it to many aspects of psi phenomena.

613 Shapin, Betty, and Coly, Lisette (Eds.). *The Philosophy of Parapsychology.* New York: Parapsychology Foundation, 1977. 295p. Chapter references; 2 figures; 1 table

This is the proceedings of a conference held by the Parapsychology Foundation in Copenhagen in August, 1976. It contains 15 papers by 16 participants, together with the discus−sion that followed each. On the whole the contributions deal with theories of specific types of psi phenomena or general dis−

cussions of philosophy and parapsychology, and paradigms in parapsychology in particular. The participants and the titles of their papers are R.G. Stanford, "Are Parapsychologists Paradigmless in Psiland?"; P.A. French, "On the Possibility of a Causal Theory of Extrasensory Perceptual Knowledge"; J. Beloff, "Backward Causation"; L. LeShan, "Philosophy and Parapsychology: Impossible Bedfellows or the Marriage of the Future?"; H. Bender, "Meaningful Coincidences in the Light of Jung—Pauli's Theory of Synchronicity"; F.C. Dommeyer, "An Acausal Theory of Extrasensory Perception and Psychokinesis"; H.L. Edge, "The Place of Paradigms in Parapsychology"; R. Haynes, "Philosophy and the Unpredictable: A Retrospective"; J.F. Nicol, "Philosophers as Psychic Investigators"; T. Penelhum, "Survival and Identity: Some Recent Discussions"; S.C. Thakur, "Parapsychology in Search of a Paradigm"; S.H. Mauskopf and M.R. McVaugh, "Parapsychology and the American Psychologists: A Study of Scientific Ambivalence"; P. Janin, "Psychism and Chance"; R. Ejvegaard, "A Philosophical Analysis of Telepathy"; and E. Servadio, "Parapsychology and the 'Ultimate Reality.'"

ChrP 3:22—23 (Dec 1978); *JASPR* 72:183—186 (Apr 1978); *JIndPsychol* 2:168 (Jul 1979); *JP* 41:215—219 (Sep 1977); *JSPR* 49:835—837 (Jun 1978).

AustlJPhil 56:184—185 (Aug 1978); *SpirFron* 10:103—106 (Spr 1979).

614 Thakur, Shivesh C. (Ed.). *Philosophy and Psychical Research.* New York: Humanities Press, 1976. 215p.
Chapter references; 1 figure; Index: 211—215; 1 table
The aim of this collection is to portray the full extent of the interest in psychical research on the part of philosophers, whether skeptical or convinced as regards the reality of psi phenomena. The chapters are by philosophers who were asked to write on whatever aspect of philosophy as related to para—psychology interested them most. The authors and the titles of their papers are as follows: Alan Gauld, "ESP and Attempts to Explain It"; Bob Brier, "The Metaphysics of Precognition"; David E. Cooper, "ESP and the Materialist Theory of Mind"; Antony Flew, "The Sources of Serialism"; Jonathan Harrison, "Religion and Psychical Research"; Pamela M. Huby, "Some Aspects of the Problem of Survival"; Hywel D. Lewis, "Religion and the Paranormal"; C.W.K. Mundle, "On the 'Psychic' Powers of Nonhuman Animals"; Michael Scriven, "Explanations of the Supernatural"; and Shivesh C. Thakur, "Telepathy, Evolution and Dualism."

ChrP 3:14—15 (Dec 1978); *JASPR* 71:304—309 (Jul 1977); *JIndPsychol* 1:81—82 (Jan 1978); *JSPR* 48:336—340 (Jun 1976);

JP 41:62—68 (Mar 1977); *PR* 8:18—19 (Jul/Aug 1977).
ExposT 87:252—253 (May 1976); *Light* 96:90—92 (Sum 1976); *Mind* 87:460—462 (Jul 1978).

615 Wheatley, James M.O., and Edge, Hoyt L. (Eds.).
Philosophical Dimensions of Parapsychology. Springfield, IL: Charles C Thomas, 1976. 483p. Bibliography: 464—478; 1 figure; 2 graphs; Index: 479—483; 1 table

This is an anthology of 30 previously published articles on philosophy and parapsychology. Both editors are professional philosophers. Wheatley contributes an introductory overview of the book and Edge wrote the introductions to each section, of which there are five. The titles of the sections and names of the authors of the articles in each are as follows: I. "Psi and Philosophy" (Broad, Ducasse, Scriven, and Mundle); II. "Psi and Cognition" (H.H. Price, Thalberg, Swiggart, Wheatley, and Roll); III. "Psi and Precognition" (Ducasse [two pieces], Broad, and Brier); IV. "Psi and Survival" (Wheatley, H.H. Price, Ducasse, Nayak, Penelhum, Flew, and Broad); V. "Psi and Science" (Beloff, Murphy, Meehl and Scriven, Chauvin, Shewmaker and Berenda, LeShan, and Tart). The 15—page bibliography should be very useful.

JASPR 71:103—109 (Jan 1977); *JIndPsychol* 1:81—82 (Jan 1978); *JSPR* 49:458—459 (Mar 1977); *JP* 40:315—317 (Dec 1976); *PR* 8:12—14 (Jan/Feb 1977).

Choice 14:748 (Jul 1977); *ContempPsychol* 21:900 (Dec 1976); *NewReal* 1(2):28 (1977); *Personnel&GdJ* 56:390 (Mar 1978).

Physics and Psi

INTRODUCTION

This is a new category. In *Sources* there were only two books that dealt with physics and psi to any extent: Rhine and Pratt's textbook [58] and Koestler's *Roots of Coincidence* [165].

There are eight books in this section. Those edited by Jahn, Josephson and Ramachandran, Oteri, and Puharich are conference proceedings. Physics and consciousness and physics and psi are considered by Jahn and Dunne, LeShan, and LeShan and Margenau. Zohar uses contemporary physical models to make sense of precognition.

Additional books with information on physics and psi: Physics and psi is dealt with by Holroyd [349]. LeShan [352] brings physics into his discussion of the nature of reality, includ–ing the basic limiting principles of the transpsychic mode. Cita–tions to the implications of psi for physics are cited by White [378]. Rothman's [482] *A Physicist's Guide to Skepticism* treats parapsychology as a pseudoscience. Abell and Singer [456] reprint a critical article on quantum physics and parapsychology by Martin Gardner. Rothman [482] attempts to show how psi cannot exist applying the laws of physics. Three of the books in the history section emphasize the views of the nineteenth–century and early twentieth–century physicists on spiritualism and psychical research: Brandon [549], Cerullo [550], and Oppenheim [560]. *Mind at Large* [505] is a collection of experimental reports and theoretical articles on physics and psi. There are two papers on physics and psi in Schmeidler [571]. In the philosophy and parapsychology section, Brier [607] writes on backwards causation. Loye [785] draws on quantum physics to theorize about precognition. Dossey [796] draws on physics in his theory of medicine that includes unorthodox healing.

BIBLIOGRAPHY

616 Jahn, Robert G. (Ed.). *The Role of Consciousness in the Physical World.* Boulder, CO: Westview Press, 1981. 136p. Chapter references; 15 figures; 21 illustrations; 6

tables

This book is based on a symposium held at the 1979 National Meeting of the American Association for the Advancement of Science in Houston, sponsored by AAAS sections B (Physics) and L (History and Philosophy of Science). The purpose of the symposium was to discuss the manifestations and ramifications of consciousness in various fields such as physics, psychology, parapsychology, engineering, cosmology, information science, and technology policy. The symposium "attempts to define critical experiments and to propose models that explicitly acknowledge the role of consciousness in physical reality" (p. v). Contents include: "The Extension of the Area of Science," by Eugene P. Wigner; "Psychophysical Interaction," by Charles Honorton; "Experimental Psi Research: Implication for Physics," by Harold Puthoff, Russell Targ, and Edwin C. May; "Not Consciousness but the Distinction Between the Probe and the Probed as Central to the Elemental Quantum Act of Observation," by John Archibald Wheeler; and "Broader Implications of Recent Findings in Psychological and Psychic Research," by Willis W. Harman.

JP 46:377−378 (Dec 1982); *JRPR* 10:114−115 (Apr 1987); *PR* 13:18−21 (Jul/Aug 1982).

IdealisticStud 13:79−80 (Jan 1983).

617 Jahn, Robert G., and Dunne, Brenda J. *Margins of Reality: The Role of Consciousness in the Physical World.* New York: Harcourt Brace Jovanovich, 1987. 415p. Chapter notes; 44 figures; 80 graphs; 69 illustrations; Index: 401−415; Suggested reading list: 377−398; 8 tables

This work provides both an overview of the parapsychological research conducted by the Princeton Engineering Anomalies Research Laboratory (PEAR) and an attempt to understand the role consciousness may play in the establishment of reality. Much of the material presented has already been published in the Laboratory's research reports and summarized in papers given at conferences by various members of PEAR's staff, but this is the first time an overall survey has been made of their research methodology, experimental findings, and theoretical approaches, drawing them together in a unitary approach. The first section provides the historical conceptual backgrounds of the question and draws on both physics and parapsychology. Section 2 describes the research conducted at PEAR with emphasis on human−machine interaction. The third section reports on research into precognitive remote perception. The fourth presents a theoretical approach to consciousness that would make sense of their experimental results using a quantum

mechanical approach. In the last section they assess the progress made, the problems encountered, and the implications and possibilities of their research and their theoretical approach.
JASPR 82:359–365 (Oct 1988); *JP* 52:345–350 (Dec 1988).
AmDows 28:43–50 (Sum 1988); *Artifex* 7:33–40 (Spr 1988); *Fate* 42:101–102; *NewReal* 9:65–66 (Mar/Apr 1989); *PsycholPerspec* 19:335–339 (Fal/Win 1988).

618 Josephson, B.D., and Ramachandran, V.S. (Eds.). *Consciousness and the Physical World: Edited Proceedings of an Interdisciplinary Symposium on Consciousness Held at the University of Cambridge in January 1978.* Foreword by F.J. Dyson. Elmsford, NY: Pergamon Press, 1980. 203p. Chapter references; Name index: 199–200; Subject index: 201–203
    The stated purpose of this conference was "to make a scientific study of subjective experience and to explore the relationships between subjective experience and the objective world" (p. 19). There are 11 contributions, an introduction, and an afterword. Discussants' remarks are given after each paper. The contributors were G. Vesey, R.L. Gregory, H.C. Longuet–Higgins, N.K. Humphrey, H.B. Barlow, D.M. Mackay, B.D. Josephson, M. Roth, V.S. Ramachandran, S. Padfield, and M.J. Morgan. In the Foreword, F.J. Dyson labels the authors as "animists." He explains: "They are not willing to exclude *a priori* the possibility that mind and consciousness may have an equal status with matter and energy in the design of the universe. They are trying to extend the boundaries of scientific discourse so that the subjective concepts of personal identity and purpose may come within its scope" (p. vii). Psi is only touched on in a few papers, but one contribution is by psychic Suzanne Padfield: "Mind–Matter Interaction in the Psychokinetic Experience."
    Brain/MindBul 5:4 (Feb 18, 1980).

619 LeShan, Lawrence. *The Science of the Paranormal: The Last Frontier.* Wellingborough, Northamptonshire, England: Aquarian Press, 1987. (Originally published in 1984 as *From Newton to ESP.*) 208p. Chapter notes; Index: 206–208
    LeShan explores the reasons why parapsychology has made little progress in the past 50 years and what can be done about this lack of progress. He charges that parapsychology has failed to take into account the insights of the great physicists concerning the nature of reality, but instead has chosen the model of nineteenth–century physics. He draws on the writings of Eddington, Einstein, and Planck to show that we live in a multi–

tracked universe in which what is anomalous or paranormal on one track could be perfectly normal on another track. He points out that the physicists' multi—tracked universe is corroborated by the twentieth—century social scientists' concept of alternate realities. He presents a new model of science that allows for psi and illustrates it with two applications: a model research program and a successfully completed research project in teach—ing psychic healing.

    ChrP 6:67—68 (Jun 1985); *JASPR* 81:64—67 (Jan 1987); *JSPR* 55:293—295 (Jan 1989); *PAI* 4:43 (Dec 1986); *PR* 17:8—10 (Jan/Feb 1986).

    Light 105:39—40 (Spr 1985); *SkptInq* 10:175—179 (Win 1985/86); *SpirFron* 21:49—50 (Win 1989).

620 LeShan, Lawrence, and Margenau, Henry. *Einstein's Space and Van Gogh's Sky: Physical Reality and Beyond.* New York: Macmillan, 1982. 268p. Chapter notes; 10 figures; Index: 261—268; 2 tables

    Psychologist LeShan and physicist Margenau hold that in order to progress in science and to have data that is lawful, one set of principles about how reality works is insufficient. A num—ber of alternate realities must be allowed for. They specify five: sensory realm; microcosm, or the too small to be sensed even theoretically; the macrocosm, or the too large to be sensed even theoretically; meaningful units of behavior of living things; and human inner experience. Physics has led the way in devising al—ternate interpretations of the reality of different experiential realms, but they cannot be applied to inner experiences or mean—ingful behavior. "Here, constructions of reality are needed that make the data from *these* realms lawful, not constructions bor—rowed from other realms" (p. 21). Our organization of reality is what we observe of it and different organizations of it work best for different purposes. However, it must be organized law—fully, not whimsically. "It ('reality') is somehow there, but we alloy it into being with our consciousness" (p. 26). There is a chapter on parapsychology in which they discuss the attitudes toward psi and offer suggestions for research into need—determined psi as opposed to flaw—determined psi.

    BestSell 42:319 (Nov 1982); *Kirkus* 50:779 (Jul 1, 1982); *LibrJ* 107:1996 (Oct 15, 1982); *Ms* 14:81 (Dec 1985); *Nature* 300:385 (Nov 25, 1982); *NewAge* 8:65 (Oct 1982); *PublWkly* 222:37 (Jul 9, 1982); *PublWkly* 224:124 (Sep 16, 1983); *SciBooks&Films* 18:253 (May 1983); *Zygon* 21:125 (Mar 1986).

621 Oteri, Laura (Ed.). *Quantum Physics and Parapsychology: Proceedings of an International Conference Held in Geneva, Switzerland, August 26—27, 1974.* New York:

Parapsychology Foundation, 1975. 283p. Chapter
references; 18 figures; 2 graphs; 14 illustrations; 4 tables
    The papers in this Parapsychology Foundation conference
dealt primarily with quantum physics and parapsychology as the
title suggests, but there also were papers on both science and
physics in general and in relation to parapsychology. A paper
by Targ and Puthoff describes their experiments with remote
viewing. The other authors and titles are: "Foundations of
Paraphysical and Parapsychological Phenomena," by E.H.
Walker; "Precognition––A Memory of Things Future," by G.
Feinberg; "Parapsychology, Quantum Logic, and Information
Theory," by C.T.K. Chari; "Quantum Paradoxes and Aristotle's
Twofold Information Concept," by O. Costa de Beauregard;
"Life and Quantum Physics," by V.A. Firsoff; "Physics,
Entropy, and Psychokinesis," by Puthoff and Targ;
"Parapsychology as an Analytico–Deductive Science," by J.H.M.
Whiteman; "A Logically Consistent Model of a World with Psi
Interaction," by H. Schmidt; and "Connections Between Events
in the Context of the Combinatorial Model for a Quantum
Process," by T. Bastin. There is a comment on the papers by
C.T.K. Chari, and the proceedings closes with a roundtable dis-
cussion by all the participants.
    JASPR 70:309–312 (Jul 1976); *JP* 41:60–62 (Mar 1977);
JSPR 48:402–404 (Dec 1976); *T* 6:16–17 (1978).
    Encounter 54:90 (Jan 1980).

622 Puharich, Andrija (Ed.). *The Iceland Papers: Select Papers
    on Experimental and Theoretical Research on the Physics
    of Consciousness.* Amherst, WI: Essentia Research As-
    sociates, 1979. 191p. Chapter references; 47 figures; 13
    graphs; 16 illustrations; 7 tables
    This is a compilation of five of the papers that the editor,
a physician and inventor, deemed most important of those given
at a conference held in 1977 in Iceland under the sponsorship of
the Orb Foundation. Engineers Harold Puthoff and Russell Targ
present remote–viewing ESP data and physicist John Hasted
reports on his metal–bending PK experiments. Physicist
Elizabeth Rauscher presents an eight–dimensional model that
shows how information from the future or past can be obtained
as though it all existed in the now of the observer. Physicists
Richard Mattuck and Evan Harris Walker postulate that the
wave function of matter can be collapsed by willpower alone.
Finally, physicist Costa de Beauregard "argues that the data
and the theory of the physical sciences alone *demands* that ESP
and PK exist" (p. 13).
    AmDows 21:43–44 (Nov 1981); *JASPR* 77:254–261 (Jul
1983).

Specula 2:16—18 (Oct/Dec 1979).

623 Zohar, Danah. *Through the Time Barrier: A Study of Precognition and Modern Physics.* London: Heinemann, 1982. 178p. Bibliography: 168—172; 3 figures; 1 graph; 11 illustrations; Index: 173—178; 1 questionnaire; Suggested reading list: 172

Zohar, whose training is in physics, surveys the historical and contemporary evidence for precognition and examines whether it can make sense in terms of current physical models. In particular, she relates the implications of acausality in physics to that of precognition. Part I consists of four chapters dealing with the evidence for precognition. Part II contains three chapters on "Who has Precognition," and the last part consists of five theoretical chapters on physics and precognition. In the absence of repeatable precognition experiments, she concludes that there exists no hard and fast proof of precognition. However, twentieth—century physics has at least weakened the case against precognition, and physics may even help hypothesize how precognition could work, "but evidence that something is possible is still not evidence that it exists" (p. 165).

JASPR 77:254—261 (Jul 1983); *JSPR* 52:200—203 (Oct 1983); *PR* 15:10—12 (Jan/Feb 1984).

BritBkN p221 (Apr 1983); *Economist* 286:90 (Feb 12, 1983); *Light* 103:29—31 (Spr 1983).

Psychokinesis

INTRODUCTION

There were only three books devoted to PK in *Sources* [118-120], but all are still useful today.

There are 14 books in this section dealing with a variety of aspects of PK. Overviews of the subject are presented by Randall [630], Robinson [632], and Rogo [633, 634]. The books by Gauld and Cornell [625], Rogo [635], and Roll [636] are on poltergeists. PK associated with sitter groups is the concern of Owen and Sparrow [628] (the "Philip" group) and Richards [631] (the SORRAT group). The books by Ebon [624] and Panati [629] are about Uri Geller and those by Hasted [627] and Taylor [637] report on their experiments with mini–Gellers. Gettings [626] writes about spirit photography.

Additional books with information on psychokinesis: Fraudulent ways of producing Geller–type phenomena are described by Fulves (pseudonym) [465, 466]. Gardner has criticisms of PK in two books [467, 468]. Grof [348] theorizes about PK and RSPK experiences. Discussions of human–machine equipment interactions by Morris and PK research by May are in Jones [373]. A number of books in the section "Biographies and Autobiographies of Healers, Mediums, and Psychics" contain accounts of macro–PK: Geller [386, 387], Jenkin's book on D.D. Home [391], Dean Kraft [392], and Matthew Manning [393]. Several books in the section on books aimed at young people are devoted wholly or in part to PK or aspects of it. Feola [435] provides a nontechnical overview of PK. Aylesworth [418] is on metal–bending. There are chapters on PK in Cohen [422, 424], Curtis [429], Hall [436], Heaps [437] and L.E. Rhine [448]. Poltergeists are covered by chapters in Cohen [421, 424], Knight [443], Myers [446, 447], and Williams [456]. Levitation is considered by Aylesworth [418] and Edmonds [432]. Several spontaneous forms of PK with a teleological aspect are described by Inglis [809]. Booth [458] has a chapter in which he criticizes D.D. Home's famous levitation, and Hall's book [471] is also critical of it. Randi [480] is critical of Uri Geller, and points out how he could have produced his phenomena fraudulently. Taylor [483] is critical of metal–

bending and other PK phenomena. The PK research at the Princeton Engineering Anomalies Laboratory is described by Jahn and Dunne [617]. Subject Suzanne Padfield contributed a chapter in Josephson and Ramachandrar [618] entitled "Mind–Matter Interaction in the Psychokinetic Experience." Braude [605, 606] deals with the philosophical foundations of PK. In the former, he reviews the best evidence for PK; in the latter, he emphasizes macro–PK. For an extensive bibliography on poltergeists see Goss [666]. The investigation of hauntings and poltergeists comprises a large portion of Auerbach's book [723] and MacKenzie [733] studies several haunting cases in depth. Rogo and Bayless [771] survey telephones and other electronic instruments that appear to be associated with psi (presumably PK).

BIBLIOGRAPHY

624 Ebon, Martin (Ed.). *The Amazing Uri Geller.* New York: New American Library, 1975. 168p. 2 figures; 1 graph; 12 illustrations; 3 tables
    Journalist and editor Martin Ebon has put together a book of popular readings on Uri Geller that presents both pro and con views of the controversial Israeli while providing a psychological portrait of Geller. Several chapters present firsthand observation of his feats, three deal with his psychic photography, one chapter covers Geller imitators, two chapters cover the SRI experiments, and there is even a chapter by Geller on Geller. Ebon himself contributes a lengthy introduction on Geller and the man who brought him to the U.S. (Andrija Puharich), and a summary chapter entitled "Everybody has His Own Geller!," in which he likens Geller to a Rorschach test.
    P 6:30–31 (Nov/Dec 1975); *PS* 2:153 (F 1977).
    PublWkly 207:52 (Mar 31, 1975).

625 Gauld, Alan, and Cornell, A.D. *Poltergeists.* Boston, MA: Routledge & Kegan Paul, 1979. 406p. Case index: 399–401; Citations in text; 1 graph; 16 illustrations; Name index: 403–406; 3 tables
    The authors, two British parapsychologists, have aimed this book at the general reader but it should be very useful to parapsychologists as well. Part I is by Cornell and "is a survey of representative cases of poltergeists and hauntings" (p. ix). His intent was to select as many different kinds of well–authenticated cases as possible. Also included is a statistical analysis of 500 sample cases. Based on this analysis, a

proposed classification of the cases is presented. Both authors contributed to Part II, in which they describe their own case investigations and discuss their findings and theories. They consider alternative explanations that can account for some, but not all, of the cases. One chapter is devoted to some polter—geists they have studied and another to some haunts they in—vestigated. In the last two chapters they "discuss some of the main theories of the nature and causes of poltergeist phenomena" (p. 319). The last chapter is devoted to the ques—tion of whether poltergeists are living or dead. An Appendix to Part I consists of a "Chronological List of 500 Cases, with Sources and Case Characteristics." The list is to make it pos—sible for anyone interested in a specific case to locate the original printed sources and to provide data for those interested in carrying out their own statistical analyses.

ChrP 3:246—247 (Jun 1980); *JASPR* 74:454—459 (Oct 1980); *JP* 43:326—340 (Dec 1979); *JSPR* 50:301—303 (Mar 1980); *PR* 12:12—13 (Mar/Apr 1981); *T* 8:20—21 (Fal 1980).

Alpha 5:29 (Nov/Dec 1979); *Booklist* 76:462 (Nov 15, 1979); *ContempRev* 236:53 (Jan 1980); *Encounter* 54:89 (Jan 1980); *LibrJ* 104:2577 (Dec 1, 1979); *Light* 99:186—188 (Win 1979); *Listener* 102:315 (Sep 6, 1979); *SpirFron* 15:54—55 (Win 1983).

626 Gettings, Fred. *Ghosts in Photographs: The Extraordinary Story of Spirit Photography.* New York: Harmony, 1978. 152p. Chapter notes; 125 illustrations

This is a profusely illustrated history of spirit photog—raphy, or the phenomenon of images which appear on photographic plates or film purportedly through the agency of deceased persons. Most of the historical cases of importance are described. Regrettably, however, cases of fraud——notoriously prevalent in this area——are excluded, so it cannot serve as a history of the subject. Also excluded are instances of the related phenomenon, psychic photography.

LibrJ 103:1644 (Sep 1, 1978); *PopPhot* 83:81—83+ (Aug 1978); *PublWkly* 213:94 (May 15, 1978).

627 Hasted, John. *The Metal—Benders.* Boston, MA: Routledge & Kegan Paul, 1981. 279p. Bibliography: 258—261; 19 figures; 21 graphs; 21 illustrations; Index: 263—279; 12 tables

Experimental physicist Hasted reports on his research with metal—benders. He deals with the physics of metal—bending and various ways of validating the phenomenon He also describes the research of others, and how his subjects exhibited other types of phenomena described, such as levitation and teleporta—

tion. He makes observations on the psychology of psychokinesis and discusses a quantum theoretical explanation.

JASPR 76:59–67 (Jan 1982); *JP* 45:355–356 (Dec 1981); *JSPR* 51:154–156 (Oct 1981); *T* 9:15–16 (Aut 1981); *ZS* No.9:108–110 (Mar 1982).

BritBkNews p398 (Jul 1981); *SkptInq* 6:63–64 (Fal 1981) [reprinted from *Spectator*]; *Spectator* 246:24 (Apr 11, 1981).

628 Owen, Iris M., with Sparrow, Margaret. *Conjuring Up Philip: An Adventure in Psychokinesis.* New York: Harper & Row, 1976. 217p. Bibliography: 215–217; 4 figures

This is an account of how eight men and women, members of the Toronto Society for Psychical Research, set out to create their own ghost or postmortem communicator. They combined the old–time method of the home circle, in which several persons sat in a circle with their hands on a table, with the psychological techniques developed by Kenneth J. Batcheldor and Colin Brookes–Smith to produce physical phenomena. They then manufactured their "ghost," named him Philip, and invented a personality and history for him. They tried to communicate with Philip and the table responded. This book recounts the work with Philip as well as that of a second group that was able to repeat the experiment by creating a different personality. It is suggested that this method of obtaining communications and producing physical phenomena is repeatable, and by ordinary persons who do not consider themselves psychic.

JASPR 71:201–212 (Apr 1977); *JP* 40:326–327 (Dec 1976); *JSPR* 49:460–464 (Mar 1977); *P* 7:44–45 (Nov/Dec 1976); *PR* 7:15–17 (Jul/Aug 1976); *T* 7:15–16 (Spr 1979).

Choice 13:726 (Jul 1976); *HumanBeh* 5:74 (Sep 1976); *Kliatt* 12:5 (Win 1978); *LibrJ* 101:1643 (Aug 1976).

629 Panati, Charles (Ed.). *The Geller Papers: Scientific Observations on the Paranormal Powers of Uri Geller.* Boston, MA: Houghton Mifflin, 1976. 317p. Chapter references; Citations in text; 19 figures; 9 graphs; 53 illustrations; 1 questionnaire; 18 tables

Whereas Ebon's *The Amazing Uri Geller* [624] is a collection of the popular writings on Geller, this book is a compilation of scholarly reports on Geller's phenomena. As Panati describes it: "The purpose of this book is to present firsthand observations on the talents of Uri Geller and, in doing so, to bring to light and offer for public scrutiny much material that has either never before appeared in print or has surfaced only piecemeal in the popular press. The book is written––through papers, reports, diary entries, and letters––by the scientists and professionals who, is various ways, have scrutinized Geller's talents" (p. vii). Physicists, mathematicians, and engineers contributed 14 of the papers, and there are three by parapsychologists, four by magicians (but not the better known magician critics of Geller such as Christopher and Randi), and one by a professional photographer. In a 29—page introduction, Panati, a physics—trained *Newsweek* science editor, reviews and evaluates, for the layperson, the major highlights of the research reported in the remainder of the book. Although this collection could be said to represent the best "scientific" work with Geller, it was severely criticized by Hyman (see review in *Zetetic* listed below under its present title of *Skeptical Inquirer* [*SkptInq*]). Those disposed to take Geller seriously should read Hyman's review.

> *JP* 40:321—325 (Dec 1976); *JSPR* 49:537—540 (Jun 1977); *P* 7:45—46 (Sep/Oct 1976); *PS* 2:153—154 (F 1977).
> *Brain/MindBul* 1:4 (Jul 5, 1976); *Choice* 13:892 (Sep 1976); *Humanist*[U.S.] 37:22—24 (May/Jun 1977); *Kirkus* 44:375 (Mar 15, 1976); *LibrJ* 101:1126 (May 1, 1976); *PsycholToday* 10:93—94 (Jul 1976); *PublWkly* 209:53 (Mar 15, 1976); *SkptInq* 1:73—80 (Fal/Win 1976); *VillVoice* 21:42 (Mar 29, 1976).

630 Randall, John L. *Psychokinesis: A Study of Paranormal Forces Through the Ages.* London, England: Souvenir Press, 1982. 256p. Bibliography: 239—250; 18 figures; Glossary: 234—238; 18 illustrations; Index: 251—256

In the first part of this book, British biologist Randall recounts and examines reports of spontaneous PK phenomena from a variety of cultures and countries through the ages. Next, he considers the scientific investigations of PK over the past 50 years. He then attempts to classify the various kinds of PK and discusses their implications for our views of matter, space, and time. He calls for a *physical* theory of psi——and offers theoretical speculations based on hyperspace and quantum theory.

> *ChrP* 5:24—25 (Mar 1983); *JASPR* 78:157—161 (Apr 1984); *JSPR* 52:141—142 (Jun 1983); *PR* 15:10—13 (Sep/Oct 1984).
> *BritBkNews* p78 (Feb 1983); *ContempRev* 242:51 (Jan 1983); *Light* 102:175—177 (Win 1982); *Nature* 300:119 (Nov 11, 1982).

631 Richards, John Thomas. *SORRAT: A History of the Neihardt Psychokinesis Experiments, 1961—1981.* Foreword by Peter Philips. Afterword by W.E. Cox. Metuchen, NJ: Scarecrow Press, 1982. 338p. Bibliography: 312—323; Chapter references; 31 illustrations; Index: 325—338

This is an account of the macro—PK phenomena occurring

in the home circle that was organized by author and critic John G. Neihardt, who called it called the Society for Research on Rapport and Telekinesis (SORRAT). It is written by an English professor and student of Neihardt who was, and still is, one of the principals of the group. The book is based on tapes, diaries, and photographs. A chapter each is devoted to the founding of the group; a representative year; the environments of PK activity; the participants; descriptions of unconfined PK; confined PK (within a sealed container); apports, odors, and apparitions; communicating with entities; interrelated phenomena (the occurrence of different types of psi in a meaningful pattern); and the influence of Neihardt on the group, after, as well as before, his death. The Afterword is by the major investigator of SORRAT, W.E. Cox.

JP 46:373−376 (Dec 1982); *JRPR* 7:55−57 (Jan 1984); *JSPR* 51:389−392 (Oct 1982); *PsiR* 1:122−126 (Dec 1982); *Psychoenergetics* 5:177−181 (1983); *T* 11:89−90 (Win 1983).

LibrJ 107:1000 (May 15, 1982); *Light* 103:135−136 (Aut 1983).

632 Robinson, Diana. *To Stretch a Plank: A Survey of Psychokinesis.* Chicago, IL: Nelson−Hall, 1981. 277p. Bibliography: 247−266; Name index: 267−270; Subject index: 271−277

Ms. Robinson is a psychologist with a long−term interest in parapsychology, and her book is a comprehensive survey of PK phenomena and theories, in three sections. The first reviews laboratory research, including a discussion of Uri Geller, Ingo Swann, Matthew Manning, Felicia Parise, and Jean−Pierre Gerard. The second reviews spontaneous and mediumistic PK. There is a chapter on psychic photography and electronic voice phenomena, and one on miracles as PK. It concludes with an interesting chapter on unexplained phenomena and accounts which demand further investigations. The final section provides an excellent review of theories, with chapters on Stanford's PMIR and the conformance model of psi, and one recapping and classifying the more traditional theory. The last chapter is devoted to the implications of PK.

JP 45:251−254 (Sep 1981); *PR* 12:23−24 (Nov/Dec 1981); *T* 10:63−65 (Aut 1982).

Choice 18:1487 (Jun 1981); *LibrJ* 106:459 (Feb 15, 1981); *NewReal* 4:45−47 (Sep 1981); *SpirFron* 14:58−59 (Win 1982).

633 Rogo, D. Scott. *Mind Over Matter: The Case for Psychokinesis; How the Human Mind Can Manipulate the Physical World.* Wellingborough, Northamptonshire, England: Aquarian Press, 1986. 160p. Bibliography:

151–156; 17 illustrations; Index: 157–159

This nontechnical introduction to PK has chapters on psychokinesis; poltergeists; the "golden age of mediumship" (latter nineteenth century and early twentieth), covering D.D. Home, Eusapia Palladino, Rudi Schneider, and Stella C.; modern PK breakthroughs (Nina Kulagina, Alla Vinogradova, Julius Krmessky, Felicia Parise); the Geller effect; PK by committee (the Philip group and other table–tipping groups such as the Goligher circle); PK and healing (emphasizing experiments); and a summary chapter assessing the evidence for PK, which reviews laboratory evidence and offers various theories to account for PK. The book is published in conjunction with ASSAP: The Association for the Scientific Study of Anomalous Phenomena, a British organization. It is part of a series of books on the paranormal under the general editorship of Hilary Evans.

PAI 5:44 (Dec 1987)
Light 107:139 (Win 1987).

634 Rogo, D. Scott. *Minds and Motion: The Riddle of Psychokinesis.* New York: Taplinger, 1978. 271p. Chapter notes: 259–266; Index: 267–271

Journalist/parapsychologist Rogo attempts, not to prove the existence of PK, although he does cite what he considers to be the best laboratory evidence for it in the first chapter, but to describe and make sense of the clues we have that might shed light on how PK operates. The literature of mediumship, spontaneous PK effects such as poltergeists, and studies of current PK stars such as Geller, Parise, Kulagina, and the Toronto "ghost," Philip, are emphasized. The final chapter is an interesting discussion of PK theories.

JIndPsychol 2:151–152 (Jul 1979); *JP* 43:153–156 (Jun 1979); *PR* 10:9–10 (Jan/Feb 1979); *T* 9:16–18 (Spr 1981).
Booklist 75:582 (Dec 1, 1978); *ContempPsychol* 24:534–535 (Jun 1979); *LibrJ* 104:198 (Jan 15, 1979); *PublWkly* 214:73 (Aug 7, 1978); *SpirFron* 15:51–53 (Win 1983).

635 Rogo, D. Scott. *The Poltergeist Experience.* New York: Penguin Books, 1979. 301p. Chapter notes; Citations in text; Index: 293–301

This is an introduction to poltergeists by a parapsychological investigator and author. Much of the impetus for writing it comes from a firsthand experience of poltergeists that occurred when Rogo investigated a case in 1974. He writes: "In one respect this book is a result of that case, for it prompted me to reevaluate and restudy everything I had ever learned about the phenomena" (pp. 12–13). Rogo surveys various types

of poltergeist activity with illustrations from actual cases, but emphasis is placed on the psychology of the poltergeist. He presents a theory of poltergeist agency and takes pains to make the point that "the poltergeist is a terrible force" (p. 284).
T 8:20—21 (Fal 1980).
LibrJ 104:1345 (Jun 15, 1979).

636 Roll, William G. *The Poltergeist.* Metuchen, NJ: Scarecrow Press, 1976. 208p. Chapter notes; 1 figure; Index: 203—208; 1 questionnaire; Suggested reading list: 199—202; 1 table

Roll's book was originally published by Nelson Doubleday in 1972 as a book club edition, but it was very difficult to obtain, as was the NAL paperback. It was reprinted by Scarecrow Press in 1976 and, in view of its prior inaccessibility, is being included here. It is one of the best books written on poltergeists and is by a parapsychologist who, among other things, has specialized in poltergeist investigation. Roll describes several cases from the literature, but primarily presents those he has observed himself. In addition there are theoretical chapters, a chapter on *PK* and consciousness, and one entitled "The Poltergeist: Psychopathology or Human Potential?" An important appendix tells how to investigate poltergeists and hauntings and a five-page suggested reading list is provided.
ChrP 1:156—158 (S 1977); *JASPR* 68:101—104 (Jan 1974); *JP* 38:94—97 (Mar 1974); *P* 6:10—11 (Nov/Dec 1974); *T* 43—44:17 (Win/Spr 1975).
Choice 12:807 (Sep 1975).

637 Taylor, John. *Superminds: A Scientist Looks at the Geller Metal—Bending Phenomena and the Paranormal.* New York: Viking, 1975. 183p. Bibliography: 178—179; 22 figures; 145 illustrations; Index: 180—182; 4 tables

Although much has been written about Uri Geller, very little has appeared concerning the mini—Gellers, or persons who develop metal—bending abilities, usually after seeing or listening to Geller on radio, TV, or in person. He feels that if the phenomena of even a few mini—Gellers are genuine, this fact indirectly vouches for Geller's phenomena. Professor John Taylor, a physicist of King's College, London, has investigated both Geller and the mini—Gellers. He describes the results of his investigations in *Superminds*, but unfortunately provides insufficient information concerning the test conditions and controls. Half the book is devoted to his investigations and the remainder to his efforts to explain psi by the known principles of physics——electromagnetism, to be specific——a theory which parapsychologists find lacking. An outstanding feature of the

book is its many illustrations. Taylor [**483**] was later to become
critical of parapsychology.

IJParaphys 9:82–83 (1975); *JP* 39:242–250 (Sep 1975);
JSPR 48:168–171 (Sep 1975).

Books&Bkmn 21:52 (Dec 1975); *Booklist* 72:479 (Dec 1,
1975); *BestSell* 35:325 (Jan 1976); *Brain/MindBul* 1:4 (1975);
Choice 12:1504 (Jan 1976); *Kirkus* 43:988 (Aug 15, 1975); *Kliatt*
11:28 (Spr 1977); *LibrJ* 100:1933 (Oct 15, 1975); *Light* 96:46–47
(Spr 1976); *Listener* 93:587 (May 1, 1975); *NYRevBks* 22:14
(Oct 30, 1975); *NYTimesBkRev* p8 (Jan 18, 1976); *Observer* p.
30 (May 11, 1975); *PublWkly* 208:114 (Aug 11, 1975); *PublWkly*
210:48 (Nov 8, 1976); *SciAm* 234:134 (Feb 1976); *Spectator*
235:83 (Jul 19, 1975); *TimesLitSuppl* p846 (Jul 25, 1975).

Psychology, Psychiatry, and Parapsychology

INTRODUCTION

In *Sources* there was a section of books on "Psychiatry and Psi Phenomena" (12 books) and one on "Psychology and Psi Phenomena" (10 books). The same total number of books (22) is included in this section, but in many instances a clear differentiation between psychology and psychiatry could not be made and so all the books were lumped in one section.

Burt [639], Eisenbud [645], and the book by Jung [647] are basically collections of the authors' papers. The papers in Dean [641] and in the two Shapin and Coly [656, 657] volumes were originally given at conferences. Corliss [640], and Ehrenwald [643], are anthologies of previously published papers by many authors. The remaining monographs are on a variety of subjects: Bolen [638] is on the psychology of synchronicity, Ehrenwald [642] presents a summation of a lifetime of thought about psychiatry and psi, Eisenbud [644] on psychoanalysis and precognition, Irwin [646] writes about ESP and PK from the viewpoint of information processing, and Krippner explores various aspects of psychology and parapsychology in the Soviet Union, Eastern Europe, and other countries. The Leaheys [649] write on the psychology of pseudoscience, including parapsychology; Leeds and Murphy [650] culled Murphy's writings on parapsychology in order to formulate a theory of psi functioning; Mindell [651] theorizes about a "subtle body" that may serve as a catalyst for psi; Mintz and Schmeidler [652] write about transpersonal psychotherapy and psi; Schatzman [653] presents a case history of a patient who had frequent apparitions of her father; Schmeidler [654] presents a perceptual theory of psi; Schwarz [655] describes a number of cases of psi occurring in his life or practice; Taub—Bynum [658] writes about the family as a vehicle for psi; and Zusne and Jones [659] present a textbook of anomalistic psychology that includes parapsychology.

Additional books with information on psychology, psychiatry, and parapsychology: See also the section "Altered States of Consciousness and Psi." Wolman [361] discusses

schizophrenia and psi, Grof [348] describes his psychedelic
therapy and his newer method of "holotropic therapy." Hunt
[807] examines automatisms including a discussion of their role
in both inducing and curing psychoses. Mavromatis [812] reviews
hypnagogic phenomena. Both Neher [355] and Reed [813] deal
with anomalous psychological experiences and states. Agor [366]
reports on his research into intuition and management using the
Myers—Briggs Type Indicator. Dean et al. [369] report on at—
titudes toward time and results of precognition tests. White's
bibliography [378] has sections on psi and personal growth and
on psychiatric and psychological applications of psi. Kautz and
Branon [374] emphasize the psychotherapeutic uses of channel—
ing. Tart [412] deals with consciousness and psi. Chapters on
the psychology of belief may be found in the book by Abell and
Singer [456]. Alcock [457] offers a critical psychological
perspective of parapsychology. Hines [474] emphasizes the
psychology of belief in the paranormal, as do Marks and
Kammann in a section of their book [477]. A view of parapsy—
chology and psychology from the Soviet approach is presented
by Dubrov and Pushkin [511]. Psychology and parapsychology
is covered in several chapters in *Advances in Parapsychological
Research 4* [523]. There are four papers on parapsychology and
psychiatry or psychology in Schmeidler [571]. Greyson and
Flynn [587] have a section on clinical aspects of NDEs. A
psychological approach to OBEs is taken by Blackmore [597],
Gabbard and Twemlow [598], and Irwin [600]. Whitton [702]
describes a technique of using hypnotic age regression to explore
the bardo state for therapeutic purposes. Perry [716] has
compiled an anthology offering guidelines on exorcism and on
counseling persons involved with the occult and paranormal.
Kelsey offers counsel to persons who have had psi experiences
using a Jungian approach. Hartmann [730], in writing on the
psychobiology of the nightmare, notes that ESP experiences tend
to be reported by nightmare sufferers. Neppe [737] writes about
the psychology of déjà vu, and Persinger [739, 740] studies
several physiological and psychological correlates of ESP ex—
periences. Ebon [778] has collected many articles by
psychologists and psychiatrists who counsel about dabbling with
the occult and psychic powers. Mishlove [782] views para—
psychology as a discipline for self—actualization, and he dis—
cusses the relationship of humanistic psychology and para—
psychology. White's bibliography on psychics [791] has a sec—
tion on the psychology of psi and on counseling persons who
have had psi experiences. Achterberg [792] covers the role of
imagination and imagery in healing. Irwin's textbook [810] em—
phasizes the psychology of parapsychology.

BIBLIOGRAPHY

638 **Bolen, Jean Shinoda.** *The Tao of Psychology: Synchronicity and the Self.* New York: Harper & Row, 1979. 111p.
Chapter notes; Index: 109—111
Each chapter presents a view of Jung's concept of synchronicity and the Oriental concept of the Tao. Bolen, who is a Jungian analyst, proposes that synchronicity and the Tao are the same thing. She holds that "much of the value of synchronicity lies in its ability to connect us to a meaning—giving, intuitively known principle in our lives by which we can find a 'path with a heart,' a *tao*, a way to live in harmony with the universe. Synchronicity can provide us with confirma—tion that we are on the right path, as well as let us know when we are not" (pp. xii—xiii). She likens synchronistic events to dreams and suggests that they be interpreted after the manner of dreams. She also conceives of synchronicity as shed—ding light on parapsychological research. In a chapter devoted to parapsychology she points out that the "archetype of the miracle," or the "archetype of 'magic effect'," is at the core of psi.
PAI 1:85 (Dec 1983).
AntR 39:393 (Sum 1981); *Choice* 17:143 (Mar 1980).

639 **Burt, Sir Cyril.** (Anita Gregory, Ed.). *ESP and Psychology.* New York: Wiley, 1975. 179p. Chapter notes; Index: 175—179; Suggested reading list: 173
British parapsychologist Anita Gregory compiled this chronological selection of Cyril Burt's lectures, reviews, and papers on parapsychology. Since Burt's death, incontrovertible evidence of fraud in his psychological research and writing has been uncovered. However, Burt's parapsychological contribution was primarily theoretical, so his views can be judged on their own merits. Although all the selections are scholarly, an effort was made to choose those pieces that were addressed more to the general reader than to professional scientists. Included are a 1953 lecture entitled "Psychical Research," "The Soul in 1966," "Parapsychology and its Implications," and reviews of Gardner Murphy's *Challenge of Psychical Research*, Jung's autobiography, and Alister Hardy's *Divine Flame*. Previously unpublished are "Cross Correspondences in Random Material" and "The Next Steps in Psychical Research."
JASPR 71:213—216 (Apr 1977); *JP* 40:74—75 (Mar 1976); *JSPR* 48:179—182 (Sep 1975); *PR* 7:15—17 (Mar/Apr 1976); *T* 7:15—16 (1979).

ContempPsychol 21:565—566 (Aug 1976); *Light* 95:186—187
(Win 1975); *NewHum* 91:213—215 (Dec 1975); *TimesLitSuppl*
p976 (Aug 29, 1975).

640 **Corliss, William R. (Comp.).** *The Unfathomed Mind: A
Handbook of Unusual Mental Phenomena.* Illustrated by
John C. Holden. Glen Arm, MD: The Sourcebook
Project, 1982. 754p. Chapter references; 7 figures; 9
graphs; 10 illustrations; Index: 749—754; 25 tables

The compiler, a physicist and freelance writer on science
and technology, presents this collection of source materials that
indicate that "the mind has powerful, subtle, often bizarre in—
fluences on the human body, human behavior, and perhaps even
the so—called objective external world" (p. v). Most of the
material has been excerpted from medical and psychological
journals. Parapsychological material is included, but Corliss
admits to being "somewhat negative on 'psi' and ESP" (p. v).
Criteria for including material were: (1) current scientific
theories are contradicted by it; or, (2) it raises questions not
answered sufficiently by the accepted literature. There are six
main sections: (1) "Dissociative Behavior: Other Control
Centers," (2) "The Possible Acquisition of Hidden Knowledge,"
(3) "Anomalous Modes of Information Processing," (4)
"Hallucinations: Sensing What is Not," (5) "Remarkable
Mind—Body Interactions," and (6) "Mind Over Matter." Each
chapter has its own detailed table of contents.
JP 46:285 (Sep 1982).

641 **Dean, Stanley R. (Ed.).** *Psychiatry and Mysticism.* Chicago,
IL: Nelson—Hall, 1975. 424p. Bibliographical footnotes;
Chapter references; 12 figures; 7 graphs; 45 illustrations;
Index: 399—424; 2 tables

At the time he edited this book, Dr. Dean was clinical
professor of psychiatry at the University of Miami in Florida.
He is the founder of the American Metapsychiatric Association.
He coined the term "metapsychiatry" for "the developing branch
of psychiatry that deals with psychic phenomena" (p. 1) and,
one could add, with mysticism. Dean defines the latter as a
by—product of the fact "that very exceptional kinds of
knowledge and awareness may reach consciousness through
channels other than those known to us at present" (p. 1). This
book is a collection of 26 papers read at three symposia on
psychic phenomena plus some additional invitational papers that
were organized by Dean and given at the 1972, 1973, and 1974
annual meetings of the American Psychiatric Association. Some
papers are just on parapsychology, some just on psychiatry, but
most deal with aspects of both. Some of the major subject

areas are altered states, psychic healing, psychosomatic medicine, psi and psychotherapy, and mysticism. The contributors are Dean himself, Francis Braceland, Jules Masserman, Edgar Mitchell, W.G. Roll, Jan Ehrenwald, Berthold Schwarz, Shafica Karagulla, Jule Eisenbud, J.B. Rhine, J.N. Emerson, John Pier—rakos, James Beal, Thelma Moss, Paul Adams, Gary Schwartz, Herbert Benson, Arthur Gladman, E. Fuller Torrey, Lawrence LeShan, Elmer and Alyce Creen, Carl Simonton, Stanislav Grof, Malcom Bowers, Jr., and Julian Silverman.

JP 40:247—248 (Sep 1976); *JSPR* 49:550—551 (Jun 1977); *JSPR* 49:550—551 (Jun 1977); *P* 7:42—43 (Jul/Aug 1976); *PR* 7:13—14 (Jul/Aug 1977); *T* 5:14—15 (1977).

AmJPsychiat 133:1104—1105 (Sep 1976); *Brain/MindBul* 1:4 (Dec 1, 1975); *Choice* 13:140 (Mar 1976); *LibrJ* 101:723 (Mar 1, 1976); *Light* 96:89—90 (Sum 1976).

642 Ehrenwald, Jan. *The ESP Experience: A Psychiatric Validation.* New York: Basic Books, 1978. 308p. Bibliog—raphy: 289—298; 9 figures; Index: 299—308; 1 table

This volume can be viewed as the summation of a lifetime of experience and thinking about the interrelation of psi and psychiatry. Psychotherapist/parapsychologist Jan Ehrenwald ex—plains how many forms of psi phenomena are corroborated by psychoanalytic and psychiatric findings. He emphasizes the psychodynamics and psychopathology of psi in daily life. A sec—tion consisting of seven chapters is devoted to his theory of the nature of psi.

JASPR 73:67—71 (Jan 1979); *JIndPsychol* 2:162—163 (Jul 1979); *JP* 43:238—240 (S 1979); *JSPR* 50:120—122 (Jun 1979); *PR* 9:18—20 (May/Jun 1978).

ContempPsychol 23:935 (Nov 1978); *Choice* 15:611 (Jun 1978); *Kirkus* 46:31 (Jan 1, 1978); *LibrJ* 103:371 (Feb 1, 1978); *SciBooks&Films* 15:65 (Sep 1979).

643 Ehrenwald, Jan (Ed.). *The History of Psychotherapy: From Healing Magic to Encounter.* New York: Jason Aronson, 1976. 589p. Bibliography: 577—583; Index: 585—589; 1 table

The editor, a psychiatrist long associated with para—psychology, calls this massive volume a "guided tour through the history of psychotherapy" (p. 6). The format consists of chronologically arranged excerpts from the basic texts on mental healing from magic to encounter, each one being introduced by Ehrenwald. Having immersed himself in this intensive, as well as vast, array of source materials, Ehrenwald concludes: "if there is a lesson to be learned, here and now, from the past, it is that man is the most important therapeutic agent for man, and that psychotherapy is the time—tested accretion and codification of a social expedient to meet universal human needs" (p. 5). This book is included here because many selec— tions and even whole sections are relevant to parapsychology. Unorthodox healing and miraculous cures are considered in several chapters and one chapter devoted to parapsychology covers Freud, Servadio, Ehrenwald, Eisenbud, and Ullman. Of special interest are Ehrenwald's comments on the personality and role of the healer and on what he has called "doctrinal compliance," which may at least partially be psi—mediated.

ContempPsychol 22:717 (Sep 1977); *Choice* 14:123 (Mar 1977); *PsychoanalRev* 65:659—666 (Win 1978).

644 Eisenbud, Jule. *Paranormal Foreknowledge: Problems and Perplexities.* New York: Human Sciences Press, 1982. 312p. Bibliographical footnotes; Bibliography: 289—302; 2 illustrations; Name index: 309—312; Subject index: 303—308

This is a clinical and theoretical approach to the problem of precognition by a psychiatrist who has wrestled with it for over 30 years. Part I, "Cases," consists of six chapters in which he presents precognitive case material drawn from his psychoanalytic practice, delving into the psychodynamics in— volved. Part II, "Hypotheses," consists of 10 chapters. He dis— cusses the "passive" and "active" approaches to precognition, the former demanding nothing beyond cognitive abilities of humans, whereas the latter utilizes "the possibilities of active extrasensory and psychokinetic influences in the development of future events" (p. 117). Eisenbud presents his own compound theory of precognition, which is an active approach. In essence, he proposes that humans create the future they "foresee." Several appendices serve as additional histories and discussions supplementing the text.

JASPR 76:288—294 (Jul 1982); *JP* 46:153—157 (Jun 1982); *JSPR* 51:386—387 (Oct 1982); *PR* 13:19—21 (Nov/Dec 1982); *PS* 5:89—91 (1983); *T* 11:72 (Aut 1983).

AmJPsychiat 140:804 (Jun 1983); *Choice* 19:1638 (Jul 1982).

645 Eisenbud, Jule. *Parapsychology and the Unconscious.* Berkeley, CA: North Atlantic Books, 1984. 251p. Bibli— ographical footnotes; Chapter references; 26 illustrations; Index: 245—251

The author remarks on the irony of the fact that the two disciplines having most to do with unconscious factors in human affairs, psychoanalysis and parapsychology, pay so little heed to one another. This collection of essays, only one of which has not been previously published in whole or in part, is offered as a tentative step in bringing psychoanalysis and parapsychology together. Some papers have been revised and rewritten to the point where they are almost entirely new. The chapter on mediumship was written specifically for this volume for, as Eisenbud points out, it is an almost unique refracting prism through which various aspects of the relationship between the unconscious and psi can be explored. The chapter titles and sources are "Four Views of Man" (originally published in Spanish and reprinted in *Fate*), "Psi and the Problem of the Disconnections in Science" (*Journal of the American Society for Psychical Research*), "Two Approaches to Psi Phenomena" (*Journal of the American Society for Psychical Research*), "The Psychology of the Paranormal" (*Journal of the American Society for Psychical Research*), "Differing Adaptive Roles of Psi in Primitive and Non—Primitive Societies" (*Psychoanalytic Review*), "How to Influence Practically Anybody (but Fellow Scientists) Extrasensorially at a Distance" (in *Psi and Psychoanalysis* and *Psychic*), "The Problem of Resistance to Psi" (*Proceedings of the Parapsychological Association* and in *The World of Ted Serios*), "Psychic Photography and Thoughtography" (in *Psychic Exploration*, edited by J. White), "Postscript: Ganymede Observed" (new), "Anthropology and Parapsychology" (in *Extrasensory Ecology*, edited by J. Long), "Psi and the Nature of Things" (*International Journal of Parapsychology*), "Why Psi?" (*Zeitschrift für Parapsychologie und Grenzgebiete der Psychologie*; also *Psychoanalytic Review*), "Chance and Necessity: Is There a Merciful God in the House?" (in his *Paranormal Foreknowledge*), "The Case of Florence Marryat" (*Journal of the American Society for Psychical Research*), and "Cagliostro Revisited: The Magic of Mediumship" (new).

JASPR 79:554—557 (Oct 1985); PAI 3:39 (Dec 1985).

646 **Irwin, Harvey J.** *Psi and the Mind: An Information Processing Approach.* Metuchen, NJ: Scarecrow Press, 1979. 173p. Author index: 167—169; Bibliography: 149—165; 1 figure; Subject index: 170—173

An Australian psychologist explores ESP and PK in terms of information processing theory, which is based on a model of the human mind as a system which processes information: it accepts inputs, extracts information from them and stores it, and when required acts upon it and generates outputs. The first three chapters describe the mind as an information—processing system. The next three present an information—processing approach to ESP and the seventh chapter presents an approach to PK. The eighth and final chapter "concerns those characteristics of psi phenomena that may reflect certain holistic principles governing the operation of the human information processing system" (p. vi). Irwin concludes that the approach to psi as an information—processing system is viable and that it provides a theoretical basis for generating productive research hypotheses. *JP* 45:163—174 (Jun 1981); *JSPR* 50:535—538 (Dec 1980). *ContempPsychol* 25:943 (Nov 1980).

647 Jung, C.G. *Psychology and the Occult.* Translated by R.F.C. Hull. Princeton, NJ: Princeton University Press, 1977. 167p. Bibliographical footnotes; Bibliography: 159—167; 4 figures

This is a collection of nine of Jung's writings on the oc—cult, including spiritualism and parapsychology. It performs a useful service, considering the fact that otherwise they would have to be extracted from the 19 volumes of his collected works, three volumes of his letters, or from reviews in peri—odicals or introductions to books. Included are Jung's disserta—tion on the psychology of automatism and mediumship, "On Spiritualistic Phenomena," "The Psychological Foundations of Belief in Spirits," "The Soul and Death," "Psychology and Spiritualism," "On Spooks: Heresy or Truth?," his Foreword to Jaffe's *Apparitions and Precognition*, and "The Future of Parapsychology." The bibliography is a general one supplement—ing the footnotes in the text.

648 Krippner, Stanley. *Human Possibilities: Mind Exploration in the USSR and Eastern Europe.* Garden City, NY: Anchor/Doubleday, 1980. 348p. Chapter references; 2 figures; Index: 335—348

This is an informal yet informative review of Soviet psychology, medicine, and parapsychology. Psychologist/para—psychologist Krippner says: "It is an account of my trips to the USSR and Eastern Europe and the people I have met who are engaged in what Americans would call . . . 'consciousness studies'——and what many foreign scientists would refer to as 'hidden reserves' or 'latent human possibilities'"(p. vii). Subjects covered are PK in Leningrad (Kulagina), Soviet parapsychology, hypnosis and creativity, suggestopedia, Kirlian photography, self—regulation, acupuncture, bioplasma, and innovative Soviet therapies, such as fasting.

Booklist 77:429 (Nov 15, 1980); *Kirkus* 48:1333 (Oct 1, 1980); *LibrJ* 106:62 (Jan 1, 1981); *SciBooks&Films* 16:245 (May 1981).

649 Leahey, Thomas Hardy, and Leahey, Grace Evans. *Psychology's Occult Doubles: Psychology and the Problem of Pseudoscience.* Chicago, IL: Nelson—Hall, 1983. 277p.

Bibliography: 267—272; Chapter notes; 3 figures; Index: 273—277; 1 table

The authors, two psychologists, write: "Our aim in this book is to understand why pseudosciences exist, why they at—tract followers, why they are rejected. We will also study the traditional grounds for distinguishing between science and pseu—doscience, to see how far they are justified and what role they have played in the historical rejection of 'real pseudosciences,' if such there be. Finally, we will reflect on the nature of science, its relation to human cognitive and emotional needs, and how far it may go in replacing other human endeavors such as religion and art" (p. viii). Pseudoscience is examined in Part I. Part II consists of four chapters on specific "pseudosciences": alchemy, astrology, phrenology, mesmerism, spiritualism, psychi—cal research, and contemporary therapeutic cults. After delving into the nineteenth—century antecedents of pseudoscience, the authors conclude that the motivating force behind pseudoscience is the need to find "a universe of meaning, rather than one of mere order" (p. 245). They hold that "there is nothing un—scientific about phrenology, mesmerism, or parapsychology——which conclusion does not make them sciences. They are not sciences because they try to delve deeper than any science can" (p. 245).

JASPR 79:89—92 (Jan 1985); *JP* 47:254—257 (S 1983); *PAI* 5:41 (Jun 1987).

AmJPsychol 97:314 (Sum 1984); *BestSell* 43:230 (Sep 1983); *Choice* 21:354 (Oct 1983); *ContempPsychol* 29:202 (Mar 1984); *LibrJ* 108:1263 (Jun 15, 1983); *PsycholRec* 34:443 (Sum 1984); *RelStRev* 10:261 (Jul 1984); *SciBooks&Films* 19:126 (Jan 1984).

650 Leeds, Morton, and Murphy, Gardner. *The Paranormal and the Normal: A Historical, Philosophical and Theoretical Perspective.* Metuchen, NJ: Scarecrow Press, 1980. 239p. Bibliography: 209—219; 29 figures; Glossary: 220—228; 1 graph; Index: 229—235; 9 tables

This book was half written by the world renowned psychologist/parapsychologist Gardner Murphy and half by Morton Leeds, a former student of Murphy's and now a social administrator. Their intention was to review and condense Murphy's writings on psi with a view to developing a theory of so—called paranormal functions. Part I is a selective historical review of parapsychology. Part II deals with philosophical im—plications. Part III formulates the relation between psi and psychology, and they theorize as to how psi is likely to occur. Part IV offers some theoretical bases for paranormal functioning, and Part V provides final reflections, with the emphasis on repeatability. The book is aimed primarily at currently practic—

ing parapsychologists with the hope of enlisting the aid of addi—
tional researchers.
    JP 44:362—365 (Dec 1980); JSPR 51:26—27 (F 1981);
PJSA 5:101—105 (Dec 1984); PR 11:19—20 (Nov/Dec 1980); T
9:15—16 (Spr 1981).
    LibrJ 105:1315 (Jun 1, 1980).

651 **Mindell, Arnold (Sisa Sternback-Scott and Becky Good-
    man, Eds.).** *Dreambody: The Body's Role in Revealing
    the Self.* Introduction by Marie—Louise von Franz. Santa
    Monica, CA: Sigo Press, 1982. 219p. Bibliographical
    footnotes; Bibliography: 201—207; 52 illustrations; Index:
    209—219
    Jungian analyst Mindell's concept of the "dreambody"
roughly corresponds to the "subtle body," which he defines as
"a higher energic field intensity, that is, a patterned experience
without definite spatial or temporal dimensions." The dream
body oscillates between matter and psyche. He holds that it is
the dreambody that is experienced in NDEs, OBEs, shamanism,
and mediumistic trance.   It can serve as a catalyst for
synchronistic, and possibly psychic, events. The main emphasis
of the book is on the role of the dreambody in psychotherapy.
Mindell sees illusions as opportunities to commence the process
of individuation accompanied by dreambody awareness.
    T No. 13/14:23—24 (Spr 1985/86).
    Choice 20:650 (Dec 1982); *LibrJ* 108:56 (Jan 1, 1983);
Light 105:43—44 (Spr 1985).

652 **Mintz, Elizabeth E., with Gertrude R. Schmeidler.** *The
    Psychic Thread: Paranormal and Transpersonal Aspects
    of Psychotherapy.* New York: Human Sciences Press,
    1983. 232p. Chapter notes; Index: 227—232
    Addressed primarily to the practicing psychotherapist, this
book "discusses the psychodynamic significance and therapeutic
management of paranormal and transpersonal experiences
reported by the patient or taking place in the therapeutic
situation" (p. 11). The clinical data presented is taken from the
author's own clinical experiences or that of her colleagues. The
author's stance is pragmatic: Seeming psychic and transpersonal
or mystical experiences *do* occur in people's lives, and whatever
their reality may be, unless responsible practitioners begin to
regard them as matters of legitimate concern for psychologists,
untrained entrepreneurs will try to answer the need. In the In—
troduction, paranormal and transpersonal experiences are defined
and discussed within a psychotherapeutic context, setting the
stage for the ten chapters to follow. The subjects covered are:
the nature of psi; apparent psi experiences occurring in

psychotherapy and their therapeutic management; patients' ap—
parent psychic awareness of the therapist's personal life; how
the therapist's hunches can help the patient; patients who seem
to be psychic; apparent psi among patients in therapy or su—
pervisory groups; transpersonal experiences occurring in the
context of group therapy; the relationship between mystical ex—
periences and schizophrenia; an overview of therapeutic ap—
proaches that allow for transpersonal dimensions of experience;
and the therapist's role in the management of bereavement and
serious illness, including a discussion of near—death experiences.
The focus of discussion throughout is on therapeutic manage—
ment.
    JP 50:72—75 (Mar 1986); *JRPR* 7:261—262 (Oct 1984);
PAI 3:41 (Jun 1985).
    ImagCogPers 4:313—315 (1984/85); *JTranspersPsychol*
16:244—245 (1984); *SciTech* 8:17 (Aug 1984).

653 Schatzman, Morton. *The Story of Ruth.* New York:
    Putnam's, 1980. 306p. 6 figures; 9 illustrations; Notes:
    293: 293—306; 5 tables
    This, in one sense, is a case history of a psychiatric
patient told by the psychiatrist. But as Schatzman points out in
the Introduction, it is more. It is concerned with such questions
as "What are apparitions? Are they real? Can one trust the
evidence of one's senses?" (pp. 9—10). The patient, Ruth, was
haunted by the apparition of her father, who was living. Some—
times the apparition could be seen by others. Schatzman ex—
amines the apparition's role as well as its reality.
    JASPR 75:186—191 (Apr 1981); *JSPR* 51:94—96+ (Jun
1981); *T* 10:38—41 (Sum 1982).
    FronSci 3:39+ (Mar/Apr 1981); *Kirkus* 48:428 (Mar 15,
1980); *PublWkly* 217:38 (Mar 28, 1980).

654 Schmeidler, Gertrude R. *Parapsychology and Psychology:
    Matches and Mismatches.* Jefferson, NC: McFarland,
    1988. 236p. Bibliography: 205—220; Glossary: 201—204;
    Index: 221—236
    This work is in three parts: "Introduction: A Theory"
(four chapters), "Examining the Theory" (nine chapters), and
"Speculation" (four chapters). In Chapter 2, "A Theory of
Parapsychology," Schmeidler notes that parapsychology is often
criticized by both proponents and opponents for (a) not having
a theory and (b) for not having a repeatable experiment. In
this book she denies the first and questions the second. She
proposes that parapsychology has had an implicit theory since
its beginning 100 years ago: it is that "psi is a psychological
function." Furthermore, psi responses should be processed in

the same manner as other psychological responses; and variables that affect how other abilities are used will also affect how psi is used. This work, then, reviews Schmeidler's own experiments and those of others that test the consequences of the above as— sumptions. The question of repeatability is also examined in Section I. In Section II, Schmeidler examines the experiments that test her theory, with a chapter each devoted to "The Ex— perimenter Effect," "Social Psychology," "Personality," "Moods," "The Longer Altered States of Consciousness," "Perception," "Cognition," "Brain and Other Body Changes," and with a closing chapter entitled "Evaluation of Replicability and of the Theory." The last section reviews theories of psi, including a chapter on "Psi and the Self," and closes with a "General Overview." Each chapter has a helpful summary overview at the end. The glossary is experimentally oriented and includes general scientific and psychological experimental terms as well as parapsychological.

ContempPsychol 26:1412 (Apr 1989); *Fate* 42:102,104 (May 1989).

655 Schwarz, Berthold Eric. *Psychic—Nexus: Psychic Phenomena in Psychiatry and Everyday Life.* New York: Van Nostrand Reinhold, 1980. 308p. Bibliography: 291—299; 21 figures; 1 graph; 14 illustrations; Index: 300—308; 12 tables

A psychiatrist, Schwarz presents many examples of the psychodynamics of psi (he coins the term "psychic—dynamics" for it) from his practice, his personal life, and his experiments. It is the aim of the book not only to describe the psychology and psychopathology of psi but to enable the reader to develop his or her own ability to be aware of and make use of psi. With the exception of a Postscript, the first 15 chapters were previously published in journals. However, in regard to the cases used, new or expanded cases have been added to avoid duplica— tion or to augment abbreviated examples. In a useful 13—page Introduction, Schwarz provides an informative autobiographical account of how his interest in psi developed and how he fol— lowed up on it, thus providing a broader framework that helps to integrate the selections. In the 16th and last chapter, "Postscript: Possible 'Impossibilities'," he makes some observa— tions about the nature of psi research and psi investigators.

Choice 18:458 (Nov 1980).

656 Shapin, Betty, & Coly, Lisette (Eds.) *Brain/Mind and Parapsychology: Proceedings of an International Con— ference Held in Montreal, Canada, August 24—25, 1978.* New York: Parapsychology Foundation, 1979. 252p.

Chapter references; 6 figures; 1 table
This is a compilation of ten papers and discussions given at a Parapsychology Foundation conference in Montreal. As the title suggests, most papers dealt with the mind/body problem and psi. Edward F. Kelly started the conference with "Converging Lines of Evidence on Mind/Brain Relations," followed by Charles Honorton's "A Parapsychological Test of Eccles' 'Neurophysiological Hypothesis' of Psychophysical Interaction." J. Bigu presented "A Biophysical Approach to Paranormal Phenomena," John Beloff wrote on "Voluntary Movement, Biofeedback Control and PK," and Thomas H. Budzynski discussed biofeedback and brain lateralization. Karl H. Pribram contributed "A Progress Report on the Scientific Understanding of Paranormal Phenomena," and Charles T. Tart gave "An Emergent–Interactionist Understanding of Human Consciousness." "Subliminal Perception and Parapsychology: Points of Contact" was presented by Norman F. Dixon, "The Right Hemisphere: Pathway to Psi and Creativity" was given by Jan Ehrenwald, and the conference closed with "The Mind–Body Problem, Reality, and Psi" by Emilio Servadio. A roundtable discussion completed the conference.
ChrP 3:205 (Mar 1980); *Light* 99:190–191 (Win 1979).

657 Shapin, Betty, & Coly, Lisette. *Communication and Parapsychology: Proceedings of an International Conference Held in Vancouver, Canada, August 9–10, 1979.* New York: Parapsychology Foundation, 1980. 230p. Chapter references; 8 figures; 2 graphs; 6 maps; 12 tables

This book contains ten papers plus discussions, and a final roundtable discussion, given at a Parapsychology Foundation convention in Vancouver. Each paper dealt with a specific aspect of communication and psi. Robert Morris discussed "Psi Functioning Within a Simple Communication Model," Harvey Irwin related information processing theory to psi, Carroll Nash reviewed the characteristics of psi communication, and "Nonverbal Communication and ESP" was discussed by Paul Byers. Robert Rosenthal discussed the implications of the experimenter effect for nonverbal communication research, Edward Storm's paper was entitled "Attention and Computation in the Brain," "Neutrino Theory of Psi Phenomena" was Martin Ruderfer's contribution, Luther Rudolf discussed "The Psi Channel Coding Problem," David R. Barker dealt with "Psi Information and Culture," and Montague Ullman described "Psi Communication Through Dream Sharing."
Phoenix 5:117–121 (1981).

658 Taub-Bynum, E. Bruce. *The Family Unconscious: "An Invisible Bond."* Wheaton, IL: Theosophical Publishing House, 1984. 230p. Glossary: 225—226; Index: 227—230; 1 table

There are five major themes evident in this book: quantum physics, family psychodynamics, yoga psychology, body energy research, and psi. Starting from a holistic model of the family as a unit, the binding agent sometimes being psi, the author, a psychotherapist, goes on to an experimental and theoretical discussion of the nature of the self. He draws out many connections between the insights of Eastern mysticism and the findings of modern physics. His insights and theories are firmly rooted in biology and physiology, but they spiral outward and inward to the limits of both the physical and the subjective worlds. The clinical observations and cases he presents are used to illustrate ideas concerning personal and interpersonal growth and to provide a context for psi. The author has obviously practiced what he preaches, and his theories are based not only on logic but on experience. There are 20 chapters in six sections. The outline of the book is provided by the six section titles: "The Enfolding Field"; "Family Dreams"; "The Radiant Organism"; "Psi, Dreams, and the Family Unconscious"; "Toward a New Paradigm: Psi, Physics, Psychology, and the Self"; and "The Common Stream."

JASPR 80:451—454 (Oct 1986); *PAI* 3:42 (Jun 1985).

Brain/MindBul 11:4 (Mar 24, 1986); *SpirFron* 18:46—47 (Win 1986).

659 Zusne, Leonard, and Jones, Warren H. *Anomalistic Psychology: A Study of Extraordinary Phenomena of Behavior and Experience.* Hillsdale, NJ: Lawrence Erlbaum Associates, 1982. 498p. Bibliography: 465—485; 13 figures; 2 graphs; 21 illustrations; Index: 487—498; 3 tables

This book by two psychologists is designed as "a teaching aid in a course on the scientific approach to all those psychological phenomena that do not fit the current scientific world view by the criteria of most psychologists, as well as paranormal phenomena . . . that at least in part can be explained in terms of known psychological principles" (p. vii). The authors try to present the psychological approach to paranormal phenomena, relate this approach to the scientific approach in general, and show what the scientific approach reveals when applied to the paranormal. There are 15 chapters, the first being introductory. Anomalistic psychophysiology is dealt with in two chapters, in part on miraculous healing and injury and pain control. Single chapters are on perception and the paranormal, anomalous imagery and hallucinations, anomalous memory, extraordinary beliefs, psychopathology and magic, the social psychology of the extraordinary, and the psychology of bad science. Two chapters are devoted to personality in occult—tism and two to parapsychology. "Parapsychology: I" covers the experimental method in parapsychological research and ESP ex—perimentation in historical perspective. "Parapsychology: II" in—cludes "Explanations, Models, and Theories of ESP"; "Psychokinesis"; and "Parapsychology: An Art or Science." A retrospective and prospective chapter on anomalistic psychology ends the book.

JASPR 77:171—180 (Apr 1983); *JP* 47:251—254 (S 1983); *PAI* 4:70 (Jun 1986); *ZS* No.12/13:166—170 (Aug 1987).

Choice 20:352 (Oct 1982); *ContempPsychol* 28:351—352 (May 1983); *PsycholRec* 33:143—144 (Win 1983); *SkptInq* 7:63—64 (Spr 1983).

Reference Works on Parapsychology

INTRODUCTION

There were 13 books in this category in *Sources*: five bib—liographies, four library catalogs, one biographical dictionary, two encyclopedias, and one index. The library catalogs are still of historical interest and the index [207] and biographical dictionary [200] are still useful, although dated. The two en—cyclopedias, however, have been absorbed by Shepard [685], the bibliography by Naumov and Vilenskaya has been superseded by White and Vilenskaya [680], and *Sources* plus the present volume have taken the place of Zorab's bibliography.

This category has become one of the largest, with 36 titles. Because there are so many books, this section has been further subdivided into "Bibliographies" (23), "Dictionaries and Encyclopedias" (8), and "Indexes" (5). There are a number of general bibliographies, the best ones being Clarie's bibliographies [663, 664], White [679], and White and Dale [820]. There are several specialized bibliographies: Goss's [666] poltergeist bibliog—raphy is excellent; Eberhart [665] provides a geographic direc—tory of spontaneous cases, and White and Vilenskaya [680] have compiled the most exhaustive bibliography of Soviet, East European, and Chinese psi research to date. Of the eight en—cyclopedias and dictionaries, Thalbourne's [687] is the most authoritative dictionary, with the emphasis on experimental parapsychology. Wolman [688] is a standard reference in the field. The indexes included are to individual journals, the best one being the *Ten—Year Index of the Skeptical Inquirer* [692]. Much useful reference information is included in the *Ashby Guidebook* [804].

Additional books with information on reference works on parapsychology: Bibliographies listed in other sections are Crabtree [344] on animal magnetism, White [378] on psi ap—plications (an earlier version appeared in Jones [373]). There is also an annotated bibliography of the survival problem in Jones [373]. White [791] also has compiled a bibliography on psychics listing works by and about individual mediums, psychics, and healers, and covering methods of training psi and the psychol—ogy of psi. There is a bibliography of remote—viewing research

1973—1982 in Targ and Harary [785]. The Kautz and Branon [374] book includes "A Directory of Channels." Because it presents both pro and con views of parapsychology, Kurtz's *Handbook* [476] could serve as a companion volume to Wolman. Schmeidler [654] includes an excellent experimentally—oriented glossary.

BIBLIOGRAPHIES

660 *American Book Publishing Record Cumulative 1876—1949: An American National Bibliography.* New York: R. R. Bowker, 1980. 15 vols. Author index: Vol. 13; Title in—dex: Vol. 14; Subject index: Vol. 15

This is a cumulation of American book production for the years 1876—1949 as cataloged by the Library of Congress and the National Union Catalog. A total of 625,000 books are ar—ranged by subject according to Dewey Decimal classification and indexed by author and title with a separate subject guide in—cluding Library of Congress subject tracings, Dewey Decimal Classification numbers and Library of Congress classification numbers. This provides access, by author, title, or Dewey Decimal number, to books on psychical research from 1876—1949. For example, in the Dewey Decimal numbers 133—136 there are nearly 3,000 titles. Because of the time span covered, it is an easier bibliography to use than, for example, *Cumulative Book Index* or the Library of Congress Catalog.

AmRefBksAnn 12:3 (1981); *Choice* 18:1067 (Apr 1981); *RQ* 20:303 (Spr 1981).

661 *American Book Publishing Record Cumulative 1950—1977: An American National Bibliography.* New York: R.R. Bowker, 1978. 15 vols. Author index: Vol. 13; Title in—dex: Vol. 14; Subject index: Vol. 15

This is a sequel to the foregoing work [660] covering the years 1950—1977. It lists the majority of the books on para—psychology published in the U.S. from 1950—1977. The basic arrangement is by Dewey Decimal Classification, so the bulk of the books on parapsychology will be found in Volume 1 listed under Dewey numbers 133—133.9.

Coll&ResLibr 40:358 (Jul 1979); *Choice* 16:1147 (Nov 1979); *LibrJ* 104:1442 (Jul 1979); *RQ* 19:79 (Aut. 1979).

804 Ashby, Robert H. (Frank C. Tribbe, Ed.). *The Ashby Guidebook for Study of the Paranormal* (rev. ed.). [See Addenda for annotation]

662 Chicorel, Marietta (Ed.). *Chicorel Index to Parapsychology and Occult Books.* New York: Chicorel Library Publishing Corp., 1978 (Chicorel index series, v. 24). 354p. Glossary: 9–12

Chicorel is a librarian, editor and publisher of indexes. She says: "This book provides a subject guide to the available books, regardless of their relative value to adherents of particular systems, beliefs, or groups" (p. 3). A total of 3,229 books are listed under 126 headings. Only English–language books published since 1908, out of print and in print, are included. About one–sixth of the headings are aspects of "mainline parapsychology," although most headings are tangentially relevant. The arrangement is alphabetical by subject, and within that by author's last name.

JP 43:350–352 (Dec 1979).

663 Clarie, Thomas C. *Occult/Paranormal Bibliography: An Annotated List of Books Published in English, 1976 Through 1981.* Metuchen, NJ: Scarecrow Press, 1984. 561p. Author index: 495–511; Subject index: 539–561; Title index: 512–538

This is a sequel to Clarie's earlier work [**664**], and it covers the years 1971 through 1975. Together they provide an extensive and insightful compendium of information on a decade of all aspects of the paranormal presented in book form. Included in this volume are "books on the occult (witchcraft, demonology, black and white magic, occult philosophy and occult traditions), books on occult–related topics (strange natural phenomena, astrology, prophecy), and books on the paranormal (including supernatural phenomena and the large body of work on psychical research/parapsychology)" (p. xi). The arrangement is alphabetical by author. There is also an index to authors, joint authors, editors, translators, and illustrators and a useful title index, which includes variant titles and foreign titles as well as English translations. The subject index contains 1,456 terms, 44 "see" and 91 "see also" references. It is arranged by five broad topics, followed by subheadings. The five topics are occult, psychical phenomena, psychical research, spiritualism, and magic. There are also useful form categories: reference; juvenile literature; autobiography; biography; bibliographies, major; and "bibliographies" (over three pages long). Not only does Clarie provide very useful and sometimes quite lengthy annotations, he also quotes the opinions of other reviewers. Emphasis in the annotations is given to content before all else.

JASPR 80:232–235 (Apr 1986); *PAI* 3:40 (Jun 1985).

ABBookman 76:3128 (Oct 28, 1985); *AmLibr* 15:748 (Nov 1984); *AmRefBksAnn* 16:245 (1985); *AmBkCollector* 6:50 (Jul

1985); *Choice* 22:532 (Dec 1984); *LibrJ* 109:1313 (Jul 1984); *RefBkRev* 6(2):25 (1985).

664 Clarie, Thomas C. *Occult Bibliography: An Annotated List of Books Published in English, 1971 through 1975.* Metuchen, NJ: Scarecrow Press, 1978. 454p. Author index: 391—409; Subject index: 440—454; Title index: 411—439

This is an annotated bibliography of books in English published from 1971—1975 dealing with the "occult," which the compiler defines as "an event or series of events supernatural or extra—natural to man, the sources or reasons for which can be determined by any human gifted or sensitive enough, and particularly, by any person willing to devote the necessary patience to fathoming the system of the universe that is behind all events" (p. xvii). The primary arrangement is alphabetical by authors' last names. There are supplementary author, title, and subject indexes. The latter is especially important, and contains a number of parapsychological terms. The annotations are excellent.

AmRefBksAnn 11:669 (1980); *BCM* 4:40 (Nov 1979); *Booklist* 76:1084 (Mar 15, 1980); *LibrJ* 104:616 (Mar 1, 1979); *Phoenix* 3:53 (Sum 1979); *WilsonLibrBul* 53:724 (Jun 1979).

665 Eberhart, George M. (Compiler). *A Geo—Bibliography of Anomalies: Primary Access to Observations of UFOs, Ghosts, and Others Mysterious Phenomena.* Westport, CT: Greenwood Press, 1980. 1114p. Glossary: xxv—xxxvii; Ethnic group index: 1113—1114; 1 map; Observer index: 1055—1109; Ship index: 1111—1112; Subject index: 1039—1053

The aim of this impressive work is to serve as a "detailed case index—bibliography for all scientific and some historical anomalies, and to make an attempt at standardization of terms for anomalous events" (p. xiii). The entries are arranged geographically by state or province, and within that by city or town. Each item consists of a brief characterization of the type of event, the date it was observed, the name of the observer, the specific location (e.g., a firehouse, an apartment), plus any further information of relevance (such as illustrations), and the reference source. Sources drawn upon are books, newspapers, and the complete contents of 28 journals plus other journal references. The parapsychological journals checked are the *Journal of Parapsychology* and the *Journal of the American Society for Psychical Research.* The magazine *Fate* is also included. An outline of anomalies is given that covers nine broad groups, three of them parapsychological: "Parapsychology," "Survival," and "Phantoms." This is a valuable tool for persons searching for sources of spontaneous cases, and its geographic approach appears to be unique.

JSPR 51:106+ (Jun 1981); *T* 10:18–19 (Spr 1982).

AmRefBksAnn 12:702 (1981); *Booklist* 78:462 (Nov 15, 1981); *Choice* 18:774 (Feb 1981); *LibrJ* 105:1847 (Sep 15, 1980); *WilsonLibrBul* 55:299 (Dec 1980).

666   Goss, Michael (Compiler). *Poltergeists: An Annotated Bibliography of Works in English, circa 1880–1975.* Metuchen, NJ: Scarecrow Press, 1979. 351p. Index: 283–331; Geographical index: 332–351

This is an annotated bibliography of 1,111 books, articles, reviews, research papers, and printed correspondence dealing with poltergeists from 1880–1975. Compiled by a librarian, it is aimed at parapsychologists, professionals in other fields, and general readers. The citations are listed alphabetically by author (or title, if no author is named). Technical material is marked "Research Level." A geographical index presents the cases cited in the text under standard geographical divisions.

JP 43:350–352 (Dec 1979); *JSPR* 50:303–304 (Mar 1980); *PR* 10:17 (Sep/Oct 1979).

LibrJ 104:1441 (Jul 1979).

667   Kies, Cosette N. *The Occult in the Western World: An Annotated Bibliography.* Hamden, CT: Shoe String Press, 1986. 233p. Glossary: 191–200; Name index: 201–211; Title index: 212–233

The aim of this work is to provide a degree of bibliographic control over the books on the occult. Emphasis is placed on the occult as it is viewed today, although a number of classic, historic works are also included. Annotations range from one to six sentences in length. There are 13 main headings: "General Works on the Occult" (broken down into nine form categories such as "Bibliographies," "Library Catalogs," "Biographies," etc.); "Traditional Witchcraft and Satanism" (broken down by form and geographically); "Modern Witchcraft and Satanism" (broken into three groups: "General Overviews," "Favorable (Advocacy) Approaches," and "Anti–Witchcraft Works"); "Magic and the Hermetic Arts" (broken into eight subcategories, mainly by subject, e.g., "Gypsy Lore," "Alchemy," "Numerology," "I Ching"); "Secret Societies, Exotic Religions, and Mysticism" (grouped in six areas mainly by subject, e.g., "Druids and Druidism," "Freemasonry," "Rosicrucianism," etc.); "Psychics and Psychical Research" (eight categories: "Reference Sources," "General Works," "Out–of–the Body Experiences," "Psychokinesis," "Clairvoyance and Dowsing," "Auras and the

Kirlian Effect," "Telepathy," and "Altered States of Consciousness"); "Ghosts, Poltergeists, and Hauntings" (a general, and three geographic categories); "Primitive Magic and Beliefs" ("General," "African," "Santeria"); "Myths, Legends, and Folklore" ("General," plus six subject categories such as "Lost Worlds," "Manimals," "Vampires"); Ancient Astronauts, Disappearances, and UFOs" ("General," "Bermuda Triangle," "UFOs"); "Prophecy and Fortune–Telling" (General, plus sub– ject categories: "Crystal Gazing," "Palmistry," "Phrenology," "Tarot"); "Astrology" (five form subheadings); and "Skeptics and Debunkers."

JASPR 82:396–397 (Oct 1988); *PAI* 6:44–45 (Jun 1988).

668 Naumov, E.K., & Vilenskaya, L.V. *Bibliography of Para– psychology (Psychotronics, Psychoenergetics, and Psychobiophysics) and Related Problems.* Translated from the Russian. Alexandria, VA: Parapsychological Associa– tion, 1981. 171p.

This is a comprehensive bibliography of works in Russian on parapsychology and related subjects. The titles of the books and articles have been translated into English, making it pos– sible for non–Russian readers to obtain a topical overview of Soviet parapsychology. The material is arranged by subject. There are two main sections: "I. Parapsychology, Psychotronics, Psychoenergetics, and Psychobiophysics," which includes general material and publications on dowsing (biophysical effect); and "II. Related and Boundary Problems (Parapsychology and Medicine, Living Organisms and Energetic–Environmental Fac– tors, Radiation of Living Organisms, Methods of Investigation of Bio– and Psychoenergetic Processes, and Miscellaneous)." An appendix lists articles by Soviet authors in proceedings of inter– national conferences, and two supplements update the bibliogra– phy through the beginning of 1980. This bibliography is updated in additional publications by Larissa Vilenskaya [673-677].

JP 45:249–251 (Sep 1981); *PR* 7:21–22 (Jul/Aug 1976); *T* 10:95–96 (Win 1982).

669 Nester, Marian L., & O'Keefe, Arthur S.T. (Compilers). *Selected Bibliography in Parapsychology for Instructors and Students.* New York: American Society for Psychical Research, 1979. 27p.

This is a subject arrangement (44 categories) of 358 ar– ticles selected from the *Journal of the American Society for Psychical Research, Journal of Parapsychology, Journal of the Society for Psychical Research, Parapsychology Review,* and *Research in Parapsychology* from January 1970 through July 1979. The compilers indicate whether each article is introductory

(I) or advanced (A). The types of articles selected are over—
views, interdisciplinary studies; fringe areas, e.g., auras, Kirlian
photography, magic; and instrumentation.

670 Nester, Marian L., & O'Keefe, Arthur S.T. (Compilers).
*Selected Bibliography in Parapsychology for Instructors
and Students.* New York: American Society for Psychical
Research, 1981. 32p.
This is a subject listing of 334 articles published in the
*Journal of Parapsychology, Journal of the Society for Psychical
Research,* and *International Journal of Parapsychology* from 1960
through 1969 and in the *Journal of the American Society for
Psychical Research* from 1950 through 1969. It is a selective bib—
liography, with those articles chosen that were especially
readable and that were on subjects likely to be covered in
parapsychology courses. Approximately 1/20th of the listings con—
tain annotations. Items are listed under 39 subject headings.
Each citation is listed as being Introductory, Advanced, or both.

671 Sable, Martin H. *Exobiology: A Research Guide.* Brighton,
MI: Green Oaks Press, 1978. 324p. Author index: 265—
311; Subject index: 312—319
A bibliographic guide to books and articles on extrater—
restrial life and UFOs, compiled by a professor of library
science. The main arrangement is by subject and there are sup—
plementary author and subject indexes. It is included here be—
cause two of the subjects covered in this well—researched guide
are relevant to parapsychology: "Occult Aspects, Including ESP,
Mental Telepathy, Psychological—Physical Effects, etc." and
"Teleportations."
AmRefBksAnn 10:663 (1979); *Choice* 15:1502 (Jan 1979);
LibrJ 103:1970 (Oct 1, 1978).

672 Society for Psychical Research. *Catalogue of the Library of
the Society for Psychical Research.* Boston, MA: G.K.
Hall, 1976. 341p.
The library of the Society for Psychical Research, which
was founded in 1882, is one of the largest (8,000 volumes) and
best—organized parapsychology libraries in the world.   This
book reproduces the cards in the SPR Library's catalog.   Al—
though the bulk of the books are in English, titles in other lan—
guages are also listed.   Part 1 is the "Title Catalogue" and
Part 2 is the "Author Catalogue."
PR 7:21—22 (Jul/Aug 1976).

673 Vilenskaya, Larissa (Compiler). *Parapsychology in the
USSR, Part I: Psychoregulation and Psychic Healing.* San

Francisco, CA: Washington Research Center, 1981. 27p.

Russian emigre Vilenskaya, an engineer and parapsy—chologist, is also a prolific bibliographer. The first of the three parts of this work contains translations of eight articles originally published in Russian reference books, newspapers, and magazines. Of note are Vilenskaya's own "Some Impressions Concerning Healing in the USSR," selected documents on the healing of Juna Davitashvili, and another by Vilenskaya on Bar—bara Ivanova's research. Part II is a 181 item bibliography of Soviet parapsychology published in English or translated into English. Part III is a bibliography of 217 publications on Soviet parapsychology by western writers. Vilenskaya concludes with a brief article on the practical applications of psi ability.

674 Vilenskaya, Larissa (Compiler). *Parapsychology in the USSR, Part II: The Biofield: Its Nature, Influences, and Interactions.* San Francisco, CA: Washington Research Center, 1981. 57p.

In this volume, bibliographer Vilenskaya provides 18 trans—lations of articles from the Russian, including three under "Living Organisms and Energetic Environmental Factors," seven under "Radiation and Living Organisms," four under "Application without Understanding," and four under "The Un—explored Around Us." The fifth section is bibliographical, and it contains additional items in the three classes established in Part I: "Bibliography on Parapsychology in the U.S.S.R. (Recent Publications)," "Bibliography on Parapsychology in the USSR in English," and "Bibliography on Parapsychology in the USSR: Foreign Language Publications."

675 Vilenskaya, Larissa (Compiler). *Parapsychology in the USSR, Part III: Observations and Experimental Studies in ESP, PK, and Psychic Healing.* San Francisco, CA: Washington Research Center, 1981. 71p. 2 figures; 3 graphs; 7 illustrations; 3 tables

In this compilation, there are 13 articles translated from the Russian, five under "Studying ESP and PK in Animals and Humans," and eight under "Psychic Diagnosis and Healing: Ex—perience and Research." The third section consists of new items regarding parapsychology and related subjects in the USSR in English. In a brief concluding section she discusses research in parapsychology and psychic healing.

676 Vilenskaya, Larissa (Compiler). *Parapsychology in the USSR, Part IV: Approaches and Findings in Studying Psi Phenomena and Related Subjects.* San Francisco, CA: Washington Research Center, 1981. 80p. 5 figures;

10 graphs; 2 tables
With this volume, Vilenskaya gives recognition to the fact that this series was becoming a periodical. This issue consists of translations of 19 items under five headings: "Experimental Studies" (5), "Biofield and Its Role in Interactions with Living Systems" (6), "In the pages of Soviet and Eastern Press" (3), "Uses and Abuses in Mind Research" (3), and "The Unexplored Around Us" (3). The bibliography is a continuation of those in earlier issues. She closes with "Parapsychology at the Crossroads," in which she classifies and discusses methodology in parapsychology and likens the present situation in the field to that of chemistry before the periodic table was discovered. She calls for "a comprehensive non—contradictory theory . . . which could generalize the existing findings and indicate new directions of research" (p. 80).

677 **Vilenskaya, Larissa (Compiler).** *Parapsychology in the USSR, Part V: Approaches and Findings in Studying Psi Phenomena and Related Subjects.* San Francisco, CA: Washington Research Center, 1982. 67p. 6 figures; 2 graphs; 10 tables
This volume consists primarily of Soviet studies in fields that interface with parapsychology. The first three parts consist of translations of articles on I. "Living Organisms and Energy Environmental Factors" (3); II. "The Biofield and its Role in Interactions of Living Organisms" (5); and III. "Latent Percep—tual Abilities" (3). Part IV is a bibliography listing new material on Soviet parapsychology. The concluding part consists of translations of Soviet controversies over healing research with the healer Juna Davitashvili as subject.

678 **White, Rhea A.** *Kirlian Photography Bibliography* (3rd ed.). Dix Hills, NY: Parapsychology Sources of Information Center, 1988. 45p.
This work consists of an author listing of English—language books; chapters; articles in journals, magazines, and newspapers; theses; conference proceedings; research notes; let—ters; and abstracts. It includes both popular and technical references.

679 **White, Rhea A.** *Parapsychology: A Reading and Buying Guide to the Best Books in Print.* Dix Hills, NY: Parap—sychology Sources of Information Center, 1987. 99p. Glos—sary: 90—97; Name index: 68—72; Subject index: 85—89; Title index: 73—84
This work is a listing by author and subject of English—language books in print on parapsychology deemed worthy of

note either because of the quality of the presentation, authority, or presence of an unusual feature such as a bibliography, or because it is one of the few works available on a particular aspect of parapsychology. A total of 421 books are included. The listing is in four major sections: "Author Listing: Current Books" (published 1960 to date); "Author Listing: Classics" (books originally published prior to 1960 that have remained in print or have been reprinted because they are still considered to be of value); "Subject Listing: Current Books"; and "Subject Listing: Classics." The books are listed alphabetically by author's last name. Each entry contains the author's full name, the book title, place of publication, publisher, date of publication, number of pages (p.) or volumes (vols.) and price in dollars ($) or in pounds (£).

JP 52:242−244 (Sep 1988); *JSPR* 55:160−161 (Jul 1988).

AmRefBksAnn 20:283 (1989); *Archaeus* 7:47 (Spr/Sum 1988); *SpirFron* 20:186−187 (Sum 1988).

821 White, Rhea A., and Dale, Laura A. (Compilers). *Parapsychology: Sources of Information.* [See Addenda for annotation]

680 White, Rhea A., and Vilenskaya, Larissa (Compilers). *Parapsychology in the Soviet Union, Eastern Europe, and China: A Compendium of Information.* Metuchen, NJ: Scarecrow Press, in press. Author index; Glossary; Subject index

This is the most comprehensive bibliography of books, articles, reports, dissertations, conference papers, popular articles, and news items on parapsychology in the USSR, East European countries, and China. Both compilers are bibliographers and parapsychologists. The first section consists of "Selected Writings about Soviet, East European, and Chinese Psi Research" in order to provide the reader with firsthand acquaintance with the types of materials cited. The second section gives the texts of reference articles on parapsychology drawn from Soviet, Chinese, and a Western source. Section III is an annotated bibliography of earlier bibliographies. Section IV begins the bibliography proper and is a "Bibliography of Western Publications in English." Section V is a "Bibliography of Publications by Soviet, East European, and Chinese Authors in English." Section VI is a "Selected Bibliography of Publications by Soviet, East European and Chinese Authors in Their Native Language." Section VII is a "Selected Bibliography of Publications in Finnish, French, German, Italian, and Spanish." Also included is a "Glossary of Terms Used in Soviet, East European, and Chinese Research on Psi and Related Subjects," a "Biographical Diction−

ary of Persons Associated With Psi Research in the USSR, East
European Countries, and China," chronologies of events indicating interest in the USSR and in the People's Republic of China,
directories of research institutes where psi research is being conducted in the Soviet Union and in China, and a list of "Soviet,
East European and Chinese Journals and Newspapers Publishing
Articles on Psi Research."

DICTIONARIES AND ENCYCLOPEDIAS

681 Brookesmith, Peter (Ed.). *The Unexplained: Mysteries of
Mind, Space and Time.* Freeport, NY: Marshall Caven-
dish, 1983. 8 vols. 1120p. Copious illustrations; Index:
Vol. 8, 1103–1120

This popular encyclopedia is on "Curiosities and Wonders."
The consultants were A.J. Ellison, J. Allen Hynek, Brian Inglis,
and Colin Wilson. There is no discernible arrangement of subjects in each volume, but each one has a contents listing, a list
of major contributors, and an "A to Z Quick Reference." Most
articles are signed. There is a subject index to the set in the
last volume. The set is lavishly illustrated, with many illustrations in color. Entries range from a paragraph to five pages.
Cross–references are sometimes included to other articles and a
few citations are given to sources of further information. Although not a source of definitive information, *The Unexplained*
provides a "first look" and may well cover individual subjects
not readily available elsewhere. About a third of the subjects
covered deal with parapsychology. As a representative sample,
Volume 1 has 37 articles, of which 10 could be said to deal
with aspects of parapsychology, with the emphasis on practical
considerations.

682 Brookesmith, Peter (Ed.). *The Unexplained II: Mysteries
of Mind, Space and Time.* Freeport, NY: Marshall
Cavendish, 1985. 8 vols. (pp. 1121–2212). Copious illustrations; Index: Vol. 8, 2221–2212

This is a continuation of the preceding set, and the pagination continues where the first set left off. The articles are independent of those in the first set but in all other respects the
arrangement is the same. As a representative sample, Vol. 2
has 35 articles, of which nine are on aspects of parapsychology.

683 Cavendish, Richard (Ed.). *Encyclopedia of the Unexplained:
Magic, Occultism and Parapsychology.* (Consultant: J.B.
Rhine.) New York: McGraw–Hill, 1974. 304p. Bibliography: 286–297; 7 figures; 175 illustrations; Index: 300–
304; 4 maps

This is the most authoritative of recent attempts to provide an encyclopedic approach to parapsychology (although, as the subtitle indicates, it is by no means devoted only to parapsychology). J.B. Rhine served as "Special Consultant on Parapsychology," and is also a contributor. The arrangement is alphabetical, and the majority of the entries are less than a page in length, but there are articles of several pages for major subjects such as Mediums, Parapsychology, and Spiritualism. The longer articles are signed and indicate further sources of information. In order to facilitate finding information, the entries are alphabetically arranged; there are "see" and "see also" references within the text; there is an index; and there is a 521—item bibliography.

JP 38:249—250 (Jun 1974); *JSPR* 47:450—452 (Sep 1974); *T* 4:15 (Win 1976).

684 Cavendish, Richard (Ed.). *Man, Myth & Magic: An Illustrated Encyclopedia of Mythology, Religion and the Unknown.* Freeport, NY: Marshall Cavendish, 1983. 12 vols plus Index. 3376p. Bibliography: 3268—3314; Copious illustrations; Index: Vol. 13

This encyclopedia is edited by British author Richard Cavendish, an authority on magic and occultism, and is a valuable tool for students and scholars. The basic arrangement is alphabetical. Some entries are limited to a paragraph but most are a page to several pages in length. Longer articles are signed and cite bibliographic references. Contributors are identified in Vol. 13, among them parapsychologists M.R. Barrington, C.D. Broad, E.R. Dodds, A. Gauld, K.M. Goldney, C. Green, A. Gregory, R. Heywood, R.G. Medhurst, A.R.G. Owen, J.B. Rhine, L.E. Rhine, L. Rose, and W.H.C. Tenhaeff. Major subjects covered in detail are psychical research, occultism, magic, supernatural, religion, spiritualism, and mythology. The volumes are profusely illustrated in color and the last volume contains a detailed index, a classified bibliography and brief information about the contributors. J.B. Rhine wrote the articles on extrasensory perception and psychokinesis; Alan Gauld contributed "Psychical Research"; L.E. Rhine, the article on spontaneous psi; G.K. Nelson wrote "Immortality and Spiritualism"; and a long article on "Mediums" was written by R.G. Medhurst and M.R. Barrington.

685 Shepard, Leslie A. (Ed.). *Encyclopedia of Occultism & Parapsychology* (2nd ed.). Detroit, MI: Gale Research Company, 1984—1985. 3 vols. 1617p. General index: 1487—1576; Topical indexes: 1577—1617

The first edition of this work was based on a reprint of

the entries in the old encyclopedias of Fodor and Spence and the *Biographical Dictionary of Parapsychology* [200] with some new categories added by the editor. According to the publisher, the entries in the first edition have been rewritten and revised and many new entries have been added, covering recent phenomena, terms, persons, organizations, publications, trends, and cults. An effort has been made to present both sides of controversial topics with references to further information. A strong point of the work is its extensive indexing, which consists not only of a standard alphabetical index but also several topi‐ cal indexes such as "Paranormal Phenomena," "Geographical," "Periodicals," "Societies and Organizations." Space does not per‐ mit listing the errors here, but many have been noted by reviewers. Nonetheless it represents a monumental effort and it is a unique reference tool for psychical research. As far as parapsychology is concerned, only persons, organizations, and spontaneous phenomena are covered. Not even such common terms as "psi—missing," "decline effect," and "random event gen‐ erator" appear in the index.

   JASPR 73:315—319 (Jul 1979); *JASPR* 81:396—402 (Oct 1987); *JP* 42:328—330 (Dec 1978); *JSPR* 53:398—399 (Jul 1986); *PAI* 4:69—70 (Jun 1986); *ZS* No.1:54—55 (May 1978).

   AmRefBksAnn 17:289 (1986); *Choice* 23:1654 (Jul 1986); *Explorer* 2:6 (Jan 1985); *FantRev* 7:13 (Sep 1984); *FantRev* 8:26 (Oct 1985); *RefBkRev* 7(2):24 (1986); *RefServRev* 14:53 (Sum 1986); *SanFranChron* 7:36 (Mar 1986); *SFictR* 15:16 (Feb 1986); *SkptInq* 3:51—58 (Sum 1979); *WilsonLibrBul* 60:64 (Dec 1985).

686 **Shepard, Leslie A. (Ed.).** *Occultism Update: First Supple—* *ment to the Second Edition of Encyclopedia of Occultism* *& Parapsychology.* Detroit, MI: Gale Research, 1987. 102p. Index: 95—102

   This is the first of two supplements to the three—volume *Encyclopedia of Occultism and Parapsychology,* also edited by Shepard. These "supplements cover new events, trends, organiza‐ tions, personalities, and discoveries . . . as well as research material not included in the *Encyclopedia* and revaluations of existing topics" (p. vii). The format follows that of the present encyclopedia [685]. Updates or modifications to entries already in the *Encyclopedia* cite the page of the article they are updat‐ ing. Most of the material is new: thus these supplements are valuable. They are useful even without the main volumes, but of course serious persons and libraries should own the complete set. Although there will always be omissions and not all the entries are accurate, it is the best encyclopedic work available.

   JASPR 81:402—403 (Oct 1987).
   AmRefBksAnn 19:310 (1988).

687 Thalbourne, Michael A. (Compiler). *A Glossary of Terms Used in Parapsychology.* Foreword by John Beloff. London, England: Heinemann, 1982. 91p. Bibliography: 89–90.

In his Foreword, John Beloff discusses the controversy about terminology in parapsychology, and he concludes that "it is a mistake to spend too much time and effort worrying about particular terms. What *is* important is that these terms should be clearly understood and their application be made consistent and this is something which a glossary such as this can help bring about" (p. ix). In his Preface, Thalbourne states that the "present glossary contains approximately 100 terms which will not be found in any previous compilation" (p. xi). It also contains some neologisms proposed by Thalbourne. Two special features of this glossary are that some etymological information is included and that citations to the original source material where a term or expression was first used. There are many dictionaries that claim to cover parapsychological terms, and they do to an extent, with emphasis on terms used in the older literature of psychical research, but this book, written as it is by a working parapsychologist, serves as the best source available of terms in use in parapsychology today, especially those dealing with the experimental approach.

JASPR 77:347–349 (Oct 1983); *JP* 47:75–76 (Mar 1983); *JSPR* 52:205–207 (Oct 1983); *PsiR* 2:104–106 (Jun 1983); *PsiR* 2:106–108 (Jun 1983).

BritBkNews p221 (Apr 1983); *Choice* 21:1593 (Jul 1984); *ContempPsychol* 23:644–646 (S 1978); *ContempRev* 242:112 (Feb 1983); *Light* 103:29–31 (Spr 1983).

688 Wolman, Benjamin B. (Ed.). *Handbook of Parapsychology.* New York: Van Nostrand Reinhold, 1977. 967p. Chapter references; 3 figures; Glossary: 921–936; 2 graphs; 29 illustrations; Name index: 937–951; Subject index: 953–967; Suggested reading list: 907–920; 9 tables

This scholarly handbook is aimed at the serious reader and attempts to review in some depth all of the major approaches to parapsychology and to delineate the interface of parapsychology with other disciplines. Wolman has specialized in the compilation of handbooks for various fields of the behavioral sciences, and in compiling this volume he was joined by three associate editors who have had a long association with parapsychology: Laura A. Dale, Gertrude R. Schmeidler, and Montague Ullman. The first of 11 parts consists of three historical chapters: an "Historical Overview" by J. Beloff, a history of experiments by J.B. Rhine, and an article on "William James

and Psychical Research" by G. Murphy. Part II has three chap—
ters on research methods: L.E. Rhine on spontaneous cases, D.S.
Burdick and E.F. Kelly on statistical methods, and G.R.
Schmeidler on "Methods for Controlled Research on ESP and
PK." Part III contains review articles on four key aspects of
psi research: ESP, by J.B. Rhine, attitudes and personality
traits and ESP by J. Palmer, intrasubject and subject—agent ef—
fects by J.C. Carpenter, and the role of the experimenter by
R.A. White. The fourth part has four chapters on physical sys—
tems and psi: a history by J.F. Nicol, PK experiments by R.G.
Stanford, "Poltergeists," by W.G. Roll, and "Paranormal Photog—
raphy" by J. Eisenbud. Part V contains three chapters on Al—
tered States and Parapsychology: C. Honorton covers "Psi and
Internal Attention States," R.L. Van de Castle covers "Sleep
and Dreams," and C.T. Tart covers "Drug—Induced States of
Consciousness." Healing is the subject of Part VI, which has a
chapter on psychotherapy and psi by J. Ehrenwald, another by
him on the healing arts, and one by M. Ullman on
psychopathology and psi. Survival is the subject of Part VII,
which has a chapter by A. Gauld on survival and one by I.
Stevenson on reincarnation. Part VIII, on the relation of parap—
sychology to other fields, has chapters by R.L. Van de Castle
on anthropology and parapsychology, by R.L. Morris on biology
and anpsi, by J. Ehrenwald on psi and brain research, by
J.H.M. Whiteman on physics and psi, by J. Beloff on
philosophy and parapsychology, one on religion and para—
psychology by W.H. Clark, and one on literature and para—
psychology by J.M. Backus. Part IX is about models and
theories of parapsychology and consists of four chapters by
C.T.K. Chari, R.G. Stanford, A. Koestler, and B.B. Wolman.
The tenth part is a single chapter on Soviet parapsychology by
J.G. Pratt, and the last part consists of an annotated list of
suggested readings by R.A. White and a glossary by L.A. Dale
and R.A. White.

   ChrP 3:20—21 (Dec 1978); *JASPR* 72:267—274 (Jul 1978);
JIndPsychol 2:157—160 (Jul 1979); *JP* 41:352—358 (Dec 1977);
JSPR 49:892—893 (Sep 1978); *PR* 9:9; *PN* 1:4 (Jul 1978); *T*
7:15—16 (Fal 1979).
   Choice 14:1568 (Jan 1978); *ContempPsychol* 23:644—646
(Sep 1978); *ContempPsychol* 23:644 (Sep 1978); *Encounter* 54:86
(Jan 1980); *LibrJ* 102:2505 (Dec 15, 1977); *LibrJ* 103:512 (Mar
1, 1978); *Light* 101:176 (Win 1981); *NewScientist* 77:590 (Mar 2,
1978); *PsycholToday* 11:162 (Dec 1977); *RefServRev* 14:54 (Sum
1986).

INDEXES

689 Ford, Clyde W. (Compiler). *Brain Mind Bulletin: Keyword Index, Volume 1–Volume 5.* Los Angeles: Interface Press, 1981. 111p.

This index covers all major articles appearing in the first five volumes of *Brain/Mind Bulletin.* Not included are book reviews, listings of upcoming events, obituaries, and brief quotations. Personal names have not been indexed unless they have a larger meaning, e.g., the Pribram/Bohm theory. Only a fraction of *Brain/Mind Bulletin* covers parapsychological subjects, but easily half of the material is nonetheless relevant to parapsychology: articles on altered states, hemispheric lateralization, yoga, trance, relaxation, psychedelics, mystical experience, near–death experience, meditation, hypnosis, holographic theory, and many more. Those terms directly relevant are clairvoyance, psychic phenomena, ESP, PK, precognition, parapsychology, synchronicity, and telepathy.

690 Morton, Diane. *Common Ground Index.* San Antonio, TX: Author [1982]. 24p.

Common Ground was a British publication with variant subtitles: (1) *Studies at the Fringe of Human Experience* and (2) *The Radical Journal of the Paranormal.* A total of nine issues were published from 1981–1982. This index covers Numbers 1–6. There is a subject index (including persons and organizations), an author index, an index to book reviews by author, and an index to book reviews by title.

691 Morton, Diane. *Psychic Magazine Index.* San Antonio, TX: Author [1983]. 75p.

Psychic, which began publication in 1969, was a popular bimonthly magazine devoted to parapsychology. It had feature articles, interviews with prominent persons associated with parapsychology, and news items on the paranormal. It broadened its coverage when it became *New Realities* in 1977. This is the first index compiled by Ms. Morton, who is the librarian of the Mind Science Foundation. The index is in four sections, the first of which is the subject index, which includes names of people, cases, and places. Part 2 is an author index, Part 3 is a list of books reviewed arranged by author, and Part 4 lists books reviewed by title. Reviewing her own efforts, Ms. Morton indicates that if she were doing it again she would change some of the subject headings as well as use more of them, and she would also have a reviewer list for the book reviews. She points out that she omitted some of the smaller articles. Nonetheless, this is a useful index and makes it possible to locate needed information that otherwise would take considerable time to find.

691a White, Rhea A. (producer, editor). *PsiLine.* Dix Hills, NY:
     Parapsychology Sources of Information Center, 1984—
     Although it is not a book, a fairly complete index to the
journals and books on parapsychology exists and merits mention
in this section.   It is   the *PsiLine* database, which abstracts
and indexes   all the major journals from 1940 on and some for
earlier years.   It also includes many books from the early twen—
tieth century to date, dissertations, chapters, and conference
proceedings.   About one third of the contents of *PsiLine* is pub—
lished and indexed in great detail in *Parapsychology Abstracts
International* [828].   For further information contact the author
at the Parapsychology Sources of Information Center [890].

692 *The Ten—Year Index of* The Skeptical Inquirer: *Volume 1
     Through Volume 10, 1976–1986.* Buffalo, NY: Committee
     for the Scientific Investigation of Claims of the Paranor—
     mal, 1987. 101p.
     This index covers the *Skeptical Inquirer* (formerly the
Zetetic) [849], the official journal of the Committee for the
Scientific Investigation of Claims of the Paranormal, from
Volume 1 through Volume 10. A third consists of an "Author
Index," which also gives the title of each author's contributions
after his or her name. Another third is a "Book Reviews
Index," which consists of three separate listings: by author, title,
and reviewer. The final third is the "Subject Index," which also
lists names of persons, places, and organizations.   Citations to
standard features of the *Skeptical Inquirer* are indicated by one
of the following capital letters: F = Forum, L = From Our
Readers, N = News and Comment, and P = Psychic Vibra—
tions.   The index was adapted from an earlier one compiled by
Diane Morton of the Mind Science Foundation.

Reincarnation Research

INTRODUCTION

This is a new section, necessitated primarily by the pioneering research of Ian Stevenson of the University of Virginia. There were four books on reincarnation in *Sources* Guirdham [250], Head and Cranston [252], Head and Cranston [253], and Stevenson [266], but they were listed under the category of "Survival." Only one is still of any major relevance, and that is Stevenson's first book on the subject [266].

The 12 books in this section can be categorized as research reports: Ryall [696], Stevenson [697-700, 818], and Whitton [702]); anthologies of writings about reincarnation: Head and Cranston [694], and surveys of the evidence for reincarnation: Christie—Murray [693], Rogo [695], Walker [701], and Wilson [703].

Additional books with information on reincarnation: Grof [348] discusses reincarnation experiences. Reincarnation is touched on in the book by Barbara Ivanova [390]. Some of the books for young people have chapters on reincarnation: Andreae [415], Ebon [430, 431], and Kettlekamp [441]. Hines [474] is critical of reincarnation. Paul Beard [747] discusses reincarnation, which is frequently alluded to in mediumistic communications.

BIBLIOGRAPHY

693 Christie-Murray, David. *Reincarnation: Ancient Beliefs and Modern Evidence.* Newton Abbot, England: David and Charles, 1981. 287p. Bibliography: 267—274; Chapter references; Index: 283—287
The author, a retired assistant master at Harrow School and a member of the Council of the Society for Psychical Research, has written this survey of the evidence for and theories of reincarnation in order to enable the reader to draw his or her own conclusions concerning the phenomenon. The types of evidence described are déjà vu, the life—readings of Edgar Cayce, hypnotic age regression, the Christos phenomenon, the research of Ian Stevenson, and the experiences of Arthur

Guirdham. The last two chapters provide a summary of the ar—
guments for and against reincarnation and a theology of rein—
carnation that tries to summarize the many differing views
about the progress of the entity after death.
    Choice 19:1573 (Jul 1982); *Light* 101:88—89 (Sum 1981).

**694   Head, Joseph, and Cranston, S.L. (Comps. & Eds.). *Rein—
        carnation: The Phoenix Fire Mystery.* New York: Julian
        Press/Crown, 1977. 620p.** Chapter notes; Index: 611—619
    This anthology is the successor to two previous volumes
compiled by these editors and published in 1961 and 1967. The
authors retain the best of their previous works, while expanding
and revising them extensively, adding new material on reincar—
nation, particularly in Oriental thought. A new section, "Stories
of 'Remembrances' of Past Lives," has been added. Material
drawn from Western thinkers on reincarnation has been
presented in chronological order by author's birth date. Included
are philosophical, religious, theosophical, and scientific views on
reincarnation, together with some case histories and personal
experiences.
    ChrP 3:206—207 (Mar 1980); *JASPR* 73:305—314 (Jul
1979).
    Choice 14:1664—1665 (Feb 1978).

**695   Rogo, D. Scott. *The Search for Yesterday: A Critical Ex—
        amination of the Evidence for Reincarnation.* Englewood
        Cliffs, NJ: Prentice—Hall, 1985. 241p.** Annotated bibli—
        ography: 225—231; Chapter notes; Index: 233—241
    The author conceives of this work as a critical and scien—
tific examination of the reincarnation question. It also provides
an overview of the various kinds of evidence indicative of rein—
carnation.   There are chapters on the historical evidence——
primarily cases of past life recall, spontaneous cases of past life
recall (dreams, waking visions, déjà vu); some of the cases in—
vestigated by Ian Stevenson; Stevenson's research methodology;
hypnotic age regression; xenoglossy; past—life therapy; and
psychedelic drugs and past—life recall.   In the final chapter
Rogo presents some explanatory models that can account for the
evidence.   He grants that "some people can tap into the
memories of other people, cultures, and places long removed in
time" (p. x), but proposes that what the evidence is pointing to
might be something more complex than our traditional views of
reincarnation.
    ChrP 6:192 (Mar 1986); *JASPR* 81:193—197 (Apr 1987);
PAI 6:47 (Jun 1988).

696   Ryall, Edward W. *Born Twice: Total Recall of a*

Seventeenth—Century Life. Introduction and appendix by Ian Stevenson. New York: Harper & Row, 1974. 214p. Bibliography: 204—205; Glossary: 185—189; 6 illustra— tions; Index: 208—214; 3 maps

Ryall relates the story of the life he recalls leading as John Fletcher, an English yeoman of the seventeenth century. Ian Stevenson, in the Introduction, compares Ryall's memories of his previous life with the Asian cases Stevenson has studied. He discusses the various hypotheses other than reincarnation that might account for this case. The Introduction includes a bibliography, and Stevenson also provides an appendix, "Notes on Verification of Details in *Born Twice*," with its own list of references. Whatever the explanation of Ryall's memories, in wealth of detail they surpass other accounts of presumed reincar— nation. However, see the criticisms of this case by Wilson [**703**], and Stevenson's later modified position (*JSPR*, 1983, *52*, 27).

JASPR 70:423—427 (Oct 1976); *P* 6:46 (Jan/Feb 1976). *Choice* 12:1372 (Dec 1975); *SpirFron* 9:61 (Win 1977).

697 **Stevenson, Ian.** *Cases of the Reincarnation Type, Volume I. Ten Cases in India.* Charlottesville, VA: University Press of Virginia, 1975. 374p. Bibliographical footnotes; Glossary: 361—367; Index: 369—374; 16 tables

Ian Stevenson, former head of the Department of Neurol— ogy and Psychiatry at the University of Virginia, has long been the world's foremost authority on the scientific investigation of reincarnation. This is the initial volume in a series aimed at providing detailed reports of reincarnation—type cases in Asia and the Mideast. Each case was investigated personally by Stevenson, sometimes over a long period of time. The account of each case begin with a summary of the case and of the in— vestigation of it, followed by complete background material based primarily on firsthand testimony. Each account ends with Stevenson's discussion of its weak and strong features. The first volume has a long General Introduction in which Stevenson gives an updated description of his methods of investigation and, as he puts it, "an analysis of the ineradicable——but, I trust, not fatal——sources of error they retain" (p. ix). There is a chapter describing the Hindu belief in reincarnation and a summary of the cases from India.

JASPR 70:231—235 (Apr 1976); *JSPR* 48:306—309 (Sep 1976); *PR* 7:12—15 (Sep/Oct 1976).
AmJPsychiat 133:868—869 (Jul 1976); *Choice* 12:808 (Sep 1975).

698 **Stevenson, Ian.** *Cases of the Reincarnation Type, Volume II. Ten Cases in Sri Lanka.* Charlottesville, VA: Univer—

sity Press of Virginia, 1977. 373p. Bibliographical foot—
notes; Glossary: 363—367; Index: 369—373; 18 tables
This second volume of reincarnation case studies presents
ten thoroughly investigated examples in Sri Lanka (formerly
Ceylon), together with a long essay on the belief in reincarna—
tion among the Sinhalese people, and of the recurrent features
of Sinhalese cases. Psychiatrist Stevenson discusses the strong
and weak features of each case.
    JASPR 73:71—81 (Jan 1979); *JSPR* 49:894—895 (Sep
1978).
    AmJPsychiat 135:1447 (Nov 1978); *Brain/MindBul* 3:4
(May 1, 1978).

699 Stevenson, Ian. *Cases of the Reincarnation Type, Volume
    III. Twelve Cases in Lebanon and Turkey.* Charlottes—
    ville, VA: University Press of Virginia, 1980. 384p. Bibli—
    ographical footnotes; Bibliography: 15—16, 370—372; Glos—
    sary: 375—377; Index: 379—384; 15 tables
University of Virginia psychiatrist/parapsychologist Ian
Stevenson presents 12 detailed case histories of reincarnation, six
in Lebanon and six in Turkey. Most of them were investigated
over a period of eight to ten years. Introductions to the chap—
ters describe the reincarnation beliefs of the Druses in Lebanon
and the Alevis in Turkey. In addition to the case histories
described, Stevenson has summarized the main features of many
other cases of reincarnation reported in these two countries. Of
particular interest is the final chapter in which Stevenson
provides a reevaluation of his interpretation of the reincarnation
evidence since his first book on the subject was published in
1966 [266]. He explains why he no longer considers ESP to be
an alternative explanation, and why he considers normal means
of obtaining information, and reincarnation itself to be the
likeliest interpretation of these cases.
    JRPR 5:57—59 (Jan 1982); *T* 9:20—22 (Win 1981).

818 Stevenson, Ian. *Cases of the Reincarnation Type: Volume
    IV. Twelve Cases in Thailand and Burma.* [See Addenda
    for annotation]

700 Stevenson, Ian. *Children Who Remember Previous Lives:
    A Question of Reincarnation.* Charlottesville, VA:
    University Press of Virginia, 1987. 354p. Bibliography:
    325—341; Chapter notes; Index: 343—354
The author, a psychiatrist and pioneer investigator of rein—
carnation, has written this book about his research for the
general reader. Rather than present detailed evidence of reincar—
nation, he provides a summary of his research methods, the

more important of his results, and his conclusions based on the research. Summaries of 12 typical cases are given and when available, references to detailed reports of the cases are provided. The significance of the title is that rather than being a book about reincarnation, this is about children who claim to remember previous lives. As Stevenson puts it: "From studying the experiences of such children some understanding about reincarnation may eventually come. Before that can happen, however, we must become confident that reincarnation offers the best explanation for these children's apparent memories" (p. 1).

ChrP 7:293−294 (Dec 1988); *JRPR* 12:57−60 (Jan 1989); *JSPR* 55:227−229 (Oct 1988); *PAI* 6:47 (Jun 1988); *PR* 19:8−10 (Jul/Aug 1988).

Choice 25:1761 (Jul/Aug 1988); *JSciExpl* 2:241−244 (1988); *SciBook&Films* 23:274 (May 1988).

701 Walker, Benjamin. *Masks of the Soul: The Facts Behind Reincarnation.* Wellingborough, England: Aquarian Press, 1981. 160p. Bibliography: 145−156; Index: 157−160; 1 table

This is a thorough study of the theory of reincarnation. Walker describes the doctrine of reincarnation as it appears in various religions and also deals with the scientific evidence for reincarnation. Alternate explanations are also discussed. Walker points out that the reincarnation hypothesis is favored by the testimony of tradition while the scientific evidence rules against it. "We therefore cannot be so skeptical as to reject it outright, nor so dogmatic as to accept it without qualification. This book presents the evidence from both sides, and it is left to readers to draw their own conclusions" (p. 8). He does not seem to be concerned with the possibility of doctrinal compliance. The lengthy bibliography leads to other studies.

ChrP 4:95−96 (S 1981); *JSPR* 51:170−171 (Oct 1981); *Psychoenergetics* 4:283−284 (1982).
ContempRev 239:112 (Aug 1981).

702 Whitton, Joel L., and Fisher, Joe. *Life Between Life: Scientific Explorations into the Void Separating One Incarnation from the Next.* Garden City, NY: Doubleday, 1986. 198p. Bibliography: 189−194; Index: 195−198

In the Introduction, Whitton, an M.D., says that in this book, as in his psychotherapy, he proceeds from the assumption that reincarnation is a fact. He says he uses "past−life therapy" to study the spiritual dimension of human beings. Specifically, he presents several case histories in which he used hypnotic age regression to discover that souls do reincarnate and that between incarnations, in the bardo state, they choose

the script for their next lifetime. Whitton proposes that by
looking at our past lives it is possible to learn how specific be—
havioral and emotional patterns developed and so are in a posi—
tion to change them.

AIPRB No11: p19 (Jan 1988); *ChrP* 7:214—215 (Jun 1988).

Booklist 83:308 (Oct 15, 1986); *Kliatt* 22:43 (Apr 1988);
LibrJ 8:102 (Nov 1, 1986); *PubWkly* 230:64 (Aug 8, 1986); *Spir
Fron* 21:58—60 (Win 1989); *WestCoastRevBks* 12(5):41 (1987).

703 Wilson, Ian. *All in the Mind: Reincarnation, Hypnotic
Regression, Stigmata, Multiple Personality, and Other
Little—Understood Powers of the Mind.* Garden City, NY:
Doubleday, 1982. 267p. Chapter notes; 18 figures; 29 il—
lustrations; Index: 261—267; 14 tables

Historian/journalist Wilson provides an overview of claims
of reincarnation and the recall of past lives under hypnosis. He
discusses the research of Ian Stevenson, the Bridey Murphy
case, and several cases resulting from the current wave of inter—
est in hypnotic age—regression. He uncovers fresh facts in the
old cases and presents some new ones. He proposes that the
phenomena of cryptomnesia, multiple personality, and Hilgard's
"hidden observer" can account for most cases, but admits that
there is a residue defying explanation. The book has been
meticulously researched and referenced and is as readable as it
is instructive.

JP 46:377 (Dec 1982); *T* 13/14:20—23 (Spr 1985/86).

BestSell 42:312 (Nov 1982); *Booklist* 79:82 (Sep 15, 1982);
Kirkus 50:727 (Jun 15, 1982); *LATimesBkRev* p4 (Oct 10, 1982);
LibrJ 107:1666 (Sep 1, 1982); *PublWkly* 221:55 (Jun 11, 1982);
SciBooks&Films 18:119 (Jan 1983); *SpirFron* 15:187—188 (Fal
1983).

Religion and Parapsychology

INTRODUCTION

In *Sources*, there were 10 books on religion and psi. The ideas expressed in them are still relevant, although the most recent thoughts of Hardy [214, 215] have since been given in the *Biology of God*, and Thouless's views [219] were updated in a chapter in *Advances in Parapsychological Research 1* [520].

Interest in the religious aspects of psi has burgeoned, as evidenced by the fact that there are 19 books in the present category, nearly twice as many as in *Sources*. Each one of these books makes a unique contribution——it is difficult to group them into any subcategories. Tart [721] deals with all the major religions, but the emphasis is on spiritual tenets, not parapsychology, save for an important essay by Tart himself. The Parapsychology Foundation proceedings [720] provides papers on several interfaces between parapsychology and religion. Rossner [719] is the most extensive effort yet to set forth the connections between religion and psi, and Whiteman [722] is perhaps the most intensive attempt yet made to relate aspects of psi to the spiritual path. The theological implications of psi are brought out by Heaney [709], Kelsey [713], and Perry [717]. Religious miracles as psi phenomena are dealt with in general by Ebon [704] and by Rogo [718], Gowan [705] provides a taxonomy of miraculous phenomena, and Haraldsson [706] reports on his firsthand investigation of the miracles associated with Sai Baba. Hardy [707] proposes a natural theology in which psi plays a major role. Hardy [708] also provides a col—lection of religious experiences, including some psi phenomena, and Heron [710] has culled all the psi phenomena in the Bible. The book edited by Higgins [711] is an anthology containing chapters on all aspects of religion and parapsychology. Hoffman [712] writes from the Jewish viewpoint. Kurtz [714] provides a critique of religion and parapsychology, holding that both are based on revelation, which he attacks. Perry [716] sets forth principles for Christian counseling of people who have had negative occult or psychic involvements. Melton's bibliography [715] is a thorough and scholarly work covering spiritual healing within the context of the Christian church.

Additional books with information on religion and para-

psychology: Barber [343] has several chapters on meditation, Zen, yoga, and mystical experiences, and the Greens [347] dis— cuss meditation and yoga.   Kelly and Locke [350] review the literature of comparative religion for parapsychological insights, and they cover psi phenomena associated with saints and yogis. Neher's book [355] is on experiences of transcendence, including mystical experiences.   Parker [386] deals with meditation and mysticism. B. Bennett contributed a paper on Biblical prophecy in the volume edited by Shapin and Coly [358].   There are chapters on meditation and mystical states in Wolman and Ullman [361].   White's bibliography [378] has a section on the religious implications of psi.   Accounts of healers with religious orientations are found in Cerutti's biography of Olga Worrall [382], DiOrio [384, 385], and Ivanova [390].   Ebon [431] has a chapter on religion and parapsychology.   Rogo [481] describes a number of religious miracles and relates them to psi.   Four of the books in the section on history discuss the relationship of spiritualism and psychical research to religion, or deal with psychical research as a pseudoreligion: see Brandon [549], Cerullo [550], Loewe and Blacker [557], and Oppenheim [560]. In a report of a survey on the paranormal [564], mystical ex— periences were included as well as questions related to psi and survival.   Shamanism is the main subject of the books by Har— ner [565] and Nicholson [569].   All the books in the section on NDEs discuss the religious quality of aftereffects of NDEs, but this aspect is emphasized in those by Flynn [585], Grey [586], Moody [590], Ring [592], and Zaleski [595].   The latter contrasts the modern NDE with the medieval Christian "otherworld journey."   Flynn describes his Love Project, which attempted to induce NDE aftereffects in small groups without the element of near death.   Thakur [614] has several chapters on religion and parapsychology and Stanley Dean's anthology [641] has several on mysticism and parapsychology. Religious aspects of reincar— nation are considered by Head and Cranston [694], Walker [701], and Whitton [702].   Forman [727] notes that people who have anomalous experiences involving time, such as precognition, also report mystical components, and many of the examples of al— ternate states presented by Evans [545a] take place in a religious context. Grosso [759], Ryzl [772], and some of the authors in Toynbee [777] deal with religious aspects of psi and survival.

BIBLIOGRAPHY

704   **Ebon, Martin (Ed.).** *Miracles.* New York: New American Library, 1981. 199p.

This is an anthology of miraculous events drawn from different religious and geographic sources. A wide range of viewpoints are presented——religious, anthropological, psychological, psychiatric, and parapsychological. There are 19 contributions, of which eight were previously published, the remainder having been written especially for this volume. Ebon, who has compiled many anthologies on parapsychology, sets the stage with an introductory chapter entitled "Man's Need for Miracles."

T No. 13/14:44—46 (Sum 1985/86).
Kliatt 16:43 (Spr 1982); *SpirFron* 14:57 (Win 1982).

705 **Gowan, John Curtis.** *Operations of Increasing Order and Other Essays on Exotic Factors of Intellect, Unusual Powers and Abilities, etc. (as Found in Psychic Science).* Westlake Village, CA: Author, 1980. 384p. Bibliography: 373—383; Subject index: 383—384; 7 tables

Educator Gowan presents some facts and theories concerning exceptional human abilities, many of them psychic. Formerly he has written extensively on creativity, but he thinks that there are more powerful levels of cognitive thought that surpass creativity. "Some very spectacular effects occur in these higher and rarer states; indeed mystics of all ages and cultures have told us so for centuries" (p. ix). In reviewing accounts of these experiences, he feels he detects operations of increasing order in the vast mass of evidence. In this book, (1) he catalogues these exotic powers and provides a rationale for them so as to make them easier to believe; (2) he presents independent testimony that gives the phenomena some reliability, thus establishing a basis for validity; he proposes that (3) the taxonomy is evidence of order; and he observes that (4) the direction of this order is toward growth and positive integration; (5) the scope of human capacity revealed is on a grand scale; (6) the various abilities and powers described are interconnected; and (7) the nature of the structure revealed is numinous. In his words: "There is an El Dorado; the map is outlined in these pages; the way to it is through the inner reaches of the latent powers of man's mind; and, knowing that it exists, having the map and the road, it behooves us all to start on the immense journey" (p. xx). The testimony is in nine major sections, of which the third deals almost exclusively with parapsychological phenomena, covering ESP and PK (including the physical phenomena of mysticism). The fourth covers "Cosmogonic Mental Abilities," including psychometry, clairvoyance, laying—on—of—hands, and psychic surgery. Both in subject matter and in treatment this is a modern—day version of Myers' monumental *Human Personality.*

JP 46:76—77 (Mar 1982); *JRPR* 5:127 (Apr 1982).

Light 101:85—88 (Sum 1981).

706 Haraldsson, Erlendur. *Modern Miracles: An Investigative Report on Psychic Phenomena Associated with Sathya Sai Baba.* Foreword by Karlis Osis. New York: Fawcett Columbine, 1988 (c1987). 300p. Bibliography: 293—295; Index: 296—300; 3 tables

Record of a ten—year investigation of the psychic abilities of Indian guru, Sathya Sai Baba. Haraldsson, a psychologist, presents his firsthand observations, supported by careful inter—rogation of witnesses, of a wide variety of phenomena. He also interviewed persons who had known Sai Baba in the various periods of his life, and tried to corroborate evidence and trace critical statements. Haraldsson made eight visits to India in all, on several of which he was accompanied by other para—psychologists. In Part 1, Haraldsson describes his encounters with Sai Baba and gives verbatim responses of witnesses. In Part 2, he evaluates the statements made in the interviews and discusses the major phenomena reported about Sai Baba: ap—ports, raising the dead, dazzling light, teleportation, bilocation, and ESP.

Fate 41:105—107 (Dec 1988); JSPR 55:225—227.

Light 107:130—132 (Win 1987); NewReal 8:77—78 (Jul/Aug 1988).

707 Hardy, Alister. *The Biology of God: A Scientist's Study of Man the Religious Animal.* New York: Taplinger, 1975. 238p. Bibliographical footnotes; Citations in text; Index: 235—238

This book may be considered an extension of Hardy's ear—lier works, *The Living Stream* [215] and *The Divine Flame* [214]. Hardy argues on behalf of consciousness and purposiveness not only in humans but in higher animals. He proposes that both humans and animals can initiate evolutionary changes through altering habit patterns by means of purposive, exploratory ac—tivity. A chapter is devoted to a discussion of the relevance of parapsychology to religion and to biology. The last part of the book is on human religious emotions. Hardy reports on the work of the Religious Experience Research Unit (now the Alister Hardy Research Centre) at Oxford, which he founded. He proposes a new natural theology that would, as the title of the book indicates, be based on the fact that man is a religious animal. Although he doesn't make the connection, there appears to be much in common between Rex Stanford's theoretical and experimental approach, which he calls "psi—mediated in—strumental response," and Hardy's suggestion that "it may well be that the various separate solutions to our individual problems are always within us if only we could reach them, and that the act of prayer brings them to the surface" (p. 229). It can be said that PMIR makes it possible to reach anywhere in the world to solve our individual problems, and as Hardy suggests, prayer may provide the optimum psychological state for achieving results. In line with Hardy's evolutionary hypothesis, PMIR may be instrumental in solving species—wide problems, not just individual ones.

JASPR 71:220−222 (Apr 1977); *JP* 41:359−362 (Dec 1977); *JSPR* 48:271−274 (Mar 1976); *T* 10:15−17 (Spr 1982).

Brain/MindBul 2:4 (Feb 7, 1977); *ChristCent* 94:282 (Mar 23, 1977); *Choice* 14:698 (Jul 1977); *JRelig* 58:221 (Apr 1978); *Kirkus* 44:169 (Feb 1, 1976); *LibrJ* 101:720 (Mar 1, 1976); *Listener* 94:125 (Jul 24, 1975); *Observer* p19 (Dec 14, 1975); *PublWkly* 209:89 (Feb 9, 1976); *Spectator* 235:49 (Jul 12, 1975); *TimesLitSuppl* p976 (Aug 29, 1975).

708 Hardy, Alister. *The Spiritual Nature of Man: A Study of Contemporary Religious Experience.* Oxford, England: Clarendon Press, 1979. 162p. Bibliography: 154−157; Index: 159−162; 7 tables

This book presents the research carried out during the first eight years of the Religious Experience Research Unit at Oxford founded by biologist Hardy. (Since his death it has become the Alister Hardy Research Centre.) Basically, it gives the results of the analysis of the first 3,000 replies to a survey of religious experience conducted by the unit. The experiences have been classified into 92 categories primarily describing the varieties of spiritual awareness. Many examples are given in the recipient's own words. Dynamic patterns, triggers, and consequences of the experiences are also discussed. Several investigations based on the records are described as well as additional quantitative research conducted by D. Hay and A. Morist.

ChrP 3:283 (S 1980); *T* 10:15−17 (Spr 1982).

AngTheolRev 63:216−217 (Apr 1981); *Gregorianum* 64(1): 157−158 (1983); *JAmAcadRelig* 49:720 (Dec 1981); *JEcumStud* 18:308 (Spr 1981); *JSciStudRelig* 20:93 (Mar 1981); *Light* 100:189−191 (Win 1980); *Observer* p37 (Dec 23, 1979); *ReligEduc* 76:101−103 (Jan/Feb 1981); *SpirFron* 14:115−117 (Fal 1982); *Zygon* 17:417−418 (Dec 1982)

709 Heaney, John J. *The Sacred and the Psychic: Parapsychology and Christian Theology.* Ramsey, NJ: Paulist Press, 1984. 286p. Chapter notes

The author, a Catholic theologian, seeks to present a balanced view of the implications of parapsychology for theology——fundamental theology in particular. By the latter he

refers first to the study of the experience and doctrine of the
foundational Christian category, revelation, and, second, to the
study of the presuppositions and assumptions, both conscious
and unconscious, that are involved in the commitment to
Christian revelation. In each chapter he presents the evidence
for the type of paranormal ability that serves as the subject of
that chapter, then the criticisms, the response to the criticisms,
and the theological implications. The titles of the chapters dealing with phenomena are "Telepathy and Clairvoyance";
"Psychokinesis, Poltergeist Manifestations and Demonic
Possession"; "Paranormal Healing"; "Precognition and
Prophecy"; "Retrocognition"; "Out−of−the−Body Experiences";
"Recent Studies of Near Death Experiences"; "Apparitions and
the Resurrection of Jesus"; "Mediums: The Unconscious or Life
Beyond?"; and "Reincarnation." In addition there are three
more theoretical and explicitly theological chapters. One deals
with Jung's concept of the collective unconscious, coupled with
the idea of a psychic universe. Another is on theological and
parapsychological aspects of eschatology. The last chapter is en−
titled "Consciousness, Parapsychology and Theology," and it
presents a holistic, transpersonal view. Throughout, Heaney
singles out the moments of transcendence associated with psi
phenomena and attempts to explicate their meaning for theol−
ogy. He sees psychic phenomena as a kind of bridge between
the secular and the sacred. Psi phenomena indicate that there
are dimensions of reality that we can contact in which we are
not the isolated egos we generally assume we are. Finally, he
proposes that the most coherent view of the cosmos reveals that
we are enfolded in a psychic universe in which we operate
together with a mysterious Otherness, Absolute Mind, and Love.

  ChrP 5:277−278 (Dec 1984); *PAI* 3:39 (Dec 1985); *PR*
17:12−14 (Sep/Oct 1986).

  BestSell 44:275 (Oct 1984); *ReligStudRev* 12:55 (Jan 1986);
SpirFron 17:45−46 (Win 1985); *TheologStud* 46:183−184 (Mar
1985); *VentIn* 1:49 (Nov/Dec 1984).

710 Heron, Laurence Tunstall. *ESP in the Bible.* Garden City,
      NY: Doubleday, 1974. 212p. Chapter notes; Suggested
      reading list: 211−212

  Heron is a newspaper reporter and editor who became interested in parapsychology and active in the Spiritual Frontiers
Fellowship after the death of his 19−year old son. This book is
based on the premise that "efforts to discern the will of God
have depended, in savage cultures and in high civilizations,
upon the exercise of man's extrasensory capacities. Great
religions arise from prophetic experience in extrasensory percep−
tion (ESP). Prophets are persons who see realities that most

men cannot see, and who hear truths that most men cannot sense" (p. 17). This is not only a useful survey of spontaneous psi experiences described in the Old and New Testaments. Heron also relates Biblical psi experiences to similar cases in the literature of psychical research, current findings concerning altered states, and modern instances of inspiration and creativity. He suggests that prophecy and divine inspiration continues, and that it is communicated to man via ESP. There is an appendix which lists the Bible passages cited.

    P 6:38 (May/Jun 1975); *PR* 5:7—8 (Sep/Oct 1974).

    LibrJ 99:1830 (Jul 1974); *SpirFron* 6/7:12 (Aut 1974/Win 1975).

711 Higgins, Paul Lambourne (Ed.). *Frontiers of the Spirit: Studies in the Mystical and Psychical Areas in Observance of the Twentieth Anniversary of the Founding of Spiritual Frontiers Fellowship.* Minneapolis, MN: T.S. Denison, 1976. 133p.

    This volume was prepared under the auspices of the Spiritual Frontiers Fellowship (SFF) in honor of its twentieth anniversary. SFF was "formed for the purpose of studying supernormal experiences in the churches and elsewhere, as they relate to personal immortality, spiritual healing, and prayer and meditation" (p. 6). Higgins, a Methodist minister and a co-founder and first president of SFF, contributes a chapter on the history of the Fellowship. There are three chapters that stress parapsychology in relation to religion, one each on Christianity and psi phenomena and on mediumship, two each on mysticism and on spiritual healing, and four on prayer and meditation. In addition to Higgins, who wrote two chapters, the authors are William V. Rauscher, J. Gordon Melton, Ross E. Sweeny, Marguerite H. Bro, Frank C. Tribbe, Carol Ann Purdy, Kenneth G. Cuming, Ambrose A. Worrall, Irene F. Hughes, Curtis Fuller, Robert S. Slater, and L. Richard Batzler. The emphasis is on the implications of parapsychology for religion. Suggestions are based on personal experiences and intuitive suggestions rather than laboratory evidence, but the research—minded reader is likely to pick up some ideas worth investigating.

    SpirFron 9:39—44 (Win 1977).

712 Hoffman, Edward. *The Way of Splendor: Jewish Mysticism and Modern Psychology.* Boulder, CO: Shambhala, 1981. 247p. Bibliography: 235—243; Chapter notes; 9 figures; Glossary: 232—234; 3 illustrations; Index: 244—247

    Psychologist Hoffman's purpose is "to introduce the Kaballah's psychological insights to the general public, unfamiliar with the intricacies of . . . Jewish . . . mysticism" (p.

vii). He feels that the Kaballah has key insights to offer that complement the psychological insights from Eastern religions. There are chapters on the Kaballah and its key scholars, the unity of the cosmos, the sacred nature of the body, techniques of inner peace, awakening ecstasy, the relevance of dreams and music, paranormal phenomena, and the nature of death and immortality. The last chapter summarizes the key psychological insights found in the Kaballah.

AREJ 17:226–228 (Sep 1982); *LibrJ* 106:2244 (Nov 15, 1981); *NewAge* 7:58 (Feb 1982); *NewReal* 4:45–46 (Jul 1982); *ReVision* 5:108 (Spr 1982); *SpirFron* 15:41–42 (Win 1983).

713 Kelsey, Morton T. *The Christian and the Supernatural.* Minneapolis, MN: Augsburg Publishing House, 1976. 168p. Annotated bibliography: 159–168; Chapter notes; 3 figures

This is an introduction to parapsychology by an Episcopal clergyman in a context meaningful to Christians. The overview of parapsychology is very cursory, but it is not the important part of the book, whose primary value lies in the way he re-lates psi phenomena to the Christian tradition. He points out that although psi phenomena were an accepted part of early Christianity, for centuries mainline churches have dissociated themselves from psi, so that in effect the Christian view of reality is back in the Middle Ages. Kelsey attempts to redress this one-sided emphasis on physical reality, and points out that psi experiences are often associated with deep experiences of God. If religion does not provide a legitimate place for these experiences, people will seek them outside the church. But, as he also points out in a section on psi theology, the value of psi experiences is questionable when sought as ends in theories and not as part of the experience of God. Kelsey has had training as a Jungian, and he offers many useful insights to those who would counsel others regarding psi phenomena. An appendix lists psi experiences occurring in the New Testament.

LutheranQuart 29:99–100 (Feb 1977); *NewRevBksRelig* 1:8 (Mar 1977); *ReformedRev* 30:223–224 (Spr 1977).

714 Kurtz, Paul. *The Transcendental Temptation: A Critique of Religion and the Paranormal.* Buffalo, NY: Prometheus Books, 1986. 500p. Bibliographical footnotes; Index: 485–500; 1 table

Kurtz is a leading secular humanist philosopher and founding chair of the Committee for the Scientific Investigation of the Paranormal. He aims this book at the general reader and argues on behalf of secular humanism as a replacement for religion and paranormal belief systems. The first part is on

"Skepticism and the Meaning of Life." The second is on "Mysticism, Revelation, and God," and the third is on "Science and the Paranormal." Kurtz argues that the major religions are based on revelation, and that the phenomena of parapsychology share a similar base. He feels that these fields have a large following because their adherents have succumbed to the "transcendental temptation," or a proclivity for magical thinking, which undermines critical judgment. He discusses the prospects for developing a humanistic society based on a truly scientific foundation. The areas of parapsychology dealt with are mediumship, ESP experiments, and survival.

    JASPR 83:180−183 (Apr 1989); *PR* 19:8−11 (Mar/Apr 1988).

    Choice 24:777 (Jan 1987); *Humanist* (U.S.) 47:37 (Jan 1987); *JSciStudRel* 27:151−152 (Mar 1988); *NoeticSciRev* No.2:25−26 (Spr 1987); *SkptInq* 12:194−199 (Win 1987/1988).

715 **Melton, J. Gordon.** *A Reader's Guide to the Church's Ministry of Healing.* Evanston, IL: Academy of Religion and Psychical Research, 1973. 78p.

    This mimeographed bibliography by the founder/director of the Institute for the Study of American Religion is for "those who are interested in furthering their knowledge of paranormal healing in the context of the historic church" (p. 4). Melton provides a short history of the modern healing movement followed by the bibliography proper, which is aimed at being "an exhaustive listing of books concerning spiritual healing as it relates to the mainline Western religious tradition in the 19th and 20th centuries. Supplementary and selective listings are given for various related topics" (p. 10). An added feature is that the location of at least one copy of each book has been attempted. Books are listed under the following headings: "Twelve Recommended Books on Spiritual Healing" (12), "Phenomenological Studies of Spiritual Healing" (21), "The Emmanuel Movement" (15), "Works Written Prior to 1920" (69), "Works Written 1920−1929" (36), "Mainline Works on Spiritual Healing" (277), "Churches' Reports on Spiritual Healing" (17), "New Thought" (60), "Christian Science" (31); "The Spiritualist Tradition" (29), "The Pentecostal Deliverance Tradition" (62), "Roman Catholic" (21), "Periodical Literature" (57), "Non−Allopathic Medicine" (22), and a "Miscellaneous Category" (62). There are addenda by chapter supplying an additional 25 titles. A total of 816 books and articles are listed.

716 **Perry, Michael (Ed.).** *Deliverance: Psychic Disturbances and Occult Involvement.* London, England: SPCK, 1987. 143p. Bibliographical footnotes; Index: 141−143

The papers presented here are the result of the work of
the Christian Exorcism Study Group and the Anglican
Churchmen. They share their training and experiences in helping
people who have become involved in a problem that has an oc—
cult or psychic component and need help. The Group is as—
sociated with the Churches' Council for Health and Healing.
One chapter offers counsel to the counselors. Twelve guidelines
are set forth to follow when confronted with a case. There are
also 12 guidelines to aid the counselor from becoming involved.
Advice is also offered on practical aspects, such as taking notes.
The remaining chapter headings make clear the types of
phenomena dealt with: "Poltergeist Phenomena"; "Ghosts and
Place Memories"; "The Unquiet Dead"; "The Occult and the
Psychic"; "Occultism, Witchcraft and Satanism, Sects and
Cults"; "Possession Syndrome"; "Possession"; and "Exorcism."
Four appendices offer background information on the Christian
tradition of exorcism.
      JASPR 83:183—184 (Apr 1989); *PAI* 6:46 (Jun 1988).
SpirFron 19:181 (Sum 1987).

717 **Perry, Michael.** *Psychic Studies: A Christian's View.* Wel—
      lingborough, Northamptonshire, England: Aquarian Press,
      1984. 224p. Chapter references; Index: 221—224
      This is a collection of Michael Perry's writings, most of
them previously published in many British and American peri—
odicals with small circulations. The overall purpose of the book
is to provide a way of looking at reality from a position be—
tween two disciplines, Christianity and parapsychology, that for
too long have been ignorant and distrustful of one another.
Many of the pieces have been modified and adapted to this
purpose. Some recurring themes are standards of scientific
evidence, the role of faith, and the extent to which the psychi—
cal and the spiritual overlap or diverge. The role of ethics in
Christianity and also in parapsychology is pointed out. About
half of the book deals with theological aspects of survival, in—
cluding resurrection and reincarnation. Throughout, the author
succeeds in showing how parapsychology may help Christianity
and vice versa.
      ChrP 5:163—164 (Mar 1984); *PAI* 3:40 (Dec 1985); *PR*
16:13—14 (May/Jun 1985).
      Churchman 98:275—276 (1984); *ExposT* 96:57 (Nov 1984);
Light 104:40—41 (Spr 1984); *ModChurchman* 27:68 (1984);
Theology 88:51—53 (Jan 1985); *VentIn* 1:49 (Jan/Feb 1985).

718 **Rogo, D. Scott.** *Miracles: A Parascientific Inquiry into
      Wondrous Phenomena.* New York: Dial Press, 1982.
      332p. Bibliographical footnotes; Chapter notes; 40 il—

lustrations; Index: 323—332

Rogo presents a modern introduction to miraculous phenomena. Miracles are described under three major groups: "Miraculous Talents" (levitation, stigmata, bilocation); "Miraculous Events" (divine images, miraculous hailstones, bleeding statues and weeping madonnas); and "Miraculous Interventions" (Marian apparitions; miraculous healings). In the final chapter, "Psyche and the Miraculous," Rogo draws on the findings of parapsychology and other sciences. Rogo theorizes about the nature of miracles in a fresh way.

JASPR 77:327—345 (Oct 1983); *JRPR* 7:195—197 (Jul 1984); *Psychoenergetics* 5:87—88 (1983); *T* 13/14:44—46 (Sum 1985/86).

Kirkus 49:1570 (Dec 15, 1981); *LibrJ* 107:183 (Jan 15, 1982); *PublWkly* 221:40 (Jan 1, 1982); *ReligStudRev* 8:353 (Oct 1982).

719 Rossner, John L. *Toward Recovery of the Primordial Tradition: Ancient Insights & Modern Discoveries.* Lanham, MD: University Press of America, 1979—1984. 5 vols. Chapter notes; Chapter references

The rationale behind this impressive series of books is that primitive Christianity was rooted in a "Primordial Tradition of psychic intuition and spiritual insight which has been shared with other great religions of the East and West in both ancient and modern times. This background of shared conceptions, enshrined in myths, doctrines, and ritual practices arose out of a common, or universal phenomenology of human psychic and spiritual experience. It was thus rooted in the human psyche and in perennial modes of consciousness and experience which cut across traditional, cultural and religious barriers" (Vol. II, Bk 3, p. vi). In this series, the Reverend Canon Rossner, a professor of religion at Concordia University, shows how recent findings in parapsychology, consciousness studies, and paraphysics, may assist in uncovering the "experiential origins of human religious beliefs in such things as ritual magic, prayer, miraculous healings, post—mortem contacts with the dead, apparitions of angels and saints, revelatory dreams and visions, resurrections and ascensions into heavenly spheres, and psychic identification with archetypical saviours, masters, teachers, or gurus, through initiatory rites" (Vol. II, Bk. 3, p. vi.). Volume I of the series is entitled "Toward a Parapsychology of Religion: The Convergence of Images of Man & the Cosmos from Ancient Religious & Metaphysical Traditions with Insights from Recent Studies in Consciousness." It consists of two books: 1. *From Ancient Magic to Future Technology*, and 2. *From Ancient Religion to Future Science.* Volume II:

"Essays in the Parapsychology of Religion: The Primordial Tradition in Contemporary Experience——Explorations in Psyche, Sacrament, Spirit," is in three books: 1. *Religion, Science, Psyche*, 2. *Spirits & Cosmic Paradigms*, and 3. *The Psychic Roots of Ancient Wisdom*. The series is designed to serve as a textbook for the use of students and instructors. The chapter bibliographies are extensive and an outline of the contents of each chapter is listed in the beginning. This is certainly the most ambitious attempt yet made to examine the relationship of parapsychology and religion and the religious implications of parapsychology.

720 Shapin, Betty, and Coly, Lisette (Eds.). *Parapsychology, Philosophy and Religious Concepts: Proceedings of an International Conference Held in Rome, Italy, August 23– 24, 1985.* New York: Parapsychology Foundation, 1987. 215p. Chapter references; 5 figures; 2 illustrations
The papers given at this conference on religion and parapsychology were varied. The most general piece is probably the appendix, which is a reprint of J.B. Rhine's "The Parapsychology of Religion——A New Branch of Inquiry." A theoretical approach is taken by Steven M. Rosen ("Psi and the Principle of Non–dual Duality"), Michael Perry ("Understanding and Explanation"), and John Rossner ("The Psychic Area as the Missing Link Between the Sacred and the Profane in Modern Western Civilization"). Emphasis on mystical experiences and psi is given in the papers by Emilio Servadio ("Mysticism and Parapsychology"), Sergio Bernardi ("Shamanism and Parapsychology"), and Leslie Price ("Theosophy as a Problem for Psychical Research"). The implications of psi for religion are dealt with by James A. Hall ("A Jungian Perspective on Parapsychology: Implications for Science and Religion"), and Michael Grosso ("The God Idea: A Parapsychological Perspective"). Rhea A. White emphasizes ways in which religion can assist parapsychology ("Meaning, Metanoia and Psi"). The discussions following the papers are included.
ChrP 7:217–218 (Jun 1988); *JASPR* 83:271–273 (Jul 1989); *JP* 52:171–175 (Jun 1988); *JRPR* 11:170–172 (Jul 1988). *SpirFron* 20:57–58 (Win 1988).

721 Tart, Charles T. (Ed.). *Transpersonal Psychologies.* New York: Harper & Row, 1975. 502p. Bibliographical footnotes; Bibliography: 475–483; 29 figures; 2 graphs; Name index: 489–493; Subject index: 495–502; 11 tables
The authors of the chapters in this anthology have attempted to present the spiritual disciplines with which each is familiar as a psychology. In a very real sense this book does for

various spiritual traditions (Buddhism, Yoga, Gurdjieff, Arica training, Sufism, Mystical Christianity, Western magic, and Zen Buddhism) what Lindzey and Hall's *Theories of Personality* does for Freud, Jung, Allport, etc. Psychologist/parapsychologist Tart is basically concerned with the development of sciences for various altered states. The first 150 pages of the book consist of three seminal chapters by Tart: "Science, States of Consciousness, and Spiritual Experiences: The Need for State−Specific Sciences"; "Some Assumptions of Orthodox, Western Psychology"; and "The Physical Universe, the Spiritual Universe, and the Paranormal." The various contents of the book are tied together by an index and a useful 11−page bibliography.

ChrP 7:33−34 (Mar 1987); *JASPR* 70:235−239 (Apr 1976); *JP* 39:339−341 (Dec 1975); *JSPR* 48:312−313 (Sep 1976); *P* 6:45−46 (Jan/Feb 1976); *PS* 1:243−244 (1976).

AmJPsychiat 33:468−469 (Apr 1976); *Choice* 12:1372 (Dec 1975); *ChristCent* 92:857 (Oct 1, 1975); *ChristToday* 20:35−36 (May 7, 1976); *JAmAcadRelig* 45:115−116 (Mar 1977); *JRelig* 57:107 (Jan 1977); *JTranspersPsychol* 7:Sup 197−202, 197−198 (1975); *SpirFron* 9:50−52 (Fal 1977).

722 Whiteman, J.H.M. *Old and New Evidence on the Meaning of Life: The Mystical World−View and Inner Contest. Volume I. An Introduction to Scientific Mysticism.* Gerrards Cross, England: Colin Smythe, 1986. 267p. Chapter notes; 3 figures; Name index: 255−257; Subject index: 259−267

The author, both a mathematician and a mystic, in this work attempts to systematize the inner world and the mystical life. Part I, "The Groundwork of Mystical and Psychical Awareness," contains three chapters that provide a classification of the inner life (or the nonphysical domain), set forth three foundational skills used in developing the inner life, and four creative functions which operate cyclically, and present a practical method for objectively evaluating any stabilized nonphysical experience on a six−point scale or "Index of Reality." Part II, "Varieties of 'Other−World' Experience: Beyond Personality" consists of five chapters that classify separation experiences and describe the first stages of a method of detachment called "objectification of influences." Part III, "Attainment and Teachings of the Mystical Life," consists of three chapters dealing with the difficulties in the description of mystical states and with states of mystical transformation. Part IV, "The Mystical Structure of Psychology and Theoretical Physics," consists of three chapters in which Whiteman presents "the answers provided by modern theoretical physics to problems of mind−matter dualism" (p. 243), expands on the classification of the four—fold cycle of creative functions, and describes the mystical significance of spatial perceptions, including experiences of ESP.

JASPR 83:173—177 (Apr 1989); *JSPR* 54:146—151 (Apr 1987); *PAI* 6:54 (Jun 1988).

Light 107:137 (Win 1987).

Spontaneous Psi

INTRODUCTION

There were 13 books in this category in *Sources*, all of which are still relevant. Data from one of them, *Phantasms of the Living* [229], written in 1886, has been used in research surveys conducted in recent years.

A total of 26 books are listed in this chapter, including books on hauntings and poltergeists, which were listed separately in *Sources* (nine titles, all of which still have value). Several authors deal with "ghosts" (i.e., apparitions, haunts, poltergeists): Auerbach [723], Davidson [724], Evans [725, 726], Finucane [726a], Goss [728], Green and McCreery [729], Hufford [732], MacKenzie [733, 735], Moss [736], and Rogo [815]. Spontaneous precognition experiences are covered by Behe [805], Forman [727], Hearne [731a], and by MacKenzie [734]. Retrocognition is the subject of Coleman's book [723a]. Spontaneous phenomena in general are described and discussed by Haynes [731] and by Persinger [738, 739]. The approach to spontaneous cases employed at the Duke University Parapsychology Laboratory is the subject of the book by L.E. Rhine [741] and also the book in her honor edited by Rao [740]. The psychodynamics of spontaneous phenomena related to psi are considered by Hartmann [730] (nightmares), Neppe [737] (déjà vu), and Vaughan [743] (synchronicity).

Additional books with information on spontaneous psi: The books in the section on animal psi [362-365] describe many spontaneous cases involving animals. Some cases involving psi or intuition in business are described by Agor [366]. Hibbard and Worring [371] write about synchronistic happenings in daily life that may be helpful in locating criminals. All of the books in the section of "Biographies and Autobiographies of Healers, Mediums, and Psychics" contain accounts of experiences. Rogo [411] describes many of the pitfalls involved in investigating spontaneous phenomena. Spontaneous psi and other nonordinary experiences of children are emphasized in the books listed in the section "Children and Psi." Considerable space also is devoted to spontaneous phenomena in several books in the section on young people: [414, 420, 424 (child ghosts), 429a (apparitions and haunts), 434 (precognition), 435 (PK), 437, 440

(PK and poltergeists), **441**, **443** (poltergeists), **444**, **445**, **446**, **447**, **448**, and **450**. Inglis [**809**] surveys a broad range of spontaneous psi experiences that suggest they are products of some
larger design. Schouten discusses the quantitative analysis of
spontaneous psi experiences in Shapin and Coly [**817**]. Gardner
[**460**] is critical of precognitions reported concerning the wreck of
the *Titanic*. Several well—known cases are criticized by Harris
[**473**]. Spontaneous phenomena are covered in Fairley and Welfare [**515**]. Emmons [**563**] compares a survey of psi experiences
in Hong Kong with answers to a 1978 Gallup poll in the U.S.
Greeley [**564**] reports on a U.S. poll by the National Opinion
Research Center. All of the books in the sections on NDEs
[**585-595**] and OBEs [**596-604**] deal with spontaneous cases of
those specific types. Of special note, as regards spontaneous psi
phenomena, are the books by Greenhouse [**599**] and Moody [**589**,
590]. The former attempts to use specific cases to illustrate all
facets of the OB experience, whereas Moody does the same
thing for the near—death experience. Braude [**606**] has a chapter on apparitions, with emphasis on collective apparitions.
Spontaneous forms of PK are covered by Randall [**630**] and
Robinson [**632**]. Three books are on poltergeists and hauntings:
Gauld and Cornell [**625**], Rogo [**635**], and Roll [**636**]. The
volume compiled by Corliss [**640**] contains descriptions of many
types of spontaneous anomalous phenomena, some of them in—
volving psi. Eberhart's *Geo—Bibliography of Anomalies* [**665**] is a
unique geographic guide to information on cases of all types of
anomalous phenomena, including psi. Goss's [**666**] poltergeist
bibliography is extremely exhaustive and also has a geographic
index. Several of the books in the reincarnation section present
histories: Ryall [**696**, Stevenson [**697-700**], and Whitton [**702**].
Wilson [**703**] critically reviews some of the cases discussed.
Spontaneous psi experiences associated with religious and mysti—
cal experiences are dealt with by Ebon [**704**], Gowan [**705**],
Haraldsson [**706**], Hardy [**708**], Heaney [**709**], Heron [**710**, Hof—
fman [**712**], Kelsey [**713**], Rogo [**718**], and Rossner [**719**]. Moody
[**767**] surveys many types of spontaneous phenomena associated
with Elvis Presley after his death. Vaughan [**789**, **790**] gives
examples of many spontaneous precognitive experiences. White
and Anderson's [**791**] bibliography on psychics has a section on
psychic experiences in daily life. Irwin's textbook [**810**] em—
phasizes surveys of spontaneous psi, and he discusses the
phenomenology of spontaneous psi.

BIBLIOGRAPHY

723 Auerbach, Loyd. *ESP, Hauntings and Poltergeists: A*

Parapsychologist's Handbook. New York: Warner Books, 1986. 463p. 2 questionnaires; Suggestions for further reading: 443–456

Written in a popular style, this handbook is aimed at persons who have had or want to know more about psi experiences: hauntings, apparitions, and poltergeists, in particular. Auerbach attempts to inform readers as to how parapsychologists conduct scientific investigations and what they have learned about these experiences, their forms, why they occur, and what can be done about them. The first of five sections is about the parapsychological approach to psi experiences in the field and in the laboratory. Section 2 presents views of psi taken by various mass media––films, radio, literature, and television. The third section describes actual cases, some well-known, some investigated by the author or other parapsychologists, and some cases highlighted by the media (e.g., *The Amityville Horror*). Section 4 tells the reader how he or she can handle a reported psi experience––one's own or someone else's. It attempts to show "how to differentiate between a paranormal experience and a 'normal' one that's been misinterpreted, and how to bring the situation under control, to understand and live with or stop the phenomenon" (p. 10). The last section consists of several appendices: A list of questions to ask in looking into a case; an annotated listing of reputable parapsychological research centers and organizations and parapsychologists associated with them; a bibliography of books, magazines, and newspaper articles arranged by subject and aimed at the general reader; and two sets of survey questions for the reader to answer and send to researchers.

JASPR 82:298–300 (Jul 1988); *PAI* 5:42 (Dec 1987).
FantR 10:46 (Jan 1987); *VoiceYouthAdv* 10:42 (Apr 1987).

805 Behe, George. *Titanic: Psychic Forewarnings of a Tragedy.* [See Addenda for annotation]

723a Coleman, Michael H. (Ed.). *The Ghosts of the Trianon: The Complete An Adventure by C.A.E. Moberly and E.F. Jourdain.* Wellingborough, Northamptonshire, England: Aquarian Press, 1988. 160p. Bibliography: 153–156; 9 figures; 20 illustrations; Index: 159–160

This book is listed as being a new printing of the most famous case of presumed retrocognition originally published as *An Adventure* by the two percipients, the Misses Moberly and Jourdain. It is true that Coleman has brought together all the accounts of the case, including a reprint of the original draft, and lists all of the editions in an appendix. However, this material, although important, comprises only 58 pages of the

text. It is the remaining 102 pages that seem to me to be
even more important, and these pages were written by Coleman,
who really merits being listed as the author, and so I have put
his name first. What Coleman has done is to reprint, sum—
marize, and discuss all of the pertinent writing about the case
in the last three—quarters of a century, aiming to convey the
essence of all the arguments fairly and give each one its proper
weight. Then he discusses reports and investigations of
analogous cases, mostly taken from the records of the SPR,
noting especially those cases on which normal explanations were
found, thus testifying to the difficulty in establishing that a case
is actually paranormal. His own explanation of the case is that
"there seems little justification for regarding the story as any—
thing other than wish fulfillment based on confused memories"
(p. 151).

724 Davidson, Hilda R. Ellis, and Russell, W.M.S. (Eds.). *The
Folklore of Ghosts.* Cambridge, England: Published for
the Folklore Society by D.S. Brewer, 1980. (Distributed
in the United States by Biblio Distribution Services,
Totowa, NJ.) 271p. Chapter notes
      The approach to apparitions taken in this book is that of
the folklorist. The concern here is not with the extent to which
the manifestations can be taken to be genuine, but to inves—
tigate how such stories develop and become embroidered as they
are retold. H.R.E. Davidson, in her introduction, states that "in
this book an attempt has been made . . . to examine a number
of ghost traditions, and to attempt to analyze them, and to see
what characteristics emerge when the various studies are put
together" (p. viii). The book consists of 11 papers that were
part of a conference on the folklore of ghosts held at the Univer—
sity of York in July, 1980. They are divided into three sections,
the first consisting of five papers under the heading, "Ghosts in
Recent Times." The second, "Ghosts in Perspective," consists of
a paper by Claire Russell, and the third section, "Ghosts
Through the Ages," contains five papers on ghosts of the past.
The paper by Russell points to an intimate connection between
the experience of haunts and apparitions and the process of
dreaming. She emphasizes the important role played by associa—
tions aroused by places and of relationships of the living persons
involved. Other papers attest to this observation as well while
still others emphasize the role creative imagination plays in
descriptions of ghosts and hauntings.
      ChrP 4:265—266 (Dec 1982); *PAI* 3:39 (Dec 1985).
      BritBkNews p343 (Jun 1982); *JAmFolklore* 96:357—359
(Jul/Sep 1983); *Listener* 108:21 (Jul 15, 1982); *Times LitSuppl*
p97 (Jan 29, 1982).

725 Evans, Hilary. *Gods—–Spirits—–Cosmic Guardians: A Com-—parative Study of the Encounter Experience.* Wel-—lingborough, Northamptonshire, England: Aquarian Press, 1987. 287p. Bibliography: 271—281; 23 illustrations; In-—dex: 283—287

This is a sequel to Evans' earlier survey, *Visions, Appari-—tions, Alien Visitors* [**726**], which was a comparative study of sightings. The present book surveys actual encounters with other beings. He says: "By *contact* or *encounter* I mean an event in which a person believes himself in direct and personal contact with an entity. Generally I use *encounter* when there seems to have been a face—to—face meeting as between two persons, and *contact* where somewhat less is involved, for instance where no words are spoken, or the entity is sensed but not seen" (p. 14). He presents representative cases from 15 types of encounters: with divinities, God and other gods, angels, the Virgin Mary and other saintly beings, Satan and other demonic beings, folklore entities, spirits, crisis apparitions, seance—room spirits, dark forces, extraterrestrials, Martians, contact encounters, and cosmic guardians. Next he examines the process of encounter, including sections on hallucinations and altered states, triggers and catalysts, and projection and psychodrama. Then he con-—siders the context of the encounter, followed by a chapter on understanding the encounter experience. In the final chapter, Evans presents his model of what happens in the encounter experience. There is a 273—item bibliography consisting primarily of case material.

PAI 6:43 (Jun 1988).

BritBkNews p589 (Sep 1987).

726 Evans, Hilary. *Visions, Apparitions, Alien Visitors: A Cooperative Study of the Entity Enigma.* Wellingborough, Northamptonshire, England: Aquarian Press, 1984. 320p. (Distributed in the United States by Sterling, New York.) Bibliography: 311—318; 4 figures; 17 illustrations; Index: 319—320; 1 table

The author is Director of the Mary Evans Picture Library in London and is concerned with the many different kinds of experiences people have had "in which they seem to see a more—or—less human—like figure which there are good reasons for believing is not as 'real' as it seems to be" (p. 11). He asks of an experience of this type: "If it is not real in the accepted sense, is it real in any other sense, or totally unreal? And are all of these varied experiences real/unreal in the same way and to the same degree? Are they, indeed, the same kind of ex-—perience in different forms, or a variety of experiences with a superficial similarity?" (p. 11). The variety of experiences he considers as types of the "entity" experience is wide: religious visions, apparitions, demons, dream images, imaginary playmates, tulpas, extraterrestrial visitors, and the double. The first part ʳ is a general discussion of the meaning of "reality," "identity," and "entity." The second part covers various means employed to deliberately induce the manifestation of an entity. The last part discusses relevant studies and offers several hypotheses to account for the entity experience.

    JASPR 80:332−336 (Jul 1986); *JSPR* 53:462−463 (Oct 1986); *PAI* 3:40 (Jun 1985).

    Light 104:133−134 (Aut 1984); *Light* 106:93 (Sum 1986); *SchLibrJ* 31:92 (Feb 1985); *SkptInq* 9:288−290 (Spr 1985).

726a Finucane, R.C. *Appearances of the Dead: A Cultural History of Ghosts.* Buffalo, NY: Prometheus Books, 1984. 232p. Chapter notes; 17 illustrations; Index: 225−232.

    In this work a professor of history considers three main problem areas: "how the dead have been perceived in Western European traditions; what changes have occurred in these perceptions through the centuries; and why these perceptions have altered" (p. 1). He studied descriptions of apparitions−−their function as well as their form−−from the classical age to Medieval times and every century thereafter to today (from the 16th century on he limits the survey to England and English reports). He found that "changes in social assumptions, particularly those associated with theological opinions and scientific accomplishments" (p. 223) had an effect on the way in which the living perceived the dead in Europe and England. "Each epoch has perceived its spectres according to specific sets of expectations, as those change, so do the spectres" (p. 223). He concludes that ghosts have their origin in this world.

    JASPR 79:442−446 (Jul 1984); *JSPR* 52:330−333 (Jun 1984).

    PublWkly 224:47 (Dec 23, 1983); *SkptInq* 13:196−198 (Win 1989); *TimesLitSuppl* 617:863 (Aug 3 1984); *WestCoastRevBks* 11:48 (Jan 1985).

727 Forman, Joan. *The Mask of Time: The Mystery Factor in Timeslips, Precognition and Hindsight.* London, England: Corgi Books, 1981. 302p. Bibliography: 289−292; 3 figures; Glossary: 293−294; Index: 295−302

    This is an inquiry into oddities of time, or timeslips such as precognition and retrocognition, and experiences such as déjà vu and time stopping. It is aimed at the general reader, though the author, who has written several books on hauntings in England, hopes it might provide researchers food for thought.

She says that "primarily, it is a book of search." She attempts "to look again at time, the familiar companion, and see it with fresh eyes" (p. 13). She made an appeal through the British press for anomalous time experiences to which there was an unexpectedly large response from many parts of the world. Accounts of these cases are liberally interspersed throughout the text. She theorizes about time, drawing on physics, religion, and parapsychology. Many of the experiences described in the book contain a mystical element, and this is incorporated in her hypotheses about time slips, although she also attempts to align it with known facts and current scientific investigation.

> *JSPR* 49:956—958 (Dec 1978).

728 Goss, Michael. *The Evidence for Phantom Hitch—Hikers.* Wellingborough, Northamptonshire, England: Aquarian Press, 1984. (Distributed in the United States by Sterling, New York.) 160p. Bibliography: 155—156; Chapter notes; 16 illustrations; Index: 157—160

Survey of the "phantom hitch—hiker" phenomenon in all its stages. Stories of "road ghosts" extend back to antiquity——it is an archetypal motif that is part of the tradition of folklore and literature. Goss, a freelance writer specializing in the paranormal, describes all the variants of the ghostly hitch—hiker theme. In addition to classifying the phenomenon, emphasis is placed on finding normal explanations for it, whether they be instances of mistaken identity or outright lies. Goss makes a distinct contribution to the folklore of a specific type of apparition, the origin of which is multiply-determined.

729 Green, Celia, and McCreery, Charles. *Apparitions.* London, England: Hamish Hamilton, 1975. (Distributed in United States by Transatlantic Arts.) 218p. Bibliography: 213; Glossary: 211—212; Index: 215—218; 2 tables

Celia Green is Director and Charles McCreery is Research officer of the Institute of Psychophysical Research in Oxford, England. In 1968 and 1974 the Institute made newspaper and radio appeals "for first—hand accounts of experiences of perceiv—ing apparitions, this term being intended to cover experiences involving any of the senses and not just visual ones" (p. viii). A total of 1,800 persons responded and were sent a question—naire, 850 of which were completed. This book is based on the experiential accounts, questionnaire answers, and previously published firsthand accounts of apparitions. The authors classify and analyze the characteristics of the apparitions, and in—dividually describe many cases. One such classification is autophany, in which a person sees an apparition of himself, as opposed to autoscopy, in which a person apparently sees his own body as if from outside it. Apparitions of animals and ob-jects are also included.

    ChrP 3:15—16 (Dec 1978); *JSPR* 48:239—241 (Dec 1975); *T* 5:11—13 (1977).

    Choice 12:1371 (Dec 1975); *ContempRev* 227:222 (Oct 1975); *Guardian Wkly* 113:19 (Jul 26, 1975); *Kirkus* 44:44 (Jan 1, 1976); *Light* 96:39—42 (Spr 1976); *TimesLitSuppl* p976 (Aug 29, 1975).

730 Hartmann, Ernest. *The Nightmare: The Psychology and Biology of Terrifying Dreams.* New York: Basic Books, 1984. 294p. Bibliography: 273—282; 22 graphs; 3 il-lustrations; Index: 283—294; 15 tables

    A sleep researcher and psychoanalyst, Hartmann reviews what is known about nightmares, who has them and why, and their meaning. In Part I he reviews what has been learned about nightmares, or dream anxiety attacks, in the sleep laboratory and distinguishes them from night terrors. He con-siders nightmare data from many earlier studies and the theories offered to explain the phenomenon. He attempts to find out who has nightmares using information based on his clinical experience, five years of research in interviewing adult nightmare sufferers, and the results of formal studies with 50 of them; and an examination of special groups that might be prone to nightmares, such as war veterans, schizophrenics, and artists. Based on the foregoing, he is "led to characterize persons with frequent nightmares in terms of unusual openness, defenseless-ness, vulnerability, and difficulty with certain ego functions" (p. 7). He observes that these people have "thin boundaries," in-cluding "sleep—wake boundaries, ego boundaries, and interper-sonal boundaries" (p. 7). Hartmann notes that almost half of those he interviewed reported spontaneous ESP or other paranormal experiences. He points out that many persons who do not suffer from nightmares also have these experiences. However, he notes: "In our study the nightmare sufferers had these experiences much more frequently and dramatically, al-though several of the twenty—four vivid dreamers and ordinary dreamers also reported such experiences" (p. 151). In Part II he considers clinical aspects of nightmares, tracing them to childhood fears and traumas. One chapter is devoted to post—traumatic nightmares and another to a discussion of nightmares, night terrors, and post—traumatic nightmares as regards their natural history, diagnosis, and prognosis, and their treatments, if required. The chemical and biological substrata of nightmares is examined in the last chapter, where he notes that the under-lying biology of the nightmare appears to differ somewhat from that of the dreams. He suggests that "the biology of the nightmare may also lead us to an understanding of the underlying biology of boundaries in the mind" (p. 9).

PAI 3:41—42 (Dec 1985).

BkWorld 15:4 (Feb 3, 1985); *Booklist* 81:667 (Jan 15, 1985); *Choice* 22:1410 (May 1985); *ContempPsychol* 31:253 (Apr 1986); *GuardianWkly* 132:18 (Feb 24, 1985); *LATimesBkRev*, p18 (Mar 3, 1985); *LibrJ* 110:66 (May 1, 1985); *NYRevBks* 32:9 (Mar 28, 1985); *NYTimes* 134:15 (Jan 2, 1985); *NYTimesBkRev* 90:23 (Jan 13, 1985); *PsychoanalRev* 75:489—491 (Fal 1988); *PsycholToday* 19:74 (Apr 1985); *PublWkly* 226:83 (Nov 30, 1984).

731 Haynes, Renée. *The Seeing Eye, the Seeing I: Perception, Sensory and Extrasensory.* New York: St. Martin's Press, 1976. 224p. Chapter references; Index: 220—224; 1 table

The author, a trained historian and a former editor of the *Journal* and *Proceedings of the Society for Psychical Research*, has had a lifelong interest in parapsychology and this book represents a summary of her thinking on that subject. She discusses the relationship between psi and sensory experience; the manner in which the human personality modifies perceptions, and the fact that the nonanalytical function of the psyche, such as imagery and emotions, can mediate the way psi is expressed. She also offers useful observations on psi—conducive states of consciousness and other variables affecting the expression of psi. Although the bulk of her examples are spontaneous cases, Haynes draws on all aspects of parapsychological research and on psychology, physiology, medicine, and biology.

ChrP 1:91 (Dec 1976); *JASPR* 71:217—220 (Apr 1977); *JIndPsychol* 2:149—150 (Jul 1979); *JP* 42:330—332 (Dec 1978); *JSPR* 49:456—458 (Mar 1977); *PR* 8:23—24 (Mar/Apr 1977).

GuardianWkly 115:22 (Sep 12, 1976); *Kirkus* 45:562 (May 15, 1977); *LibrJ* 102:1506 (Jul 1977); *PublWkly* 211:83 (May 9, 1977); *Spectator* 237:14 (Aug 14, 1976); *TimesLitSuppl* p1591 (Dec 17, 1976).

731a Hearne, Keith. *Visions of the Future: An Investigation of Premonitions.* Wellingborough, Northamptonshire, England: Aquarian Press, 1989. 143p. Chapter references; 3 figures; Index: 139—143; 1 table

Dr. Hearne is a British psychologist who has specialized in the investigation of psychophysiological states, lucid dreams, and precognition. This work is not only a report of Hearne's investigations but a survey of previous research by others, and it covers both spontaneous and quantitative studies. The first chapter is a brief history of reports of precognition from ancient times to the 19th century. There follow chapters describing precognitions associated with death, accidents, injuries, trivial events, and a small fraction which involves beneficial informa— tion. Chapter 4 consists of in—depth studies of three typical female experients and their experiences. Attempts to investigate precognition scientifically are reviewed next, followed by a chapter on explanations and counterhypotheses. The last chap— ter discusses the evidence for precognition, the characteristics of precognitive experiences, a review of questions about precogni— tion that need to be answered, proposals for future research and methodology, and attitudes toward precognition.

732 Hufford, David J. *The Terror That Comes in the Night: An Experience—Centered Study of Supernatural Assault Traditions.* Philadelphia, PA: University of Pennsylvania Press, 1982 (Publications of the American Folklore Society: New Series, Vol. 7). 278p. Bibliographical foot— notes; Bibliography: 259—265; Index of features: 267— 270; Index: 271—278; 5 tables

The author, an associate professor of behavioral science at the Pennsylvania State College of Medicine, is concerned with the experiential basis of supernatural beliefs. In this study he concentrates on one such belief: that of the nightmare, incubus experience, or supernatural assault. Known as the experience of the "Old Hag" in Newfoundland, it has the following charac— teristics: "(1) awakening (or an experience immediately preceding sleep); (2) hearing and/or seeing something come into the room and approach the bed; (3) being pressed on the chest or strangled; (4) inability to move or cry out until either being brought out of the state by someone else or breaking through the feeling of paralysis on one's own" (pp. 10—11). Through questionnaire/interview and case study approaches, Hufford found that accounts of this type of experience are consistent regardless of the culture of origin. Some of the accounts have features that border on the parapsychological. As an outcome of the research for this book, Hufford has developed and advocates an experience—centered approach to this and other types of su— pernatural (and, one could say, psychic) accounts and stories. By making a phenomenological study of the experiences, it be— comes possible "to gain a better knowledge of the experiences lying behind a particular supernatural belief and to begin to consider the role of those experiences in such belief" (p. 256).

JASPR 78:278—280 (Jul 1984); *PAI* 3:41 (Jun 1985).

Archaeus 2:91—93 (Fal 1984); *Choice* 20:1372 (May 1983); *Explorer* 3:14 (Oct 1986); *ImagCogPers* 4:225—226 (1984/85); *JSciStudRelig* 23:208—209 (Jun 1984); *LibrJ* 107:2262 (Dec 1, 1982).

733 MacKenzie, Andrew. *Hauntings and Apparitions.* London, England: Heinemann, on behalf of the Society for Psychical Research, 1982. 240p.

7 figures; 26 illustra— tions; Index: 237—240; Select bibliography: 228—236

This volume, by SPR council member MacKenzie, surveys the studies of apparitions and hauntings conducted by the SPR over the past 100 years. MacKenzie excerpts material that has appeared in the Society's *Journal* and *Proceedings*, but also in— cludes much hitherto unpublished archival material supple— mented by material from his own investigations of some classic cases, together with his own insightful observations. An overall introduction and history is given in the first chapter, followed by an ongoing discussion of the nature of apparitions provided by various SPR researchers over the years. There follow chap— ters devoted to many well—known exemplary cases: The Chel— tenham haunting, the Haunted Mill House at Willington, a haunted Shropshire vicarage, the Beavor Lodge ghost, the Snet— tisham ghost, a haunted road, the Versailles "adventure," two retrocognitive cases, the apparition of Samuel Bull, the haunting of Cleve Court, the haunting of Abbey House in Cambridge, and two cases from the 1970s. The last chapter is an insightful discussion of theories of haunts and apparitions in which MacKenzie calls for a psychological approach to these cases. In order to understand the nature of apparitions, he points out that we need to understand human nature.

ChrP 4:227—228 (S 1982); *JASPR* 78:162—166 (Apr 1984); *JSPR* 52:137—139 (Jun 1983); *Psychoenergetics* 5:84—87 (1983).

BritBkNews p536 (Sep 1982); *ContempRev* 241:109 (Aug 1982); *LondRevBks* 4:10 (Jul 1, 1982); *NewStatesmann* 104:20—21 (Aug 6, 1982); *Observer* p31 (Aug 1, 1982).

734 MacKenzie, Andrew. *Riddle of the Future: A Modern Study of Precognition.* New York: Taplinger, 1975 (c1974). 172p. Bibliography: 164—168; Index: 169—172

Along with Louisa Rhine, Ian Stevenson, and Celia Green, Andrew MacKenzie has specialized in the study of spontaneous psi phenomena. He says: "This book is an attempt to present the evidence for precognition in everyday life. I have not set out to prove that there is such a thing as precognition but simply to present some of the theories relating to it and to give a large number of cases . . . so that readers can make up their own minds on the reality (or non—reality) of precognition" (p. 12). He describes cases from the literature and fresh material he gathered himself. The literature on spontaneous precognition is reviewed in a lengthy introduction, and in a concluding chapter he considers the theories that have been offered to explain precognition.

JP 41:71—72 (Mar 1977); *JSPR* 48:99—101 (Jun 1975); *P* 7:32—33 (May/Jun 1976); *T* 5:13 (1977).

Books&Bkmn 20:79 (Feb 1975); *ContempRev* 226:53 (Jan 1975); *LibrJ* 100:1832 (Oct 1, 1975); *Light* 95:45—46 (Spr 1975); *TimesLitSuppl* p1438 (Dec 20, 1974).

735 MacKenzie, Andrew. *The Seen and the Unseen.* London, England: Weidenfeld and Nicolson, 1987. 286p. 17 illustrations; Index: 283—286; Select bibliography: 281—282
This collection of authenticated cases of apparitions by a veteran investigator is in two sections. The first contains four chapters devoted to the sense of a presence. The second section consists of 26 chapters devoted to hauntings (or hauntings plus poltergeist phenomena). Evidential aspects of the cases are emphasized. In a concluding chapter, he assesses the evidence for the preceding cases, relating them to other surveys and cases in the literature.
JASPR 83:51—53 (Jan 1989); *JSPR* 55:98—99 (Apr 1988); *PAI* 6:45 (Jun 1988).

736 Moss, Peter. *Ghosts Over Britain.* Illustrated by Angela Lewer. London: Hamish Hamilton, 1977. 173p. 60 illustrations
A collection of 60 modern British "ghost stories" selected from letters and telephone calls the author, an historian, writer, and broadcaster, received in response to ads placed in British newspapers. These cases were chosen from the hundreds of accounts received because those who contributed them were obviously sincere and because corroborative evidence of some sort was obtained. Some types included are apparitions, historical haunts, family ghosts, poltergeists, and ghostly warnings.
JP 43:354 (Dec 1979); *JSPR* 49:842—843 (Jun 1978); *PsiR* 3:196—199 (Sep/Dec 1984).
LibrJ 104:116 (Jan 1, 1979).

737 Neppe, Vernon M. *The Psychology of Déjà Vu: Have I Been Here Before?* Johannesburg, South Africa: Witwatersrand University Press, 1983. 277p. Author index: 268—270; Bibliography: 256—267; Glossary: 248—255; 6 graphs; Subject index: 271—277; 3 questionnaires; 8 tables
Based on the author's dissertation research, this is the first book—length treatment of déjà vu. There are chapters on the concept and terminology of déjà vu, and the qualitative aspects of the déjà vu experience. The author also reports on his surveys of déjà vu in schizophrenics, temporal lobe epilepsy, and normal subjects. Neppe concludes that there are different kinds

of déjà vu. The questionnaires used in his research are presented in appendices, and there is a glossary and a lengthy bibliography.
PJSA 5:50−53 (Jun 1984); *PsiR* 3:196−199 (Sep/Dec 1984).
AmJPsychiat 142:772−773 (Jun 1985); *ContempPsychol* 30:337 (Apr 1985); *Quadrant* 17:102−103 (Fal 1984).

738 Persinger, Michael A. *The Paranormal: Part I. Patterns.* New York: MSS Information Corporation, 1974. 248p. Bibliography: 241−248; 1 figure; 16 graphs; 13 tables
A psychologist at Laurentian University reports on a study in which he applied the method of operational analysis of verbal behavior concerning purported psi experiences. His database consisted of 529 reports published in *Fate* from 1965 through 1969. Given the fact that spontaneous experiences cannot provide proof of psi, the aim of the study was to show that it may be possible, through analyzing verbal behavior about psi events, "to establish consistencies about the events in the en−vironment and organisms which precede (or come after) the psi experience" (p. 43). Recognition of these patterns can facilitate laboratory investigations of psi. The types of psi phenomena considered are telepathy and clairvoyance, precognition and retrocognition, apparitions, out−of−body experiences and near−death experiences, and postmortem reports. He compares his data with other spontaneous case studies and with para−psychological and psychological laboratory findings.
PR 7:17−18 (Jul/Aug 1976); *T* 5:12−13 (1977).

739 Persinger, Michael A. *The Paranormal: Part II. Mechanisms and Models.* New York: MSS Information Corporation, 1974. 195p. Bibliography: 189−195; 4 graphs
In this volume, psychologist Persinger, after summarizing the patterns noted in his analysis of the verbal reports of spon−taneous psi presented in Part I, sets forth a number of models and mechanisms to explain those reports. The behavioralistic and physicalistic emphasis is indicated by the chapter headings: "Conditioning and Learning," "Paranormal Experiences in Terms of Normal Behavioral Mechanisms," "Interesting Speculations on Paranormal Reports Based on Extensions of Conditioning Models: Paranormal Experiences as Responses to Actual Exter−nal Stimuli," and "Possible Physical Mechanisms for Paranormal Responses." In the latter he deals with extremely low frequency electromagnetic fields, infrasonic stimuli, high voltage static fields, and an environmental peltier effect.
PR 7:17−18 (Jul/Aug 1976); *T* 5:12−13 (1977).

740 Rao, K.R. (Ed.). *Case Studies in Parapsychology: Papers Presented in Honor of Dr. Louisa E. Rhine.* Jefferson, NC: McFarland, 1986. 130p. Bibliography: 120–126; Chapter references; Index: 127–130

This volume provides an overview of field work in case studies involving parapsychological experiences. The editor provides a tribute to L.E. Rhine and her work, especially her investigations of spontaneous psi experiences and the nature of the psi process. One of Dr. Rhine's daughters, Sally Rhine Feather, provides a biographical sketch of Louisa Rhine entitled "Something Different." It emphasizes the period from her early years through the years of young motherhood. Debra H. Weiner and JoMarie Haight contributed "Charting Hidden Channels: Louisa E. Rhine's Case Collection Project," which reviews Rhine's case studies and discusses the questions and controversies raised by her work. Sybo A. Schouten's "A Different Approach for Analyzing Spontaneous Cases with Particular Reference to the Study of Louisa E. Rhine's Case Collection" contrasts the model of the psi process based on his studies of spontaneous cases and those of L.E. Rhine. In "Recurrent and Nonrecurrent Psi Effects," W.G. Roll examines L.E. Rhine's study of spontaneous cases of macro–PK in relation to poltergeist and haunting cases. K.R. Rao contributed "L.E. Rhine on Psi and Its Place," which surveys the thinking of both J.B. Rhine and L.E. Rhine on the nature of psi. In "Apparitions Old and New," Karlis Osis reviews what is known about the nature of apparitions. The present status of and future research on reincarnation–type cases is the subject of a paper by Emily Williams Cook. D. Scott Rogo reviews what is presently known about out–of–body experiences, H. Kanthamani provides a bibliography of L.E. Rhine's works arranged chronologically, with the first item a botanical article published in 1924 and the last her biography of her husband, *Something Hidden*, which was published in 1983.

JASPR 82:295–298 (Jul 1988); *JIndPsychol* 6:43–45 (Jan/Jul 1987); *JP* 57:337–352 (Dec 1987); *PAI* 6:47 (Jun 1988).

ABBookman 79:299 (Jan 26, 1987); *ContempPsychol* 32:579 (Jun 1987).

741 Rhine, Louisa E. *The Invisible Picture: A Study of Psychic Experiences.* Jefferson, NC: McFarland, 1981. 267p. Bibliography: 259–261; Index: 263–267; 5 tables

Pioneer parapsychologist Louisa Rhine studied thousands of spontaneous psi experiences for years. This is a distillation of that research in which she explains why she did it and what she found. Except for two introductory chapters and a final overview, the subjects of 15 chapters had already been dealt with by Dr. Rhine in articles in the *Journal of Parapsychology*. However, she views the material in each chapter from the perspective of the entire study and from over three decades of concentration on this material. New examples are used, and in a real sense this book provides a fresh look at the psi process as revealed in spontaneous cases.

> *JP* 44:353–357 (Dec 1980); *JSPR* 51:160–162 (Oct 1981); *T* 9:18–19 (Spr 1981).
> *Choice* 18:1344 (May 1981); *LibrJ* 106:155 (Jan 15, 1981); *NewReal* 4:42–43 (Mar 1982); *SkptInq* 6:50–53 (Win 1981/82); *Spectator* 247:20 (Nov 28, 1981); *SpirFron* 16:106–109 (Spr 1984).

742 Rogo, D. Scott. *An Experience of Phantoms.* New York: Taplinger, 1974. 214p. Chapter notes; 8 illustrations; Index: 210–214; Suggested reading list: 207–209

This is a survey of apparitions of the dead and the living, including apparitions associated with poltergeists, animals, hauntings, and out–of–body experiences. It is useful because it not only draws on the most important historical studies and cases, but it also describes current research and theories on the nature of apparitions and their implications. Rogo, a parapsychologist who specializes in the investigation of spontaneous psi experiences, suggests that the study of spontaneous psi "can offer insights into the process of psychical phenomena that may well elude us in the laboratory" (p. 10). Recent cases are included––some of which were his own and others that he investigated himself. He concludes that "apparitions, as no other phenomenon, bring to the fore that man is a spiritual being . . . . . Far from being governed by our physical universe they transcend it. It is no wonder that so many who have experienced phantoms, whether by seeing an apparition or by having an out–of–the–body experience, acquire from that experience the belief that man is a far greater mystery than is the material universe in which we live" (p. 202). An annotated bibliography of "suggested readings" is appended.

> *P* 6:36–37 (May/Jun 1975); *PR* 6:19–20 (Jul/Aug 1975).
> *Kirkus* 42:524 (May 1, 1974); *LibrJ* 99:1965 (Aug 1974); *PublWkly* 205:62 (May 20, 1974).

815 Rogo, D. Scott. *On the Track of the Poltergeist.* [See Addenda for annotation]

743 Vaughan, Alan. *Incredible Coincidence: The Baffling World of Synchronicity.* New York: New American Library, 1980. 247p. Case index: 243–247; Chapter notes; Sug–

gested reading list: 241—242

Author, editor, and psychic Alan Vaughan presents the details of over 150 cases of synchronicity, or meaningful coin—cidences that have no apparent cause. Many of the cases were previously unpublished. He also provides a classification of such cases and theorizes about both their significance and their rela—tion to psi phenomena. He discusses "The Synchronicity of Synchronicity," which is about the fact that synchronisitic events seem to occur when we need them.

JP 43:349—350 (Dec 1979); *PR* 11:18—20 (Jan/Feb 1980).

Kirkus 47:681 (Jun 1, 1979); *LibrJ* 104:1147 (May 15, 1979); *NewReal* 3:38 (Oct 1979); *PublWkly* 215:106 (Apr 30, 1979); *SpirFron* 15:58—60 (Win 1983).

Survival of Death

INTRODUCTION

In *Sources*, there were 29 titles in this category, which deals with the question of whether or not any aspect of human personality survives death.

It is still a subject of great interest as evidenced by the fact that there are 36 books in the present section. Survival is a subject that is of perennial relevance to philosophy. Seven of the books are discussions of survival by philosophers: Almeder [744], Badham and Badham [746], Flew [752], Lewis [763], Lorimer [764], Lund [765], and Moore [768]. Others survey the evidence for survival with a more empirical than theoretical orientation. These are the books by Berger [748], Ebon [750], Gauld [755], Jacobson [761], Kastenbaum [762], McAdams and Bayless [766], Rogo [770], Ryzl [772], Sibley [773], Spong [774], Wilson [777a] and the reprinted classic by Lombroso [811]. Research on survival is reported by Berger [749] (case studies), Ellis [751] (electronic voice phenomena), Osis and Haraldsson [769] (deathbed experiences), Rogo and Bayless [771] (telephone calls purporting to come from the dead), and the two books by Stevenson [775, 776] (xenoglossy). Two books review opinion polls of attitudes toward life after death: those by Gallup [754] and Greeley [757]. A crosscultural approach emphasizing religious aspects of survival is provided by Badham and Badham [745], Hick [760], and Toynbee et al. [777]. Grof and Grof [758], and Grosso [759] emphasizes the psychological im—portance and implications of survival for the living. Gill [756] has written a biography of prominent thanatologist Elizabeth Kübler—Ross who, because of her belief in survival, has turned from science to mysticism. Fuller [753] recounts the details of a classic case of purported postmortem communications through a medium, the R—101 case involving Eileen Garrett (but note that several books in the section on "Criticisms" criticize this case), and Moody [767] surveys unusual experiences and phenomena associated with Elvis Presley since his death.

Additional books on survival listed elsewhere:
LeShan [352] discusses the survival problem and alternate realities. The *Ashby Guidebook* [804] has an appendix setting forth Robert Ashby's views on survival research. LaBerge [351]

shows how lucid dreams shed light on survival, and Parker
[356] asks whether there is a postmortem altered state. Grof
[348] touches on mediumistic experiences and survival. Animal
survival is treated by Schul [364]. Finucane [726a] describes
how the way ghosts have been viewed down through the ages is
dependent on theological opinions and scientific accomplish—
ments. There is a bibliography on survival in Jones [373].
Psychic Dixie Yeterian [379] has a chapter on what she has
learned about the survival of victims of violent crimes.
Postmortem communications are reported in the books dealing
with George Anderson [394] and Ena Twigg [398]. The latter
contains accounts of five cases the authors consider cannot be
explained by the superpsi hypothesis. The survival problem
figured largely in the work of E.R. Dodds, described in his
autobiography [402] and in the biography of Sir Oliver Lodge
[403]. Young [455] describes psychic experiences in children
suggestive of survival. Chapters on survival are included in
three books in the section "Books for Young People": [415, 430
(Kidd bequest), and 448], and the volume by Deem [429a]
teaches youngsters how to become a "ghost detective." Abell
and Singer [456] have a chapter critical of survival, Booth [458]
has a chapter suggesting how the cross correspondences could
have been produced by normal means, Christopher [459] has
chapters critical of survival and has written an entire book on
the subject [460], and Hines [474] has a critical chapter on sur—
vival. The religious significance of survival is brought out in
the histories of Brandon [549], Cerullo [550[, and Oppenheim
[560]. The primary concern of psychical research was to obtain
evidence of survival. Several books in the history section
describe these early attempts: see those by Douglas [551], Hayes
[554], and Inglis [555, 556]. A 144—page survey of survival
evidence and theories by W.G. Roll appeared in *Advances in
Parapsychological Research 3* [522]. Three books in the section
on mediumship and related subjects deal with spirit healing and
spirit possession: Ebon [575], Pettiward [582], and Rogo [584].
The implications of NDEs for survival are considered in Moody
[589], in which he was equivocal, and [591], in which decides in
favor of survival, and Lundahl [588], who has chapters on
deathbed visions as well as NDEs. Philosophical aspects of the
survival question are considered in Flew [608] and in Wheatley
and Edge [615]. Spirit photography is covered by Gettings
[626]. All the books in the reincarnation section are relevant to
the survival question, especially those books by Stevenson [697-
701]. Wilson [703] reviews the survival evidence and Rogo sur—
veys the methodology [695]. About half of Perry's book [717] is
on theological aspects of survival. Apparitions and hauntings
are studied by Auerbach [723], Davidson [724], MacKenzie [733,

735], and Rogo [742].

BIBLIOGRAPHY

744 **Almeder, Robert.** *Beyond Death: Evidence for Life After Death.* Springfield, IL: Charles C Thomas, 1987. 100p. Chapter notes; Name index: 97—98; Subject index: 99—100; Suggested reading list: 20—21, 40, 52—53, 57

Almeder, a philosopher, has written this book to acquaint the general reader with the evidence supporting belief in some form of life after death. He attempts to confront "the strongest objections to all the best evidence presently available," and he concludes that "it is more reasonable to believe in some form of life after death than it is to believe in nothing after death" (p. vii). In an interesting Introduction, he describes some personal experiences with mediums that led him to abandon some long-held theories about the limitations of knowledge. There are five chapters. One is on reincarnation, in which he examines the best evidence, primarily the cases studied by Ian Stevenson. The second is on several apparitional cases. OBEs are the subject of Chapter 3 and Chapter 4 is on postmortem communications, emphasizing the mediumship of Mrs. Piper and Mrs. Willett. In a brief conclusion, he argues that it is the *aggregate* of the evidence available that compels belief rather than isolated cases or types of evidence. In an appendix, he discusses and attempts to refute "other skeptical considerations." Chapters 1—4 each end with a list of recommended readings.

JASPR 83:57—59 (Jan 1989); *PAI* 6:42 (Jun 1988).

745 **Badham, Paul, and Badham, Linda (Eds.).** *Death and Immortality in the Religions of the World.* New York: Paragon House, 1987. 238p. Chapter references; Index: 233—238

The chapters of this volume consist of some of the papers given at an international conference on "Death and Immortality in the Religions of the World" held in Korea in 1984. The contributors were concerned with answering the question of whether or not it is possible to find unity in the beliefs in forms of life after death found in the major world religions and if so, whether this united belief can "stand in the face of contemporary philosophical analyses, and can cohere with the well-established facts that the natural sciences have discovered?" (p. 1). For each religion the exponent chosen was drawn from that tradition. The traditions covered are African, Jewish, Christian, Roman Catholic, Islam, Advaita, and Theravada Buddhism. Linda Badham and Antony Flew in two chapters argue against

survival of death. Arthur Berger presents "A Critical Outline
of the Prima Facie Evidence for Survival," in which he reviews
the major types of evidence, and David Lorimer closes with a
paper entitled "Current Western Attitudes to Death and
Survival." He concludes that there will soon be "a profound
shift of preoccupations away from the exclusively material
towards a more spiritual, integrated, and compassionate perspec—
tive" (p. 229). In an introductory chapter the editors provide a
summary view of the contributions in the book.
    PAI 6:42 (Jun 1988).

746 **Badham, Paul, and Badham, Linda.** *Immortality or Extinc—
      tion?* Totowa, NJ: Barnes & Noble Books, 1982. 146p.
      Chapter notes; Index: 144–146; Select bibliography:
      142–143
      The purpose of this work by a theologian (P.B.) and a
philosopher of science (L.B.) is to examine a wide variety of ar—
guments for and against life after death. Part 1 concerns
whether or not the question of survival can be asked meaning—
fully. The second part examines the intellectual developments of
the past 400 years that have eroded the traditional religious
foundations in favor of immortality, concluding that it is neces—
sary to go beyond the traditional framework in order to find
support for the doctrine of a future life. In Part 3, the authors
consider some claims that, at face value, may be taken as
evidence for survival: the evidence, respectively, from near—
death experiences, from psychical research, and from claimed
memories of reincarnation. In the final part, they sum up the
case for and against immortality, and find that there is compell—
ing evidence on both sides. They conclude that an
experientially—based relationship with God that provides a trust
that God will hold one in being through death may tip the
scales in favor of survival, but for those who doubt the reality
of the religious dimension, a belief in survival is not tenable.
    ChrP 4:229–230 (S 1982); *Light* 104:82–83 (Sum 1984).

747 **Beard, Paul.** *Living On: How Consciousness Continues and
      Evolves After Death.* New York: Continuum, 1981. 202p.
      Chapter notes; Index: 200–202
      The President of the College of Psychic Studies attempts
to review existing evidence for survival of death to see what can
be learned about the possible continuation and evolution of con—
sciousness after death. The book is based on purported postmor—
tem communications of high caliber. The primary com—
municators cited are Oliver Lodge, Mrs. Willett, Frederic Myers,
Helen Salter, Frances Banks, persons known to C. Drayton
Thomas, and T.E. Lawrence. Beard discusses the methodology

involved and the reliability of the sources, and whether or not it is morally justifiable to approach this material. He analyzes the postmortem accounts under the headings: "What It Is Like To Die and Find One's Identity Intact"; "The Illusion of the 'Summerland'"; "The Judgment"; and "Life in the First, Second and Third Heavens." A chapter is devoted to reincarnation, which is frequently alluded to in the accounts. He concludes that a life of growth continues in the individual after death. He writes that this "spiritual task which lies before men and women on earth and those who have passed through death is a shared and common one . . . . This task is the regeneration of the world by regeneration of the individual, and the bringing to earth of the Kingdom of Heaven" (p. 194).

ChrP 4:16—18 (Mar 1981); *JRPR* 6:71—75 (Jan 1983); *JRPR* 6:250—252 (Jul 1983); *JSPR* 50:544—545 (Dec 1980); *T* 10:68—71 (Aut 1982).

Alpha No. 9:17—18 (Oct 1980); *ContempRev* 237:112 (Aug 1980); *Light* 100:129—131 (Aut 1980); *NewReal* 4:46 (Mar 1982); *Punch* 279: 71 (Jul 9, 1980); *SpirFron* 14:122—123 (Fal 1982).

748 Berger, Arthur S. *Aristocracy of the Dead: New Findings in Postmortem Survival.* Foreword by Antony Flew. Jefferson, NC: McFarland, 1987. 209p. Bibliography: 195—198; Chapter notes; Index: 199—209; 24 tables

According to philosopher Antony Flew, lawyer/parapsy—chologist Arthur Berger takes a new approach in this book to the question of whether or not there is survival after death. In the first four chapters, Berger presents the existing evidence for and against survival as found in the parapsychological literature. In Chapter 4, Berger begins the presentation of his new ap—proach. First, he proposes that we question two widely held assumptions that may have served as blocks to progress in inves—tigating survival: (1) that if any survive, all do, and (2) if one person is able to communicate after death, all should be able to communicate with equal facility. He reanalyzes key cases in the literature that indicate that some persons communicate better than others. Based on a sample of good communicators, he sets forth their characteristics and qualities. He then proposes that in looking for survival evidence, good communicators be made the focus of study. He cites an example where evidence for survival was obtained in the case of a man who met several of the criteria of good communicators. He concludes with the proposal that instead of starting the investigation of survival when a person dies, researchers should instead look among the living for possible "star" communicators, and work with them and train them in the art of communicating and establishing their identity when they eventually do die.

ChrP 7:293—294 (Dec 1988); *JSPR* 55:238—239 (Oct 1988);
PAI 6:42 (Jun 1988).
   Light 108: 84—85 (Sum 1988); *SpirFron* 20:181—182 (Sum 1988).

749 Berger, Arthur S. *Evidence of Life After Death: A
Casebook for the Tough—Minded.* Springfield, IL: Charles
C Thomas, 1988. 149p. Chapter notes; Index: 145—149
   This is a casebook of experiences, mediumistic and other,
that indicate that some component of human beings survives
biological death. Berger has culled the parapsychological litera—
ture of the past 100 years to choose 15 cases and has added 15
more investigated by the Survival Research Foundation, which
he directs. A lawyer by training, Berger follows the judicial
practice of representing both sides of a situation. Thus, each
case is followed by presenting both pro and con views in order
to "stimulate discussion of points, raise questions, [and] bring
out the real points in controversy" (p. viii). There are four
chapters. The first presents seven cases illustrating "Mental
Phenomena" (apparitions and ESP). Chapter 2, "Mental
Mediumship," considers 13 cases of communications received
through a medium and purporting to come from deceased per—
sons. Chapter 3 contains 10 cases of "Physical Phenomena,"
which may or may not occur when a medium is present. In
Chapter 4, Berger discusses "Evidence and Proof" in regard to
both mental and physical phenomena.
   ChrP 8:70—71 (Jun 1989); *JRPR* 12:115—117 (Apr 1989).
   Light 108:84—85 (Sum 1988).

750 Ebon, Martin. *The Evidence for Life After Death.* New
York: New American Library, 1977. 177p. Suggested
reading list: 175—176
   Following an historical chapter on research into survival,
Ebon reviews modern research bearing on the problem, including
the work of Raymond Moody, Elizabeth Kübler—Ross, Gardner
Murphy, Karlis Osis, Susy Smith, W.G. Roll, Stuart ("Blue"——
now Keith) Harary, J.B. Rhine, Robert Crookall, Ingo Swann,
L. Richard Batchler, Carl Wickland, and three modern
mediums: Douglas Johnson, Jane Roberts, and Rosemary Brown.
   ChrP 3:112—113 (Jun 1979); *JIndPsychol* 2:153—154 (Jul
1979); *JP* 43: 156—158 (Jun 1979).
   Light 98:34—36 (Spr 1978); *SpirFron* 10:116 (Spr 1979).

751 Ellis, D[avid] J. *The Mediumship of the Tape Recorder.*
Harlow, England: Author, 1978. 161p. Bibliography:
151—155; Glossary: 156—159; 4 graphs; 3 illustrations;
Index: 160—161; 6 tables

This is an account of "a detailed examination of the
(Jürgenson, Raudive) phenomenon of voice extras on tape record—
ings" undertaken while Ellis was the Perrott—Warrick Student
in Psychical Research at Trinity College, Cambridge, 1970—1972.
He describes and discusses all the research on tape—recorded
voices as well as his own experiments. Although he began the
study with the conviction that something paranormal was occur—
ring, his investigation led him to conclude that there is "no
reason to postulate anything but natural causes . . . to explain
the 'voice phenomenon.'" However, he suggests that use of the
tape recorder can serve as a medium for ESP: the listener can
turn out to be "more psychic than his apparatus." Ellis's in—
formed account is essential reading for anyone interested in the
"voice phenomenon."
JP 43:70—72 (Mar 1979); JSPR 50:34—35 (Mar 1979); T
8:15 (Win 1980).

752 Flew, Antony. *The Logic of Mortality: On Personal Iden—
tity.* New York: Basil Blackwell, 1987. 200p. Bibliogra—
phy: 185—195; Name index: 197—200.
This volume is based on the author's Gifford Lectures for
the academic year 1986—87, but his interest in the topic began
40 years earlier. In the first chapter he points out that there
are three possible means of survival. He characterizes these as
the reconstitutionist method, the "astral body" method, and the
"platonic criticism" approach, which he first developed in an es—
say reprinted in Wheatley and Edge [615]. These philosophical
views are discussed further in Chapters 2—6. Personal identity
is the subject of Chapters 7 and 8. He discusses the mind—
body problem in Chapter 9, and in the final chapter, "The Sig—
nificance of Parapsychology," he poses the following problem:
"The very concepts of psi are just as much involved with the
human body as are those of all other distinctively human . . .
concepts" (p. 183).

753 Fuller, John G. *The Airmen Who Would Not Die.* New
York: Berkeley, 1980. 373p. Bibliography: 343—348; Chap—
ter notes; 1 figure; 10 illustrations; Index: 369—373; 1
table
This book by a science writer provides a full—scale ac—
count of one of the major cases in the literature of psychical
research: predictions regarding the ill—fated crash of British air—
ship R—101. The predictions were purported to come through
the deceased transatlantic pilot, W.R. Hinchcliffe. His apparition
appeared after his death to a friend, a Mrs. Earl, who received
messages alleged to come from him via the ouija board. Addi—
tionally, many communications were received through the

mediumship of Eileen Garrett. After R—101 crashed, Mrs. Gar—
rett also received communications from its deceased captain,
H.C. Irwin. For criticisms of this case see: [458, 473].
   JRPR 3:143—144 (Apr 1980); *JSPR* 50:314—316 (Mar
1980); *PR* 10:15 (Mar/Apr 1979); *T* 8:22 (Spr 1980).
   Booklist 75:1530 (Jun 15, 1979); *Kirkus* 46:1394 (Dec 15,
1978); *PublWkly* 214:47 (Dec 25, 1978); *PublWkly* 216:68 (Dec
10, 1979); *SchLibrJ* 25:154 (Mar 1979); *SchLibrJ* 25:36 (May
1979).

754 Gallup, George, Jr., with Proctor, William. *Adventures in*
      Immortality. New York: McGraw—Hill, 1982. 226p. In—
      dex: 223—226; 21 tables
      Gallup, who is not only President of the Gallup Poll but
also Director of the Princeton Religious Research Center, reports
the results of several nationwide polls on attitudes and beliefs
concerning the afterlife, with emphasis on near—death ex—
periences (NDES). Many persons reporting NDES were inter—
viewed intensively and the results are reported here. He also
reports on surveys of the views of medical doctors and other
scientists concerning an afterlife. A 40—page appendix gives the
results of all the present surveys as well as Gallup polls con—
ducted in 1952 and 1966.
      ASPRN 10:4 (Jan 1984); *ChrP* 5:125—126 (Dec 1983);
JASPR 77:180—185 (Apr 1983); *T* 12:33—36 (Sum 1984).
      America 147:158 (Sep 25, 1982); *Booklist* 42:236 (Sep
1982); *BritBkNews* p221 *(Apr 1983); WestCoastRevBks* 8:58
(Nov 1982); *ChristCent* 100:375 (Apr 20, 1983); *Kirkus* 50:589
(May 1, 1982); *LibrJ* 107:1230 (Jun 15, 1982).(Spr 1983).

755 Gauld, Alan. *Mediumship and Survival: A Century of Inves—*
      tigations. London, England: Heinemann, 1982. 287p. Bib—
      liography: 268—282; 14 illustrations; Index: 283—287; 2
      tables
      Psychologist Gauld has undertaken a review of some of the
factual evidence, both old and new, which bears on the survival
problem; much of it is taken from the publications of the SPR
and ASPR. The major classes of evidence discussed are certain
types of apparitions, some instances of mental mediumship, and
some cases of ostensible reincarnation. There are seven chapters
on mediumship (including chapters on Mrs. Piper, Mrs.
Leonard, drop in communicators, manifestations of purpose and
of other characteristics, and mediumistic controls). There are
two chapters on the super—ESP hypothesis; three on apparitions
(including out—of—body experiences); and one each on obsession
and possession, reincarnation, and memory and the brain. In his
"Concluding Remarks," Gauld writes: "It seems to me that

there is in each of the main areas I have considered a sprink—
ling of cases which rather forcefully suggest some form of sur—
vival" (p. 261), but that such evidence as we have does not tell
us anything about what kind and degree of survival there may
be. He emphasizes the need for more data, and provides several
useful suggestions for further research.

    ChrP 5:20 (Mar 1983); *JASPR* 78:153—156 (Apr 1984); *JP*
48:127—148 (Jun 1984); *JSPR* 52:203—205 (Oct 1983); *PR*
14:10—12 (Sep/Oct 1983).

    Light 103:29—31 (Spr 1983).

756 Gill, Derek. *Quest: The Life of Elisabeth Kübler—Ross.*
    Epilogue by Elisabeth Kübler—Ross. New York: Harper
    & Row, 1980. 329p. 20 illustrations

    Writer Derek Gill came to write this authorized biography
of physician/pioneer thanatologist Kübler—Ross not by design
but as a result of a series of coincidences. Because the book
describes her whole life, her interest in survival of death is not
dealt with until toward the end. Some experiences described in
the book suggest that Kübler—Ross herself is psychic. This book
provides an important backdrop for understanding the teachings
and activities of Elisabeth Kübler—Ross. In the Epilogue she her—
self tells why she turned from the ways of science to those of
mysticism, and of her conviction about the meaning of life, in
which death is viewed as a stepping stone.

    PAI 1:31 (Aug 1983).

    SpirFron 19:233—234 (Fal 1987).

757 Greeley, Andrew. *Death and Beyond.* Chicago, IL: Thomas
    More, 1976. 144p. Chapter notes; 1 figure; 10 tables

    Priests have always been concerned with questions of death
and immortality, but only a handful of sociologists have been
involved professionally with such intangibles, and that only
recently. Greeley is unique in that he is both a priest and a
sociologist, and he brings his background in both disciplines to
bear on the meaning of death and what may come after. From
the sociological point of view, he presents the results of various
surveys of the belief in life after death from several countries
during the years 1936—1973. He found that in the U.S., at
least, our increasingly secularized lifestyle has not diminished
belief in survival——in fact, it has increased slightly. In a chap—
ter on mystical experiences, he presents the results of a poll he
conducted in which he found that "he who has had an ecstatic
experience is absolutely certain of the goodness and purposeful—
ness of the universe and of his own personal survival in it" (p.
87). He notes further that their testimony "forces the skeptic to
doubt his skepticism" (p. 110). He deals with the evidence for

survival from the "wonder sciences," his term for parapsychol—
ogy and transpersonal psychology. He concludes that they
"provide reassurance. They suggest that the universe is a mar—
velously more complex place than we had thought, and that the
rigid, mechanistic scientific worldview . . . is a much less than
adequate model" (p. 87). He concludes that even though it is
likely we will never obtain absolute proof of survival, there are
sufficient grounds even now for holding that belief in survival is
a highly rational position to hold.
BestSell 36:55 (May 1976); *LibrJ* 101:1127 (May 1, 1976);
RevReligious 36:494 (May 1977).

758 Grof, Stanislav, and Grof, Christina. *Beyond Death: The
Gates of Consciousness.* New York: Thames and Hudson,
1980. 96p. Bibliography: 96; 158 illustrations
This lavishly illustrated book offers traditional and current
insights into the processes of dying and death and views them
as precursors to new life and consciousness. Ancient religious
beliefs and mythologies are seen to refer to psychic realities that
are experienced in altered states and that are intrinsic to
human personality and should not be denied. "For the full ex—
pression of human nature, they must be recognized, acknow—
ledged and explored, and in this exploration, the traditional
depictions of the afterlife can be our guide" (p. 31). The death
of the body may be the door to "an adventure in
consciousness." The encounter with death is described in many
contexts: in near—death experiences, in schizophrenic and
psychedelic states, and in traditional mythological and religious
views of death, rebirth, and the afterlife.
PAI 2:119 (Jun 1984).

759 Grosso, Michael. *The Final Choice: Playing the Survival
Game.* Walpole, NH: Stillpoint, 1985. 348p. Bibliography:
341—348; 2 tables
The purpose of this "sketch of a New Age Metaphysics of
Death and Enlightenment" is to "draw a picture of Mind at
Large and to show how that concept may relate to our survival,
both here and hereafter" (p. 6). The book is in six parts, with
the first presenting an outline of the idea of a modern
deathcraft or art of dying. Part 2 deals with research on sur—
vival and stresses the importance of a synoptic view of psi, sur—
vival research, altered states, and biological theory. Part 3 at—
tempts to show that the OBE is a way we explore the fringes
of Mind at Large. Part 4, on near—death phenomena, which are
important routes to Mind at Large, shows that there is a
general mechanism triggered by near death that may work in
any crisis of transformation. In Part 5, NDEs are viewed as

"part of a larger class of apparitional phenomena whose function is to assist, reassure, guide and direct individuals, or groups of individuals, in times of crisis" (p. 8). These he calls Messengers from Mind at Large. The sixth part is on the possibility of global near—death and transformation. There is much in this book about the meaning and significance of parapsychological phenomena.

   ChrP 7:178—179 (Mar 1988); *JASPR* 82:160—165 (Apr 1988); *JNDS* 7:44—54 (Fal 1988); *JRPR* 10:108 111 (Apr 1987); *JSPR* 55:36—37 (Jan 1988); *PAI* 5:43 (Dec 1987).

   Fate 40:104,109 (Jan 1987); *ReVision* 10:39—44 (Win 1988); *SpirFron* 19:34—37 (Win 1987).

760 Hick, John H. *Death and Eternal Life.* New York: Harper & Row, 1976. 495p. Bibliography: 467—481; Chapter references; Index: 485—495

   A British theologian takes a philosophical approach to the questions of death and immortality. In the first half, Hick reviews the history of belief in an afterlife and contemporary approaches, including the contribution of parapsychology. the last half deals with past and present ideas concerning immor—tality, including survival, resurrection, reincarnation and rebirth. Hick concludes with his own conception of human destiny that integrates both Eastern and Western views.

   ChrP 1:122—124 (Jun 1977); *JASPR* 72:365—371 (Oct 1978); *T* 10:17—18 (Spr 1982).

   America 136:251 (Mar 19, 1977); *America* 136:417 (May 7, 1977); *ChristCent* 94:477 (May 18, 1977); *Choice* 14:878 (Sep 1977); *Commonweal* 104:156 (Mar 4, 1977); *CrossCurr* 28:310—317 (Aut 1978); *HeythropJ* 18:328—330 (Jul 1977); *Interpretation* 32:93 (Jan 1978); *JRelig* 58:217 (Apr 1978); *Kirkus* 44:1213 (Nov 1, 1976); *LibrJ* 101:2377 (Nov 15, 1976); *Observer* p29 (Nov 14, 1976); *PublWkly* 210:46 (Nov 22, 1976); *ReVision* 5:107—108 (Spr 1982); *TimesLitSuppl* p390 (Apr 1, 1977).

761 Jacobson, Nils O. *Life Without Death? On Parapsychology, Mysticism, and the Question of Survival.* Translated from the Swedish by Sheila LaFarge. New York: Delacorte Press/Seymour Lawrence, 1973 (c1971). 339p. Chapter notes; 8 figures; Glossary: 327—329; Index: 331—339

   In his Foreword, Swedish psychiatrist Jacobson says the book has two purposes: (1) to provide an introduction to parap—sychology, "with a survey of some of its current areas of research and findings to date" and (2) to consider the question: "Can some kind of experience be thought to continue after the death of the body? If the answer should be yes, then what might that experience turn out to be like?" (p. vii). The book is in five parts: the first presents four views of death; the second is a survey of the major areas of parapsychology; the third describes "Improbable Experiences," or "examples of paranormal phenomena [OBES, apparitions and ghosts, spiritualism, possession, tape recorded voices, reincarnation, xenoglossy] which are difficult to study in controlled experimental conditions but which are of the greatest importance to a discussion of what happens at death" (p. vii); the fourth part relates psi to various states of consciousness; and the last part is entitled "How Does It Look on the Other Side?"

JASPR 69:80–84 (Jan 1975); *JSPR* 48:114–115 (Jun 1975); *PR* 5:6–7 (Nov/Dec 1974); *T* p16 (Win/Spr 1975).

Choice 11:672 (Jun 1974); *Commonweal* 100:388 (Jul 12, 1974); *LibrJ* 99:3205 (Dec 15, 1974); *PublWkly* 204:35 (Dec 17, 1973); *PublWkly* 206:308 (Aug 26, 1974).

762 **Kastenbaum, Robert (Ed.).** *Between Life and Death.* New York: Springer, 1979. 184p. Chapter references; 3 figures; 4 tables

This is primarily a revised version in printed form of a 1977 symposium of the American Psychological Association entitled "Communication With the Dead: New Data or Same Old Story," plus additional contributions. Symposium papers include Sheldon Ruderman's account of his near–death experience (NDE), a history of survival research by Barbara Ross, Charles Garfield's discussion of NDEs as altered states of consciousness attainable in other ways, and Richard Kalish's article on the social and cultural determinations of an individual's orientation toward death. The remaining contributions are Kastenbaum's attempts to place NDEs in clinical perspective; Russell Noyes' discussion of NDEs as a depersonalization response in the face of life–threatening danger, Sandor Brent's observation that OBEs are not limited to the near–death situation and his method for the deliberate induction of OBEs; and Sandra Bertman's examination of communication between the living and the dead as expressed through literature and art. Kastenbaum concludes with a survey of the evidence, theories, and methods involved in the survival question.

JNDS 2:11–12 (F 1981); *JP* 44:180–184 (Jun 1980); *T* 8:17–18 (Sum 1980).

AmJPsychiat 137:1213 (Oct 1980); *Essence* 5:249–251 (1982); *Omega* 12:83–85 (1981/1982); *ReligStudRev* 7:141 (Apr 1981).

763 **Lewis, Hywel D.** *Persons and Life After Death.* New York: Barnes & Noble, 1978. 197p. Chapter references; Name index: 189–190; Subject index: 191–197

This collection of essays by the distinguished British philosopher of religion may be considered as a sequel to his *The Self and Immortality*. He endeavors to show that human survival is not a philosophical impossibility. Some of the essays were originally given as symposia, and he includes the papers of the other symposiasts who disagreed with him: Anthony Quinton, Bernard Williams, Antony Flew, and Sydney Shoemaker. In addition to the papers on the mind—body question and survival, he includes a chapter on religion and the paranormal (originally published in Thakur [**614**]).

ChrP 3:105—106 (Jun 1979); *JASPR* 74:227—239 (Apr 1980); *T* 8:16—17 (Spr 1980).

Choice 15:1389 (Dec 1978); *ExposT* 90:127 (Jan 1979); *ReligStud* 15:122—124 (Mar 1979); *ScotJTheol* 32:471—473 (1979).

811 Lombroso, Cesare. *After Death——What? Researches into Hypnotic and Spiritualistic Phenomena.* [See Addenda for annotation]

764 Lorimer, David. *Survival? Body, Mind and Death in the Light of Psychic Experience.* Boston, MA: Routledge & Kegan Paul, 1984. 342p. Bibliography: 322—331; Chapter notes; Case index: 341—342; Index: 332—341

Lorimer begins with a history of views about immortality and the mind—body problem. He points out that interactionism is compatible with the idea that the conscious self continues after death. In the second part he discusses the empirical evidence suggestive of survival: apparitions, OBEs, NDEs, and post—mortem communications. He concludes that although "the data surveyed are not in themselves coercive or conclusive proof that the conscious self survives bodily death, they are . . . concrete pointers which demand a coherent and comprehensive explanation" (p. 304).

ChrP 5:276—277 (Dec 1984); *JASPR* 80:98—192 (Jan 1986); *PAI* 4:44 (Dec 1986).

Brain/MindBul 10:4 (Sep 9, 1985); *BritBkNews* p587 (Oct 1984); *Light* 104:129—133 (Aut 1984); *RelStudRev* 11:173 (Apr 1985).

765 Lund, David H. *Death and Consciousness.* Jefferson, NC: McFarland, 1985. 194p. Chapter notes; Index: 191—194

In Part 1 of this book, philosopher Lund presents and defends "a view of the nature of a person and the model he encounters which is not only compatible with the claim that a person survives death but suggests that survival is more than a remote possibility" (p. x). In the second part, he considers the various kinds of empirical evidence favoring survival: OBEs,

deathbed experiences, apparitions, hauntings, mental medium—
ship, reincarnation, and resurrection.

    ChrP 7:107–108 (Sep 1987); *JASPR* 82:170–172 (Apr
1988); *JP* 50:162–166 (Jun 1986); *JSPR* 53:463–465 (Oct 1986);
PAI 6:45 (Jun 1988); *PsiR* 5:232–234 (Mar/Jun 1986).

    Choice 23:668 (Dec 1985); *Light* 106:74–76 (Sum 1986).

766 McAdams, Elizabeth E., and Bayless, Raymond. *The Case
for Life After Death: Parapsychologists Look at the
Evidence.* Chicago, IL: Nelson–Hall, 1981. 157p. Chapter
notes; Index: 155–157; Selected bibliography: 151–153

    In this nontechnical work, the authors present cases il-
lustrating some of the best types of evidence for survival. A
chapter each is devoted to NDEs, OBEs, deathbed experiences,
apparitions, animal hauntings, materializations, direct voice,
mediumistic communications, poltergeists, electronic voice
phenomena, spirit photography, and anomalous phone calls. In
a brief final chapter, the authors weigh the evidence for survival, considering the super–ESP hypothesis, but opting for the
former.

    JP 45:358 (Dec 1981); *JRPR* 5:260–263 (Oct 1982); *JSPR*
51:316–317 (Jun 1982); *T* 11:45–46 (Sum 1983).

    LibrJ 106:243 (Nov 15, 1981); *NewReal* 4:43–45 (Dec
1981).

767 Moody, Raymond A., Jr. *Elvis After Life: Unusual Experiences Surrounding the Death of a Superstar.* Atlanta, GA: Peachtree Publishers, 1987. 158p.

    This is not strictly a book on parapsychology but in part
it deals with experiences traditionally labelled parapsychological.
Moody's approach illustrates a kind of folklore in the making,
and it is included here because this approach to the study of
human experiences is one that parapsychologists might well emu-
late.   Moody interviewed at firsthand all of the persons whose
experiences are included. It is valuable because he does not rule
out the possibility of some of these experiences due to their improbability.   By studying the full range of experiences of an
unusual nature following a single focal point——the death of Elvis Presley——it becomes possible to view the experiences as
aspects of a continuum, which is needed in parapsychology.
Rather than attempting to "prove" whether the experiences
"really" happened as reported, psychiatrist Moody is interested
in them "for what they reveal about the human mind and
spirit" (p. 2 of Introduction). Some of the phenomena described
are premonitory of Elvis's death. There are apparitions and inexplicable physical effects, and many of the persons interviewed
are convinced that Elvis has been in contact with them since

his death. Moody adds a new dimension to the study of psychic experiences associated with the death of a person by his sensitivity to the "emotional context of human grief and bereave— ment in which they occur" (p. 154).

PAI 6:46 (Jun 1988).

PublWkly 231:66 (Jun 26, 1987); *SpirFron* 21:60 (Win 1989).

768 Moore, Brooke Noel. *The Philosophical Possibilities Beyond Death.* Springfield, IL: Charles C Thomas, 1981. 222p. Bibliographical footnotes; Bibliography: 206—213; Index: 214—222

The author is a professor of philosophy at California State University, Chico. The book is in four parts: in the first the various theories of survival are described and the basic assump— tions on which any view of survival must rest are explained. The second part deals with the question of whether these theories are subject to antecedent philosophical refutation. The third is a review of the early evidence for survival, including religious evidence, mediumship, cross correspondences, drop—in communicators, and apparitions. The final part examines more recent evidence such as near—death experiences, out—of—body experiences, ESP, reincarnation, and xenoglossy. In the final chapter, he concludes that survival of death is not probable.

JASPR 76:375—379 (Oct 1982); *JSPR* 51:314—316 (Jun 1982); *T* 12:44—46 (Sum 1984).

769 Osis, Karlis, and Haraldsson, Erlendur. *At the Hour of Death* (rev. ed). Introduction by Elisabeth Kübler—Ross. New York: Avon, 1986. 250p. Bibliography: 241—246; Chapter notes; 3 graphs; Index: 241—244; 1 question— naire; 13 tables

The authors, both psychologically trained parapsychologists, interviewed over 1,000 doctors and nurses for this four—year study of experiences of the dying in the U.S. and India. Al— though the heart of the book is the report of their own findings concerning deathbed experiences, they also provide a useful his— tory of research on the subject as well as some account of clini— cal death (or "near death") experiences, including a chapter on the afterlife as seen through the eyes of the dying. This revised edition contains an Introduction by Osis in which he reflects on the research reported by psychiatrists, psychologists, and physicians after the first edition was published. He points out that this book concentrates on the experiences of people who ac— tually died rather than on those of survivors, and whose ex— periences were even more intense, but on the same continuum with, those who lived. The concluding chapter has been ex—

panded to include a discussion of the bearing that new research
on NDEs and OBEs have on the moment of death and after.
Two pages have been added to the bibliography. *ChrP* 3:106−107 (Jun 1979); *JASPR* 72:375−379 (Oct
1978); *JIndPsychol* 2:69−73 (Jan 1979); *JP* 42:143−144 (Jun
1978); *JSPR* 49:885−887 (Sep 1978); *PR* 8:11−13 (Nov/Dec
1977); *T* 7:16−18 (Fal 1979).
    America 138:174 (Mar 4, 1978); *Booklist* 74:247 (Oct 1,
1977); *Choice* 15:1034 (Oct 1978); *Choice* 16: 188 (Apr 1979);
LibrJ 103: 177 (Jan 15, 1978); *Light* 99:125−126 (Aut 1979);
PublWkly 212:67 (Aug 8, 1977); *SpirFron* 10:123−124 (Spr
1979).

770 Rogo, D. Scott. *Life After Death: The Case for Survival
     of Bodily Death.* Wellingborough, Northamptonshire,
     England: Aquarian Press, 1986. 158p. Bibliography: 153−
     155; Chapter notes; 17 illustrations; Index: 156−158
    Rogo evaluates the various forms of evidence for survival
after death−−asking whether humans possess the capacity to
live beyond bodily death and whether they can communicate
with the living. He begins with a history of psychical research
and studies of mediumship. Next he considers the relevance of
out−of−body experiences for survival, followed by a chapter on
near−death experiences. A chapter is devoted to spontaneous
postmortem contacts. Electronic voice phenomena are considered
as well as phone calls from the dead. The evidence for survival
based on claimed memories of past lives, or reincarnation, is
also considered. In a final chapter, Rogo presents his personal
views on survival and the cases that he considers point most
directly to survival.
    PAI 6:47 (Jun 1988).

771 Rogo, D. Scott, and Bayless, Raymond. *Phone Calls From
     the Dead.* Englewood Cliffs, NJ: Prentice−Hall, 1979.
     172p. Chapter notes; 1 figure; Index: 169−172
    Two California psychical investigators present what is pos−
sibly a new type of psi phenomenon: telephone calls involving a
paranormal element. They investigated many of the cases per−
sonally and real names are used where possible. The phone calls
they describe purportedly came from dead persons. They also
discuss inexplicable messages received on other electronic instru−
ments such as the telegraph, gramophone, amplifiers, tape re−
corders, and telephone answering devices. The book is enhanced
by an unusual appendix containing the comments and criticisms
on the book by three parapsychologists: John Beloff, John Pal−
mer, and Gertrude Schmeidler, together with the authors'
remarks.

PR 10:11—12 (Jul/Aug 1979); *T* 8:19—20 (Spr 1980).
LibrJ 104:837 (Apr 1, 1979); *SpirFron* 11:115—117 (Fal 1979); *WestCoastRevBks* 5:36 (May 1979).

772 Ryzl, Milan. *Death . . . and After?* San Jose, CA: Author, 1980. 200p.

After a brief introduction to the basic findings and theoretical implications of parapsychology, parapsychologist Ryzl, an emigré to the U.S. from Czechoslovakia, analyzes in detail the contribution of parapsychology to the survival question. He considers mediumistic phenomena, apparitions, out—of—body experiences, deathbed visions, and reincarnation. In each case, he examines the strengths and weaknesses of the evidence and concludes that none of the traditional evidence proves survival. Rather, a living person seems always to be the source. In place of survival of an individual entity he sets forth a theory of mental impregnation to explain the data. He closes with a plea—— whether investigating ESP, survival, or even a totally unrelated field——to use ESP to advance knowledge. He proposes that we attempt to use ESP to know the World Beyond. He feels a sounder basis for knowledge than that provided by either religion or science can be developed by each individual: "There will be no need to require faith in revealed dogmas, and everybody will have a chance to verify for himself, by his own insight, the truth in the inspired teachings" (p. 182).
T 11:19—21 (Spr 1983).

773 Sibley, Mulford Q. *Life After Death?* Minneapolis, MN: Dillon Press, 1975. 160p. Chapter notes; Glossary: 152—153; 5 illustrations; Index: 158—159; Suggested reading list: 154—157

This is another volume in Dillon Press's "Psychic Explorations" series. As in the case of Feola [**435**], it uses language that high school students and uninformed laypersons can comprehend. Sibley, who is a political science professor at the University of Minnesota, where he also teaches parapsychology, begins with three introductory chapters on the nature of the survival problem. The next nine chapters deal with specific types of evidence for survival: ghosts and apparitions, dreams, out—of—body experiences, materialization, automatic writing, direct voice, cross correspondences, reincarnation, and electronic voice phenomena. A concluding chapter summarizes conflicting views of the evidence, and there are five appendices consisting of documents relevant to the survival question.
JP 40:256—257 (Sep 1976); *P* 7:45 (Jul/Aug 1976); *PR* 7:18—20 (Jul/Aug 1976); *T* 5:15 (1977).
SchLibrJ 22:73 (May 1976).

774 Spong, John S. (Ed.). *Consciousness and Survival: An Interdisciplinary Inquiry into the Possibility of Life Beyond Biological Death.* Introduction by Claiborne Pell. Sausalito, CA: Institute of Noetic Sciences, 1987. 190p. Chapter references

This is the proceedings of a symposium sponsored by the Institute of Noetic Sciences convened at Georgetown University on October 26–27, 1986. The purpose of this conference, according to Bishop Spong, was to examine anew the concept of life after death after having stripped away the traditional religious content that formally surrounded it. A total of 11 papers is included. The titles of the papers, and their authors, are as follows: "Survival of Consciousness after Death: A Perennial Issue Revisited," by Willis W. Harman; "The Cartesian Presuppositions of the Survival Hypothesis," by Antony Flew; "Altered States of Consciousness and the Possibility of Survival of Death," by Charles T. Tart; "The Survival of Consciousness: A Tibetan Buddhist Perspective," by Sogyal Rinpoche; "Can Our Memories Survive the Death of Our Brains?," by Rupert Sheldrake; "Neuropeptides, The Emotions and Bodymind," by Candace Pert; "A Possible Conception of Life After Death," by John Hick; "The Mind–Body Problem and Quantum Reality," by Paul C. Davies; "The Inner Self–Helper: Transcendent Life Within Life?," by Jacqueline Damgaard; "Survival of Consciousness After Death: Myth and Science," by Stanislav Grof; and "Near–Death Experiences: Intimations of Immortality?," by Kenneth Ring. The book closes with excerpts from the panel discussions held during the conference.
ChrP 7:211–212 (Jun 1988); *JASPR* 83:53–57 (Jan 1989).

775 Stevenson, Ian. *Unlearned Language: New Studies in Xenoglossy.* Charlottesville, VA: University Press of Virginia, 1984. 223p. Bibliography: 211–215; Index: 217–223; 7 tables

Detailed, authenticated cases of xenoglossy, or the ability to speak a foreign language not learned normally, and especially responsive xenoglossy, in which the subject engages in intelligible conversation in the foreign language, are very rare. Stevenson [776] has already published one such case, that of Jensen, in 1974. Here he presents the results of detailed investigations of two more cases, those of Gretchen and Sharada. The Gretchen case resembles the Jensen case, but in many instances the Sharada case differs from both. The two case reports comprise approximately two–thirds of the book. There follows a "general discussion" in which Stevenson deals with questions of authenticity in cases of xenoglossy, responsive xenoglossy cases as instances of incommunicable paranormal capacities, and some linguistic aspects of cases of responsive xenoglossy. A lengthy appendix contains excerpts from the Gretchen transcripts and a second consists of translated excerpts from notes and tapes of conversations with Sharada.

JASPR 79:549–554 (Oct 1985); *JP* 51:99–103 (Mar 1987); *PAI* 3:40 (Dec 1985).
AmJPsychiat 142:1218–1219 (Oct 1985); *Language* 61: 739 (1985); *ModLangRev* 80:890–891 (1985); *SkptInq* 11:367–375 (Sum 1987); *SpirFron* 17:121–125 (Spr 1985).

776 Stevenson, Ian. *Xenoglossy: A Review and Report of a Case.* Charlottesville, VA: University Press of Virginia, 1974. 268p. (Also published as Vol. 31 of the *Proceedings of the American Society for Psychical Research.*) Bibliography: 90–95; Index: 265–268; 1 table

This is the most complete account of a case of xenoglossy in the English language, if not in the literature of any language. The case under review concerns a woman whose consciousness, when under hypnosis, was taken over by a personality taking the name of "Jensen," who was able to carry on conversations in Swedish with three native born Swedish–speaking persons. Stevenson considers the counterhypotheses of fraud, cryptomnesia, and ESP and concludes "that the case offers strong evidence of the survival of physical death by some aspect of human personality" (p. 88). Nearly two–thirds of the book consists of a transcript of the tape of one session in which the taped words and phrases in Swedish are presented followed by an English translation and comments. Greatly adding to the value of the book is a review of the literature and a bibliography of 112 items.

JSPR 47:507–510 (Dec 1974); *PR* 6:10–13 (Sep/Oct 1975); *T* 5(1):11–12 (1977);
Choice 12:808 (Sep 1975).

777 Toynbee, Arnold, and others. *Life After Death.* New York: McGraw–Hill, 1976. 272p. Chapter references; Index: 261–272

Fourteen philosophers, theologians, social scientists, psychics, and historians examine whether consciousness can survive the physical death of the body and whether it is possible to communicate with the dead. In particular, religious beliefs concerning the nature of death and immortality both in the West and in other cultures, past and present, are examined. Part I consists of Toynbee's essay, from which the book takes its title. It is an historical survey of views of life after death. Part II, "The Idea of the Hereafter: Past and Present," contains nine contributions dealing with specific cultures and religions. The final part is "The Idea of the Hereafter: The Future." It consists of four essays, including "Psychedelics and the Ex—perience of Death," by S. Grof and Joan Halifax—Grof; "Illusion——or What?," by Rosalind Heywood; and "Whereof One Cannot Speak . . .?," by Arthur Koestler.

ChrP 1:87—89 (Dec 1976); *JSPR* 48:314—315 (Sep 1969); *P* 7:40,61 (Jan/Feb 1977); *PR* 8:16—18 (May/Jun 1977).

Economist 259:122—123 (May 15, 1976); *HastingsCtrRep* 7:40—42 (Oct 1977); *HeythropJ* 18:330—333 (Jul 1977); *NewStatesman* 92:183 (Aug 6, 1976).

777a Wilson, Ian. *The After Death Experience: The Physics of the Non—Physical.* New York: William Morrow, 1987. 234p. Chapter references; 10 figures; 28 illustrations; Index: 229—234; Select bibliography: 223—227

Wilson is a popular writer with training in history. He examines the evidence for survival of death, with the emphasis on near—death experiences (which he considers provide the best evidence). He considers reincarnation, hypnotic age regression, mediumship, spontaneous phenomena (apparitions), deathbed ex—periences, near—death experiences (four chapters). Having ex—amined the evidence and concluded that there may be an aspect of human nature that survives, he devotes a chapter to what that aspect may be. Another chapter is devoted to the "Physics of the non—physical." The last chapter, "On Not Being Afraid" of death.

PublWkly 234:65 (Dec 16, 1988).

Training and Development of Psi

INTRODUCTION

This category contains books that attempt to set forth methods or shed light on enabling persons to learn how to use psi abilities. There wasn't any section on training psi in *Sources*, but some of the books in other sections were on the subject of developing psi or mediumistic abilities: two of Eileen Garrett's autobiographies [103, 105], Mrs. Leonard's autobiography [108], Bendit [139], and the Parapsychological Monograph by Rush [197]. As with so many of the books in *Sources*, these volumes are still useful, especially the last two.

There are 17 books in this section, including a bibliography compiled by White and Anderson [791] that lists 1,357 sources on all aspects of developing psi ability. Works that deal with general psi training, but especially ESP, are by Goldberg [779], Huson [808], Mishlove [782], Reed [814], Rogo [783], Swann [784], the two Tart books [786, 787], and Thurston [788], with Mishlove being the most comprehensive. Methods for training specific psi abilities are discussed by Hunt [807] (ouija board), LeShan [780] (healing), Loye [781] and Vaughan [789] (precognition) and Targ and Harary [785] (remote viewing). Ebon [778] describes the potential dangers involved in developing psi and mediumistic abilities.

Additional books on training and development of psi listed elsewhere: See also the section on altered states of consciousness and psi. Some of the books in the section on applications discuss training psi. See Agor [366], Bird [368], Dean et al. [369], Hurley [372], and Kautz [374].
All the books in the section on "Biographies and Autobiographies of Healers, Mediums, and Psychics" contain chapters or passages on the development and training of psi and mediumistic abilities. All of the titles in the section on "Children and Psi" deal with development of psi in children [see 451-455]. Couttie [461] has two chapters on fraudulent and deceptive ways to "be psychic." The U.S. Army report edited by Druckman and Swets [462] is critical of parapsychology and various psychological techniques thought by many to be psi-conducive such as biofeedback, guided imagery, meditation, and split-brain effects. Targ and Puthoff [504] present a method

for developing psi ability. Eisenberg has a section on commer—
cial exploitation of psi by mind—training courses and fortune—
tellers.   Gruss and Hotchkiss [577] deal with using the ouija
board for psychic development and the dangers involved in
doing so.   Grey [586] has four chapters on the development of
psi abilities following NDEs and Ring [592] has a chapter on
"NDEs and psychic development," precognition in particular.
Psi experiences associated with OBEs are described by Black—
more [597]. In the section on out—of—body experiences, several
books deal with methods of inducing OBEs, especially Rogo
[603], but also Black [596], Blackmore [597], and Monroe [602].
Ryzl [772] advocates that parapsychologists should train them—
selves to use ESP and to apply it in looking for solutions to
parapsychological questions.   Doore [807] deals with shamanic
training and shamanism and personal growth.

BIBLIOGRAPHY

778 Ebon, Martin (Ed.). *The Satan Trap: Dangers of the Oc—*
      cult. Garden City, NY: Doubleday, 1976. 276p.
      Ebon, prolific author and editor of popular works on parap—
sychology, introduces this collection in the following words:
"Writers on such subjects as telepathy, clairvoyance, prophecy,
mind—over—matter phenomena and other psychic practices must
constantly be alert to the danger of presenting their subject
only in glowing and positive terms. There is another side, a
dark side to these phenomena, and in our time this darkness
seems to be spreading. On the one hand, scientific study of the
frontiers of the human mind is today more active than ever in
history. But on the other hand, we are also experiencing a vir—
tual epidemic of irresponsible toying with psychic powers" (p.
vii). There are 25 selections by parapsychologists, psychiatrists,
psychologists, and writers who describe the potential dangers in—
volved in the overzealous pursuit of personal psychic experience.
      JP 40:255 (Sep 1976); *P* 7:42—44 (Jul/Aug 1976); *T* 7:9
(Win 1979).
      AmJPsychiat 134:603 (May 1977); *Booklist* 72:1145 (Apr
15, 1976); *Choice* 13:804 (Sep. 1976); *LibrJ* 101:1026 (Apr 15,
1976); *PublWkly* 209:92 (Mar 1, 1976).

779 Goldberg, Philip. *The Intuitive Edge: Understanding and*
      Developing Intuition. Los Angeles: Jeremy P. Tarcher,
      1983. 241p. Bibliography: 227—237; 12 figures; Index:
      238—241; 1 questionnaire
      The author argues that science does not progress by em—
pirical observation and logical inference but by creative insights

and intuitions that are subsequently empirically verified and after a logical theory has been devised to explain the phenomenon under investigation. He observes that there has been resistance to acknowledging the important role intuition plays in our lives, but the intellectual atmosphere is changing. He proposes that although intuition is a spontaneous phenomenon and cannot be forced, much can be done to develop intuitive capacity and to create conditions that are con—ducive to it. In the first part of the book, he defines intuition and gives examples drawn from science and art. A chapter is devoted to the intuitive experience and another to the question of who is intuitive. Theories are offered to explain intuition. There are three chapters on how to develop intuition, drawing heavily on studies of meditation and brainstorming, and on East—ern sages such as Patanjali. Goldberg closes with a call to recognize the current limits of scientism and the importance of enlarging scientific methodology to include enlightened subjec—tivity. Some suggestions are offered for researching intuition.

AP 3:12—13 (Sum 1984); *PAI* 3:41 (Dec 1985).

BestSell 43:418 (Feb 1984); *Booklist* 80:381 (Nov 1, 1983); *Choice* 21:1056 (Mar 1984); *Kirkus* 51:103 (Sep 15, 1983); *LATimesBkRev* p20 (Nov 20, 1983); *LibrJ* 108:2164 (Nov 15, 1983); *Light* 106:42—43 (Spr 1986); *ReVision* 7:122—124 (Spr 1984); *SciBooks&Films* 19:273 (May 1984).

807 Hunt, Stoker. *Ouija: The Most Dangerous Game.* [See Ad—denda for annotation]

808 Huson, Paul. *How to Test and Develop Your ESP.* [See Addenda for annotation]

780 LeShan, Lawrence. *The Medium, the Mystic, and the Physicist: Toward a General Theory of the Paranormal.* New York: Viking, 1974. 299p. Chapter notes; 5 tables
    The author argues that although PK and ESP do not make sense within the framework of sense—based reality, they do fit in with an equally valid view of reality that he terms "clairvoyant reality." He describes this reality based on views of mediums, mystics, and physicists who seem to be saying the same thing. LeShan tells how he put the theory of clairvoyant reality to the test by using its postulates to become a healer. He describes his success both in healing and in teaching others to heal.

P 5:39—40 (May/Jun 1974).

781 Loye, David. *The Sphinx and the Rainbow: Brain, Mind and Future Vision.* Foreword by Willis Harman.

Boulder, CO: Shambhala/New Science Library, 1983.
236p. Bibliography: 216—225; Chapter notes; Index:
227—236; 15 figures; 3 graphs; 1 table
Loye attempts to synthesize neurophysiology, psychology,
parapsychology, and theoretical physics at a popular level in or—
der to show how all parts of the brain can function as a unit
in future prediction, including precognition. He takes a
holographic approach to precognition and attempts to show how
to develop forecasting ability.   There are four appendices, the
first offering suggestions for   improving individual and group
forecasting. The second presents some of the tenets of the new
psychophysics.   The third is a response to critics of precognition
(primarily C.E.M. Hansel), and the last is a guide to educa—
tional possibilities concerned with the forecasting mind:
workshops, seminars, discussion guides, a sample college course,
field trip possibilities, and tests.
     JASPR 79:92—94 (Jan 1985); *PAI* 4:45 (Dec 1986).
     Booklist 80:772 (Feb 1, 1984); *Futures* 16:325—326 (Jun
1984); *Futurist* 18:54—55 (Oct 1984); *Humanist* 44:33 (Sep 1984);
LaughMan 5:75 (1985); *LibrJ* 108:2164 (Nov 15, 1983); *NewAge*
9:64 (Oct 1983) *ReVision* 7:124—125 (Spr 1984).

782 Mishlove, Jeffrey. *Psi Development Systems.* Jefferson, NC:
McFarland, 1983. 299p. Bibliography: 265—290; 3 figures;
Index: 291—299; 1 table
     The author offers a revision of his dissertation in which he
attempts to present a "framework for the legitimate academic
evaluation, design and practice of programs for the cultivation
of psi" (p. 1). Mishlove also tries "to define parapsychology on
new terms as a discipline in and of itself" (p. 1), based on a
new synthesis with three characteristics: the extension of "the
self—conscious history of parapsychology further into its histori—
cal traditions" (p. 1), a critical evaluation of popular programs,
and the establishment of a disciplinary matrix that would allow
parapsychology to serve as a dual discipline for individual self—
actualization and as a means of scientific progress. Psi develop—
ment systems are surveyed within three contexts: the prescien—
tific traditions, such as shamanism, yoga, and ceremonial magic;
popular psi development systems, such as spiritualism, the
Rosicrucian Order, and Transcendental Meditation; and ex—
perimental parapsychology, represented by studies of feedback,
hypnosis, and relaxation. In the final chapter, a systems ap—
proach is applied to psi development procedures. The relation—
ship of parapsychology and humanistic psychology is also dis—
cussed.
     JASPR 79:86—89 (Jan 1985); *JP* 48:333—351 (Dec 1984);
PAI 3:41 (Jun 1985).

LibrJ 108:2164 (Nov 15, 1983); *Light* 104:41—42 (Spr 1984).

814 Reed, Henry. *Awakening Your Psychic Powers.* [See Addenda for annotation]

783 Rogo, D. Scott. *Our Psychic Potentials.* Englewood Cliffs, NJ: Prentice—Hall, 1984. 172p. Chapter references; Index: 169—172

Rogo divides this book on psychic development into two parts. In the first, he reviews the parapsychological studies that provide evidence that psi ability exists and that it can be used on demand and possibly developed, especially in association with various altered states of consciousness. In the second part, he explores specific psychic development techniques and shows the reader how he or she can use this material in a systematic program of self—testing and psi training. An effort is made throughout to only present suggestions that are based on the findings of laboratory experiments. Specific areas dealt with are dreams and ESP, mental imagery techniques, relaxation, suggestion, and feedback. There also is a critical chapter on mind training courses. There are two appendices, the first presenting a method for judging free—response material and the second discussing drugs and ESP.

PAI 5:41 (Jun 1987).
Booklist 81:462 (Dec 1, 1984).

784 Swann, Ingo. *Natural ESP: The ESP Core and Its Raw Characteristics.* Foreword by Marilyn Ferguson. Introduction by Harold E. Puthoff. New York: Bantam Books, 1987. 216p. 77 figures; 6 illustrations; Index: 209—216; Bibliography: 205—208

Swann is an artist of note as well as one of the outstanding laboratory subjects of modern times. This book is an outgrowth of his own experience with ESP and his considerable ability to theorize about it. In Chapter 1, he calls for a new stage in the study of ESP——one in which the major prerequisite for understanding ESP functioning is self—experience. In Chapter 2, he redefines ESP as "the result of an external sensing by which information and knowledge is contacted and, through subliminal processes, brought into consciousness, without the use of any of the known physical senses" (p. 16). In the third chapter, he outlines the ESP process as being one in which information required exists in the core of every individual: the job is to bring it to the level of conscious awareness. In the next chapter, he relates how he had to divest himself of all preconceptions about ESP so that "my ESP core eventually opened itself to me and my conscious mind had new

experience—oriented information with which it eventually estab—
lished new values and appreciation for raw extrasensory percep—
tion" (p. 40). The idea of the existence of an extrasensory self
is emphasized in the next chapter. Swann discovered that the
best way to eliminate the interference of the conscious mind
was to draw his response to the target. In the following seven
chapters he explains how one can contact one's natural ESP
core and how to learn from one's responses with drawings,
using the basic imaging process of the psyche. In a concluding
chapter on "ESP and the Future," he argues: "Only those in—
dividuals who have made some effort to locate their own raw—
core extrasensory capabilities will be in a position to appreciate
the age of applied ESP that is now dawning, and perhaps, to
be a part of it. One's familiarity with real ESP potential is
going to become——indeed it already has——a basis for com—
prehending applied ESP. Those who do not make an effort to
locate their own core ESP will become mere bystanders, unable
either to appreciate or take part in the ESP Age now nearly
upon us" (p. 199).

785 Targ, Russell, and Harary, Keith. *The Mind Race: Under—
standing and Using Psychic Abilities.* Foreword by Willis
Harman. Epilogue by Larissa Vilenskaya. New York: Vil—
lard Books, 1984. 294p. Bibliography: 265—269; Chapter
notes; 20 figures; 26 illustrations; Index: 279—294

According to the authors, *Mind Race* is aimed at those
who have wondered what is really happening in psi research in
the U.S. and in the Soviet Union in order to counteract the dis—
torted picture provided by the media. It also attempts to
provide compelling evidence of the existence of high—quality
psychic functioning. They tell how psychic abilities might be use—
ful to the reader in everyday life, and describe techniques that
can be used to develop one's psychic abilities. Emphasis is given
to remote viewing, including precognitive remote viewing. Three
informative and highly readable chapters educate the reader as
to the misrepresentations of psychic abilities and psi research by
various cults, in movies and TV, and by the press. Psi is
presented as a perfectly normal component of human function—
ing. An epilogue, "Psi Research in the Soviet Union: Are They
Ahead of Us?," is contributed by Larissa Vilenskaya. A bibliog—
raphy of remote—viewing research from 1973—1982 is appended.
JASPR 78:273—278 (Jul 1984).

BestSell 44:63 (May 1984); *InstNoeticSciNewsl* 12:22—24
(Spr 1984); *IntJPsychosom* 32:25 (1985); *Kirkus* 51:1253 (Dec 1,
1983); *LATimesBkRev* p14 (Jun 3, 1984); *LibrJ* 109:1245 (Jun
15, 1984); *NewReal* 6:70—71 (Nov/Dec 1984); *PublWkly* 224:60
(Dec 16, 1983); *PublWkly* 227:92 (Jan 25, 1985).

786 Tart, Charles T. *The Application of Learning Theory to ESP Performance* (Parapsychological Monograph No. 15). New York: Parapsychology Foundation, 1975. 145p. Bibliographical footnotes; Bibliography: 144—146; 14 figures; 9 graphs; Index: 147—149; 12 tables

One of the major concerns of University of California psychology professor Tart has long been the need to train subjects in order to produce a reliable flow of ESP. To this end, Tart proposed in a 1966 paper (reprinted here) that learning theory be applied to ESP performance. He also reviews studies published in the decade since that article was published. A small—scale study is described and he uses it to show some of the complexity that an expanded theory must address. The heart of the book is a report of a study to demonstrate that by providing feedback to subjects, scoring declines can be eliminated and learning can be produced in some subjects. The behavior and internal processes of the five best subjects and the most successful experimenter are described. A chapter is devoted to the implications of the research. Two appendices describe computer—based equipment for testing the role of feedback in ESP performance. There has been controversy in the parapsychological literature over the interpretation of Tart's findings. See Stanford's review article in *JASPR* (cited below) and Tart's reply on pp. 81—102 of the same issue.

JASPR 71:55—80 (Jan 1977); *JP* 40:76—81 (Mar 1976).

787 Tart, Charles T. *Learning to Use Extrasensory Perception.* Chicago, IL: University of Chicago Press, 1976. 170p. Bibliography: 157—161; Chapter notes; 18 figures; 9 graphs; Index: 167—170; 12 tables

This book revises and expands the author's earlier Parapsychology Foundation monograph [**786**]. Tart presents a survey of the literature on ESP and learning theory, and reports on a series of experiments he conducted to show that by providing immediate feedback to ESP test subjects when a response is correct, customary scoring declines may be eliminated and some subjects may even increase their performance.

JASPR 72:61—63 (Jan 1978); *JIndPsychol* 2:163—164 (Jul 1979); *JP* 41:223—226 (Sep 1977); *JSPR* 49:546—548 (Jun 1977); *T* 7:15 (Spr 1979).

Choice 13:1665 (Feb 1977); *ContempPsychol* 22:893 (Dec 1977); *LibrJ* 101:2289 (Nov 1, 1976); *NewReal* 1:28—29 (1977).

788 Thurston, Mark A. *Understand and Develop Your ESP: Based on the Edgar Cayce Readings.* Virginia Beach, VA: A.R.E. Press, 1977. 81p. Bibliography: 79—80; 20

figures
Psychologist Thurston heads the Enlightenment Services
Division of the Association for Research and Enlightenment
(A.R.E.). This book is based on the Edgar Cayce readings, as
well as some of the current findings in psychology and parapsychology. Part I consists of a chapter—length survey of major
theories and models for understanding psi operation and another
chapter describing Cayce's theory. Part II is concerned with the
development of ESP and builds on the ideas presented in Part
I. There is a chapter that describes attunement procedures that
are to be incorporated in daily living, followed by a chapter
describing exercises for developing specific aspects of psi ability.
The first draft of this book was modified by the suggestions of
400 A.R.E. members who took part in a research project on
developing psi ability.

789 Vaughan, Alan. *The Edge of Tomorrow: How to Foresee
and Fulfill Your Future.* New York: Coward, McCann &
Geoghegan, 1982. 224p. Chapter references; Suggested
reading list: 224; 1 table
Psychic and editor Vaughan is a specialist——in both writ—
ing and practice——in precognition and prophecy. To prospective
readers he writes: "My purpose in writing this book is to
provide you with tools to glimpse your future promise, your
potentials, abilities, and talents. By holding up a mirror to your
inner self, you can glimpse how best to proceed with fulfilling
your dreams" (p. 12). He offers concrete suggestions and exer—
cises for those who would like to foresee and even shape future
events in their own lives and those of others. Emphasis is on
dream interpretation and the role of imagery. In the last chap—
ter, he makes a number of predictions in 12 key areas.
LibrJ 107:264 (Feb 1, 1982); *VillVoice* 27:40 (Mar 16,
1982).

790 Vaughan, Alan. *Patterns of Prophecy.* New York: Hawthorn
Books, 1973. 244p. Chapter notes; Index: 237—240
The author, a writer and a psychic specializing in precogni—
tion, reviews many prophecies——his own and those of others——
concerning a variety of events in different contexts and states of
consciousness, in an effort to search for patterns. He describes
his own psychic development and his efforts to pinpoint
prophetic elements, especially in dreams. He surveys all the
major types of prophecy throughout history and concludes with
a discussion of destiny and foreknowledge in which he offsets a
theory of prophecy based on synchronicity and archetypal pat—
terns.
JASPR 68:434—437 (Oct 1974); *JSPR* 47:459—460 (Sep

1974); *P* 5:34 (Mar/Apr 1974).
  Books&Bkmn 19:64 (Jul 1974); *Kirkus* 41:1024 (Sep 1, 1973); *PublWkly* 204:91 (Oct 8, 1973); *SpirFron* 15:58–60 (Win 1983).

791 White, Rhea A., and Anderson, Rodger I. *On Being Psychic: A Reading Guide* (2nd ed.). Dix Hills, NY: Parapsychology Sources of Information Center, 1989. 151p.
  This work lists 1,921 information sources directed at per—sons who are interested in psychic (psi) experiences and the ex—periences of being a psychic (sensitive), healer, medium (channel), or out—of—body experient ("astral traveller"). Cita—tions are listed alphabetically under nine broad subject categories: Psychic experiences in everyday life; works on in—dividual psychics, mediums, and healers (this section is in three parts: 133 autobiographical works by 96 individuals, 534 biographical works on 238 persons, and 599 investigations of 194 persons); information sources on the psychology and physiology of being psychic; works on possible methods of inducing psi ex—periences and training psi abilities; methods of testing psi or mediumistic abilities; sources on resistance to and fear of psi and ways of coping with it; information sources on counseling persons who have had psi experiences and how to distinguish between the normal, the psychopathological, and the psychic; potential dangers involved in trying to develop psi or mediumis—tic ability and ways in which one who is interested in psi can be exploited or deceived by commercialized training courses and by professional psychics and mediums; and a section on "manufactured psi," or methods of producing psi or mediumistic effects through conjuring, illusion, or other nonpsi means.
  JP 52:242–244 (Sep 1988); *JSPR* 55:160–161 (Jul 1988).
  AmRefBksAnn 20:282–283 (1989); *Archaeus* 7:47 (Spr/Sum 1988); *SpirFron* 20:186–187 (Sum 1988).

Unorthodox Forms of Healing

INTRODUCTION

These books are about nontraditional methods of healing and psychotherapy. There were seven books in *Sources*, all still worth reading, especially that by West, which is a classic.

The rise of holistic health, psychoneuroimmunology, the development of imagery and therapeutic touch as healing modalities, and studies of shamanic healing have greatly enlarged this category. Whether or not a psi component is present has not yet been determined. The 14 books cited are some of the most notable examples representing the above categories. The most important, from a parapsychological viewpoint, is the book by Krippner and Villoldo [801]. Harvey's book [798] is a popular survey. Achterberg [792] writes on imagery and healing. Therapeutic touch is dealt with by Borelli and Heidt [794] and a holistic approach is taken by Dossey [795, 796] and Hastings et al. [799]. Anthropological studies are presented by Finkler [797], Katz [800], McGuire [801a], and Villoldo and Krippner [803]. Doore [806] and Walsh [803a] deal with shamanic healing, and Nolen's book [802] is critical.

Additional books on unorthodox healing listed elsewhere:
Gowan [346] examines psychic healing powers that occur as by—products of other pursuits. Grof discusses healing in nonordinary states of consciousness, as does Sargent [357]. Benor reviews psychic healing research in Shapin and Coly [817]. White's bibliography [378] has a section on unorthodox healing. For autobiographies of healers see DiOrio [384, 385], Kraft [392], and Tanous [397]. For biographies see the book about Olga Worrall [382]. Heaps [437] has a chapter on the new frontier of psychic healing. Abell and Singer [456] reprint a chapter from Nolen's critical book on psychic healing, and Booth [458] has a chapter critical of psychic surgery. Hines [474] devotes a chapter to the dangers of faith healing. Randi [478] is critical of faith healing, especially televangelists. Harner [565] and Nicholson [569] are concerned largely with shamanic healing. Rogo [584] deals with "rescue circles," and Grey [586] with healing and NDEs. Ehrenwald [643] has several chapters on unorthodox healing and miraculous cures. For sources on unorthodox heal—

ing in the Soviet Union, Eastern European countries, and China see the White/Vilenskaya bibliography [680] and the Vilenskaya titles [673-677]. White's Kirlian photography bibliography emphasizes healing applications[678]. Spiritual healing and miraculous cures are considered by Ebon [704], in Higgins [711], and Rogo [719]. Melton [715] compiled a spiritual healing bibliography. LeShan [780] relates how he trained himself and others to heal.

BIBLIOGRAPHY

792 Achterberg, Jeanne. *Imagery in Healing: Shamanism and Modern Medicine.* Boston, MA: New Science Library/Shambhala, 1985. 253p. Bibliography: 231−246; Chapter notes; 5 figures; Index: 247−253; 2 tables

The role of imagery is the focus of this book, beginning with shamanic healing, because "shamanism *is* the medicine of the imagination" (p. 6). In Chapter 2, the role of imagination as a healing tool in the history of Western medicine is presented. The third chapter is on current attempts to combine ancient wisdom with modern technology as in autogenics, biofeedback, and placebo research. Chapter 4 is on the physiology and biochemistry of the transition of mental images into physical changes. The fifth summarizes the findings of the behavioral and social sciences on imagination in the therapy of psychological disorders. The final chapter is about immunology and emphasizes the effect of imagery on the immune system. Several imagery scripts are presented, and there is an extensive bibliography in addition to the text references.

PAI 5:42 (Jun 1987).

Booklist 81:1284 (May 15, 1985); *LATimesBkRev* 110:135 (Jul 7, 1985); *LibrJ* 110:135 (Jun 1, 1985); *Light* 106:136−137 (Aut 1986); *SciBooks&Films* 21:218 (Mar 1986).

794 Borelli, Marianne D., and Heidt, Patricia (Eds.) *Therapeutic Touch: A Book of Readings.* Foreword by Dolores Krieger. New York: Springer, 1981. 180p. Chapter references; 6 figures; 1 graph; Index: 178−180; 4 tables

The editors view therapeutic touch within a holistic framework, and say it is "based on the assumption that the human body is an energy field which is continually affecting and being influenced by fields outside itself" (p. viii). They describe research and clinical applications used in various settings. In three sections, three chapters present theoretical foundations of therapeutic touch, nine are on clinical applications and research, and seven are on responses of patients to therapeutic touch. Of 19 chapters, only three were previously published.

    PAI 1:85 (Dec 1983).
    ContempPsychol 26:851 (1981); *LibrJ* 106:569 (Mar 1, 1981).

806 Doore, Gary (Ed.). *Shaman's Path: Healing, Personal Growth, Empowerment.* [See Addenda for annotation]

795 Dossey, Larry. *Beyond Illness: Discovering the Experience of Health.* Boston, MA: New Science Library/Shambhala, 1984. 207p. Chapter references

Dossey presents a unitary view of the universe and its im—plications for health and illness, and emphasizes the experience of health, which is seen as the capacity to accept the whole of experience and knowledge that everything is interdependent. He also criticizes holistic health. He stresses that spirit is beyond health and illness and is not subject to dualism. He makes some interesting observations on the doctor—patient relationship.

    PAI 5:44 (Dec 1987).
    LibrJ 109:1853 (Oct 1, 1984); *PublWkly* 226:80 (Sep 21, 1984).

796 Dossey, Larry. *Space, Time, and Medicine.* Foreword by Fritjof Capra. Boulder, CO: Shambhala, 1982. 248p. Chapter notes; 3 illustrations; Index: 243—248; 2 tables

Dossey, an internist, draws on quantum physics, dissipa—tive structures, Gödel, etc. to provide a theoretical underpinning for modern medicine that can accommodate unorthodox aspects. He presents a holistic, nonmechanistic view that he feels is more in line with the nature of reality than the old mechanistic one. He discusses the role consciousness plays in health and disease and emphasizes the importance of our views of time. He proposes that a linear view of time is associated with disease and a nonlinear concept is associated with health. He calls his theory the "space time model of birth, life, health and death." He proposes that the goal of the doctor is to assist the patient to realize that he or she is a "process in space time," not iso—lated from the healthy world. He quotes Bohm's "everything is alive; what we call dead is an abstraction" (p. 181).

    JRPR 7:200—201 (Jul 1984); *T* No.13/14:38—43 (Sum 1985/86).
    BestSell 42:199 (Aug 1982); *LATimesBkRev* p11 (May 30, 1982); *LibrJ* 107:1002 (May 15, 1982); *Light* 103:139—141 (Aut 1983); *NewStatesman* 104:23 (Sep 10, 1982); *PublWkly* 221:77 (Apr 2, 1982); *SocSci&Med* 17:1676—1677 (1983); *SpirFron*

16:111—113 (Spr 1984).

797 Finkler, Kaja. *Spiritualist Healers in Mexico: Successes and Failures of Alternative Therapeutics.* Foreword by Arthur Kleinman. South Hadley, MA: Bergin & Garvey, 1985. 256p. Bibliography: 228—242; Chapter notes; 6 graphs; Glossary: 243—244; Index: 245—256; 15 tables

This report by a medical anthropologist is about her investigation of the successes and failures of Spiritualist healing in rural Mexico. Finkler participated in the rituals and underwent initial training as a Spiritualist healer. She summarizes her work as follows: "(1) to demonstrate the ways in which an alternative healing system succeeds and fails to heal, as perceived by its patients, and in which one rural segment of a complex society expresses and manages illness, and (2) to describe the interaction between a therapeutic regime and the historical, social, and economic forces of the larger society of which it is a part. . . . I have unveiled a folk Mexican cognitive model of recovery, which I postulate is rooted in the history of its bearers and is fostered by current ecological conditions" (p. 193). There are useful chapters on "Spiritualist Healing Techniques" and "How Spiritualist Healers Heal." Several appendices present background information. A glossary, an extensive bibliography, and an index enhance the book's usefulness.

JASPR 83:64—67 (Jan 1989); *PAI* 5:43 (Dec 1987).

AmAnthropol 88:963—964 (Dec 1986); *AmEthnol* 13:590—592 (Aug 1986); *AmJPhysAnthropol* 67:277—278 (Jul 1985); *Choice* 22:1667 (Jul 1985); *ContempSociol* 16:404 (May 1987); *JAmFolklore* 101:121 (Jan 1988); *JPsychol&Theol* 15:84 (Spr 1987); *RevsAnthropol* 13:56 (Win 1986).

798 Harvey, David. *The Power to Heal: An Investigation of Healing and the Healing Experience.* Wellingborough, Northamptonshire, England: Aquarian Press, 1983. 192p. 4 figures; 14 illustrations; Index: 191—192; Select bibliography: 187—190

Journalist Harvey surveys psychic healing, primarily in Great Britain. He describes the work of the National Federation of Spiritual Healers, and through depictions of the life and work of individual healers past and present he provides a history of unorthodox healing (psychic, spiritual, and mediumistic). The individual healers touched on are Rose Dawson, Rose Gladden, Harry Edwards, John Cain, George Chapman, and Edgar Chase. Harvey also presents the results of a survey he conducted of people who had been treated by healers with positive results. The survey examined the circumstances under which any improvement occurred and the way the healing was ex—

perienced. There are chapters discussing how medicine can benefit from healers, a brief and selective survey of research on healing, and a summary of theories of healing. There is an appendix on "Finding a Healer and Learning to Heal," and a select, briefly annotated bibliography of 18 books on healing.
    ChrP 5:127—128 (Dec 1983); *JASPR* 80:227—231 (Apr 1986); *PAI* 6:50 (Jun 1988).
    Light 103:136—137 (Aut 1983).

799 Hastings, Arthur C., Fadiman, James, and Gordon, James S. (Eds.). *Health for the Whole Person: The Complete Guide to Holistic Medicine.* Boulder, CO: Westview Press, 1980. 529p. Chapter references; 2 figures; 5 graphs; Name and title index: 503—521; Subject index: 523—529; 12 tables
    Based on a report prepared by the Institute of Noetic Sciences for the National Institute of Mental Health, this book is aimed at both the health care professional and the layperson in order to brief them on "attitudes, information, and tools for a holistic approach to medicine, health, and mental health" (p. xi). Each chapter is written by a specialist and surveys a specific topic in six broad subject areas: "Holistic Medicine," "The Context of Health," "Modalities for Health and Mental Health," "Applications Throughout the Life Cycle," "Alternative Medical Aspects," and "Social and Political Implications." A subsection on "The Role of the Mind" contains chapters on: "Meditation and Holistic Medicine," by Deane H. Shapiro, Jr.; "Psychic Healing," by Stanley Krippner; and "The Placebo Effect," by Herbert Benson and Mark D. Epstein. A primary feature is the extensive bibliographies after each chapter, containing lengthy annotations of the most important practical and theoretical books and articles on the preceding subject.
    PAI 2:40 (Dec 1984).
    AmRefBksAnn 14:691 (1983); *Choice* 18:690 (Jan 1981); *Instructor* 91:128 (Mar 1982); *LibrJ* 105:2508 (Dec 1, 1980); *Phoenix* 5:138—139 (1981); *ReVision* 5:109 (Spr 1982).

800 Katz, Richard. *Boiling Energy: Community Healing Among the Kalahari Kung.* Cambridge, MA: Harvard University Press, 1982. 329p. Bibliography: 317—323; Glossary: 314—316; 1 figure; 19 illustrations; Index: 324—329; 1 map
    Katz reports on his field investigation of the ritual healing dance of the !Kung. The rituals are described in detail, and he shows how they are perceived at the experiential level through interviews with participants. He traces the role of healing in the healers' lives and in the community as a whole, and he

presents a sociological analysis of how the form and function of healing are shaped by the character of the !Kung social order. The book grew out of a request from the !Kung to "tell our story of healing to your people" (p. xiii). The talks of four !Kung healers are the heart of the book. Of special interest are four chapters: "At a Healing Dance," "Education for Healing," "Career of the Healer," and "Psychological and Spiritual Growth." Biographies are given of some important !Kung healers.

    JASPR 77:266−272 (Jul 1983); *PAI* 4:45 (Dec 1986).

    AmerAnthropol 86:461−462 (Jun 1984); *AmerEthnol* 10:792−793 (Nov 1983); *Choice* 20:618 (Dec 1982); *Commonweal* 109:604 (Nov 5 1982); *JTransPsychol* 14:188−190 (1982); *LibrJ* 107:2106 (Nov 1, 1982); *Man* 19:689−690 (Dec 1984); *Mankind* (Aut) 16:153−154 (Aug 1986); *NewAge* 8:65 (Oct 1982); *Parabola* 8:124 (Jan 1983); *RevsAnthropol* 11:44 (Win 1984); *VillVoiceLitSup* p16 (Sep) 1982; *Zygon* 19:514−517 (Dec 1984).

801 Krippner, Stanley, and Villoldo, Alberto. *The Realms of Healing* (3rd ed.). Introduction by Evan Harris Walker. Berkeley, CA: Celestial Arts, 1986. 210p. Author index: 205−207; Bibliography: 187−195; 25 il−lustrations; Subject index: 197−204

    This is a survey by two psychologists of unorthodox healers and their practices in various parts of the world. The work of individual healers is emphasized in the first four chap−ters. Three North American healers (Rolling Thunder, doña Pachita, and Olga Worrall) are dealt with in Chapter 1. There follows a chapter on Brazilian Spiritist healers, including psychic surgeons. Chapter 3 is on Maximo Gomez of Peru and Josef Zezulka of Czechoslovakia. Next is a chapter devoted to Josephina Sison and Filipino psychic surgeons. In all four chap−ters the world views that form the basis for the healing prac−tices are set forth and statements of individual healers are provided. The fifth chapter is a review of laboratory studies of psychic healing, concluding with a discussion of why scientific investigations of healing are needed and the role that such inves−tigations might play in revising the current world view or paradigm of science. The last chapter is entitled "How Healing Happens" and is a discussion of various models of healing, in−cluding the physical, the psychological, and the psychic.

    JP 41:68−70 (Mar 1977); *JSPR* 49:612−614 (Sep 1977); *P* 7:40 (Jan/Feb 1977); *PAI* 6:45 (Jun 1988); *T* 8:20−22 (Spr 1980).

    Booklist 73:230 (Oct 1, 1976); *Brain/MindBul* 1:4 (Jul 19, 1976); *Choice* 13:1361 (Dec 1976); *LibrJ* 101:1643 (Aug 1976); *NoeticSciRev* No.3:28 (Sum 1987); *SpirFron* 11:75−79 (Fal 1979).

801a McGuire, Meredith B, with Debra Kantor. *Ritual Healing in Suburban America.* New Brunswick, NJ: Rutgers University Press, 1988. 324p. 5 figures; Chapter notes; Index: 323–324; 1 table

Sociologist McGuire studied the roles of various alternative healing groups in suburban New Jersey. She found that a large number of college–educated, middle–class persons were involved. She interviewed over 130 healers and healees on the importance of alternative healing practices. The chapters cover middle–class use of alternative healing methods, the features of the groups studied, Christian healing groups, traditional metaphysical movements, Eastern meditation and human potential groups, psychic and occult healing, the healer's role, seeking help beyond the traditional medical model, the role of ritual and symbolism in healing, the importance of alternative healing for the self, and the social impact of alternative healing approaches. Her research methods and interview schedule are included in two appendices.

AmerJSociol 95:544 (Sep 1989); *Choice* 26:969 (Feb 1989); *ChristCent* 106:330–331 (Mar 22, 1989); *ContempSociol* 18:274 (Mar 1989).

802 Nolen, William. *Healing: A Doctor in Search of a Miracle.* New York: Random House, 1974. 308p.

The book describes the author's personal investigations of Kathryn Kuhlman, Norbu Chen, and several Filipino psychic surgeons. He found that healers could relieve symptoms and even cure some functional diseases, but he did not find a healer who could cure organic disease.

JASPR 70:101–108 (Jan 1978); *P* 7:34 (May/Jun 1976); *PR* 7:14–15 (Jan/Feb 1976); *PS* 2:158–160 (F 1977).

Booklist 71:933 (May 15, 1975); *Kirkus* 42:1244 (Nov 15, 1974); *LibrJ* 100:594 (Mar 15, 1975); *NatlObs* 14:21 (Feb 15, 1975); *NYTimesBookRev* p6 (Feb 2, 1975); *PsycholToday* 8:108 (Apr 1975); *PublWkly* 206:47 (Nov 11, 1974); *PublWkly* 209:52 (Jan 12, 1976); *SatEvePost* 247:70 (May/Jun 1975); *SchLibrJ* 21:115 (Mar 1975); *SciDigest* 77:91–93 (May 1975); *SkptInq* 2:104–110 (Fal/Win 1977); *Time* 105:86 (Mar 17, 1975); *VillVoice* 20:36 (Jan 27, 1975).

803 Villoldo, Alberto, and Krippner, Stanley. *Healing States.* Foreword by Lynn V. Andrews. New York: Simon & Schuster, 1987. 207p. Chapter notes; 38 illustrations

The aim of this book on various forms of healing, according to the two psychologist authors, is to "document our jour–neys and experiences with some of the most extraordinary healers of our time and describe techniques of healing and ecstatic trance that can be used to maintain health and for self–healing" (p. xv). Shamanic healing is emphasized, and they describe the shaman's path to power, which is also a means of achieving wholeness and healing. Part I is entitled "The Dimensions of Spiritual Healing" and consists of five chapters about a Brazilian psychologist and medium, Luis Antonio Gasparetto; psychic surgeon/spirit healer Edson de Quieroz, M.D.; Eliezer Mendes, M.D., who cures epilepsy by training patients to develop their psychic powers; spirit mediumship and spiritist healing among the Candomblé and Umbanda sects of Brazil; and the Buddhist firewalkers of São Paulo. Part II consists of six chapters that describe the experiences of Villoldo and a research group who became apprentices to don Eduardo Calderon, a Peruvian shaman and healer, who led them on an initiation journey both to key Peruvian power spots and into the world of the spirit. The last part is entitled "From Primitive Myths to Planetary Healing" and consists of four theoretical chapters on shamanic healing, ritual, and trance states. It closes with a chapter on healing traditions of the world and their integration.

PAI 6:47 (Jun 1988)

Booklist 83:1240 (Apr 15, 1987); *PsycholToday* 21:78 (Aug 1987); *PublWkly* 231:77 (Feb 20, 1987).

803a Walsh, Roger N. *The Spirit of Shamanism.* Los Angeles, CA: Jeremy P. Tarcher, 1990. 285p. Bibliography: 272–280; Index: 281–285

Psychiatrist/philosopher Walsh's aim in this book is to provide an overview of shamanism, to examine it in the light of modern psychology, and to evaluate extremist claims about shamanism. He also compares "shamanic practices, initiation, training, beliefs, and states of consciousness with those of other traditions" (p. 6). There are 6 parts: "What and Why is Shamanism?," "The Life of the Shaman," "The Shaman's Universe," "Shamanic Techniques," "Shamanic States of Mind," and "Ancient Tradition in a Modern World." Part IV is largely on shamanic healing, and deals with diagnosis, the psychological principles of shamanic healing, shamanic techniques that go beyond psychology, and how shamans heal themselves. He attributes shamanic healing results to several factors: placebo effect, spiritual insight into the patient, and possible psi ability.

Kirkus 58:641 (May 1, 1990); *PublWkly* 237:51 (Apr 27, 1990).

Addenda

INTRODUCTION

Books have been placed in the Addenda for a number of reasons. Some titles, for example, Irwin [810], Reed [813], and the volume edited by John White [819] were not yet available at the time the main bibliography of books had been completed. Some titles were omitted in error, such as Stevenson [818], and White and Dale [820]. Some were considered earlier but rejected. Upon further consideration, it was decided they should be included, such as Huson [808], Rogo [815], and Schwartz [816]. Each of these books is mentioned in the intro—duction to the section in which they belong and author, title, and book number is listed in the bibliography of each section with a note to see the Addenda for the annotation. These books are also listed in the **Additional books listed elsewhere in the bibliography** subsection of each introduction. Following the bibliographic citation in brackets is the heading of the subject section in which the book belongs, for example: [Reference Works on Parapsychology: Bibliographies] for Ashby [804]. This addenda has made it possible to extend the deadline for the in—clusion of books to December 31, 1988. Some additional books were added even later, using a method suggested by Barbara Lee of Scarecrow: that of repeating the preceding number and adding an a to it.

BIBLIOGRAPHY

804 Ashby, Robert H. (Frank C. Tribbe, Ed.). *The Ashby Guidebook for Study of the Paranormal* (rev. ed.). York Beach, ME: Samuel Weiser (in collaboration with Spiritual Frontiers Fellowship), 1987. 215p. Chapter references; Glossary: 201—215 [Reference Works on Parap—sychology: Bibliographies]
   This paperback serves a fivefold purpose. It provides an introduction to psychical research that emphasizes opportunities for personal and spiritual growth. It serves as a handbook for individuals and groups on how to develop psi abilities and how to test or investigate them in others. It has annotated bibliographies——one aimed at beginners and one at more ad—

vanced students (only books are cited). It provides a directory of organizations, including research facilities, foundations, libraries, journals, and bookstores in the United States and the United Kingdom. It has a biographical dictionary of some of the major figures of parapsychology past and present. Because it attempts so much, the *Guidebook* is necessarily limited to only the most important (and the best of the popular) information sources, but for a first glimpse it is a useful guide for beginners, and it is likely to whet the appetite for more. It is handicapped by the lack of an index. Tribbe has done an excellent job of updating and enlarging this edition.
    JASPR 83:274−277 (Jul 1989).
    SpirFron 19:234−236 (Fal 1987).

805 Behe, George. *Titanic: Psychic Forewarnings of a Tragedy.* Foreword by Edward S. Kamuda. Wellingborough, Northamptonshire, England: Patrick Stephens, 1988. 176p. Chapter references; 50 illustrations [Spontaneous Psi]
    The author, a member of the Titanic Historical Society, has culled newspaper and other accounts and interviewed survivors to compile this record of experiences that may have been precognitions of the *Titanic* disaster. Behe classifies and discusses the reports as follows: curious coincidences, mistaken accounts and deliberate hoaxes, phenomena associated with W.T. Stead (a well−known spiritualist who died in the disaster), possible psychic phenomena (accounts that lack some vital pieces of information that would warrant classifying them as probably psychic), and probable psychic phenomena (cases that cannot be explained by normal means and that may be genuine examples of psi). There were 35 cases in the last category. The total number of cases examined is 153. Each case is described and then is followed by Behe's commentary. In a final summary chapter, he discusses the possible normal explanations for seeming precognitions of the sinking of the *Titanic* and concludes that the better accounts "are numerous enough, and contain enough data, to be accepted as probable psychic foreshadowings of" (p. 162) the *Titanic* disaster. However, see the book by Martin Gardner [**469**] that criticizes possible literary precognitions of the sinking of the *Titanic*.

806 Doore, Gary (Ed.). *Shaman's Path: Healing, Personal Growth, Empowerment.* Boston: Shambhala, 1988. 236p. Chapter notes; Index: 227−236 [Unorthodox Healing]
    The editor and compiler of this book edits *Yoga Journal*. The work consists of 19 contributions written specifically for this volume by leaders of the neo−shamanism movement:

Jeanne Achterberg, Doore himself, Larry Dossey, Brook Medicine Eagle, Felicitas D. Goodman, Stanislav Grof, Joan Halifax, Michael Harner, Ake Hultkrantz, Serge King, Jürgen W. Kremer, Stanley Krippner, Frank Lawlis, Jacques Lemoine, Lewis E. Mehl, Rowena Pattee, Jim Swan, and Joan B. Townsend. Doore writes: "The book is addressed to many dif—ferent kinds of readers, both to those whose interests are theoretical as well as to those who wish to learn more about practical 'core' shamanism as a tool for healing and personal growth" (p. 2). Part I, "Shamans and Neo—Shamans," consists of seven essays about shamans and their world. The second part contains seven contributions on shamanic healing. The final part consists of five essays on personal growth and shamanism.

807 Hunt, Stoker. *Ouija: The Most Dangerous Game.* New York: Barnes & Noble, 1985. 156p. Citations in text; 1 figure; Index: 153—156 [Training and Development of Psi]

This is an overview of information about the ouija board. Hunt reports that a survey he conducted found that 30% of the respondents used the ouija board to communicate with the dead, the same percentage to communicate with the living, and the remainder to develop psychic ability or as a means of guidance. He gives some case histories illustrating that use of the ouija board can be both helpful and dangerous. Well—known cases discussed are Jane Roberts' Seth and the Toronto "Philip" group experiment. He discusses automatisms as both inducers of psychosis and cures. An interview with Barbara Honegger is the subject of one chapter. The last chapter is en—titled "How to Get the Best Results from Your Ouija Experiments."

PAI 4:45 (Dec 1986).

Booklist 81:1276 (May 15, 1985); *Kirkus* 53:320 (Apr 1, 1985)

808 Huson, Paul. *How to Test and Develop Your ESP.* New York: Stein and Day, 1975. 215p. Bibliography: 204—207; Glossary: 197—203; 12 figures; Index: 209—215 [Training and Development of Psi]

Huson, who has previously written about the occult, says "this book is a nontechnical personal record of ten years of ex—periments in developing and testing my own ESP" (p. 11). He also read widely and consulted with parapsychologists. Al—though much of the information on organizations is outdated now, this is still one of the better nontechnical books on the popular lore of developing psi abilities. In addition to ESP he

covers remembering past lives, OBEs, PK, and haunts and apparitions. There is a chapter devoted to studying parapsychology that is now outdated.
JP 39:350–356 (Dec 1975); *P* 6:45 (Jan/Feb 1976); *PR*
7:20–21 (May/Jun 1976).
Kirkus p545 (May 1, 1975).

809 **Inglis, Brian, with Ruth West and the Koestler Foundation.** *The Unknown Guest: The Mystery of Intuition.*
London: Chatto and Windus, 1987. 224p. Bibliography: 205–209; Chapter references; Citations in text; Index: 217–224 [Applications of Psi]
The authors report on the results of a survey they conducted of experiences of the "unknown guest," or episodes that
appear to transcend everyday realities and that "suggest design:
as if some prompter in the wings is operating through our subconscious minds" (p. ix). Inglis concentrated his efforts on uncovering historic cases, whereas Ruth West handled the cases
that were sent to the Koestler Foundation, which she directs.
These experiences take many forms, and some of the major
ones are covered by chapters in the book: "Daemon" (primarily
experiences of guidance in well–known people beginning with
Socrates), "The Muses" (experiences of authors, poets,
musicians, artists, and visionaries), "Eureka!" (examples of invention, scientific creativity, and prodigies), "The Superconscious
Self" (ESP, precognition, retrocognition, and clairvoyance),
"Hallucinations" (visions, doubles, apparitions, voices, olfactory
and tactile experiences), "Mind Over Matter" (levitation, weight-
lessness, and spontaneous forms of PK), "Synchronicity," and
"The Oceanic Feeling." In the next to last chapter, the pos–
sibility of tapping the superconscious self without harming the
conscious self is discussed, touching on dreams, OBEs, meditation, drugs, and divination. In the last chapter, "The Act of
Creation," an evolutionary view is taken of the nature of the
"unknown guest." The citations in the bibliography omit mention of the publishers' names.
ContempRev 251:218 (Oct 1987); *Observer* p29 (Sep 20,
1987); *Spectator* 259:23 (Aug 15, 1987); *TimesLitSuppl* p. 403
(Apr 8, 1988).

810 **Irwin, H.J.** *An Introduction to Parapsychology.* Jefferson, NC: McFarland, 1989. 321p. Bibliography: 281–
309; 16 illustrations; Name Index: 311–315; Subject index: 316–321. [Education]
Australian psychologist Irwin has prepared a textbook of
parapsychology that provides a somewhat different approach to
the subject——one that should be acceptable to many of

parapsychology's critics because of his definitions and one that should also be instructive to parapsychologists and especially to psychologists because it is systematic and places emphasis on the psychology of parapsychology. In his presentation of parapsychology, Irwin is not so much concerned with proving that psi is paranormal as with process—oriented studies. More than most other parapsychologists, he emphasizes the phenomenology of seeming psi experiences. Irwin's position is that as far as the definition of parapsychological experiences is concerned, it is the study of experiences that appear to be paranormal, but not until this can be established will he assume that the underlying processes are necessarily paranormal. Because of his emphasis on *experiences*, Irwin devotes more space to spontaneous occurrences and to surveys and attitudes toward these experiences and less space to psi experiments than is the case with other textbooks reviewed here. An initial chapter in which he defines parapsychology and seeming types of parapsychological phenomena is followed by chapters on the origins of parapsychological research, the phenomenology of ESP, ESP experiments, ESP and time, psychokinesis, special PK topics (macro PK, experimenter PK, metal—bending, EVP), theories of psi, the survival hypothesis, poltergeist experiences, NDEs, OBEs, apparitional experiences, reincarnation experiences, a chapter entitled "Matters of Relevance" (on clinical psi and some suggested practical applications of the findings of parapsychological research), and a final chapter entitled "The Evaluation of Parapsychology as a Scientific Enterprise." Each chapter closes with a list of key terms and study questions concerning the content of that chapter.

PAI 4:68—69 (Jun 1986).

811 Lombroso, Cesare. *After Death——What? Researches into Hypnotic and Spiritualistic Phenomena.* Introduction by Colin Wilson. Wellingborough, Northamptonshire, England: Aquarian Press, 1988 (c1909). 364p. Citations in text; 29 figures; 44 illustrations; Index: 359—364; 2 tables [Survival of Death]

This classic book has been reprinted as a volume in the Colin Wilson Library of the Paranormal. Lombroso was an Italian physician and psychiatrist who conducted pioneer research into hypnosis and mediumistic phenomena. Although initially he was skeptical, Lombroso's investigations (described here) convinced him that the phenomena were real, although he could not accept the spiritualistic explanation of the results. He describes his investigations of Eusapia Palladino and other mediums, phantasms and apparitions of the dead, primitive beliefs in spirits of the dead, the question of personal identity, the phenomena of doubles, spirit photography, and apports. There is a chapter on counterhypotheses, and there are conclud— ing theoretical chapters in which he attempts to construct a biol— ogy and psychology of spirits.

Bookman 30:258 (Nov 1909); *Dial* 47:284 (Oct 19, 1909); *LitDig* 39:538 (Oct 2, 1901); *NYTimesSatRev* 14:621 (Oct 23, 1909); *Outlook* 93:830 (Dec 11, 1909); *RevRevs* 40:763 (Dec 1909).

812 Mavromatis, Andreas. *Hypnagogia: The Unique State of Consciousness Between Wakefulness and Sleep.* New York: Routledge & Kegan Paul, 1987. 360p. Bibliogra— phy: 321—349; Chapter notes: 297—320; 59 figures; 1 graph; 3 illustrations; Index: 351—360 [Altered States of Consciousness and Psi]

This work is an outgrowth of Mavromatis' doctoral disser— tation, in which he performs a valuable service by pulling together what is known about hypnagogic experiences, or hal— lucinations and quasihallucinations that occur in a state between waking and sleeping. In the first part, he examines the phenomenology of the hypnagogic state (hypnagogia), considering both somatosensory or perceptual or quasiperceptual experiences and cognitive—affective phenomena. In the second part, he ex— amines the phenomenological relationships of hypnagogia with other experiences and states (dreams, schizophrenia, meditation, psi, and creativity). He notes that an important finding is that "some of these experiences are only distinguishable from hyp— nagogia by the subject's set of beliefs and the setting in which the experiences take place. Practically all of them . . . either begin with, or develop into, hypnagogic phenomena" (p. 12). In the last part he examines brain mechanisms and the function of hypnagogia within an evolutionary perspective. He concludes that hypnagogia "opens great vistas of psychological exploration. Its introspective study may furnish the individual not only with the benefits of an integrated personality but also with the means of discovering new or little known modes of experiencing which will undoubtedly enrich him/her as a psychological en— tity" (p. 13).

Choice 24:1758 (Jul 1987); *JMentImag* 11:159—160 (Sum 1987); *SciTechBkN* 11:21 (Jul 1987).

813 Reed, Graham. *The Psychology of Anomalous Experience: A Cognitive Approach* (rev. ed.). Buffalo, NY: Prometheus books, 1988. 207p. Bibliography: 199—202; Name Index: 203—204; Subject Index: 205—207; Sug— gested further reading: 195—197 [Altered States of Con— sciousness and Psi]

Reed, who chairs the Psychology Department at Glendon College, York University, Toronto, Canada, has provided what could be called a textbook of anomalous experiences, some of them considered to be psi—based. Throughout the book he attempts to show how the mind's organizing abilities, plus its information processing capability, can create seemingly strange events and phenomena. His aim is to show how unconscious choices can explain the world we perceive without recourse to the supernatural (or the paranormal). In the latter sense it is a critical book, but for the most part, it is a very useful compendium that sets forth the limits of normal perceptual behavior and shows how this behavior can account for what were thought to be anomalous experiences and behavior. There are chapters on anomalies of attention, imagery, perception, recall, recognition, experiencing the self, judgment and belief, consciousness, and on the rate of the flow of consciousness. There is a list of suggested readings arranged by chapter at the end.

814 Reed, Henry. *Awakening Your Psychic Powers.* Foreword by Charles Thomas Cayce. San Francisco, CA: Harper & Row, 1988. 226p. Bibliography: 221—222; Index: 223—226 [Training and Development of Psi]

This is a volume in the new Harper & Row series entitled "Edgar Cayce's Wisdom for the New Age," which is published in conjunction with the Association for Research and Enlightenment and is based on the teachings of the psychic Edgar Cayce. Reed, who is a psychologist, writes in the Preface that his primary aim is to make the psychic real to the reader. Part I, "The Psychic Imagination," consists of three chapters that attempt to explain the nature of reality and the role of psi in that reality. Part II, "Psychic States of Awareness," describes various psi phenomena and ways of inducing them (specifically, intuition, dreams, meditation, and hypnosis). The third part contains three chapters that theorize about the role played by the soul, the mind, and the body in psi awareness. The final part, "Developing Psychic Awareness," presents some exercises and experiments and also discusses "the ultimate purpose of psychic awareness" (p. xi), which he thinks is the development of psi ability in humanity at large, beginning with small groups and extending to the species as a whole.

815 Rogo, D. Scott. *On the Track of the Poltergeist.* Foreword by W.G. Roll. Englewood Cliffs, NJ: Prentice—Hall, 1986. 206p. Bibliography: 201; Chapter references; Index: 203—206 [Spontaneous Psi]

This book is largely an autobiographical account and documents Rogo's investigations of poltergeist cases. He also at—

tempts to provide a critical reevaluation of how poltergeists are viewed. He concludes that "different poltergeist cases have dif— ferent etiologies" (p. xiv). In addition to poltergeist cases, he has included cases that on the surface appear to be typical haunts but which, upon investigation, have often turned out to be caused by 'hidden' or 'camouflaged' poltergeists . . . . No ghosts nor outside agencies were involved" (p. xiv). Finally, he attempts to educate the reader as to how to respond when con— fronted with a poltergeist outbreak. In the last chapter he presents some formal guidelines for carrying out field investiga— tions. There are two appendices. One is an historical note on epilepsy and the poltergeist and the other is on some out—of— body poltergeists.
JASPR 82:172–177 (Apr 1988).

816 Schwartz, Stephan A. *The Alexandria Project.* New York: Delacorte Press/Eleanor Friede, 1983. 274p. 48 il— lustrations; Index: 269–274 [Applications of Psi]
The author is Project Director of the Mobius Society, an organization dedicated to conducting scientific research into con— sciousness by employing a team approach in which both psychics and scientists, right and left brain, are used. Using this technique to solve an archaeological problem——the location of Alexander's tomb——Schwartz took a team of 22 persons to Egypt. While still in the U.S., he supplied 11 psychics with a map covering 40 square miles of the site of Alexandria. Using a technique similar to remote viewing, the psychics, working in— dependently, pinpointed three small areas to be explored. In Egypt, the group worked with Egyptian archaeologists, following the leads provided by the psychics as well as the on—site assis— tance of two of the psychics, Hella Hammid and George McMul— len. Their search located the main sites for further investiga— tion and led to making several finds not considered possible by using traditional archaeological techniques. The account in itself is absorbing, and it breaks new ground methodologically.
JP 48:57–59 (Mar 1984); *PAI* 4:69 (Jun 1986).
Booklist 80:8 (Sep 1, 1983); *Brain/MindBul* 8:4 (Sep 12, 1983); *Kirkus* 51:815 (Jul 15, 1983); *LibrJ* 108:1801 Sep 15, 1983); *PublWkly* 224:56 (Jul 29, 1983); *SpirFron* 16:125–126 (Spr 1984).

817 Shapin, Betty, and Coly, Lisette (Eds.). *Current Trends in Psi Research: Proceedings of an International Con— ference Held in New Orleans, Louisiana August 13–14, 1984.* New York: Parapsychology Foundation, 1986. 282p. Chapter references; 3 figures; 1 illustration; 3 tables [General Parapsychology]

The 10 authors in the 33rd Parapsychology Foundation proceedings dealt with a wide—ranging number of topics connected by the thread that these were the subjects on which the participants were working at the time of the conference. R.L. Morris spoke on human factors and the role of psi in human—equipment interactions. Sybo Schouten dealt with the quantitative analysis of spontaneous experiences and methods of testing various hypotheses to explain the characteristics observed. W.G. Roll took a systems—theoretical approach to psi. Daniel J. Benor reviewed research in psychic healing and Carlos Alvarado surveyed research on OBEs for the period 1980—1984. Jeffrey Mishlove surveyed psi applications, and Keith Hearne covered lucid dreaming and psi research. Michael Grosso contributed a theoretical paper on "Psi, Near—Death, Mind at Large: Some Possible Connections," and Debra Weiner discussed psi errors as a clue to the boundary of psi functioning. John Palmer reviewed some common elements in criticisms of parapsychology. This Parapsychology Foundation proceedings provides a good slice of parapsychological research and thinking. The papers all represent the sort of thinking and research that make up the contemporary scene in parapsychology.

JASPR 82:384—388 (Oct 1988); *JP* 50:271—274 (Sep 1986). *ContempPsychol* 32:195 (Feb 1987).

818 Stevenson, Ian. *Cases of the Reincarnation Type: Volume IV. Twelve Cases in Thailand and Burma.* Charlottesville, VA: University Press of Virginia, 1983. 308p. Bibliographic footnotes; Bibliography: 10—11; Glossary: 297—301; 1 illustration; Index: 303—308; 14 tables [Reincarnation]

In the fourth volume of this series of detailed, scholarly case reports of persons who claim to remember past lives, seven cases from Thailand and five from Burma are presented. All cases were personally investigated in situ by the author, a psychiatrist, who conducted extensive interviews with the aid of an interpreter and made detailed field notes. Two introductory chapters provide background information on reincarnation beliefs among Burmese and Thai Buddhists, and summarize the main characteristics of the cases in Burma and Thailand. Although in many respects the cases tend to be similar to those investigated elsewhere, in several respects they differ. There is a glossary of Asian terms used in the text.

JASPR 79:80—85 (Jan 1985); *PAI* 3:42 (Jun 1985).

819 Ullman, Montague, and Krippner, Stanley, with Alan Vaughan. *Dream Telepathy: Experiments in Nocturnal ESP* (2nd ed.). Jefferson, NC: McFarland, 1989. 264p.

Chapter references; 21 illustrations; Index. [Experimental Parapsychology]

The first edition of this book went to press the same time that *Sources* did, so it missed being included in that volume. This second edition almost suffered the same fate! It is an im—portant volume for the general public as well as for scholars. It describes the history of ESP in dreams, beginning with Cicero, but the bulk of the work is devoted to reports of the dream telepathy experiments conducted by Ullman and Krippner at the Dream Laboratory of the Maimonides Medical Center in Brooklyn over a 10—year period. Part I has four chapters on spontaneous telepathic dreams, including one on dreams during psychotherapy. Part II consists of 11 reports of dream experi—ments with the emphasis on the Maimonides research. The third section presents theoretical implications. The first edition comprised the foregoing chapters plus three appendices, only one of which ("Erwin's Nocturnal Tour of the French Quarter"—— William Erwin was a star subject in the Maimonides experi—ments was repeated in the second edition. All of the older material has been amended and updated. One of the new ap—pendices is "Psychology and Anomalous Observations: The Ques—tion of ESP in Dreams," an article by Irvin L. Child reprinted from the *American Psychologist*. He reviews psychologists' treat—ment of reports of the Maimonides dream experiments. It serves as a response to criticisms of the first edition and articles describing the dream experiments. Appendix C, "ESP in Dreams: Comments on a Replication 'Failure' by the 'Failing' Subject," is Robert L. Van de Castle's account of his participa—tion as a subject in an attempted repetition of the Maimonides experiments at the University of Wyoming, which failed. (Van de Castle had also been a star subject in the Maimonides research.) The last appendix is "A Group Approach to the Anomalous Dream" by Montague Ullman. Ullman presents his general approach to dreams, and he emphasizes the importance of the metaphorical transform in dreams. In particular, he describes the experiential dream group as a psi facilitating sys—tem.

JASPR 68:299—305 (Jul 1974); *JP* 38:86—89 (Mar 1974); *P* 5:36 (Sep/Oct 1973); *PR* 5:11—13 (Mar/Apr 1974); *SF* 6:56—57 (Win 1974); *T* 42:9—11 (Fal 1974).

ContempPsychol 20:566—567 (Jul 1975); *JNerv&MentDis* 160:444—445 (Jun 1975); *PsychiatQ* 48:454—456 (1974).

820 White, John (Ed.). *Psychic Warfare: Fact or Fiction?* Wellingborough, Northamptonshire, England: Aquarian Press, 1988. 222p. Chapter notes and references [Applications of Psi]

John White, who has previously compiled several an—
thologies, here presents one on possible military applications of
psi using material drawn from books and articles. The arrange—
ment is in three parts: (1) three documents that provide histori—
cal background; (2) 11 selections that present views of the
present situation, and (3) two documents that are future—
oriented. White provides an Introduction, an Afterword, and an
appendix on psychic energy. He emphasizes peaceful and
peace—making applications of psi. He also contributes an appen—
dix on Hitler and psychic warfare. The authors and the num—
ber of documents contributed by each are John B. Alexander
(1), Thomas E. Bearden (3), Christopher Bloom (1), Martin
Ebon (1), Randy Fitzgerald (1), Anita Gregory (1), Stanley Krip—
pner and Arthur Hastings (1), Ron McRae and Sue Merrow (1),
D. Scott Rogo (1), Michael Rossman (1), Dennis Stacy (1).

821 White, Rhea A., and Dale, Laura A. (Compilers). *Para—
psychology: Sources of Information.* Foreword by J.G.
Pratt. Metuchen, NJ: Scarecrow Press, 1973. 303p. Glos—
sary: 238—249; Name and Subject index: 250—268; Title
index: 269—279 [Reference Works on Parapsychology:
Bibliographies]

White and Dale compiled this bibliographic handbook to
the literature of parapsychology [referred to in the present
volume as *Sources*] in order "to assist librarians, students, and
others interested in psi phenomena to select wisely" (p. 14)
from the large number of books and periodicals on the subject
and to provide additional information sources. The longest part
of eight is the first, which is an annotated bibliography of
books listed under 25 subject headings. The next seven parts
are on coverage of parapsychology in individual encyclopedias
(general and specialized), organizations, periodicals, scientific
recognition of parapsychology (including a chronology of events
indicating recognition by academic institutions, graduate degrees
in parapsychology, opportunities for graduate work, chronology
of events indicating recognition by other disciplines, and recogni—
tion by special issues in other professional journals), a glossary,
name and subject index, and an index of book and periodical
titles.   There are seven appendices to Parts I and II: (1)
"Books Containing Glossaries," (2) "Books and Encyclopedias
Containing Illustrations," (3) "Books with 100 or More Bibliogra—
phic References," (4) "Books by Reading Level Rating," (5)
"Books by Type of Library Indicator," (6) "Abbreviations," and
(7) "Addresses of Publishers not Readily Available."
JP 38:247—249 (Jun 1974); *JSPR* 47:337—338 (Mar 1974);
PR 4:13—14 (Nov/Dec 1973); *PS* 1(2):92—93 (1975); *T*
No.43/44:18—19 (Win/Spr 1975).

Choice 10:1850 (Feb 1974); Choice 12:808 (Sep 1975); LibrJ 99:126 (Jan 15, 1974); RQ 14:72 (Fal 1974); WLBul 48:263 (Nov 1973).

822 Houtkooper, Joop M. *Observational Theory: A Research Programme for Paranormal Phenomena.* Lisse, Nether—lands: Swets & Zeitlinger, 1983. 123p. Bibliography: 103—110; 5 figures; Author index: 111—112; Subject in—dex: 113—116; 22 tables [Experimental Parapsychology]*
Work that the author carried out for his doctoral disserta—tion at the State University of Utrecht in the Netherlands. The text is in English but the Summary and Curriculum vita are in Dutch. It contains eight chapters, four of which are experimen—tal reports (two with coauthors) published separately in the *European Journal of Parapsychology*. The first chapter is intro—ductory and the last three are methodological/theoretical in na—ture. This work is included here because it depicts a research program of experiments, theories, and attempted methodologies converging on a single goal——to test the observational theories of psi. Its main value is for researchers in the field or scien—tists in other fields who want a glimpse of ways in which psi research is relevant to the role of the observer as expressed in quantum theory. For those without access to the experimental parapsychology journals it also has value in that its chapters have been or else could have been taken from the journals. The chapter titles are (1) "Introduction," (2) "Exploratory PK Tests With a Programmable High Speed Random Number Generator," (3) "A Comment on Schmidt's Mathematical Model of Psi," (4) "A Study of Repeated Retroactive Psychokinesis in Relation to Direct and Random PK Effects," (5) "A Hierarchi—cal Model for the Observational Theories," (6) "The Hierarchical Model and the Design of Experiments," (7) "Observational Theory and Quantum Mechanics," and (8) "A Re—evaluation and Concluding Remarks."
JASPR 81:381—388 (Oct 1987).

---

*This is an addendum to the addenda, and so is not listed in alphabetical order!

II
Periodicals

QUICK INDEX TO SECTIONS IN THIS CHAPTER

ALPHABETICAL LIST OF PERIODICALS

The professional journals are the most important vehicles for the publication of original research findings, scholarly review articles, methodological papers, and, to a lesser extent, historical and theoretical papers. Listed below in five categories are English—language periodical titles that are either centrally or peripherally related to parapsychology. The categories are Major Professional Journals, Specialized Parapsychological Journals, Publications Dealing With Anomalous Phenomena, Newsletters, and Discontinued Publications. Three x's indicate a cross reference for a title change. The publications within each section are listed in alphabetical order by title.

A wide net was cast in selecting the titles included in this chapter so that it could serve as a record of periodicals dealing with all aspects of the subject. The list is limited to English—language publications. (Non—English language journals were not included because insufficient information is available concerning them. A list of names and addresses of the major journals ar—ranged by language is available from the Parapsychology Sources of Information Center. English—language abstracts of several of these journals appear regularly in *Parapsychology Abstracts International* [828].) A core collection of the English—language journals should include the titles in the first section plus the *Journal of Scientific Exploration* [846], *Parapsychology Review* [829], and the *Zetetic Scholar* [850].

Because journals tend to change, we are including the titles that were described in *Sources* with updated information. The location number in *Sources* is given in brackets following the number in this volume should it be desirable to locate ear—lier information on a given periodical. Some sources list the major journals by abbreviation only. In such cases the abbrevia—tion used is set off by parentheses following the title.

Where the information is available we list the date and fre—quency of publication, whether or not the publication is refereed, ISSN number, editor's name, publisher's address and telephone number, average number of pages per issue, size, cir—culation figure, price, a description of the contents, style, format (subjective judgment, based upon: type and paper quality, form of publication, i.e., typeset, photo offset, mimeographed, etc.), whether or not the journal is indexed and what indexing serv—ices, if any, include it, whether it contains abstracts or is abstracted and by what abstracting services, computerized databases, whether back issues are available, from whom and how much, and whether it has been reprinted or is in microform and for what volumes.

The names (with their abbreviations used here) and ad—dresses of firms that sell reprinted editions or microfilms of the journals listed are:

AMS Film Service (**AMS**), 56 East 13th Street, New York, NY 10003

Johnson Reprint Corporation (**JR**), 111 Fifth Ave., New York, NY 10003

Kraus Reprint Co. (**KR**), Route 100, Milwood, NY 10546

University Microfilms International (**UMI**), 300 North Zeeb Road, Ann Arbor, MI 48106

Note: Libraries owning periodicals were listed in *Sources*. This information is omitted here because a *Union List of Para—psychology Serials* is being compiled by the Parapsychology Sources of Information Center that will provide detailed informa—tion on volumes and sometimes even issues held by major parapsychological libraries.

The reliability of the information published in the peri—odicals included varies. The most reliable are those that are listed as being refereed. It may be of interest to note that the value of parapsychology journals for parapsychology has recently been questioned by a psychiatrist formerly identified with parapsychology, Ian Stevenson. (See his Guest Editorial in *JASPR*, 1984 (Apr), *78*, 97—104. In the same editorial Steven—son strongly criticized newsletters on parapsychology as being unreliable sources of information. For a rebuttal of Stevenson see Stillings and Duke (*JASPR*, 1986, *80*, 348—350), together with a further exchange of letters in the same issue on pages 350—352.)

Many newsletters have been described in this chapter be—cause I feel they serve important functions not otherwise being addressed. They are a major source of information on current activities and personalities, thus providing a starting point for tomorrow's histories. They serve as a barometer of popular inter—est in parapsychology, which can answer several needs; and they are usually written in nontechnical language that can interest newcomers in the field. Newsletters can acquaint persons with names to follow up in journals and books. They sometimes provide insights into the direction in which the field might be going, or present dissident views that round out the general pic—ture of the field as presented by the mainline journals.

MAJOR PROFESSIONAL JOURNALS

This section contains the core parapsychology journals deal—ing with the subject as a whole. The *Journal of Parapsychology* [825] and *Journal of the American Society for Psychical Research* [826] are generally considered to be primary. *Para—psychology Review* [829] is more popular in style but serves several useful purposes, and *Research in Parapsychology* provides the best overview of parapsychological activity. *Parapsychology Abstracts International* [828] also provides a good overview and serves as a valuable index to persons, sources, and topics. Five of the journals in this group are affiliated with the Para—psychological Association: *European Journal of Parapsychology* [823], *Journal of Parapsychology* [825], *Journal of the American Society for Psychical Research* [826], *Journal of the Society for Psychical Research* [827], and *Parapsychology Abstracts Interna—tional* [828].

823 *European Journal of Parapsychology* (*EJP*) *ISSN*: None
Vol. 1, 1975 to date
Published semi—annually in May and November

Refereed: Yes

Editor: Sybo A. Schouten   *Publisher*:
Av. pages per issue: 100   University of Utrecht
Size: 23.8 cm   Sorbonnelaan 16
Circulation: 200   3584 CA Utrecht
Price: $10.00   The Netherlands
                                    Tel.: (30) 533824

Purpose: To stimulate and enhance activity in parapsy—
chology, especially in Europe, by communicating research results
and issues related to professional parapsychology. *The European
Journal of Parapsychology* is an affiliated journal of the Parapsychological Association and welcomes contributions from all
over the world. In an attempt to avoid selective reporting of
only articles providing evidence of psi, *EJP* has instituted a
policy such that the acceptance or rejection of a manuscript is
made before the data are collected. Thus, the quality of the
design and methodology and the rationale of a study are con—
sidered more important than the significance level of the results.
Manuscripts that adhere to this publication policy are given
priority as regards publication.

Contents: An average of seven articles per issue. Experimen—
tal reports dominate, but theoretical articles are also welcome.
Also contains letters and obituaries.

Style: Scholarly. Figures; graphs; reference lists; tables.
Format: Fair. Wide margins.
Indexes: None.
Abstracts: Author abstracts for most articles. Also abstracted
in *Journal of Parapsychology*; *Parapsychology Abstracts Interna—
tional* (1975 to date); *Psychological Abstracts* (1983 to date).
Computerized databases: *PsiLine* (1975 to date); *PychINFO*
(1983 to date).
Back issues: Available from publisher.
Reprinted editions: None.
Microforms: None.

824 *Journal of Indian Psychology*   *ISSN*: 0379—3885
      Vol. 1, 1978 to date
      Published semi—annually in January and July
      Refereed: Yes

Editor: K. Ramakrishna Rao   *Publisher*:

Av. pages per issue: 60
Size: 24.8 cm
Circulation: 500
Price: Per year: $20 to
   institutions, $6 to
   individuals; $5 apiece

Dept. of Psychology and
   Parapsychology [892]
Andhra University
Waltair
Visakhapatnam 530 003
Andhra Pradesh, India
Tel: 3279

Purpose: Parapsychology does not appear in the title of this journal so as not to confuse it with the earlier *Journal of Indian Parapsychology.* Also, it contains nonparapsychological articles on psychology, and the articles on parapsychology emphasize the psychology of psi. The subtitle is *A Journal of Classical Ideas and Current Research.* Thus it is a journal of ideas as well as of hard facts. This international journal discusses classical ideas concerning human nature and current research aimed at their empirical testing and application. Emphasis is placed on the integration of research and theory. Cross—cultural and inter-disciplinary research and studies to integrate normal, abnormal and paranormal experiences so as to stimulate alternative scientific paradigms and heuristic models for the study of humans are of special interest.

Contents: Five articles per issue (theoretical, experimental, review); several book reviews as well as a number of books described under "Books in Brief," "News and Views," and "Books Received."

Style: Scholarly. Contains figures; graphs; reference lists; tables. Follows *Publication Manual of the American Psychological Association* (3rd. ed.), modified.
Format: Fair
Indexes: A contents listing for Vol. 1, titled "Index," appeared in Vol. 2(1).
Abstracts: Abstracts for experimental articles. *Journal of Parapsychology; Parapsychology Abstracts International* (1988 to date); *Psychological Abstracts* (1978 to date).
Computerized databases: PsiLine (1988 to date), *PsychINFO* (1978 to date).
Back Issues: Available from publisher at $6 apiece.
Reprinted editions: None.
Microforms: None.

825 [329]   *Journal of Parapsychology (JP)*   *ISSN:* 0022–3387
   Vol. 1, 1937 to date
   Published quarterly in March, June, September, and Decem—

ber
Refereed: Yes

Editor: K. Ramakrishna Rao
Av. pages per issue: 88
Size: 22.8 cm
Circulation: 1000
Price: $25 to individuals,
    $35 to institutions

Publisher:
Parapsychology Press [870]
Box 6847
College Station
Durham, NC 27708
    Tel.: (919) 688–8241

Purpose: "Devoted primarily to the original publication of experimental results and other research findings in extrasensory perception and psychokinesis. In addition, articles presenting reviews of literature relevant to parapsychology, criticisms of published work, theoretical and philosophical discussions, and new methods of mathematical analysis will be published as *Journal* space allows."

Contents: Each issue contains an average of four articles averaging 25 pages in length and three book reviews averaging three pages. As an affiliated organ of the Parapsychological Association, it carries announcements of that organization and, from 1958 through 1971, abstracts of its convention. Abstracts of other conventions are also included from time to time. Other features are correspondence and "Books Received." Most issues contain a valuable section entitled "Parapsychological Abstracts," initiated in 1958. This section presents abstracts (averaging half a page in length) of relevant articles published in other journals, both parapsychological and nonparapsychological, some in foreign languages, dissertations on parapsychological topics accepted at universities, and of unpublished reports on file at the Institute for Parapsychology of FRNM. The *JP* is an affiliated journal of the Parapsychological Association.

Style: Scholarly. Follows *Publication Manual of the American Psychological Association* (3rd. ed.). Figures, graphs, illustrations, reference lists, tables. A glossary is included in each volume for lay readers.
Format: Excellent.
Indexes: Each volume has a separate table of contents and a name and subject index. The table of contents for Vol. 1, 1937, through Vol. 16, 1952, was published in the September 1953, issue. It is indexed in *Mental Health Book Review Index*, *Current Contents*, *Social Sciences Citation Index*, *Social Sciences Index*.
Abstracts: Abstracts provided for experimental reports. It is abstracted in *Biological Abstracts*, *Excerpta Medica*, *Para–*

psychology Abstracts International (1937–1958; 1987 to date), *Psychological Abstracts* (1937 to date).
Computerized databases: *PsiLine* (1937 to date), *PsychINFO* (1965 to date).
Back issues: Some back issues from Vol. 13, 1949 to date may be obtained from the *Journal.*
Reprinted editions: Vol. 1–12, 1937–1948, $350.00; $280.00, paper (JR). Reprint service: UMI
Microforms: Vols. 1–46, 1937–1982, $405.00 or $9.00 per vol. (KR). Vol. 1 to date, Vols. 1–23, 1937–1959, $137.80; 1960–1987, $15.60 per Vol., 1988, $12.20; 1989– $13.20 (UMI).

826 [330]   *Journal of the American Society for Psychical Research* (*JASPR*)   *ISSN*: 0003–1070
Vol. 1, 1907 to date
Published quarterly in January, April, July, and October
Refereed: Yes

Editor: Rhea A. White
Av. pages per issue: 104
Size: 22.8 cm
Circulation: 2500
Price: free to members;
  $25 to nonmembers; $50 to
  institutions

Publisher:
American Society for
Psychical Research [869]
5 West 73rd St.
New York, NY 10023
Tel.: (212) 799–5050

Purpose: To further the advance of parapsychology by publishing reports of experimental work carried out at the ASPR and at other major research centers, case reports and discussions, theoretical and statistical papers, articles dealing with relevant trends in other disciplines, and reviews of current books and periodicals in the field.   *JASPR* is an affiliated journal of the Parapsychological Association.

Contents: Each issue contains an average of four articles averaging 15 pages in length, plus five book reviews averaging four pages. Also contains correspondence and "Books Received." As an affiliated organ, it carries announcements of the Parapsychological Association.

Style: Scholarly. Follows *Publication Manual of the American Psychological Association* (3rd ed.), modified. Figures; graphs; illustrations; tables.
Format: Excellent.
Selections: Three papers on the survival problem by Gardner Murphy, originally published in the *Journal*, are avail—

able in book form.

Indexes: The *Journal* provides a name and subject index for each volume as well as a separate table of contents. In 1985 and 1986 *Tables of Contents* of the *Journal* and *Proceedings* of the ASPR from 1885–1984 was published in the *Journal* as follows: Part I: 1885–1910 (January 1985), Part II: 1911–1925 (April 1985), Part III: 1926–1940 (July 1985), Part IV: 1941–1952 (October 1985), Part V: 1953–1963 (January 1986), Part VII: 1971–1980 (July 1986), Part VIII: 1981–1984 (October 1986). A cumulative card index is maintained at the ASPR's headquarters. It is also indexed in *Mental Health Book Review Index* and *Social Science Citation Index*. The table of contents of each issue is reproduced in *Selected List of Tables of Contents of Psychiatric Periodicals*.

Abstracts: Abstracts are provided for all articles. It is also abstracted in *Parapsychology Abstracts International* (1981 to date), and *Psychological Abstracts* (1933 to date).

Computerized databases: *PsiLine* (1940 to date; projected: 1907 to date), *PsychINFO* (1965 to date).

Back issues: Pre–1941 and post 1960: some issues available from the ASPR at $3.00 each.

Reprinted editions: Vol. 35–54, 1941–1960, $350.00, paper (JR). Reprint service: UMI.

Microforms: Vol. 1 to date (UMI).

827 [331]   *Journal of the Society for Psychical Research (JSPR)*   *ISSN*: 0037–1475
Vol. 1, 1884 to date
Published quarterly in March, June, September, and December
Refereed: Yes

Editor: John Beloff	*Publisher*:
Av. pages per issue: 64	Society for Psychical
Size: 21.3 cm	Research [874]
Circulation: 1150	1 Adam & Eves Mews
Price: free to members	London, England W8 6UG
Note: Vols. 1–34, 1884–1948,	*Tel.*: 01–937 8984
were restricted to members	
only. From September 1949	
to date available to others	
at $36 per year.	

Purpose: "To examine without prejudice or prepossession, and in a scientific spirit those faculties of man, real or supposed, which appear to be inexplicable in terms of any generally recognized hypotheses." From 1884 to 1948, when the circulation of the *Journal* was limited to members, it was used as the main vehicle for publishing spontaneous cases. Since 1949 it has also published experimental reports and various other kinds of material that serve to inform readers about what is happening in parapsychology. *JSPR* is an affiliated journal of the Para—psychological Association.

Contents: Average of three articles per issue, averaging 12 pages in length, dealing with quantitative studies, mediumistic and spontaneous case reports, theoretical papers and criticisms. Also included are correspondence, notices, and approximately six book reviews per issue averaging two pages in length.

Style: Scholarly. Figures; graphs; illustrations; tables.
Format: Good.
Indexes: A separate name and subject index and a table of contents are provided for each volume. The Society has issued a list of the principal contents of the *Journal* from Vol. 35, 1949 to date. The *Journal* is included in the index to the Society's publications (reproduced in part in 207) for Vol. 1—33, 1884—1946. Also indexed in *British Humanities Index*.
Abstracts: Contains abstracts of experimental papers. It is also abstracted in *Journal of Parapsychology*; *Parapsychology Abstracts International* (1979 to date), *Psychological Abstracts* (1954 to date).
Computerized databases: *PsiLine* (1940 to date; projected: 1884 to date), *PsychINFO* (1965 to date).
Back issues: Some copies are available from the SPR.
Reprinted editions: None.
Microforms: None.

828 *Parapsychology Abstracts International (PAI) ISSN*: 0740—7629
Vol. 1, 1983 to date
Published semi—annually in June and December
Refereed: Not applicable

Editor: Rhea A. White
Av. pages per issue: 84
Size: 28.4 cm
Circulation: 250
Price: $35.00 to individuals, $50 to institutions

Publisher:
Parapsychology Sources of
Information Center [890]
2 Plane Tree Lane
Dix Hills, NY 11746
Tel.: (516) 271—1243

Purpose: To provide summaries of the literature of and

about parapsychology from earliest times to date. *PAI* is an af—
filiated journal of the Parapsychological Association.

Contents: Each issue contains abstracts of approximately
300 documents arranged in the following categories: Para—
psychology journals (beginning with Vol. 6 the June issue
carries English—language journals and the December issue non—
English journals); articles on parapsychology in nonparapsy—
chological journals; articles on parapsychology in general interest
magazines; conference proceedings papers; book chapters; disserta—
tions and theses; monographs, pamphlets and other separate
reports; parapsychology books; and consciousness studies books.
Each issue contains a list of books received for review and most
issues have an editorial. Beginning with Vol. 6 each parapsy—
chology journal abstracted is preceded by a brief sketch of the
journal, its purpose, editor, address, and other pertinent details.

Style: Scholarly.
Format: Good. Photo offset.
Indexes: For Vols. 1—5, each issue contained an author,
title, and subject index. Beginning with Vol. 6, there are
author, title, and subject indexes for each volume. There also
are separate cumulated Author/Title/Subject Indexes for Vols.
1—2 and 3—5.
Abstracts: Not relevant.
Computerized databases: PsiLine.
Back issues: All are available at the regular subscription
prices or $20 per single issue.
Reprinted editions: None.
Microforms: None.
Note: Beginning with Vol. 8, 1990, *Parapsychology Abstracts
International* will become *Exceptional Human Experience* with
expanded coverage and a new format.

829 [336]   *Parapsychology Review (PR)*   *ISSN*: 0031—1804
          Vol. 1, 1970 to date
          Published bimonthly
          Refereed: No

Editor: Betty Shapin   *Publisher*:
Av. pages per issue: 16   Parapsychology Foundation [873]
Size: 27.8 cm   228 East 71st Street
Circulation: 2,000   New York, NY 10021
Price: $12.00; $2 per issue   *Tel.*: (212) 628—1550

Purpose: Took over the functions of the *Newsletter of the*

Parapsychology Foundation [341] and the *International Journal of Parapsychology* [857] [340]. It presents directory—type information such as the latest news coverage on persons and organizations associated with parapsychology, as well as items on events, grants, courses, lectures, conferences, symposia, degrees, etc. International in scope.

Contents: Contains an average of two or three brief articles emphasizing the current status of parapsychology or parapsychology in various countries or state—of—the—art reviews of various aspects of the subject. Also provides book reviews, conference reports, obituaries, books received, and the annual report of the Parapsychology Foundation's grants program.

Style: Nontechnical. Magazine—type format.
Format: Excellent.
Indexes: An annual index is published in the first issue of the following year.
Abstracts: *Parapsychology Abstracts International* (1988 to date), *Psychological Abstracts* (1970 to date).
Computerized databases: *PsiLine* (1988 to date; projected: 1970 to date), *PsychINFO* (1970 to date).
Back issues: Some are available from the Parapsychology Foundation.
Reprinted editions: None.
Microforms: None.
Note: *Parapsychology Review* ceased publication with the second 1990 issue.

830 [332]   *Proceedings of the American Society for Psychical Research* (*PASPR*)   *ISSN*: None.
  Vol. 1, 1885—1889; new series, Vol. 1, 1907 to date [the latest one as of 2/89 is Vol. 31, February 1974]
  Published irregularly as needed
  Refereed: Yes

Editor: Rhea A. White
Pages per issue: 32—500
Size: 22.5 or 22.7 cm
Circulation: 2500
Price: free to members; prices vary according to size of issue for nonmembers.

Publisher:
American Society for
  Psychical Research [869]
5 West 73rd St.
New York, NY 10023
Tel.: (212) 799—5050

Purpose: The publication of experimental reports, theoretical papers, or field studies too long for inclusion in the *Journal*.

Contents: Up to 1926 a varying number of articles were published per issue, but since then usually only one monograph—like paper appears in each issue. Experimental reports, large case collections, and theoretical or methodological articles are included.

Style: Scholarly.   Figures; graphs; illustrations; reference lists.

Format: Excellent.

Indexes: The Society has an unpublished card index of its *Proceedings* (1907–1960).   The index to SPR publications [207] includes Vol. 1 of the old ASPR *Proceedings*.   The contents of Vol. 1 of the old *Proceedings* and Vol. 1 through Vol. 31 (1907–1927) of the new series is given in Fodor [201] and reprinted in Shepard [685, 686]. The contents of the *Proceedings* from Vol. 1 to date was published in *JASPR* in 1985 and 1986. Part I: 1885–1910 (January 1985), Part II: 1911–1925 (April 1985), Part III: 1926–1940 (July 1985), Part IV: 1941–1952 (October 1985), Part V: 1953–1963 (January 1986), Part VII: 1971–1980 (July 1986), Part VIII: 1981–1984 (October 1986). A cumulative card index is maintained at the ASPR's head—quarters. It is also indexed in *Mental Health Book Review Index* and *Social Science Citation Index*. The table of contents of each issue is reproduced in *Selected List of Tables of Contents of Psychiatric Periodicals*.

Abstracts: Some articles have abstracts.

Computerized databases: PsiLine (1940 to date; projected 1885 to date).

Back issues: Some issues are still available from ASPR at varying prices.

Reprinted editions: None.

Microforms: None.

831 [334]   *Proceedings of the Society for Psychical Research* (PSPR)   *ISSN*: None
Vol. 1, 1882 to date
Published irregularly as needed. Each volume contains un—
   specified number of "parts" (issues)
Refereed: Yes

Editor: John Beloff   *Publisher*:
Pages per issue: 30–600   Society for Psychical
Size: 21.3 cm   Research [874]
Circulation: 1150   1 Adam and Eve Mews
Price: free to members;   London, England W8 6UG

prices vary for nonmembers   *Tel.*: 01—937 8984
according to size of issue.

Purpose: To publish the "cream" of the Society's work at
whatever length and as often as necessary as determined by the
amount of material available. The *Proceedings* contain major
research reports, the Society's presidential addresses, and papers
of an analytical or theoretical nature.

Contents: From 1882 through the 1920s each issue contained
four to 10 articles, but from 1930 on there is usually only one
article per part. Included are experimental, theoretical, critical,
and review papers on psychical research topics, as well as collec—
tions of spontaneous cases, presidential and other addresses, and
obituaries of prominent persons associated with the field. Until
Part 177, 1949, the *Proceedings* also contained from two to four
lengthy reviews of important books, but all reviews (except for
an occasional review article) are now published in the *Journal*.

Style: Scholarly. Figures; graphs; illustrations; reference lists
or footnotes; tables.
Format: Good.
Indexes: Indexed in Vols. 3—11 (1920—1949) of the *Social
Sciences and Humanities Index* (then the *International Index*),
and in *British Humanities Index*. The Society has issued an in—
dex to its publications, including the *Proceedings*, in four
volumes, covering 1882—1970. It has also issued a list of the
principal contents of *Proceedings* from 1882 to date. The con—
tents from 1882 through 1933 are included in Fodor [201],
reprinted in Shepard [685, 686].
Abstracts: Some articles have abstracts.
Computerized databases: PsiLine (1940 to date; projected:
1882 to date); *Psychological Abstracts* (1963 to date).
Back issues: Some parts are still available from the Society
at varying prices (half price to members).
Reprinted editions: None.
Microforms: None.

832 *Research in Parapsychology (RIP)**   *ISSN*: 0090—5399
     Vol. 1, 1972 — to date
     Published for the Parapsychological Association [872] by
     Scarecrow Press

_____

*Individual volumes are listed in the chapter on
"Parapsychology in General" in Part I.

Refereed: by Program Committee of Parapsychological Association

Editors: Various**   *Publisher*:
Av. pages per issue: 240   Scarecrow Press
Size: 22.3 cm   52 Liberty St.
Circulation: 500   Box 4167
Price: $25.00   Metuchen, NJ   08840
                                        Tel.: (908) 548–8600

Purpose: To publish abstracts and papers from the annual convention of the Parapsychological Association––Continues the *Proceedings of the Parapsychological Association* [863].

Contents: Each volume contains a Preface describing the annual convention of a given year, 1,500–word abstracts of papers and symposium contributions, 1,000–word abstracts of research briefs; 500–word abstracts of roundtable contributions and the full text of the presidential address and (in some volumes) of the J.B. Rhine address. These volumes comprise an important record of current psychological research, methodology, and theory.

Style: Scholarly. Figures; graphs; illustrations; reference lists; tables for addresses only. Abbreviated citations embedded in text of abstracts.
Format: Good. Hardcover (library binding).
Indexes: Each volume has detailed Name and Subject Indexes.
Abstracts: Not applicable.
Computerized databases: *PsiLine* (1972 to date; bibliographic citations and descriptors only).
Back issues: Available from publisher. Listed in *Books in Print*.
Reprinted editions: None.
Microforms: None.

833 *Theoretical Parapsychology: Incorporating Psychoenergetics*
(*TP*) [**867**]   *ISSN*: 0894–2528
Vol. 6, 1988 to date
Published quarterly
Refereed: Yes
Editor: Brian Millar   *Publisher*:

**Each volume has two editors: the primary editor (starting with *1988*, Linda A. Henkel) and the Chair of that year's Program Committee (for 1988, Rick E. Berger).

Av. pages per issue: 52
Size: 21.3 cm
Circulation: Unknown
Price: $30 to individuals;
$194 to libraries

Gordon and Breach Science
Publishers
Orders to:
STBS, Ltd.
P.O. Box 786
Cooper Station
New York, NY 10276
Tel.: (212) 243−4411/4543

Purpose: Publishes papers dealing with matters of theoretical interest to parapsychology, which is treated from varying viewpoints, and is defined as "the scientific study of those effects known to our ancestors as 'magic.'" Although the existence of statistical anomalies in parapsychology is granted, it is considered "uncertain whether such results reveal a new kind of (psi) interaction or whether they are due to undetected experimental errors. The keynote of this journal is the elucidation of mechanism, the mechanism of pseudo−psi effects as well as the mechanism of real psi. This is the common interest of skeptic and believer alike and contributions are welcome from both sides of this traditional great divide. The majority of papers on quasi−psi will be psychological, while those on real psi will be predominantly mathematical/physical in nature. Apart from theory in a strict sense, theory−based empirical studies are particularly welcome. Methodology and instrumentation are also areas of interest. Aspects of the philosophy of science and the sociology of the field can be legitimate areas of inquiry for *Theoretical Parapsychology*." The editor notes also that *TP* aims to complement existing parapsychology journals, not compete with them.

Contents: Five to six articles per issue averaging eight pages. Correspondence and book reviews are planned. Descriptors are listed for each contribution.

Style: Scholarly. Figures; graphs; illustrations; reference lists.
Format: Very good.
Indexes: Unknown.
Abstracts: Journal abstracts for articles. *Parapsychology Abstracts International* (1988 to date).
Computerized databases: *PsiLine* (1988 to date).
Back issues: Individual issues are not available but the publisher maintains a fee−based document delivery service for individual articles.
Reprinted editions: None.
Microforms: Subscriptions are available from the publisher in microform editions.

Note: As we go to press we have learned that Brian Millar resigned as editor after the first issue. Apparently a new editor has not been assigned as of July, 1990.

SPECIALIZED JOURNALS

The publications in this category are scholarly journals that are of a narrower range than the journals in the preceding group. Most are devoted to specific subjects such as religion and parapsychology [**834, 836, 838, 839**], near—death experiences [**835**], survival [**837, 841**], or a specific geographic area, e.g., the Netherlands [**840**].

xxx   *Anabiosis: The Journal of Near—Death Studies.* Superseded by the *Journal of Near—Death Studies* [**835**]

834 *The Christian Parapsychologist*   *ISSN*: 0308—6194
     Vol. 1, 1975 to date
     Published quarterly in March, June, September, and Decem—
     ber
     Refereed: No

Editor: The Venerable
   Michael Perry
Av. pages per issue: 40
Size: 20.2 cm
Circulation: 1,500
Price: $8.00

Publisher:
Churches' Fellowship for
   Psychical and
   Spiritual Studies [**920**]
The Priory
44 High Street
New Romney, Kent TN28 8B2
England
Tel.: 0679—66937

Purpose: It attempts to facilitate communication among Christians of different views engaged in the interface between parapsychology and Christianity, and to report from an ecumeni— cal viewpoint on Christian publications, conferences, and special groups concerned with psi.

Contents: Each issue contains an average of three articles averaging five pages, letters to the editors, an average of six book reviews a page in length, an editorial, and news about recent books, periodical articles, events, conferences, organiza— tions, etc.

Style: Scholarly. Reference notes.
Format: Good.
Indexes: A cumulative index to the first two volumes is available from the Churches' Fellowship for Psychical and Spiritual Studies. An author, title, and subject index to subsequent volumes is published in the last number of each volume.
Abstracts: *Parapsychology Abstracts International* (1987 to date).
Computerized databases: *PsiLine* (1987 to date; projected: 1975 to date).
Back issues: Some issues available from publisher.
Reprinted editions: None.
Microforms: None.

xxx   *Gateway.* Superseded by *Spiritual Frontiers* [839]

835   *Journal of Near—Death Studies*   *ISSN*: 0891—4494
      Vol. 1, 1981 to Vol. 5, 1987 published semiannually as
      Anabiosis: A Journal of Near—Death Studies.
      Published quarterly in Fall, Winter, Spring, and Summer
      Refereed: Yes

Editor: Bruce Greyson, M.D.
Av. pages per issue: 64
Size: 22.8 cm
Circulation: 1000
Price: Free to members of the Research Division of the International Association for Near—Death Studies.
$30 per year for individuals; $76 for libraries.

Publisher:
Human Sciences Press for the International Association for Near—Death Studies [903]
233 Spring Street
New York, NY
10013—1578
Tel.: (212) 620—8000

Purpose: To publish articles concerned with near—death experiences and allied phenomena. In particular it welcomes scientific studies of human consciousness as it is affected by the prospect of or near—occurrence of death. In addition to articles on near— death experiences it publishes articles on out—of—body experiences, deathbed visions; experiences of dying persons or those in contact with them prior to the onset of death; and experiences of persons following the death of another. It may also publish articles on other topics which make a definite contribution to the understanding of the experience and meaning of death (e.g., experiences suggestive of reincarnation).

Contents: An average of four experimental, methodological, and theoretical articles per issue average 14 pages in length and one or two book reviews averaging four pages. A new practice was introduced with Vol. 6 of publishing a controversial lead article followed by several peer commentaries expressing varying viewpoints.

Style: Scholarly. Follows *Publication Manual of the American Psychological Association* (3rd ed.). Figures; graphs; references lists; tables.
Format: Good.
Indexes: Table of contents for volume in last issue of that volume. Vol. 5, No. 2 contained a cumulative listing of the first five volumes (published as *Anabiosis*). Also indexed in *International Bibliography of Periodical Literature*.
Abstracts: Journal abstracts for most articles. Some articles abstracted in *Journal of Parapsychology*; *Parapsychology Abstracts International* (1981 to date); *Psychological Abstracts* (1981 to date); *Social Work Research and Abstracts*; *Sage Family Abstracts*.
Computerized databases: *PsiLine* (1981 to date), *PsychINFO* (1981 to date).
Back issues: Available from publisher.
Reprinted editions: Available from IANDS [905].
Microforms: None.

836 *Journal of Religion and Psychical Research (JRPR) ISSN*: 0272−7188
Vol. 1 1978 to date
Published quarterly in January, April, July, and October
Refereed: No

Editor: Mary Carman Rose
Av. pages per issue: 64
Size: 21.5 cm
Circulation: 350
Price: Free to members of the Academy. $8.00 per year by subscription.

Publisher:
Academy of Religion and Psychical Research [916]
Box 614
Bloomfield, CT 06002
Tel.: (203) 242−4593

Purpose: Established as a vehicle and forum for dialogue and exchange of ideas among Academy members and others in regard to the interface between religion and psychical research.

Contents: An average of six articles per issue eight pages in length plus book reviews averaging three pages, and abstracts of research, reports of conferences and symposia, editorials, news, comments, and correspondence.

Style: Nontechnical. Reference lists; tables.
Format: Fair.
Indexes: Separate table of contents for each volume. Name index and Index of Topics and Titles in the October issue. Also in *Religion Index One: Periodicals*.
Abstracts: Brief abstracts for each article. Also in *Journal of Parapsychology*; *Parapsychology Abstracts International* (1988 to date).
Computerized databases: *PsiLine* (1978 to date).
Back issues: Available from publisher at $2.00 each.
Reprinted editions: None.
Microforms: None.

837 *Light: A Review of Psychic and Spiritual Knowledge and Research* ISSN: 0047 4649
Vol. 1, 1881 to date
Published three times a year in Spring, Summer, and Winter
Refereed: No

Editor: Brenda Marshall
Av. pages per issue: 48
Size: 21 cm
Circulation: 2,500
Price: Free to members;
$12.00 ($18.00 air mail)

Publisher:
College of Psychic Studies [**912**]
16 Queensberry Place
London SW 7 2EB
England, U.K.
Tel.: 071−589−3292/3

Purpose: Originally the oldest British Spiritualist weekly, *Light* is now open−ended and "exists for the reasoned and cour− teous examination of all aspects of spiritual and psychic explora− tion and experience."

Contents: Each issue has an editorial, seven to eight ar− ticles averaging five pages, which can include spontaneous cases, post−mortem communications, or essays on aspects of medium− ship and survival, spiritual aspects of psi and survival, or on various altered states. Also publishes news concerning the Col− lege of Psychic Studies and other related organizations. There are 20 or so books reviewed per issue ranging in length from a paragraph to several pages. Also carries letters to the editor.

Style: Nontechnical.
Format: Good.

Selections: A collection of articles edited by Neville Armstrong under the title *Harvest of Light* was published by Neville Spearman in 1976.
Indexes: None, but it is projected.
Abstracts: *Parapsychology Abstracts International* (1987 to date).
Computerized databases: *PsiLine* (1987 to date).
Back issues: Some issues available from publisher at £1.50 each.
Reprinted editions: None.
Microforms: None.

838 *Proceedings of the Academy of Religion and Psychical Research*   *ISSN*: None
Vol. 1, 1980 to date
Published annually
Refereed: Yes. By Program Committee

Editor: Mary Carman Rose   *Publisher*:
Av. pages per issue: 88   Academy of Religion and
Size: 21.5 cm   Psychical Research [916]
Circulation: 350   Box 614
Price: Free to members of   Bloomfield, CT 06002
  the Academy (dues: $20.00)   *Tel.*: (203) 242−4593

Purpose: To publish the papers presented at the annual con−ference of the Academy of Religion and Psychical Research. Each conference is usually devoted to a specific theme. The themes thus far have been:
Mysticism and Other Transformative States of Consciousness (1980)
Psychical Research and Mysticism: An Interdisciplinary Sym−biosis (1981)
Holism in Religion and Psychical Research (1982)
The Synergy of Religion and Psychical Research: Past, Present and Future (1983)
Mysticism, Creativity and Psi: A Search for a New Science (1984)
Explorations in Psycho−Spiritual Transformations (1985)
The Relevance of Religion and Psychical Research to Planetary Concerns (1986)
Psychical Research and Spirit (1987)
Kundalini: Biological Basis of Religion and Genius? (1988)
Kundalini and the Paranormal (1989)

Contents: An average of 13 papers about 12 pages in length plus an editorial.

    Style: Nontechnical. Reference lists; tables.
    Format: Fair. Typed; photo offset.
    Indexes: Volumes for 1980–1982 had Proper Name and Sub—
ject indexes.
    Abstracts: There are journal abstracts for each paper. Also
abstracted in *Parapsychology Abstracts International* (1985 to
date).
    Computerized databases: *PsiLine* (1980 to date).
    Back issues: All available from publisher at $5 each. Cas—
settes also available.
    Reprinted editions: None.
    Microforms: None.

839 *Spiritual Frontiers* (Published as *Gateway* 1962–1968)
    Vol. 1, 1956 to Vol. 13, 1968.   *ISSN*: None
    n.s. Vol. 1, 1969 to date
    Published quarterly in Winter, Spring, Summer, and Fall
    Refereed: No

Editor: Frank C. Tribbe
Av. pages per issue: 62
Size: 22.8 cm
Circulation: 4,800
Price: Free to members; $12
  for nonmembers

Publisher:
Spiritual Frontiers Fellowship
  International [**924**]
P.O. Box 7868
Philadelphia, PA 19101
Tel.: None until 1990

    Purpose: To publish material in line with the purposes of
Spiritual Frontiers Fellowship, which is "to encourage and inter—
pret to the Churches, and receive interpretation from the
Churches of the rising tide of interest in mystical, psychical and
paranormal experience. SFF is **spiritual** in that it is concerned
with nonphysical phenomena which relate to God, the human
spirit, and the future life. It is **frontier** because it explores mat—
ters beyond the usual range of church worship and activity, the
paranormal. It is a **fellowship** of those who, having accepted
the validity of one or more of these phenomena, would en—
courage each other and ultimately the whole church to seek for
further light and greater reality in the spiritual life."

    Contents: Eight to ten articles per issue averaging seven
pages, some poetry, a research report by the editor, ap—
proximately 15 book reviews an average of two pages in length,
and letters to the editor.

Style: Nontechnical. Reference lists.
Format: Good.
Indexes: None.
Abstracts: None.
Computerized databases: None.
Back issues: Recent issues available at $5.00 apiece from publisher.
Reprinted editions: None.
Microforms: None.

840 *SRU Bulletin: Report of the Synchronicity Research Unit*
Vol. 1, 1976 to date   *ISSN*: None
Published quarterly in March, June, September, and Decem—ber (month may vary)
Refereed: No

Editors: Jeff C. Jacobs   *Publisher*:
       and J.A.G. Michels   Synchronicity Research
Av. pages per issue: 36   Unit [885]
Size: 29.6 cm   P.O. Box 7625
Circulation: 50   5601 JP Eindhoven
Price: $12.00   The Netherlands
                                   Tel.: (40) 413171

Purpose: Primarily a regular progress report of the research and activities of the Synchronicity Research Unit and is written by SRU staff members, with an occasional contribution by an invited guest author. Originally it was in the Dutch language with only occasional articles in English, but beginning in 1988 it became primarily an English—language journal.

Contents: Usually contains one theoretical paper and four to five research reports an average of six pages long. Also carries news items concerning the Synchronicity Research Unit and parapsychology in general. Lists of books and journals received, contents list of journals, one or two book reviews.

Style: Scholarly. Figure; graphs; reference lists; tables.
Format: Fair.
Indexes: None.
Abstracts: There are journal abstracts for most experimen—tal reports and some other articles. Also *Journal of Parapsychol—ogy*; *Parapsychology Abstracts International* (1988 to date).
Computerized databases: *PsiLine* (1988 to date; projected: 1981 to date).
Back issues: Some available from publisher at $3 each,

postage included.
    Reprinted editions: None.
    Microforms: None.

841 [337]   *Theta: Contemporary Issues in Parapsychology*
    No. 1, 1963 to date   *ISSN*: 0040–6066
    Published three times a year in Winter, Spring–Summer,
    and Autumn
    Refereed: No

Editors: William George Roll   *Publisher*:
Av. pages per issue: 48   Parapsychological Services
Size: 28 cm   Institute [**907**]
Circulation: 900   Department of Psychology
Price: Free to PSI members; $15.00 West Georgia College
   per year for nonmembers   Carrollton, GA 30118–0001
                     Tel.: (404) 836–6510

Purpose: Named after the first letter of the Greek word *thanatos* [death], *Theta* is oriented to the application and meaning of psychical and transpersonal experiences. It serves as a forum for studies of such topics as clinical parapsychology, outof–body and near–death experiences, reincarnation, haunting cases, and transcendent experiences.

    Contents: Each issue usually contains three or four articles discussing current research, parapsychological counseling, and other activities touching on the meaning and practical uses of parapsychology, plus four or five book reviews averaging three pages. A new feature, beginning in 1987, is a critical section entitled "Forum," which is comprised of an initial article written to provoke further reflection on a basic theme in parapsychology, followed by several brief responses.

    Style: Scholarly, yet nontechnical. Follows *Publication Manual of the American Psychological Association* (3rd ed.). Figures; illustrations; reference lists; tables.
    Format: Good.
    Indexes: None.
    Abstracts: *Journal of Parapsychology*; *Parapsychology Abstracts International* (1987 to date).
    Computerized databases: *PsiLine* (1963 to date).
    Back issues: Some issues available from publisher.
    Reprinted editions: None.
    Microforms: None.

PUBLICATIONS ON
ANOMALOUS PHENOMENA

In this category are journals, magazines, and newsletters that deal with anomalous phenomena in general (*Explorer* [845], *Journal of Scientific Exploration* [846], *Skeptical Inquirer* [849], *Zetetic Scholar* [850]) or areas of anomalous behavior [*Archaeus*, *Artifex* (bioenergetic fields), *Brain/Mind Bulletin* [844] (neurosciences), *New Realities* [847], and *Noetic Sciences Review* [848] (new age subjects)]. The *Journal of Scientific Exploration* and *Zetetic Scholar* [850] belong in any core collection. The *Skeptical Inquirer* is useful for its critical stance, and *Brain/Mind Bulletin* is an excellent information source on new findings. *Noetic Sciences Review* is aimed at the educated reader, whereas *New Realities* is aimed at the general public.

842 *Archaeus*   *ISSN*: 0895–125X
        Vol. 1, 1982 to date
        Published annually
        Refereed: Most articles refereed

Editor: Dennis Stillings   *Publisher*:
Av. pages per issue: 80   Archaeus Project [917]
Size: 28 cm   2402 University Ave.
Circulation: 450   St. Paul, MN 55414
Price: $20 per year [subscribers *Tel.*: (612) 781–5012 and
    also receive *Artifex* (843)]   641–0177

Purpose: To publish longer articles of a more serious and scholarly nature than those published in *Artifex* [843]. The articles deal with current and historical processes for the detection and measurement of bioenergic fields——electric, magnetic, and psychical——as reflected in holistic medicine, parapsychology, and Jungian thought.

Contents: Editorial, six or seven articles per issue averaging 20 pages, an occasional book review, letters to the editor, and notes on contributors.

Style: Scholarly. Bibliographic footnotes; figures; illustrations; tables.
    Format: Good.
    Indexes: None.
    Abstracts: Infrequent abstracts in journal. Also in *Para*–

psychology Abstracts International (1985 to date).
Computerized databases: *PsiLine* (1982 to date).
Back issues: Available from the publisher at $8.00 apiece ($10, overseas).
Reprinted editions: None.
Microforms: None.

xxx   *Archaeus Project Newsletter.* Superseded by *Artifex* [843]

843 *Artifex*   *ISSN*: 0895–125X
Vol. 1, 1982 to date (Vols. 1–3 were titled *Archaeus Project Newsletter*)
Published quarterly in Spring, Summer, Fall, Winter
Refereed: Some articles refereed

Editor: Dennis Stillings
Av. pages per issue: 56
Size: 27.6cm
Circulation: 400
Price: $20, with subscription to *Archaeus* [842]

Publisher:
Archaeus Project [917]
2402 University Ave.
St. Paul, MN 55414
Tel.: (612) 781–5012 and 641–0177

Purpose: To publish articles, reviews, and news concerning current and historical processes for the detection and measure—ment of bioenergic fields— -electric, magnetic, and psychical——as reflected in holistic medicine, parapsychology, and Jungian thought. Initially published as a newsletter, *Artifex* became a quarterly journal in 1988. Also aims to chronicle the activities of the Archaeus Project. Editorial policy published in *Artifex*, 1987, *6*(6), 31–32.

Contents: Three—four articles per issue averaging 10 pages. There is usually a lengthy editorial essay, detailed information on recent library acquisitions, conference news and reviews, notes of interest, annotated bibliographies, obituaries, and let—ters.

Style: Scholarly. Advertising; bibliographic footnotes; figures; illustrations; tables.
Format: Excellent.
Indexes: A list of contents of back issues is available.
Abstracts: *Parapsychology Abstracts International* (1987 to date).
Computerized databases: *PsiLine* (1987 to date).
Back issues: Available from publisher at $4 per issue ($5,

overseas).
  Reprinted editions: None.
  Microforms: None.

844 *Brain/Mind Bulletin*   *ISSN*: 0273—8546
  Vol. 1, 1975 to date.
  Published monthly
  Refereed: Not relevant

Editor: Marilyn Ferguson   *Publisher*:
Av. pages per issue: 8   Interface Press
Size: 27.8 cm   P.O. Box 42211
Circulation: 10,000   4717 N. Figueroa St.
Price: $35 per year   Los Angeles, CA 90042
                                  Tel.: (213) 223—2500

Purpose: To serve as a clearing house for information about individuals, organizations, publications, and meetings dealing with breakthrough findings and theories in brain science. Covers consciousness research, psychology, learning, creativity, holistic healing, parapsychology, culture, arts, the mind/body interface, human consciousness, and transformation.

Contents: An average of six brief articles usually summarizing recent published research reports, conference papers, or interviews, one or two book reviews per issue plus several brief mentions. Regular features are "Upcoming," a listing of upcoming programs, courses, and lectures dealing with aspects of consciousness research, and "Tools and Resources," a listing of books, films, periodicals, special issues, newsletters, and pamphlets on consciousness studies.

Style: Nontechnical. Bibliographic citations in text; figures; illustrations.
  Format: Excellent.
  Selections: Themepacks of articles available on "Parapsychology and Mysticism" covering Vols. I—II, III, and IV, respectively.
  Indexes: Five—year index available [689]. *New Periodicals Index* (ceased).
  Abstracts: Not applicable.
  Computerized databases: *PsiLine*.
  Back Issues: Available from publisher.
  Reprinted editions: None.
  Microforms: Vol. 1— (microfilm, microfiche) (UMI).

845 *The Explorer*   *ISSN*: None
Vol. 1, 1983 to date.
Published semi—annually in Spring and Autumn.
Refereed: No

Editor: Angela Thompson
Av. pages per issue: 16
Size: 28 cm.
Circulation: 400
Price: Free to members;
  subscriptions available to
  nonmembers via the
  Secretary, Lawrence Frederick

Publisher:
Society for Scientific
  Exploration [905]
ERL 306
Stanford University
Stanford, CA 94305
Tel.: (804) 924—4905

Purpose: To serve as the "newsletter of the Society for Scientific Exploration." It publishes news of the Society and of other organizations devoted to the scientific investigation of anomalous phenomena. In a position paper published in the first issue the Society (whose full name is the Society for Scientific Exploration Formed for the Study of Anomalous Phenomena) states that it uses "the term anomalous to characterize a phenomenon which appears to contradict existing scientific knowledge and/or which is generally regarded by the scientific community ... as being outside their established fields of inquiry." Some of the anomalies covered in the newsletter are cryptozoology, UFOs, and parapsychological phenomena such as PK and precognition.

Contents: Issues contain abstracts of papers given at the Society's annual convention, news items, four or five book reviews, obituaries, and biographical sketches of members, and, with *4*(3), a diary page.

Style: Scholarly. Illustrated.
Format: Excellent. Brown print on beige.
Indexes: None.
Abstracts: *Parapsychology Abstracts International* (1988 to date).
Computerized databases: *PsiLine* (1988 to date).
Back issues: Available at $2.50 per copy.
Reprinted editions: None.
Microforms: None.

xxx *Institute of Noetic Sciences Newsletter.* Superseded by *Noetic Sciences Review* [848]

846   *Journal of Scientific Exploration*   *ISSN*: 0892–3310
Vol. 1, 1987 to date.
Published semi–annually
Refereed: Yes

Editor: Ronald E. Howard   *Publisher*:
Av. pages per issue: 92   Pergamon Press, Inc. for the
Size: 22.8 cm   Society for Scientific
Circulation: 400   Exploration [905]
Price: Free to members;   Fairview Park
   $40, individuals; $75,   Elmsford, NY 10523
   institutions.   *Tel*.: (914) 592–7700

Purpose: A publication of the Society for Scientific Explora–
tion, the *Journal of Scientific Exploration* publishes scientific
papers "aimed at advancing the study of anomalous phenomena.
. . . Papers are considered for publication if they focus on any
aspect of anomalous phenomena, including phenomena outside
the current paradigms of one or more of the sciences, such as
the physical, psychological, biological, or earth sciences;
phenomena within the scientific paradigms, but at variance with
current scientific knowledge; the scientific methods used to study
anomalous phenomena; and the study of the impact of
anomalous phenomena on science and society in general."

Contents: Contains an average of five articles averaging 15
pages in length; correspondence, and an occasional book review.
Also contains information on the Society for Scientific Explora–
tion.

Style: Scholarly. Follows *Publication Manual of American
Psychological Association* (3rd ed.). Advertising; figures; graphs;
illustrations; reference lists; tables.
   Format: Excellent.
   Contents listing: Contents of Vols. 1 and 2 in Vol. 2, No. 2.
   Indexes: Author index to Vols. 1 and 2 in Vol. 2, No. 2.
   Abstracts: Abstracts for all articles. Also *Journal of Para–
psychology*; *Parapsychology Abstracts International* (1987 to date).
   Computerized databases: *PsiLine* (1987 to date).
   Back issues: Available from publisher.
   Reprinted editions: None.
   Microforms: Back issues in microform are available from
UMI.

847 *New Realities* (formerly *Psychic*, q.v.)   *ISSN*: 0147—7625
Vol. 1, 1977 to date
Published bimonthly.
Refereed: No

Editor: Neal Vahle
Av. pages per issue: 80
Size: 27.2 cm
Circulation: 20,000
Price: $22 individuals;
$25 institutions

Publisher:
Heldref Publications
4000 Albemarle St. NW
Washington, DC 20016
Tel.: (202) 362—6445

Purpose: This is one of the major "new age" journals as indicated by its subtitle: "Oneness of Self, Mind, and Body." It is not parapsychological but parapsychological subjects are often included. It serves as a useful barometer of new age trends.

Contents: Six or seven articles on new age topics, biographies of prominent persons, interviews. Features are en—titled "Exploring Your Dreams," "Tools for Transformation," "Profiles: Visions and Visionaries," and "Sights and Sounds" (reviews of new books, tapes, and videos).

Style: Popular. Advertising; bibliographic citations in text; figures; illustrations.
Format: Excellent. Glossy magazine.
Indexes: *New Periodicals Index*, *Magazine Index*.
Abstracts: *Parapsychology Abstracts International* (1988 to date).
Computerized databases: *PsiLine* (1977 to date).
Back issues: Some available from publisher.
Reprinted editions: Reprints (orders of 100 copies or more) of specific articles are available from publisher.
Microforms: Vol. 7 to date (UMI).

848 *Noetic Sciences Review*   *ISSN*: 0897—1005
No. 1, 1986 to date (Formerly the *Institute of Noetic Sciences Newsletter*)
Published quarterly in Spring, Summer, Autumn, Winter
Refereed: No

Managing Editor:
David A. Johnson
Av. pages per issue: 32
Size: 27.8 cm
Circulation: 20,000
Price: Free to members

Publisher:
Institute of Noetic Sciences [**923**]
475 Gate Five Road
Suite 300
Sausalito, CA 94965
Tel.: (415) 331—5650

(Associate membership: $35.00)

Purpose: To provide a high quality publication with broad appeal to encompass and express the ever−expanding spheres of activity and influence of the *Institute of Noetic Sciences* which addresses "what are deemed to be the most promising areas of the still unexplored frontiers of the human mind and spirit." Broadly speaking, these areas are exceptional abilities, health and healing, and societal transformation.

Contents: Four to six articles per issue averaging five pages, special reports on institute−sponsored research, an annual report on the Temple Award for Creative Altruism, and three to four book reviews, four paragraphs to two pages in length. Also covers news of the Institute and Books Received.

> *Style*: Nontechnical. Figures; illustrations; reference lists.
> *Format*: Excellent. Glossy magazine.
> *Indexes*: None.
> *Abstracts*: Brief summaries in Table of Contents. *Parapsychology Abstracts International* (1986 to date).
> *Computerized databases*: *PsiLine* (1986 to date).
> *Back issues*: Most issues available from publisher.
> *Reprinted editions*: None.
> *Microforms*: None.

849 *Skeptical Inquirer* (*SI*)   *ISSN*: 0194−6730
  Vol. 1, 1976 to date (Vol. 1−2[1] published under title *Zetetic*)
  Published quarterly in Spring, Summer, Fall, and Winter
  Refereed: Yes

Editor: Kendrick Frazier
Av. pages per issue: 112
Size: 23 cm
Circulation: 44,000
Price: $25.00 a year

Publisher:
Committee for the Scientific
  Investigation of Claims
  of the Paranormal [902]
P.O. Box 229
Buffalo, NY 14215−0229
Tel.: (716) 834−3222

Purpose: To serve as the official journal of the Committee for the Scientific Investigation of Claims of the Paranormal. It is intended to communicate scientific information about the many esoteric claims that have shown a growing influence upon the general public, educational curricula, and scientific institutions themselves, and to reflect a growing intelligent dialogue between those making claims for the paranormal and their critics.

Contents: An average of seven main articles per issue, some—times on different aspects of the same topic; several book reviews. Regular features are "News and Comment," "Notes of a Fringe Watcher" (Martin Gardner), "Psychic Vibrations" (Robert Sheaffer), "Forum," "From Our Readers," and two an—notated bibliographic listings, one covering recent books and the other articles. Approximately half of each issue is devoted to parapsychological topics.

Style: Scholarly. Figures; graphs; illustrations; reference lists; tables.
Format: Good.
Indexes: Published a Ten—Year Index [692]; *Popular Periodicals Index*. There is a partial list of the contents of all past issues in each issue.
Abstracts: *Linguistics and Language Behavior Abstracts*; *Parapsychology Abstracts International* (1987 to date); *Sociological Abstracts*.
Computerized databases: *PsiLine* (1987 to date).
Back issues: Available from publisher at $6.00 each (Vol. 1(1) through 2(2): $7.50 each).
Reprinted editions: None.
Microforms: None.

xxx   *Zetetic*. Superseded by *Skeptical Inquirer* [849]

850 *Zetetic Scholar: Journal of the Center for Scientific Anomalies Research (ZS)*   *ISSN*: 0741—6229
Vol. 1, 1978 to date
Published irregularly, but approximately once a year.
Refereed: Unknown

Editor: Marcello Truzzi
Av. pages per issue: 148
Size: 12.5 cm
Circulation: 600
Price: $15 for two issues to individuals; $20 for institutions and foreign

Publisher:
Center for Scientific Anomalies Research [901]
c/o Marcello Truzzi
Dept. of Sociology
Eastern Michigan University
Ypsilanti, MI 48197
Tel.: (313) 663—8823

Purpose: To provide an independent scientific review of

claims of anomalies and the paranormal. Emphasis is placed on independent opinion. Opinions expressed are not necessarily those of the Center. It attempts to create a continuing dialogue between proponents and critics of the paranormal, and it is con— cerned mainly with enhancing communications, but not only with the adjudication of the claims but with the sociology of the disputes themselves. *Zetetic Scholar* seeks to balance science's proper skepticism towards extraordinary claims with its need for objectivism and fairness. Finally, it seeks to help the scientific community reach rational judgments based upon the empirical facts.

Contents: Average of four articles per issue averaging nine pages. A special feature in each issue is a ZS Dialogue which consists of a position paper with commentaries by other authors with response by the original author. Dialogues are carried over to succeeding issues. Each issue has an editorial, letters, a ran— dom bibliography on the occult and paranormal, average of two book reviews a page in length, short notes on approximately 60 books and information on contributors.

Style: Scholarly.
Format: Fair. Photo offset.
Indexes: None.
Abstracts: *Parapsychology Abstracts International* (1987 to date); *Sociological Abstracts*.
Computerized databases: *PsiLine* (1987 to date).
Back issues: $8 apiece from publisher in reduced photocopy
Reprinted editions: None.
Microforms: None.

NEWSLETTERS

Except for *Applied Psi* [853], which is devoted to the sub— ject specialty of its title, the remaining newsletters in this group are house organs that also contain materials aimed at or of in— terest to the wider parapsychology community. The *Anthropol— ogy of Consciousness Newsletter* [852a] is written by and for the members of the Society for the Anthropology of Consciousness, for whom the interface between anthropology and parapsychol— ogy is a central concern. The *ASPR Newsletter* [854] and the *SPR Newsletter* [856] contain news items about their respective parent organizations, but the former also carries articles and news items about the field of parapsychology as a whole, and the latter also carries news items about the United Kingdom. The same can be said of the *A.I.P.R. Bulletin* [852], which is also the major source of information about parapsychology "down under" as well as information about the Institute. *FRNM Bulletin* [855] provides information on what is happening at the Foundation for Research on the Nature of Man, but it also has news of general parapsychological interest.

xxx *AASC Quarterly.* Superseded by *Anthropology of Consciousness Newsletter* [852a]

852 *A.I.P.R. Bulletin*   *ISSN*: 0813-2194
No.1, 1983 to date
Published triannually since 1988
Refereed: No

Editor: Michael Hough   *Publisher*:
    (Nos. 1-12)   Australian Institute of Para-
    Harvey Irwin   psychological Research [909]
    (No. 13 on)   P.O. Box 445
Av. pages per issue: 16-20   Lane Cove NSW 2066
Size: 28cm   Australia
Circulation: 300   *Tel.*: (02) 660 7232
Price: Free to members;
    $12 per year

Purpose: As the official organ of the Australian Institute of Parapsychological Research, the *A.I.P.R. Bulletin* contains information about the Institute and its activities in particular, and more broadly, about parapsychology in Australia. It also informs readers about current subjects of parapsychological interest. Beginning with No. 13, more articles by parapsychologists overseas will be published.

Contents: One or two lead articles, "Research Notes," "Recommended Reading," "Notes & News," "Letters," and two to four book reviews.

Style: Nontechnical. Figures; illustrations; reference lists.
Format: Good (from No. 11 on).
Indexes: Retrospective index projected.
Abstracts: *Parapsychology Abstracts International* (1987 to date).
Computerized databases: *PsiLine* (1987 to date; projected: 1983 to date).
Back issues: Available from publisher.
Reprinted editions: None.

852a *Anthropology of Consciousness Newsletter* [Formerly *AASC Quarterly*, *AAAS Newsletter*, *Newsletter of the Association for Transpersonal Anthropology*, and before that the *Newsletter for the Anthropological Study of Paranormal and Anomalous Phenomena*]
Vol. 1, 1990 to date
Published quarterly in March, June, September, and December
Refereed: No

Editor: Joseph K. Long   *Publisher*:
Av. pages per issue: 20   Society for the Anthropology
Size: of Consciousness [905a]
Circulation: 150   Dept. of Anthropology
Price: Plymouth State College
  (Dues: Members, $75; Associ- Plymouth, NH   03264
  ates, $45; includes membership
  in the the American Anthropo-
  logical Assn. and the *Newsletter*;
  Subscription only: $25.)

Purpose: To publish articles about and disseminate information on the anthropological study of consciousness, or transpersonal anthropology. Subjects covered are altered states of consciousness, the ethnography of shamanic, spiritual and magical training and initiations, indigenous healing practices, divination, linguistic and symbolic studies of myth and consciousness, and psychic archaeology, other applied aspects of parapsychology, and anomalous human abilities.

Contents: One or two articles, one of them sometimes being an autobiographical statement of a member's interest and activities in transpersonal anthropology. Some are brief research reports. News on transpersonal anthropology in general and news concerning the Society for the Anthropology of Consciousness, There are occasional book reviews, conference notes, bibliographies, and other lists of information sources of interest.

Style: Nontechnical. Follows *Publication Manual of the American Psychological Association* (3rd ed.), modified. Illustrations; tables.
Format: Fair.
Indexes: Contents for each volume published in December issue as "Index."
Abstracts: *Parapsychology Abstracts International* (1988 to date).

Computerized databases: *PsiLine* (1988 to date; projected: 1985 to date).
Back issues: Available from publisher at $5 apiece.
Reprinted editions: None.
Microforms: None.

xxx *Applied Psi Newsletter*. Superseded by *Applied Psi* [856]

853 [335]   *ASPR Newsletter*   *ISSN*: 0044—7919
No. 1, 1968 to date
Published quarterly in Spring, Summer, Autumn, and Winter
Refereed: No

Editor: Donna L. McCormick   *Publisher*:
Av. pages per issue: 6   American Society for
Size: 28 cm   Psychical Research [869]
Circulation: 2,000   5 West 73rd St.
Price: free to members; $10   New York, NY 10023
  to nonmembers   *Tel.*: (212) 799—5050

Purpose: To keep members and laypersons informed of the activities and opportunities of the Society and of scientific parap—sychology in general. Also, to make research reports more easily understood by readers lacking the necessary technical vocabulary. To this end, short rewrites of JASPR articles in layman's language occasionally appear in the *Newsletter*.

Contents: Each issue usually contains a brief lead article, fol—lowed by news of ASPR activities and parapsychological news in general.

Style: Nontechnical.
Format: Good.
Selections: Two anthologies of articles from the *ASPR Newsletter* have been published by the ASPR: *Exploring ESP and PK: Selections from the Newsletter with Added Material* (1976) and *Exploring Parapsychology: A New Collection from ASPR Newsletter Including a Section on Education in Parapsy—chology* (1980), compiled by Marian Nester and Arthur S.T. O'Keefe.
Indexes: Cumulative table of contents 1976 to date.
Abstracts: *Parapsychology Abstracts International* (1988 to date).
Computerized databases: *PsiLine* (1988 to date; projected:

1968 to date).
Back issues: Some available from publisher.
Reprinted editions: None.
Microforms: None.

854 [339]   *FRNM Bulletin*   *ISSN*: None
No. 1, Summer 1965 through No. 14, Autumn 1969 [No. 15
published as *Parapsychology Bulletin*]; see (**861**)], Nos. 21–
26 published in *Journal of Parapsychology* [**825**]. No. 27,
1982 to date.
Published quarterly in Spring, Summer, Autumn, Winter.
Refereed: No

Editor: Linda Vann   *Publisher*:
Av. pages per issue: 4   Parapsychology Press [**870**]
Size: 27.8 cm   Box 6847, College Station
Circulation: 500   Durham, NC 27708
Price: Free to members of   *Tel.*: (919) 688–8241
the FRNM Association

Purpose: Similar to that of the *Parapsychology Bulletin*, but
geared to communicate news of FRNM rather than of the Duke
University Parapsychology Laboratory. It also provides "glimpses
of research development, interim reports, and special items that
give some indication of the trend and scope of current
inquiries."

Contents: An average of five news items or articles per
issue.

Style: Nontechnical.
Format: Excellent.
Indexes: None.
Abstracts: *Parapsychology Abstracts International* (1988 to
date).
Computerized databases: *PsiLine* (1988 to date; projected:
1965 to date).
Back issues: Some are available from FRNM.
Reprinted editions: None.
Microforms: All issues available (UMI).

xxx   *Newsletter for the Anthropological Study of Paranormal
and Anomalous Phenomena.* Superseded by *Anthropology of
Consciousness Newsletter* [**852a**]

xxx *Newsletter of the Association for Transpersonal Anthropology.* Superseded by *Anthropology of Consciousness Newsletter* [852a]

855 *SPR Newsletter*   *ISSN*: None
No. 1, 1981 to date
Published quarterly in January, April, July, and October
Refereed: No

Editor: Susan J. Blackmore   *Publisher*:
Av. pages per issue: 6   Society for Psychical
Size: 29.6 cm   Research [874]
Circulation: 1150   1 Adam and Eve Mews
Price: Free to members   London   W8 6UG
                                     England, U.K.
                                     Tel.: 071–937–8984

Purpose: To serve as a channel of communication within the Society for Psychical Research, with emphasis on current and future activities and developments. A *Supplement* presents experiences sent by members and others and correspondence con— cerning them.

Contents: Occasional one—two page articles. Mostly paragraph length notices of events under "Conferences," "Societies," "Grants," "Obituaries," "New publications," etc. The SPR Newsletter Supplement edited by Renée Haynes was initiated in 1985 and consists of two or three pages of verbatim case accounts plus comments by the editor and others.

Style: Informal. Illustrations.
Format: Fair. Beginning in 1989, very good.
Indexes: None.
Abstracts: None.
Computerized databases: *PsiLine* (Projected: 1981 to date).
Back issues: Some issues available from publisher.
Reprinted editions: None.
Microforms: None.

DISCONTINUED PUBLICATIONS

The journals and newsletters in this section are for the most part quite different, their only point in common being

that they are no longer published. They are included, nonethe—
less, because they are still referred to and because the standard
reference books on periodicals do not include discontinued publi—
cations. Therefore, a specialized work such as this is the only
likely place to find out about them. (Most of them are cited in
the book review references in the first chapter on books.)

Because they are no longer published, information on the
ISSN, circulation, price, and telephone has been omitted. An ex—
ception is made in the case of *Psi Research* [865], which will
continue publication if funding permits. The Parapsychological
Association has announced that it will revive *Psi News* [864]
with Charles Honorton as editor, but as we go to press it has
not yet become a reality. Also, as we go to press, notice been
received that *Parapsychology Review's* [829] last issue was No.
3—4, 1990.

The *International Journal of Parapsychology* [857] was an
innovative journal that gave full recognition to the interdiscipli—
nary and international scope of parapsychology. Authors of its
articles were from many countries and it carried lengthy
abstracts of the main articles in French, German, Italian, and
Spanish. It also stressed theoretical papers (as did another jour—
nal in this group: the *Journal of Research in Psi Phenomena*
[858]), as well as historical, anthropological, sociological
biographical, and literary approaches to the paranormal.
Whereas the *IJP* had emphasized Western European parapsy—
chology, *Psi Research* extended international coverage to Eastern
Europe and Asia with the emphasis on the U.S.S.R., other East
European countries, and the People's Republic of China. *Psychic*
[866] was also a pioneer publication——the first glossy magazine
aimed at reaching the general public with articles on mainline
parapsychology. The *Research Letter* [868] emphasized research
reports as did the *Parapsychological Journal of South Africa*
[860] (which was discontinued when its editor and a major con—
tributor, Vernon Neppe, moved to the United States).

Another regional journal, *New Horizons* [859], served as a
vehicle for reporting the research of the Toronto Society for
Psychical Research and New Horizons Research Foundation.
Phoenix [862] pioneered as a journal about anthropological
parapsychology (or parapsychological anthropology), but the im—
petus for it vanished with the untimely death of Philip
Staniford, the dynamic leader of the association that published
it. (It is continued in part by the *Anthropology of Consciousness
Newsletter* [852a].)

The remaining publications also became reincarnated in
other publications: *Applied Psi* will be replaced by *Intuition*; the
Newsletter of the Parapsychology Foundation became *Para—
psychology Review* [829], the *Proceedings of the Parapsychological*

Association [863] became *Research in Parapsychology* [832], *Parapsychology Bulletin* [861] became the *FRNM Bulletin* [855] and then *Parapsychology Bulletin* again and is now once more *FRNM Bulletin* [855]. *Psychoenergetics* [867] has become *Theoretical Parapsychology* [833].

856 *Applied Psi*   *ISSN*: None
    Vol. 1, 1982 to 1987 (Vols. 1−2 were published under the
    title *Applied Psi Newsletter*). To be superseded by a
    magazine entitled *Intuition*.
    Published quarterly.
    Refereed: No

Editor: William H. Kautz   *Publisher*:
Av. pages per issue: 16   Center for Applied
Size: 28 cm   Intuition [920]
Circulation: 300   2046 Clement Street
Price: $18, individuals; $32,   San Francisco, CA 94121
    institutions   *Tel.*: (415) 221−1280

Purpose: To explore "the practical applications and social implications of psi phenomena in science, technology, business and the helping professions" and to serve as a vehicle for disseminating and exchanging information about practical applications of psi.

Contents: Three−four article two−three pages long. Occasional book reviews. Also carries interviews and occasional bibliographies. Some news items.

Style: Popular. Figures; illustrations; reference lists.
Format: Good.
Indexes: Brief index to preceding volume in first issue of a new volume.
Abstracts: *Parapsychology Abstracts International* (1986 to 1987).
Computerized databases: *PsiLine* (1986 to 1987; projected: 1982 to 1987).
Back Issues: Available from publisher.
Reprinted editions: None.
Microforms: None.

857 [340]   *International Journal of Parapsychology*
    Vol. 1, 1959 to Vol. 10, 1968
    Published quarterly in Winter, Spring, Summer, and

Autumn
Refereed: Yes

Editor: Martin Ebon
Av. pages per issue: 108
Size: 22.5 and 28 cm

Publisher:
Parapsychology Foundation [873]
228 East 71st Street
New York, NY 10021

Purpose: Established to act "as a forum for scholarly in—quiry, linking parapsychology with psychology, physics, biochemistry, pharmacology, anthropology, ethnology, and other scientific disciplines, the *International Journal of Parapsychology* publishes contributions dealing with the total nature of the mind."

Contents: An average of six articles per issue averaging 15 pages in length, consisting of general and critical review papers, theoretical contributions, exploratory essays, some experimental reports, and some book review articles. About six book reviews per issue, averaging three pages.

Style: Scholarly. Figures; graphs; illustrations; reference lists; tables.
Format: Good.
Selections: *The Psychic Force* edited by Allan Angoff [81] is composed of articles originally published in the *International Journal*.
Indexes: A resume of each article is given in French, German, Italian, and Spanish. Separate indexes were published for Vols. 1–3, Vol. 4, Vol. 5, Vol. 6, Vol. 7, and Vols. 8–10.
Abstracts: *Journal of Parapsychology*; *Parapsychology Abstracts International* (complete).
Computerized databases: *PsiLine* (complete).
Back issues: None.
Reprinted editions: None.
Microforms: Vols. 1–10, microfilm (AMS).

858 *Journal of Research in Psi Phenomena*
Vol. 1, no. 1, 1976 – Vol. 2, no. 1, 1977.
Three issues published.
Refereed: Unknown

Editor: J. Bigu
Av. pages per issue: 76
Size: 12 1/2 cm

Publisher:
Kingston Association for
    Research in Parascience
P.O. Box 141

Kingston, Ontario
K7L 4V6   Canada

Purpose: Intended (1) to allow and encourage the publica—
tion of fundamental theoretical research, leading to experimen—
tally testable predictions, concerned with the nature of extrasen—
sory perception, psychokinesis and other psi phenomena, (2) to
stimulate experimental work aimed at the testing of these
theoretical predictions, (3) to encourage a multi—disciplinary
scientific approach to the study of psi phenomena.

Contents: Average of four full—length theoretical/
methodological/experimental articles per issue an average of 18
pages long, an editorial, and occasional book reviews.

Style: Scholarly. Figures; graphs; reference lists; tables.
Format: Fair.
Indexes: None.
Abstracts: Journal abstract for each article.
Computerized databases: *PsiLine* (complete).
Back issues: No response.
Reprinted editions: None.
Microforms: None.

859   *New Horizons*
Vol. 1, 1972 — Vol. 2, 1977.
8 issues published.
Published irregularly.
Refereed: No

Editor: A.R.G. Owen
Av. pages per issue: 44
Size: 23 cm

Publisher:
New Horizons Research
   Foundation [904]
P.O. Box 427
Station F
Toronto, Ontario
Canada M4Y 2L8

Purpose: To publish reports on some of the investigations
made by the New Horizons Research Foundation or by persons
working in close association with it. The New Horizons Research
Foundation is a nonprofit organization whose purpose is to
promote research on the frontiers of science and to disseminate
information. It incorporates the Transactions of the Toronto
Society for Psychical Research (terminated in 1987).

Contents: Four—12 articles per issue averaging four pages, primarily research reports on all aspects of parapsychology and paraphysics, with emphasis on psychokinesis. Also an editorial and news of the Toronto Society for Psychical Research and New Horizons Research Foundation.

Style: Scholarly. Figures; graphs; illustrations; reference lists; tables.

Format: Good.

Special issues: Vol. 1(5), 1975 consisted of the Proceedings of the First Canadian Conference on Psychokinesis and Related Phenomena held in June, 1974.

Indexes: None.

Abstracts: Brief abstracts of each article.

Computerized databases: *PsiLine* (complete).

Back issues: Available from New Horizons Research Foundation for postage and fee (please inquire).

Reprinted editions: None.

Microforms: None.

xxx *Newsletter of the Parapsychology Foundation.* Superseded by *Parapsychology Review* [829]

860 *Parapsychological Journal of South Africa (PJSA)*
        Vol. 1, 1980 to Vol. 7(1) 1986
        Published semiannually in June and December. Publication
        suspended——perhaps permanently.
        Refereed: Yes

Editor: Vernon M. Neppe
Av. pages per issue: 80
Size: 21 cm
Circulation: 400

Publisher:
South African Society for
    Psychical Research [911]
    in Conjunction with the
    South African Institute for
    Parapsychological Research
P.O. Box 23154
Joubert Park 2044
Republic of South Africa

Purpose: The first regular publication complying with international standards in the field of parapsychology in South Africa, this journal reflects the development of ideas and research in this area, allowing local progress in this scientific discipline to become known throughout the world. Contributions, therefore, will derive predominantly from within Southern Africa, but suitable papers from further afield dealing with any aspect of parapsychology are welcomed, as well as correspondence. Any well—attested information bearing on the area of parapsychology whether describing positive or negative results is considered for publication.

Contents: Each issue contains an editorial, an average of five articles, averaging 20 pages. Also contains correspondence and an occasional book review. The "Catalogue of the Parapsy—chological Collection in the University of Witwatersrand Library" (books and journals) was published in 2(1), 1981, and it is updated by supplements in later issues.

Style: Scholarly. Figures; graphs; illustrations; reference lists; tables. Key words supplied for all articles. Follows the rules of the Instructional Steering Committee of Medical Editors, May 2, 1979.
Format: Fair.
Indexes: None.
Abstracts: Journal abstracts for most articles. Vols. 4—5 in *Journal of Parapsychology*. Also *Psychological Abstracts* (1980 to 1984).
Computerized databases: *PsiLine* (complete), *PsychINFO* (1980 to date).
Back issues: Limited number of copies available from pub—lisher.
Reprinted editions: None.
Microforms: None.

861 [342]   *Parapsychology Bulletin (PB)*
   No. 1, 1946 through No. 72, 1965 and new series No. 15, 1970 through No. 20, 1971. Between 1965 and 1970 pub—lished as *FRNM Bulletin* [855].
   Published quarterly in February, May, August, and Decem—ber.
   Refereed: No.

Editor: Dorothy H. Pope   *Publisher*:
Av. pages per issue: 4—6   Parapsychology Press [870]
Size: 22.8 and 28 cm   Box 6847, College Station
                                                Durham, NC 27708
                                                Tel.: (919) 688—8241

Purpose: Initiated to bring parapsychological research closer to the general reader and "to help bring together the family of parapsychologists everywhere, many of whom ... know [each other] only through the formal medium of their scientific publications." These functions have now been taken over by a new section in the *Journal of Parapsychology* entitled: "News and Comments."

Contents: An average of six articles a paragraph or two long. Regular features are "Events," "Personnel," and very brief reviews of books and periodical articles (mostly in non—parapsychological publications).

  Style: Informal. Bibliographic citations in text; illustrations.
  Format: Good.
  Indexes: None.
  Abstracts: None.
  Computerized databases: PsiLine (projected: 1966−1971).
  Back Issues: Some are available from FRNM.
  Reprinted editions: None.
  Microforms: The first series is available from University Microfilms (Ann Arbor, MI 48106) with the *Journal of Parapsychology* [825].

862 *Phoenix: Journal of Transpersonal Anthropology*
  Vol. 1, 1977 to Vol. VIII, nos. 1−2, 1985
  Published semi−annually.
  Refereed: Unknown

Managing   *Publisher*:
  Editor: Shirley W. Lee   Association for
Av. pages per issue: 144   Transpersonal Anthropology
Size: 27.8 and 20.8 cm   (defunct)

Purpose: To explore and examine the paranormal, the psychic, and consciousness itself—and to some extent their rela—tion to culture in accord with the new field of transpersonal anthropology, which combines anthropology, parapsychology, and transpersonal psychology. As such it is the official journal of the Association for Transpersonal Anthropology (defunct). To promulgate the position that in the study of transpersonal anthropology objective science must be coupled—as far as possible—with experience.

Contents: An average of eight articles averaging 10 pages per issue; an editorial; several book reviews, occasional poetry, and information concerning the Association for Transpersonal Anthropology.

Style: Scholarly. Figures; graphs; illustrations; reference lists; tables.
Format: Good.
Indexes: None.
Abstracts: Abstracts for some articles.
Computerized databases: *PsiLine* (Vol. 1, 1977 to Vol. 8, 1985).
Back issues: Available from publisher at 2/3 original price.
Reprinted editions: None.
Microforms: None.

863 [333]   *Proceedings of the Parapsychological Association*
No. 1, 1966 to 1971. In 1972 became *Research in Parapsychology* [**832**]
Published annually
Refereed: by program committee of the Parapsychological Association.

Editors: W.G. Roll,   *Publisher*:
   R.L. Morris, J.D. Morris   Parapsychological
Av. pages per issue: 124   Association [**872**]
Size: 21.5 cm   P.O. Box 12236
                                       Research Triangle Park, NC
                                       27706

Purpose: To provide a published record of the annual convention of the Parapsychological Association (323), the professional parapsychological society.

Contents: No. 1 summarizes the conventions held from 1957–1964. Beginning with No. 2, each volume contains shortened versions of the papers delivered, arranged by subject; the complete text of the presidential address; and the complete text of the invited dinner address, usually given by an outstanding scholar from another field.

Style: Scholarly. Illustrated.
Format: Excellent.
Indexes: Each number has a combined subject index and glossary, and a name index. The last volume, No. 8, 1971 has cumulative indexes for all 8 volumes.
Abstracts: Not relevant.
Computerized databases: *PsiLine*.
Back issues: Inquiries to W.G. Roll, Parapsychological Services Institute [**902**].
Reprinted editions: None.
Microforms: None.

864 *Psi News: Bulletin of the Parapsychological Association*
Vol. 1, July 1978 to Vol. 5, January 1982.
Published quarterly in January, April, July, and October.
Note: The Parapsychological Association has announced
plans to publish this newsletter again.
Refereed: No

Editor: *Publisher*:
Vols. 1(1) − 3(2):   Parapsychological
   William G. Roll   Association [872]
Vols. 3(3) − 5(1):   P.O. Box 12236
   Hoyt Edge.   Research Triangle Park, NC
Av. pages per issue: 4   27709
Size: 11 cm

Purpose: To serve as the official bulletin of the Para−
psychological Association, the international society of professional
parapsychologists. To provide current information on the scien−
tific and educational parapsychological activities.

Contents: Brief articles, 1/2 to 1&1/2 pages in length on
issues and activities in parapsychology; news about education in
parapsychology; information on the Parapsychological Associa−
tion: its origin, goals, purpose, current activities; news releases
of current research; upcoming events; book reviews; correspon−
dence.

Style: Nontechnical.
Format: Good. Illustrations; reference lists.
Indexes: None.
Abstracts: None.
Computerized databases: PsiLine.
Back issues: Vol. 1, No. 2−Vol. 4, No. 4 available at $7.00
for entire set from Hoyt Edge, Rollins College, Dept. of
Philosophy and Religion, Campus Box 2659, Winter Park, FL
32789−7499.
Reprinted editions: None.
Microforms: None.

865 *Psi Research: An East−West Journal on Parapsychology,*
Psychotronics and Psychobiophysics   *ISSN*: 0749−2898
Vol. 1 1982 to Vol. 4(3/4) Sep/Dec 1986.
Published quarterly in March, June, September, and Decem−
ber. Publication suspended pending funding.

Refereed: Some articles refereed

Editor: Larissa Vilenskaya
Av. pages per issue: 124
Size: 21.2 cm
Circulation: 450
Price: $28

Publisher:
Psi Research [**891**]
484—B Washington St., #317
Monterey, CA 93940
Tel.: (408) 655—1985

Purpose: *Psi Research* is an outgrowth of Larissa Vilenskaya's *Parapsychology in the USSR*. It was initiated by a group of researchers affiliated with the Foundation for Human Science and Washington Research Center to encourage an inter— national exchange of information, views, opinions, and research results in parapsychology, psychotronics, and psychobiophysics. It is now produced by the editor. Emphasis is on practical ap— plications resulting from advances in research directed to the understanding of human nature, consciousness, psi, and intercon— nections in biological systems. New approaches, methodologies, and research paradigms are encouraged. A special effort is made to present an international array of authors, but all material is in English.

Contents: Contains an average of eight articles per issue, five pages in length and an average of three book reviews averaging three pages in length. several book reviews; section of "news and views," "suggestions for further reading," and often a section of translations from the Soviet press or summaries of material published in foreign language parapsychology journals.

Style: Semitechnical. Figures; graphs; illustrations; reference lists; tables.
Format: Fair.
Indexes: Annual.
Abstracts: *Parapsychology Abstracts International* (1986 to date); *Psychological Abstracts* (1982 to date).
Computerized databases: *PsiLine* (1986; projected: 1982 to 1986); *PsychINFO* (1982 to 1986).
Back issues: Available from publisher at $8 apiece.
Reprinted editions: None.
Microforms: None.

866 [338]   *Psychic*. Became *New Realities* [**847**]
Vol. 1, 1969 to Vol. 7(6), 1976
Published bimonthly.
Refereed: No

Editor: James G. Bolen   *Publisher*:
Co—editor: Alan Vaughan   James G. Bolen
Av. pages per issue: 48   680 Beach Street
Size: 27.5 cm   San Francisco, CA 94109

Purpose: Initiated, writes Bolen, because of "the field's publi—
cation gap at the popular level...." The function of *Psychic* is to
"bring forward interesting and assorted material of every perspec—
tive and from every corner, fairly and forthrightly. The conclu—
sions, if any, will be left to the venturesome reader."

Contents: Regular features include an interview with a
prominent person associated with parapsychology. In addition
there are three or four articles; a section entitled "Phenomena"
(mostly case reports); "News Ambit" (brief items of interna—
tional scope); "Comments" (readers' opinions); book reviews, and
a list of books and articles relevant to the interviews and ar—
ticles in each issue.

Style: Popular. Figures; illustrations; references listed at
back.
Format: Excellent. Glossy magazine.
Indexes: None.
Abstracts: None.
Computerized databases: *PsiLine* (Vol. 1, 1969 to Vol. 7(6),
1976).
Back issues: None.
Reprinted editions: None.
Microforms: None.

867 *Psychoenergetics: The Journal of Psychophysical Systems*
Vol. 1, 1974 to 1985   *ISSN*: 0278—6060
Vols. 1—3 were under the title *Psychoenergetic Systems: An
International Journal*. Four issues published per volume as
material is available. Continued as *Theoretical Parapsy—
chology* [**833**]
Refereed: Yes

Editor: Stanley Krippner   *Publisher*:
   followed by Derek Lawden   Gordon & Breach Science
Av. pages per issue: 120   Publishers Ltd.
Size: 22.6 cm   PO. Box 197
Circulation: Unknown   London WC2E 9PX
Price: $164 individuals; $266   England, U.K.
   institutions   *Tel.*: 071—836—5125

Purpose: This journal was initially edited by Stanley Kripp—ner under the title *Psychoenergetic Systems*, and under that title it explored the relationship among consciousness, energy, and matter using a general systems theory approach. Under the present title and the editorship of Derek Lawden, it published selected original papers in English describing contemporary research into the behavior of psychophysical systems. A psychophysical system is defined as any assembly of elements whose behavior cannot be explained by appeal to the known physical modes of interaction and where the presence of mind or consciousness or other psychic factors within the system sug—gests the existence of an additional psychic mode of interaction between its elements. All papers were refereed by one of the consulting editors or by other authorities appointed by the editor.

Contents: Averages five articles per issue averaging 10 pages in length, which may be theoretical papers or accounts of ex—perimental or other scientific investigations of psychic phenomena. Occasional commissioned articles survey the current state of knowledge over a restricted field. Each issue contains several book reviews, an average of two pages in length, primarily by the editor. Published papers, or progress reports on research currently being conducted, are presented in a special section and given high priority. It is an affiliated journal of the Parapsychological Association.

Style: Scholarly. Figures; graphs; illustrations; reference lists; tables.
Format: Excellent.
Indexes: None.
Abstracts: Abstracts for most articles.
Computerized databases: PsiLine (1974 to 1985).
Back issues: None.
Reprinted editions: None.
Microforms: Microfiche and microfilm available from pub—lisher.

868 *Research Letter*
   No. 1, 1971 to No. 12, 1984
   Published irregularly
   Refereed: Yes

Editor: Martin Johnson   *Publisher*:
Av. pages per issue: 64   Parapsychology Laboratory
Size: 23.8 cm   University of Utrecht [881]

Price: Sent free to subscribers   Sorbonnelaan 16
to the *European Journal*   3584 CA Utrecht
of Parapsychology and   The Netherlands
members of the Parapsycho‐
logical Association.

Purpose: To present information about ongoing research ac‐
tivities in the Utrecht Parapsychology Laboratory or from other
research groups, and to provide an opportunity for researchers,
who for some reason are not able to publish a manuscript in
one of the regular journals. In this way it is hoped to stimulate
research activities as it is often frustrating, especially for
younger workers, when they are not able to communicate
reports of their activities because they do not meet the stan‐
dards required by professional journals. The *Research Letter* is
also a means of communicating deviant or controversial opinions
or theories which, although we might not agree with them,
merit publication because some people might be interested in
them. The editors are responsible only for contributions that
state University of Utrecht after the author's name.

Contents: An average of six experimental or methodological
papers, an average of 10 pages in length.

Style: Scholarly. Figures; graphs; reference lists; tables.
Format: Fair.
Indexes: None.
Abstracts: Abstracts for some articles. Some items abstracted
in *Journal of Parapsychology*.
Computerized databases: *PsiLine* (complete).
Back issues: Some issues available from publisher.
Reprinted editions: None.
Microforms: None.

III

Parapsychological Organizations

QUICK INDEX TO SECTIONS IN THIS CHAPTER

ALPHABETICAL LIST OF ORGANIZATIONS

Unitarian Universalist Psi Symposium (UUPS), **925**
West Georgia College, **897**

A total of 58 organizations are described here. Fourteen of the 15 that were described in *Sources* have been updated for purposes of this volume. The 15th organization was the Dream Laboratory of the Maimonides Medical Center, which was reborn as Psychophysical Research Laboratories [**884**]. However, by the time this book appears, PRL may no longer exist due to lack of funding. The same fate may soon befall the Parapsychology Laboratory of the University of Utrecht. Another organizations removed from this list because it will close in two months (September 1989) is the Science Unlimited Research Foundation, which has conducted several worthwhile research projects reported by Gary Heseltine and Rick Berger.

No attempt has been made to list all the parapsychological organizations in existence. The primary criterion for the inclusion of an organization is that its main purpose must be the advancement of parapsychology as a science. To this end, research or the results of research must be the focal point of a group's interest, although its expression need not be in engagement in actual research but can range from theorizing about it, supporting it financially, educating others about it, or criticizing it. The list consists primarily of organizations in English—speaking countries because of space and time limitations and availability of information, although a few of the better known research centers are included. To reflect the fact that the organizations were formed for a variety of purposes, they are arranged in nine categories: General Parapsychology, Research Institutions, Information Centers, Educational Institutions, Anomalies Research Institutions, Counseling Centers, Regional Organizations, Organizations Interested Principally in Survival of Death, Religious/Human Potential Organizations with Interests in Psi. An asterisk appears after the name of those organizations in which at least one member of the Parapsychological Association [**872**] has an important role as an officer or board member. An exception is the Committee for the Scientific Investigation of Claims of the Paranormal, many of whose members have a negative bias toward parapsychology. Nevertheless, CSICOP was included because I feel that it is important for the advancement of parapsychology as a science to attend to all sides of a given question. The organizations are arranged alphabetically within each of the nine subcategories.

GENERAL PARAPSYCHOLOGY

869   American Society for Psychical Research (ASPR)*

President: C.B. Scott Jones, Jr., Ph.D.
Director of Administration: Donna L. McCormick
Director of Public Information & Education: Patrice Keane
5 West 73rd Street
New York, NY 10023

Telephone: (212) 799—5050/51

Under the leadership of William James, the ASPR was organized in 1885 with Simon Newcomb, the astronomer, as
its first president.   Later it became a branch of the Society
for Psychical Research (London), with Richard Hodgson,
working out of Boston, directing its research.   Hodgson died
in 1905, at which time an independent ASPR was organized
in New York with James H. Hyslop as its secretary—
treasurer.   He initiated the Society's publications program in
1907.
The purpose of the ASPR, a nonprofit institution, "is to
advance   the   understanding   of   phenomena   alleged   to   be
paranormal: telepathy,   clairvoyance,   precognition,
psychokinesis,   and   related   occurrences   that   are   not   at
present thought to be explicable in terms of physical,
psychological, and biological theories."   At one time the
Society maintained an active research department under the
direction of Karlis Osis.   Currently it has limited funds avail—
able to sponsor research. It publishes a quarterly *Journal*,
an occasional *Proceedings*, and the *ASPR Newsletter*. A use—
ful directory entitled *Courses and Other Study Opportunities
in Parapsychology* [486] is compiled by Donna McCormick as
needed.   The ASPR also disseminates information by lec—
tures, forums, seminars, and workshops, and provides coun—
sel to research workers, writers, and educators under the
guidance of Patrice Keane, Director of Public Information
and Education. Students and others are supplied with infor—
mation and materials upon request. The Society maintains
an 8,000—volume library. Use of the library is restricted;
and only members may borrow books.
There are no membership requirements. In addition to
regular members ($35.00), student members and senior
citizens ($20.00), and libraries/institutions ($50.00), higher
classes of membership are available with benefits beyond the
regular ones. The total membership averages 2,000.

870 Foundation for Research on the Nature of Man (FRNM)*

Director: K. Ramakrishna Rao, Ph.D.

Box 6847
College Station
Durham, NC   27708

Telephone: (919) 688–8241

Established in 1962, the founding Board of Directors of FRNM was drawn primarily from the Duke University Parapsychology Laboratory. Its aim was to make the transition over a period of time from the Duke Laboratory to an international world center for the advancement of parapsychology and its integration with other scientific studies of human nature. The Foundation's stated purpose is to "explore fully and carefully all the unusual types of experiences that suggest underlying capacities or principles as yet unrecognized in the domain of human personality." In addition to research, FRNM serves as an international clearinghouse for information on parapsychology and as a training base for persons interested in pursuing empirical studies of psi. It also promotes bridging the gap between the academic community and isolated researchers, between the sources of support and needy projects, and between an intelligently interested public and the experimental laboratory.

FRNM houses the Institute for Parapsychology [877] and the Parapsychology Press, its publication arm, which publishes the quarterly *Journal of Parapsychology* [825] and a newsletter, the *FRNM Bulletin* [855]. The Press also publishes a limited number of books, especially collections of papers originally appearing in the *Journal of Parapsychology* [329] which have been rewritten in less technical language [55 and 57]. FRNM also houses the FRNM Clinical Interest Group [907]. It offers a summer study program which has come to serve an important educational function in parapsychology. Instructors include FRNM staff members and visiting research scholars. The course, which lasts 8 weeks, includes lectures, discussions, and participation in research meetings and projects.   FRNM also offers qualified persons with a career interest in parapsychology an opportunity to receive specialized training in its year–round residential Advanced Program. Research fellowships are sometimes available that cover tuition costs plus small stipends. It is also

possible to pursue a graduate degree while at FRNM
through Andhra University [892] and other colleges and
universities. The Foundation maintains a 2,600—volume
library on scientific parapsychology. It also sells several
dozen of the best available books on psi research as well as
ESP cards and testing instructions. Its book list is available upon request as are a variety of complimentary informational items and reprinted articles.

871 Institut für Grenzgebiete der Psychologie (Institute for Bor-
der Areas of Psychology)

Director: Hans Bender, Ph.D., M.D.

Eichhalde 12
D 7800 Freiburg im Breisgau
West Germany

Telephone: (0761) 55035

The Institute was founded in 1950 by Hans Bender and is
subsidized chiefly by a private foundation ("Fanny—Moser—
Stiftung"). The Institute's activities consist of information services for the general public; investigations of spontaneous psi
phenomena, especially field investigations of RSPK cases; counsel—
ing persons with alleged psi or "anomalous" experiences; and
serving as a historical/archival center of documents, correspon—
dence, and papers relevant to the development of German
psychical research. The Institute maintains a 23,000—volume
library (under the direction of a half—time professional librarian)
on scientific parapsychology and related areas, including occultism, spiritualism, magic, witchcraft, dowsing, unorthodox healing, ufology, conjuring, Forteana, and "anomalistics." The
library is funded as a "special library" for parapsychology by
the Deutsche Forschungsgemeinschaft (German Foundation for
Research). The library currently receives approximately 200 periodicals from around the world, including the major journals of
scientific parapsychology published in English, French, Italian,
Spanish, and other European languages. A bibliography of the
Institute's periodical literature was published as a special issue
(Nos. 2&3/1987) of the *Zeitschrift für Parapsychologie und
Grenzgebiete der Psychologie*. The library is open to the public
Monday through Thursday 1:00 to 5:00 p.m. Books can be
loaned for home (and also for interlibrary loan) directly at the
Freiburg University Library. Staff members include Eberhard
Bauer and Rolf Streichhardt, Research Associates, and Micaela

Brunner, Librarian.

872 Parapsychological Association (PA)*

President: elected yearly
Business Mgr: Richard Broughton, Ph.D.

Box 12236
Research Triangle Park, NC 27709

Telephone: (919) 688–8241

International in scope, the PA is the professional society of parapsychology. Formed in 1957, its Constitution states that the aim of the Association is "to advance parapsychology as a science, to disseminate knowledge of the field, and to integrate the findings with those of other branches of science." It aims to enlarge further the number of active research workers by improving working conditions in the field and by fostering communication both inside parapsychology and with the scientific community as a whole. One of its major activities is to hold an annual convention (in August or September), the results of which are circulated, unedited, in a proceedings volume and published in lengthy abstracts in *Research in Parapsychology* [832]. It also encourages research, fosters interdisciplinary communication, and collaborates with psychical research societies and other groups. Five journals are affiliated with the PA: *EJP* [823], *JP* [825], *JASPR* [826], *JSPR* [827], and *PAI* [828].

In 1969 the Association was granted affiliation with the American Association for the Advancement of Science. Membership is restricted. There are five membership categories: full, associate, affiliate, student affiliate, and honorary. In 1988 there was a total of 267 members, of which 110 were full members, 150 associate members, and 7 honorary members.

873 Parapsychology Foundation (PF)

President: Eileen Coly

228 East 71st Street
New York, NY 10021

Telephone: (212) 628–1550

Founded in 1951 as a "non—profit educational organization to support impartial scientific inquiry into the total nature and working of the human mind and to make available the results of such inquiry." The Foundation's first president, Eileen J. Gar—rett, played an active role, until her death in 1970, in encourag—ing scientific investigation through independent studies as well as laboratory controlled experimental programs, by providing financial support to pursue such research. The Foundation's con—cern is not only with parapsychology as such, but also with the relation of the field to other disciplines as they touch on the nature of mind such as psychology, psychiatry, physics, religion, philosophy, medicine, and psychopharmacology. Education, together with support of scientific research, remain the primary thrust of the Parapsychology Foundation.

One of the Foundation's major functions is to provide financial support for research and education in the form of grants to individuals and to institutions. An annual scholarship award, the Parapsychology Research Scholarship, is offered for students continuing college level studies in parapsychology. The publication program maintained by the Foundation includes scholarly *Parapsychological Monographs*, yearly *Proceedings* of its conferences (edited by Betty Shapin and Lisette Coly), and the bimonthly *Parapsychology Review* [**336**]. Participants in the Foundation's annual international and interdisciplinary con—ferences are invited to present formal papers and discuss a selected parapsychological theme. *Parapsychology Review* pub—lishes articles, book reviews, and news notes from around the world devoted to parapsychological research and activities. All publications are available from the Foundation. The Eileen J. Garrett Library of the Parapsychology Foundation is a reference library devoted to psychical research and related subjects and is open to students and researchers to facilitate studies in para—psychology. The librarian is Wayne Norman.

874 Society for Psychical Research (SPR)*

President: 1989/1990, Alan Gauld, Ph.D.
Secretary: Eleanor O'Keeffe

1, Adam & Eve Mews
London W8 6UG
England, U.K.

Telephone: 071—937 8984

The SPR was organized in 1882 by a group of distin—

guished scholars and scientists, with Henry Sidgwick of Cambridge University as its first president. The object of the Society was initially to investigate "that large group of debatable phenomena designated by such terms as mesmeric, psychical and spiritualistic." Its purpose as stated today is "to examine without prejudice or pre—possession and in a scientific spirit those faculties of man, real or supposed, which appear to be inexplicable on any generally recognized hypothesis." Since its inception a number of well—known thinkers have occupied the presidential chair, among them Oliver Lodge, William F. Bar—rett, Charles Richet, Eleanor Mildred Sidgwick, Andrew Lang, Henri Bergson, Gilbert Murray, William McDougall, Hans Driesch, Camille Flammarion, C.D. Broad, and Alister Hardy.** The Americans who have served as president are William James, Walter Franklin Prince, Gardner Murphy, J.B. and Louisa Rhine, and Ian Stevenson. The presidential addresses of these and many others provide a valuable contribution to psychi—cal research. (They are published in the Society's *Proceedings* or *Journal*.)

In addition to conducting research, mainly by six com—mittees of members (research advisory, experimental ESP, hyp—nosis, mental mediumship, physical phenomena, and spontaneous cases), the SPR maintains a library and valuable archives. It has published its *Proceedings* [334] since 1882, the *Journal* [331] since 1884, and the *Newsletter* [856] since 1981. The Myers Memorial Lectures and a number of pamphlets on various aspects of psychical research have also been published. It spon—sors lectures and provides speakers upon request from groups and organizations. It has a small studentship trust fund to sup—port graduate work in parapsychology in British universities.

Members or Associates pay $45 but the former must be sponsored by two SPR members. The latest membership figure available is 812 members in September 1988.

RESEARCH INSTITUTES

875 Cognitive Sciences Program* (Ceased in 1990)

Executive Director: Edwin C. May, Ph.D.
(Program Manager)

Geoscience and Engineering Center
SRI International

**For a complete list see R. Haynes' history of the SPR [554].

333 Ravenswood
Menlo Park, CA 94025—3493

Telephone: (415) 859—3765

The Cognitive Sciences Program investigates the relation—
ship between certain cognitive processes and the environment,
and determines the degree to which this relationship suggests
applications.   In the past, the Cognitive Sciences Program
research direction was oriented toward establishing the existence
of and applications for certain phenomena defined as direct inter—
actions between human consciousness and the environment.
These human capabilities fall into two main categories: (1) the
aquisition of information, and (2) the production of physical ef—
fects.   These can be further defined as: (1) remote viewing, or
the ability to gain access, by mental means alone, to informa—
tion which is otherwise blocked by distance, time, or shielding,
and (2) remote action, or the ability to interact with the physi—
cal environment by mental means alone.

Currently, the effort is to determine possible physical,
physiological, and psychological mechanisms for remote viewing.
The research area, therefore, spans a wide variety of disciplines,
including general relativity and multidimensional spaces, informa—
tion theory, linguistics, neurophysiology, perceptual psychology
and physiology, philosophy, psychology, quantum measurement
theory, and statistics and set theory.

Basic and applied research related to these special forms of
cognitive processes——remote viewing and remote action——has
been conducted at SRI International (formerly Stanford Research
Institute) since 1972 and within the Geoscience and Engineering
Center since 1977.   Although support has been primarily from a
variety of federal agencies, it also has consisted of industrial,
private foundation, and individual sources.   Because the Cogni—
tive Sciences Program is not hardware oriented, there are no
unique physical facilities.   However, a number of unique applica—
tions have been developed.   They include (1) computer
programs designed to maintain double—blind protocols, and (2)
a real—time UNIX application to drive special perceptual testing
hardware.

There are five staff members in addition to May: Nevil
Lantz, Ph.D., senior research psychologist; Beverly S. Humphrey
and Virginia V. Trask, senior research analysts; Thane J.
Frivold, software engineer; Wanda Walsh Luke, research analyst;
and Frances Saucedo, secretary. The Program also uses the serv—
ices of approximately 30 consultants in the fields of art, en—
gineering, medicine, philosophy, physics, physiology, psychology,
and statistics.

In addition to publishing reports in *Nature* and in physics and parapsychology publications, the program has published 30 technical reports by various staff members since 1976.

876 Institute for Advanced Psychology*

Director: Keith Harary, Ph.D.

2269 Chestnut Street, #875
San Francisco, CA 94123

Telephone: (415) 931—0419

The Institute for Advanced Psychology is a nonprofit, tax—exempt California corporation that conducts pioneering investigations in perception, cognition, communication, creativity, and other emerging areas of psychology. The research effort focuses on finding the psychological and physical correlates that will bring hitherto poorly understood areas of individual behavior, experience, and interaction into the mainstream of modern scientific research. Experimental and field investigations are supplemented by theoretical studies. Results of the Institute's research are published in scientific journals and disseminated to the general public through popular publications, educational workshops, and the media. Psi research is conducted in the areas of remote viewing and in studies of human communication and human—animal communication.

877 Institute for Parapsychology*

Director: K. Ramakrishna Rao, Ph.D.

Box 6847
College Station
Durham, NC 27708

Telephone: (919) 688—8241

The designated successor to the Duke University Parapsychology Laboratory is the division of FRNM known as the Institute for Parapsychology. Among several research units envisaged as part of the larger body, it was the first to be created because "as a branch of inquiry already far advanced in methods, and with a body of confirmed knowledge well tested by decades of controversy, parapsychology is the subject of

study qualified to lead the advance further into the territory of
what identifies man." The Institute houses the Duke University
collection of more than 10,000 accounts of spontaneous psi ex—
periences initiated by Louisa E. Rhine. The founders state that
"the increasing responsibilities of the center have come to re—
quire an independent, self—governing and self—sustaining institution, inter—university in character, and international in scope."
The broad objective of the Institute's entire research effort is to
assure that the answers to questions about psi experiences——
Who has them and why? How do they relate to specific
physiological processes and subjective mental states? Do they
represent abilities that can be controlled and applied to the solution of human problems?——rest on a firm foundation of systematic and dispassionate observations rather than on shifting
metaphysical speculations. In addition to its own research, the
Institute extends supportive services to other research centers
throughout the world. Many parapsychologists working at
universities and institutions not specializing in psi research are
dependent on the Institute's staff for consultation and some
times even actual evaluation of collected data.

The Institute conducts research, serves as an international
forum, and holds weekly meetings to which research workers
and other interested persons from other centers are invited.
The resulting papers, as well as reports of the Institute's own
research, are published primarily in the *Journal of Parapsychol—
ogy* [825]. *See also* the Foundation for Research on the Nature
of Man (FRNM) [810].

878 Kairos Foundation*

Director of Research: Norman Don, Ph.D.
Contact Person: Ann Wilmot

405 North Wabash Avenue
Chicago, IL 60611

Telephone: (312) 236—4432

Incorporated in 1981, the primary aim of this research cen—
ter is to support laboratory and phenomenological research in
the area of consciousness studies and, to a lesser extent, mental
health, with the main focus on unusual nonclinical mental
states. The brainwave activity and event—related potentials of
unusual states (including psi) have been emphasized.

879 Mind Science Foundation (MSF)*

Executive Director: Catherine Nixon Cooke

8301 Broadway, Suite 100
San Antonio, TX 78209

Telephone: (512) 821–6094

When the Foundation was organized in 1958 by Tom Slick, research projects were undertaken in several locations sponsored by the Mind Science Foundation. In 1975 research operations were consolidated in a facility located in San Antonio. By bringing together a team of researchers from physics and psychology, the Mind Science Foundation promulgates a multidisciplinary approach to the scientific investigation of the human mind with projects in psychoneuroimmunology, parapsychology, and human potential.

In addition to research, the MSF performs educational functions such as giving college and university courses, seminars, workshops, public lectures, film programs, and media exposures. It also has established a specialized research library that may be used by the public. The results of the Foundation's research are published in scientific journals. The MSF publishes a newsletter entitled the *Mind Science Foundation News*, and in 1982 it published *Psi Notes*, a book by W. Braud [509]. It is available from MSF. Another staff member, D. Morton, has compiled indexes to *Psychic* [691], *Common Ground* [690], and *Skeptical Inquirer* [692].

Mind Science Foundation is headed by a Board of Trustees. The staff presently consists of a director: Catherine Nixon Cooke; assistant–to–the–director: Katie McGee; two senior research associates: Helmut Schmidt, Ph.D. and William Braud, Ph.D.; a research associate, Marilyn Schlitz, M.A.; three research assistants: Steve Dennis, Donna Shafer, and Susan Straus; information coordinator Diane Morton; and a secretary: Elisa Ruiz.

880 The Mobius Society*

Chairperson: Stephan A. Schwartz

4801 Wilshire Blvd., Suite 320
Los Angeles, CA 90010

Telephone: (213) 933–9266

This independent nonprofit research public research founda—
tion is a multidisciplinary society for the scientific study of con—
sciousness founded in 1977 by Stephan A. Schwartz. Primarily
the research program has investigated remote viewing in practi—
cal situations using a consensual methodology consisting of inter—
disciplinary teams of experts and psychics individually selected
for specific projects. Then, by means of a consensus or majority
vote technique, an attempt is made to solve practical problems
in such areas as archaeology, historical verification, and criminol—
ogy. Statements are obtained from psychics giving their impres—
sions of the needed answers and these are analyzed for several
types of patterns: internal patterns, conceptual agreement, com—
mon linguistics, and common geometric forms. Steps suggested
by the consensus views are followed up and verified by experts
in the subject area under investigation. The Mobius Society
also has conducted two nationwide mass ESP experiments via
the science magazine *Omni* and research into unorthodox heal—
ing.

The Mobius Society engages in approximately 11 small ex—
periments and one large probe per year. Results are published
in various forms. For example, both a film and a book
describe a major archeological probe at Alexandria, Egypt. It
also sponsors seminars, compiles statistics, and operates a
speaker's bureau. It publishes a quarterly newsletter and *Current
Research*, a semi—annual. It also issues periodic annotated sub—
ject bibliographies and *Mobius Reports*. It has a library of 3,500
volumes.

881 Parapsychology Laboratory, University of Utrecht

Director: Sybo A. Schouten, Ph.D.

University of Utrecht
Sorbonnelaan 16
3584 CA Utrecht
The Netherlands

Telephone: (30) 533824

After the inception of a special professorship (which in—
volves only teaching and limited or no funding by the univer—
sity) for parapsychology by the Dutch S.P.R. at the University
of Utrecht in 1953, the first incumbent, W.H.C. Tenhaeff, estab—
lished the Parapsychology Institute. When the university estab—
lished a full professorship (research and teaching fully supported

by the university), the Parapsychology Laboratory was founded as part of the Psychology Department of the Faculty of Social Sciences and the university terminated funding the Parapsychology Institute, which later was taken over by a private organization not related to the university. After Tenhaeff retired, the special professorship went to H. van Praag who occupied the chair from 1978 to 1988.

The Parapsychology Laboratory functions primarily as a research institute, but educational courses are provided for all levels. The Laboratory has invited many foreign parapsychologists to carry out research at the laboratory for periods ranging from a few months to four years. The Laboratory published a *Research Letter* [868] at irregular intervals, and it still publishes a bi–annual professional journal, the *European Journal of Parapsychology* [823].

The Laboratory was headed by Martin Johnson, a full professor in parapsychology from 1974, until his retirement in 1988, when Sybo A. Schouten took over. Due to extensive nationwide cuts in university budgets in 1987, the Faculty of Social Sciences cancelled the professorship and reduced funding for the Laboratory. It was closed after we went to press. For a history of its work see *European Journal of Parapsychology* [823], 1988/1989, 7, 95–116.

882 Princeton Engineering Anomalies Research Laboratory (PEAR)

Director: Robert G. Jahn, Ph.D.
Laboratory Manager: Brenda J. Dunne

C131
School of Engineering/Applied Science
P.O. Box CN 5263
Princeton University
Princeton, NJ   08544–5263

Telephone:

This laboratory was organized by Robert G. Jahn in 1979 to explore anomalous phenomena in human factors engineering. The program has three components: experiments in low–level psychokinesis; analytical studies in precognitive remote perception and theoretical modelling of the phenomena. Research in these areas is addressed to an engineering assessment of the influence of human consciousness on physical systems or processes. Primary emphasis is on the engineering aspects and implications of the phenomena under investigation.

The program of the laboratory is under the direct supervision of Dr. Robert G. Jahn, Dean Emeritus of the School of Engineering and Applied Science, Princeton University. Developmental psychologist Brenda J. Dunne serves as laboratory manager, and Dr. Roger D. Nelson, a cognitive psychologist, oversees protocol and analytical methodology and maintains the interface with psychology. Other members of the staff are John Braddish, electrical engineer; York Dobyns, Ph.D., theoretical physicist; and Angela Thompson, developmental psychologist. In addition to the full—time research personnel, the program has the part—time assistance of several technical and administrative staff persons.

PEAR itself publishes reports of its research and also publishes them in professional journals. An overview of PEAR research and theory—building is presented in a book entitled *Margins of Reality* by Jahn and Dunne [617]. The research laboratory consists of a five—room complex centrally located on the first level of the School of Engineering/Applied Science near various supporting facilities such as the machine shop, electronics shop, stock rooms, and design and reproduction facilities.

Although no formal academic program is associated with the project, a number of undergraduate and graduate students are involved in various capacities. Through its professional contacts and publications the program performs an educational function.

As of mid—1989 PEAR has published 15 technical reports as well as the book by Jahn and Dunne [667], and papers in various physics, engineering, psychology, and parapsychology publications.

883 Psychical Research Foundation (PRF)*

Director: William George Roll, Ph.D.

P.O. Box 217
Carrollton, GA 30117

Telephone: (404) 836—6510

Founded in 1960, the PRF serves as a scientific and educational research center to investigate the possibilities of the continuation of consciousness after the death of the physical body. It has conducted research on OBEs, meditation, haunting and poltergeist activity, and mediumship. It publishes the quarterly

Theta [841] and holds an annual meeting.

884 Psychophysical Research Laboratories (PRL)* (Closed in 1989)

Director: Charles R. Honorton

301 College Road East
Princeton, NJ 08540

Telephone: (609) 452—8144

Located in the Forrestal Research Center in Princeton, New Jersey, the Psychophysical Research Laboratories (PRL) is the successor to the Division of Parapsychology and Psychophysics of the Maimonides Medical Center in Brooklyn (formerly the Maimonides Dream Laboratory). PRL was founded in June 1979 by James S. McDonnell (President of McDonnell Douglas Aircraft) and Charles Honorton, former Director of the Division of Parapsychology and Psychophysics.

The research program at PRL is oriented toward increasing the reliability of laboratory ESP and PK. Computers are emphasized at every stage of the research process: in planning experiments, reviewing literature, monitoring psi tests, using computer games as psi tests, measuring physiological states, tabulating data, and in writing research reports.

885 Research Institute for Psi Phenomena and Physics (RIPP)*

Director: Dick Bierman, Ph.D.

Alexanderkade 1
1018 CH Amsterdam
The Netherlands

Telephone: 431 20262764

The Research Institute for Psi Phenomena and Physics was organized in 1979 to provide an interdisciplinary professional approach to research on psi phenomena. The research policy of the Institute is interdisciplinary in nature and emphasizes interactions between psychology, biology, physics and parapsychology. It also stresses the interaction between experiment and theory. Experiments are viewed primarily as tests of theories and theories are seen as predictors of experimental out—comes. The relevance of spontaneous cases for the evaluation of theories is emphasized, and attempts are made to reconcile cur—rent physical theories with psi phenomena.

In order to optimize exchange of ideas the Institute invites visiting research fellows at regular intervals to stay for a period of three months. Young researchers receive special encourage—ment.

Since 1981 RIPP has been a member of The Foundation for Fundamental Research into Man & Matter, whose primary goal is to coordinate psi research in The Netherlands and to for—mulate criteria for the evaluation of research plans in order to obtain government support for parapsychological research.

886 Synchronicity Research Unit (SRU)*

President: Jeff C. Jacobs
Secretary: J.A.G. (Hans) Michels

P.O. Box 7625
5601 JP Eindhoven
The Netherlands

Telephone:

SRU was initiated in 1969 as the Studiegroep Synchronici—teiten. The founders (J.C. Jacobs, N.V. Luxemborg, and T. Henst) were primarily interested in studying synchronicity, or remarkable coincidences that make sense on a psychological level but that in terms of current knowledge in the behavioral and physical sciences are noncausal. In 1975 the group adopted its current name and had five core members plus 20 associates who were free—lance researchers. In 1979 SRU became a foundation. Currently it has six core members: philosopher G.H. Hövelmann (aasociate); statistician Jeff C. Jacobs (president); psychologist J.A.G. Michels (secretary); parapsychologist Brian Millar, Ph.D. (vice—president and librarian); graphics designer R. Paré (associate); and engineer A.C.M. Verbraek (treasurer). The primary activity of SRU is experimental psi research. It also col—lects and studies spontaneous cases. It publishes the results of its research in its journal, *SRU Bulletin* [840]. A laboratory is maintained with special equipment (computers, random number generators, and video, audio, film, and photo equipment). It also has a library of 1,800 books and the major parapsychology jour—nals. SRU maintains contact with several other European para—psychological research centers.

INFORMATION CENTERS

888   Mary Evans Picture Library

Contact Person: Hilary Evans

59 Tranquil Vale
Blackheath
London SE3 OBS
England

Telephone: 081–318–0034   *FAX*: 01–852 7221

This unusual library consists of historical pictorial material covering all aspects of life and the universe. The library also owns many thousands of illustrated books and bound periodicals of the last two centuries, including foreign ones. These materials may be borrowed for a fee. The Library is listed here because of some special collections that are administrated by the Library on behalf of private owners: the collection of the Society for Psychical Research, the Harry Price collection of London University, the Sigmund Freud family photograph archive, and historical items from the *Psychic News* relating to spiritualism and psychical research in the 19th and 20th centuries. Hilary Evans is a member of the SPR and has written books on spontaneous psi and related experiences [see **552, 725, 726,** and **345a**].

889   Parapsychology Information Network*

Director: William Tedder, Ph.D.

P.O. Box 10456
Denver, CO 80210

Telephone: (303) 722–9635
Computerline: (303) 751–8653

The purpose of the Parapsychology Information Network is "to provide the public with a balanced and understandable view of the paranormal, from both scientific and experimental aspects." This goal is fostered by means of a microcomputer–based network; making available current information on local and national events; encouraging interaction and the exchange of ideas; promoting scholarly and scientific research on psi phenomena; sponsoring educationally oriented activities such as classes, lectures, seminars, and workshops; maintaining a database of parapsychological information; and assisting individuals in understanding their psychic experiences. The Network provides up—to—date information on parapsychology. Because the Network is a message—based system, user interaction is possible through the question and answer board, the open forums, and the recording of personal experiences; users can also engage in a dialogue with authors and provide feedback on articles presented via the Network. The public has access to only specific sections of the Network: a tutorial, current Network offerings, subscription information, testing for ESP abilities, eight parapsychological surveys, the open forum (editorial comments), read privileges on the question and answer board, 10 categories of definitions, local library offerings, and books and materials for purchase from parapsychological organizations. The full member has access to the following additional sections: 15 categories of parapsychological bibliographies; descriptions of parapsychological organizations located throughout the world; the parapsychology digest, which contains abstracted or original articles presented in layman's terms (new articles are provided at approximately three—week intervals); both read and write privileges on the question and answer board; and current local and national events. Two classes of membership are offered: Full (to those who can access the Network by computer) and Associate (to individuals who cannot access the Network by computer).

890   Parapsychology Sources of Information Center (PSIC)*

Director: Rhea A. White

2 Plane Tree Lane
Dix Hills, NY 11746

Telephone: (516) 271—1243

The Parapsychology Sources of Information Center (PSI Center) was established in 1981 to serve as a clearinghouse for information on research findings, theories, organizations, publications, and personnel involved in parapsychology and consciousness studies.

The PSI Center collects, catalogs, and indexes books, journals, articles, pamphlets, cassette tapes, and unpublished manuscripts dealing with parapsychology and consciousness studies. Its holdings include 4,500 books, 100 periodical titles, and a collection of 12,500 articles, book reviews, and chapters from nonparapsychological periodicals and books. It maintains alphabetical information files on organizations, persons, and pub— lications and has a collection of bibliographies. The Center also sells secondhand books and back issues of parapsychological peri— odicals.

The services of the PSI Center are fee—based and consist of searching the *PsiLine* [691a] database system, manually com— piling bibliographies on request; answering specific questions deal— ing with parapsychology and consciousness studies, and consult— ation. The Center maintains a publications program that in— volves compiling and publishing several parapsychology reference works, including bibliographies on specific subjects [e.g., see **318**, **454**, **491**, and **679**], directories, and a journal, *Parapsychological Abstracts International* [**828**]. It also produces the computerized database *PsiLine*, which eventually will contain abstracts of the parapsychological literature from earliest times to date.

891 Psi Research*

Director: Larissa Vilenskaya

484—B Washington Street, #317
Monterey, CA 93940

Telephone: (408) 655—1985

Promotes research and the international exchange of infor— mation on parapsychology, folk healing, and fire walking with emphasis on the USSR, Eastern Europe, and China. Psi Research also offers occassional seminars and lectures and main— tains a speaker's bureau and small library. From 1982 to 1986 it also published *Psi Research* [**865**], which has been suspended.

EDUCATIONAL INSTITUTIONS

891a Abteilung für Psychologie und Grenzgebiete der Psychologie des Psychologischen Institut der Universität Freiburg i. Br. (Department of Psychology and Border Areas of Psychology, Psychological Institute, University of Freiburg i. Br.)

Director: Johannes Mischo, Ph.D.

Belfortstrasse 16

D 7800 Freiburg im Breisgau
Germany

Telephone: (0761) 203—4154 (Secretary)

Between 1954 and 1967 there existed a personal chair for
"Grenzgebiete der Psychologie" (Border Areas of Psychology) at
the University of Freiburg held by Professor Hans Bender; in
1967 this personal chair was changed into a full professorship
for "Psychology and Border Areas of Psychology." Since
Bender's retirement in 1975, the chair holder is Professor
Johannes Mischo, who collaborates closely with Bender's private
Institute [871]. The staff consists of a fulltime research as—
sociate and a halftime secretary. Since 1978, at the University
of Freiburg it has been possible to choose
"parapsychology/border areas of psychology" as optional, with
additional study for the final examination for the German de—
gree of "Diplompsychologe" within the normal curriculum for
psychology. Students of psychology have the chance, moreover,
to treat parapsychological problems in their diploma or doctoral
theses in psychology (Ph.D.). There are regular courses and/or
lectures on current topics of parapsychological research. Because
only 30% of the teaching and research facilities of the chair are
devoted to parapsychology, experimental psi research is possible
only on a very limited scale.

Since 1957 the Institute has published a scholarly journal,
the *Zeitschrift für Parapsychologie und Grenzgebiete der
Psychologie*, which contains English summaries of most articles.
In 1990, a thirty—year index of the *Zeitschrift* (Volume 1, 1957
— Volume 30, 1988) will be available. The editorial staff of the
Zeitschrift is composed of Eberhard Bauer, Hans Bender, Walter
von Lucadou, Johannes Mischo, and Ulrich Timm. The
Zeitschrift is published quarterly by AURUM publishers, POB
5204, D 7800 Freiburg im Breisgau, Telephone (0761) 36409.
The subscription rate is DM 80.00.

For a detailed chronology and documentation of the parap—
sychological work done in Freiburg, see Eberhard Bauer and
Walter von Lucadou: "Parapsychologie in Freiburg——Versuch
einer Bestandsaufnahme," in the *Zeitschrift für Parapsychologie
und Grenzgebiete der Psychologie*, 1987, *29*, 214—282.

892 [315] **Andhra University, Department of Psychology and
Parapsychology***

Director: P.V. Krishna Rao, Ph.D.

Waltair
Visakhapatam 530 003
Andhra Pradesh, India

Telephone: 3279

The Department was established in 1967 by the University Grants Commission with K. Ramakrishna Rao as director. In addition to Rao there were two research scholars on the staff. Regular students for graduate study were first admitted in the fall of 1968. It was reported in 1970 that "the Department restricts admissions to eight students every year . . . ." The minimum eligibility requirement for admission to Ph.D. in Parapsychology is a Master's degree in arts or science with satis— factory evidence of competence to carry on research in parap— sychology. The minimum required residence at the University is two years. The Ph.D. degree is awarded entirely on the basis of the thesis. Ph.D. students have no required course work. Note: Andhra University recognizes FRNM's Institute for Parap— sychology as an approved center for research. Therefore, it is possible for a student enrolled in the Ph.D. program of Andhra University to spend all or part of the period of study at the In— stitute for Parapsychology which would count toward the residential requirement. The M.A. Degree in Psychology in a two—year course of study. Minimum eligibility for admission is a bachelor's degree in arts or science. Parapsychology is offered as a subject for the final year students. Also the student may select to write a dissertation on a parapsychological topic.

Currently there are ten staff members.

The department publishes the *Journal of Indian Psychology* [824].

893 Atlantic University

Executive Director: Charles Thomas Cayce, Ph.D.
Contact Person: Douglas G. Richards, Ph.D.

67th Street and Atlantic Avenue
P.O. Box 595
Virginia Beach, VA 23451

Telephone: (804) 428—3588

The mission of the university is expressed in the offering of graduate—level programs which focus on the nature of humanity, the nature of the universe, and holistic living. Cur—

rently the university offers the Master of Arts degree in Transpersonal Studies. The curriculum includes courses which provide an opportunity to integrate knowledge and experience.

The university sponsors a research program directed by Douglas G. Richards, which serves as a source of information on anomalous phenomena such as perception beyond the traditional five senses. It also studies areas of metaphysics such as meditation, spiritual growth, and healthful living.

The charter of Atlantic University, which was originally granted in 1930, was reactivated in 1985. The University is associated with the Edgar Cayce Foundation and the Association for Research and Enlightenment [918]. It is in the process of pursuing accreditation from the Southern Association of Colleges and Schools.

The faculty of Atlantic University is composed primarily of adjunct professors. Visiting scholars are also available to teach in their field of specialization. The Faculty for 1989—1990 includes Charles Thomas Cayce, Ph.D.; Eileen Connelly, Ph.D.; Michael Dennis, Ph.D.; Richard Drummond, Ph.D.; Margaret Irby, Ph.D.; Robert Jeffries, Ph.D.; Henry Reed, Ph.D.; Douglas Richards, Ph.D; Scott Sparrow, Ed.D.; Frances Sporer, M.A.; Mark Thurston, Ph.D.; and James C. Windsor, Ed.D., L.L.D.

894   Consciousness Research and Training Project

President: Joyce Goodrich, Ph.D.

315 East 68th Street, Box 9G
New York, NY 10021

Telephone: Unlisted

This organization is devoted to research and education on the healing method developed by Lawrence LeShan, Ph.D. [For a description of LeShan's method see **352** and **780**]. Introductory and advanced training seminars are taught by a small group of people trained at length to teach. The seminars are offered in central locations throughout the U.S., are five days long, and residential. There are two important provisos: no person thus trained may charge or receive donations of money or gifts from persons they do healing work with, and any person whom they attempt to heal must also be under the care of an appropriate professional. There are four ongoing groups that meet in Manhattan, Brooklyn, Syracuse, and San Francisco. A research group meets every three months. Double—blind research with humans, based on the work of its trainees, is being developed.

The Project also publishes a privately circulated newsletter. In 1989, 12 years' editions of the *Newsletter* are being condensed, edited, and made available for use in the seminars.

In addition to LeShan's books, persons seeking more information may consult a chapter on the Consciousness Research and Training Project by Dr. Goodrich in J.L. Fosshage and P.L. Olsen (Eds.), *Healing: Implications for Psychotherapy* (pp. 84–108). New York: Human Sciences Press, 1978 (a publication of the National Institute for the Psychotherapies). A healing meditation tape is available for people who have taken the introductory seminar.

895 Edinburgh University

Koestler Chair: Robert L. Morris, Ph.D.*

Department of Psychology
7 George Square
Edinburgh EH8 9JZ
Scotland, U.K.

Telephone: (031) 667–1011

Under the leadership of John Beloff, the Department of Psychology at the University of Edinburgh has been actively involved in parapsychology since 1964. Not only do students and staff conduct psi research but several doctorates in psychology with dissertations on parapsychology have been granted by the University of Edinburgh. The department acts as a clearing house and meeting place for all those interested in parapsychology, staff or students, from any department of the university. Regular seminars and research meetings are held to bring the parapsychological community together and facilitate the exchange of information and views.

In 1985, at the bequest of Arthur and Cynthia Koestler, the Koestler Chair of Parapsychology was established at Edinburgh. The well–respected American psychologist/parapsychologist Robert L. Morris is the first holder of the Chair. It is possible to do graduate research under the direction of Dr. Morris leading to a doctorate in psychology with a concentration in parapsychology. Under Morris, the research activity at Edinburgh includes studies of anomalous interactions between humans and computers, the use of imagery in enhancing PK test scores, the evaluation of psychic development techniques in free–response ESP tasks, spontaneous psi experiences within family units, exploring the similarities between psychic develop–

ment techniques and those of sports psychology, and the psychology of conjuring and deception.
     The staff consists of Dr. Morris, two postdoctoral research
fellows, Drs. Deborah Delanoy and Julie Milton; a research associate, Caroline Dow; a secretary, Helen Sims; and four
postgraduate students.

896 John F. Kennedy University*

   Dean of Graduate School for the Study of Consciousness:
   David Komito, Ph.D.

   12 Altarinda Road
   Orinda, CA 94563

   Telephone: (415) 254—0200

     The Graduate School for the Study of Human Conscious-
ness offers master's degree programs in a variety of facets of
human experience, including transpersonal psychology, holistic
health, and interdisciplinary consciousness studies. The latter
has four areas of specialization, one of them being parapsychology (24—units covering an in—depth examination of the full
range of psi phenomena). The curriculum is designed to
provide professional skills in experimental laboratory research,
methodologies, field investigations, and clinical, phenomenological, and survey approaches. A theoretical examination is
provided of historical and modern models for mapping consciousness and psi phenomena in psychology, physics, and allied fields
to facilitate the student's understanding of the process of inquiry. Students may select courses that focus directly on practical applications of intuition and psi in business, industry, and
social services, as well as daily living. The parapsychology
specialization prepares students to work as researchers, consultants, and teachers in innovative, challenging, and multidisciplinary fields. The list of Master's degrees in Chapter VII contains 11 degrees granted by JFK for work in parapsychology
starting in 1981.
     The current faculty in parapsychology includes Julian
Isaacs, Ph.D., Jerry Solfvin, Sharon Franquemont, Loyd Auerbach, and Cynthia Siegel. The Program Director for Interdisciplinary Consciousness Studies is Vernice Solimar, and the
Program Advisor is Stephanie Austin.

897 Saybrook Institute*

Contact Person: Stanley Krippner, Ph.D.

1550 Sutter Street
San Francisco, CA 94109

Telephone: (415) 441—5034

Founded in 1970 as the Humanistic Psychology Institute, Saybrook Institute offers external degree graduate education and research projects in humanistic psychology and human science. The graduate program consists of an accredited correspondence course requiring residence of only two weeks annually. There is also a small on—site program with regularly scheduled classes. Degrees offered are Master of Arts in Psychology and in Human Science, Doctor of Philosophy in Psychology, and Doctor of Philosophy in Human Sciences. There are four interest areas: Clinical Inquiry, Human Systems, Health Studies, and Conscious—ness Studies. The latter program, in particular, lends itself to investigation of aspects of parapsychology. It provides a focus for descriptive, experimental, and theoretical approaches to con—sciousness research. Topics include theory and research in hyp—nosis, meditation, dreams, creativity, anomalous phenomena, and altered states of consciousness. Emphasis is placed on tech—niques and methods used to study various aspects of conscious—ness, with particular attention to scientific data collection, phenomenological description, and abstract model building. A course in experimental parapsychology is offered. Thus far five persons have been granted doctorates for work in parapsychol—ogy and others are currently enrolled.

Independent studies can also be arranged with Institute faculty members and, on occasion, volunteer assistantships can be arranged for graduate students at the Washington Research Institute in San Francisco. Institute seminars are held peri—odically with well—known scholars, including some parapsy—chologists.

897a West Georgia College*

Chairman: Mike Arons, Ph.D.

Department of Psychology
Carrollton, GA 30118

Telephone: (404) 836—6510

Accredited by the Southern Association of Schools and Col—
leges, the Psychology Department at West Georgia College offers
a B.A. and M.A. in psychology in a humanistic program that
teaches not only traditional psychology but also explores the fur—
ther reaches of personal and human experience. The Psychology
Department also has a holistic orientation including the follow—
ing approaches; humanistic, phenomenological, transpersonal, and
experiential. Both undergraduate and graduate coursework for
credit and opportunity for independent study are offered.
Courses include developmental and clinical parapsychology; life
after death; and neuropsychology, system theory and psi. Also
in Carrollton are the Parapsychological Services Institute [907]
and the Psychical Research Foundation [883].

ANOMALIES RESEARCH INSTITUTES

xxx   Association for the Anthropological Study of Conscious—
      ness (AASC) *see* Society for the Anthropology of Con—
      sciousness [905a]

899   Association for the Scientific Study of Anomalous
      Phenomena (ASSAP)

      Executive Director: Hugh Pincott

      20 Paul Street
      Frome Somerset BA11 1DX
      England, U.K.

      Telephone: 0373—51777

      This association was formed in 1981 to study a wider field
of anomalous phenomena than was being investigated by exist—
ing British societies, including the SPR. The Association
studies anomalous phenomena such as UFOs, psi phenomena,
Forteana, earth mysteries, cryptozoology, etc. by means of the
scientific method. Activities of AASAP includes investigating
spontaneous cases, research programs, education and archives,
meetings and conferences, and publishing research and news. It
publishes a journal, *Anomaly*, twice a year and a quarterly
newsletter, *ASSAP News*. Although ASSAP is not housed in
separate quarters, it maintains a central pool of research equip—
ment for use by members.

900   Center for Applied Anomalous Phenomena

Contact Person: C.B. Scott Jones, Ph.D.

6435 Shady Lane
Falls Church, VA 22042–2335

Telephone: (703) 534–2305

This nonprofit institution was established in 1982 in order to conduct research and analysis of anomalous phenomena, including ESP and survival phenomena, and to explicate their implications and applications; to conduct conferences, give lectures, and publish information relating to implications and applications of anomalous phenomena; to promote a professional and open atmosphere between researchers in anomalous phenomena and those involved in research of more commonly recognized and accepted phenomena; to remove any impediments to research and open discussion of anomalous phenomena; and to raise funds to accomplish the above goals. The Center takes full advantage of the fact that it is located in the metropolitan Washington, DC area where it is in a position to interact with the federal system and the professional organizations of conventional science and research.

901 Center for Scientific Anomalies Research (CSAR)*

Director: Marcello Truzzi, Ph.D.

P.O. Box 1052
Ann Arbor, MI 48106

Telephone: (313) 663–8823

This private center was organized in 1981 by Marcello Truzzi, a professor of sociology at Eastern Michigan University, in order to bring together scholars and researchers concerned with furthering responsible scientific inquiry into and evaluation of claims of anomalies and the paranormal.

The primary focus of CSAR is on the study and evaluation of bodies of anomalous observations rather than upon esoteric theories seeking to explain already known phenomena. The orientation of the Center is exclusively scientific, places the burden of proof on the claimant, and recognizes the need for a degree of proof commensurate with the extraordinary character of the phenomenon claimed. But the Center also wishes to promote open and fair–minded inquiry that will be *construc–*

tively skeptical.   Critically and constructively approached,
legitimate anomalies should be welcomed by science rather than
perceived as ill—fitting nuisances.

In order to do so CSAR fosters the interdisciplinary scien—
tific study of alleged and verified anomalies, acts as a clearin—
ghouse for scientific anomaly research, and publishes the *Zetetic
Scholar* [850], research reports, and bibliographies.   It promotes
the dissemination of information about scientific anomalies
research. It sponsors scientific research into anomalies, paranor—
mal, and related phenomenona. It has created a public network
of experts on anomalies research, and attempts to promote im—
proved communication between critics and proponents of scien—
tific anomalies.

Persons associated with CSAR do so as consultants or
members. Consultants are those who have demonstrated exper—
tise in some area of anomalies research.   (There are two types
of consultants: Research Consultants, who give evidence of ex—
pertise in research and methodology, and Resource Consultants,
who give evidence of special information or knowledge of pos—
sible use to others doing scientific research into anomalies.)
Membership is open to all who agree with the scientific goals of
CSAR. CSAR is run by a governing board. As of March 1989
there were 16 members and 96 consultants.

CSAR has sponsored two research projects——one on the
use of alleged psychics by law enforcement agencies, and the
other on industry engineers' attitudes toward and experiences
with unidentified flying objects.   It is also compiling a CSAR
directory of consultants.

**902   The Committee for the Scientific Investigation of Claims of
the Paranormal (CSICOP)**

Chairman: Paul Kurtz, Ph.D.

P.O.Box 229
Buffalo, NY   14215—0229

Telephone: (716) 834—3222

This nonprofit scientific and educational organization was
founded April 30, 1976 at the State University of New York at
Buffalo. Its purpose is "to encourage the critical investigation of
paranormal and fringe—science claims from a responsible, scien—
tific point of view and to disseminate factual information about
the results of such inquiries to the scientific community and the
public."   In order to carry out these objectives, the Committee

maintains a network of people interested in critically examining claims of the paranormal; prepares bibliographies of published materials that carefully examine such claims; encourages and commissions research by objective and impartial inquirers in areas where it is needed; convenes conferences and meetings; publishes articles, monographs, and books that examine claims of the paranormal; and does not reject claims on a priori grounds antecedent to inquiry, but rather examines them objec—tively and carefully.

In 1988 there were more than 100 Fellows and scientific consultants of CSICOP, mainly psychologists, philosophers, astronomers, and science writers. CSICOP has a staff of 15 per—sons. There are more than 35 local and regional CSICOP groups in the U.S. and 30 national groups throughout the world with aims similar to those of CSICOP.

The Committee has active subcommittees for UFOs, Educa—tion, Paranormal Health Claims, Astrology, and Para—psychology, respectively. It publishes a quarterly journal, the *Skeptical Inquirer* [849], and it holds an annual convention.

903 International Association for Near-Death Studies, Inc. (IANDS)*

Director: Elizabeth W. Fenske, Ph.D.
Director of Research: Bruce Greyson, M.D.

P.O. Box 7767
Philadelphia, PA 18101—7767

Telephone: None listed

The main purpose of this not—for—profit international educational and research organization incorporated in 1981 is to encourage and promote the scientific study of near—death ex—periences and related phenomena. It is a direct outgrowth of the Association for Scientific Study of Near—Death Phenomena founded in 1977. IANDS differs from the former association in that its scope is international, it has a larger range of programs and activities, and its publications have been upgraded and ex—panded. The specific objectives of IANDS is to import knowledge concerning NDEs and their implications; to encourage and support research dealing with NDEs and related phenomena; to further the utilization of NDE research findings in such settings as hospitals, hospices, nursing homes, funeral establishments, and death education programs; to form local chapters of near—death survivors and others interested in NDEs;

to sponsor international symposia, conferences and other programs concerned with NDEs and related phenomena; to main— tain a library and archives of material pertaining to NDEs; and to provide reliable information through its printed materials, media work, and speaker services.

IANDS is governed by a Board of Directors. It also has an international advisory board. Membership is open to all per— sons, lay or professional, with an interest in NDEs and related phenomena. In addition to General Membership, members may elect to join the Research Division. All general members receive the Association's quarterly newsletter *ReVitalized Signs*. All Research Division members receive, in addition, the quarterly *Journal of Near—Death Studies* [835] (formerly *Anabiosis*).

904 New Horizons Research Foundation

Executive Director: A.R.G. Owen, Ph.D.

P.O. Box 427, Station F
Toronto, Ontario
Canada M4Y 2L8

This is a small nonprofit organization founded by Donald C. Webster in 1970 to investigate parapsychological and other frontier science phenomena and to disseminate information about it. The Foundation at one time published eight issues of the journal *New Horizons* [859], but it has suspended publication, and there are no plans at present to activate it. In association with the Toronto Society for Psychical Research (now defunct), the Foundation was involved in the sitter PK group that produced Philip, an account of which may be found in Owen and Sparrow [628]. Seminars and conferences are held from time to time.

In 1974 the Foundation and the Toronto S.P.R. held a Canadian Conference on Psychokinesis and Related Phenomena and published the papers in *New Horizons* (Jan 1975). The New Horizons Research Foundation also has put together an ex— tensive library of frontier research subjects: history and philosophy of science; psychology, psychedelics, parapsychology, archaeology, anthropology, ethnopharmacology, primitive medicine, earth sciences, strange phenomena, UFOs, radiesthesia, dowsing, unorthodox healing, acupuncture, etc. It maintains contact with various other institutions, societies, and individuals throughout the world, as well as with faculty members of several Canadian universities.

The emphasis of New Horizons Research Foundation is on

research, and it is run by unpaid volunteers who are willing to consult with other researchers but do not have the facilities to respond to requests from the general public.

905 Society for Scientific Exploration (SSE)*

Secretary: Laurence W. Frederick, Ph.D.

Department of Astronomy
University of Virginia
P.O. Box 3818
Charlottesville, VA 22903

Telephone: (804) 924—4905

In 1982 an academic group called the Society for Scientific Exploration was formed to study by scientific means phenomena that are actually or potentially investigatable by established methods or by future developments of these methods, but which are generally regarded by the scientific community as being out—side their established fields of inquiry. The Society, in a posi—tion paper, notes that there is little prospect of progress toward arriving at an understanding of such topics (beginning with the basic question of their reality) until have been subjected to the normal processes of open publication, debate, and criticism that are the lifeblood of science and scholarship. The Society, there—fore, provides a forum by which research on these topics may be presented to other members of the Society and to the scien—tific and scholarly community in general. The Society has no intention of endorsing the reality or significance of any par—ticular topic, but no subject will be prohibited from discussion or publication simply because it is not now an accepted part of scientific or scholarly knowledge.

In 1988 the Society initiated a journal entitled the *Journal of Scientific Exploration* [846] and it publishes a newsletter, *The Explorer* [845]. It holds an annual meeting, and it is consider—ing the publication of a proceedings containing the papers presented at the annual meetings.

The Society's guidelines for membership qualifications state that most elicited members will have a doctoral degree or equiv—alent from a recognized institution of higher learning; an ap—pointment at a university, college, or other research institution; and a record of having made a substantial contribution to a recognized field of knowledge through publication in the scien—tific or scholarly literature. In early 1989 there were 250 mem—bers. The Society is governed by an elected Council.

905a   Society for the Anthropology of Consciousness (SAC)*

President: Geri—Ann Galanti, Ph.D.

2906 Ocean Ave.
Venice, CA 90291

Telephone: (213) 827—0937

The SAC was organized in May, 1984, as the Association for the Anthropological Study of Consciousness (AASC), when it separated from the Association for Transpersonal Anthropology. It was an interdisciplinary organization of academically oriented professionals dedicated to the diverse and growing field of the anthropology of consciousness. Among the aims of the Association was to provide a meeting place for the exploration of consciousness from crosscultural, experimental, experiential, and theoretical perspectives. Members considered phenomena that challenge the world view that underlies the traditional sciences. Areas of interest included states of consciousness (trance, possession, dreams, dissociation, psychophysiological studies of consciousness states), ethnography of shamanic, spiritual, and magical training (initiations, mediumistic communication, mystical and other transpersonal experiences); indigenous healing practices (non—Western psychotherapeutics, psychopharmacology, divination and other rituals); linguistic, philosophical, and symbolic studies (power of language, myth, literature, and oral traditions), and psi phenomena (clairvoyance, clairaudience, clairsentience, clairparlance, remote viewing, OBEs).

Another major change occurred in 1990: the Association was accepted as a unit of the American Anthropological Association, and its name was changed to the Society for the Anthropology of Consciousness. The Society "is an interdisciplinary organization of academically oriented professionals dedicated to the diverse and growing field of the anthropology of consciousness in its social and cultural context." Current 1990 officers are Geri—Ann Galanti, President; Lisa Ann Mertz, Secretary, and James G. Matlock, Treasurer. The Society publishes a quarterly newsletter, *Anthropology of Consciousness Newsletter* (852a), formerly the *AASC Newsletter*, and plans to publish a yearly volume of readings. SAC also holds an annual conference consisting of individual papers, panel discussions, workshops, slide/video presentations, and experiential sessions. Membership in SAC is "open to all who share an interest in consciousness research who are members of the American

Anthropological Association. Associate membership is available for those whose interest in anthropology is strictly avocational and who do not desire income from anthropologically related activities."

COUNSELING CENTERS

906 FRNM Clinical Interest Group

Executive Director: K. Ramakrishna Rao, Ph.D.
Contact Person: Jim Carpenter, Ph.D.

402 Buchanan Boulevard
Durham, NC 27701

Telephone: (919) 688–8241

The purpose of the FRNM Clinical Interest Group is to assist persons who approach FRNM [870] with concerns about ostensive psi experiences that also involve some measure of personal suffering, confusion, or emotional distress/disorder; to study how psi events occur in psychotherapy; and to carry out research on the facilitation of psi functioning by therapeutic techniques.

The Group meets weekly to discuss issues of concern. It conducts telephone and face–to–face counseling with persons who are troubled by ostensive psi experiences. It has conducted research to explore group process as a vehicle of psi response. Members of the group, notably James Carpenter, have published reports of research in the *Journal of Parapsychology* [825] and elsewhere.

907 Parapsychological Services Institute, Inc. (PSI)*

Chairman: William George Roll, Ph.D.

P.O. Box 217
Carrollton, GA 30117

Telephone: (404) 836–8696

This institute was formed in 1987 as a counseling, educational, and research center for those undergoing uncertainties or disturbances related to apparent psychical and spiritual experiences or who wish to explore the meaning and transforma–

tive values such experiences may have for their lives. The orientation of the PSI staff is that the psychic and transpersonal experiences of clients are accepted for what they seem to be with no attempt to reduce them to psychopathological phenomena. The counselors consist of persons with professional backgrounds in such fields as psychiatry, psychology, parapsychology, philosophy, and religious studies. Their aim is to help persons to integrate their life—changing experiences into the broader fabric of human existence.

PSI also assists families who are experiencing disturbances in their homes and businesses. Psychical household disturbances include cases suggestive of recurrent spontaneous psychokinesis (RSPK, popularly known as "poltergeist") and localized psi effects ("hauntings"). PSI also provides counseling for people suffering from intrusive personality disorders ("possession"), and individuals who are hypersensitive to psychical impressions from their social and physical environments. Services include comprehensive evaluation and treatment, including psychological, neurological, and parapsychological assessments.

PSI sponsors courses, workshops, conferences, and publications related to the practical application of parapsychological knowledge. Workshops explore how modern science and parapsychology may be revealing a matrix that connects people with one another, with their physical environment, and possibly even with persons no longer living. The program assists participants to experience themselves and the universe around them with an expanded awareness and fresh consciousness, and helps them to integrate these experiences into daily life. They learn how to use psychic development as a step to spiritual maturity.

There are three classes of membership: associate, professional, and sustaining. In early 1989 there were 125 members.

907a Parapsychologische Beratungsstelle der Wissenschaftlichen Gesellschaft zur Förderung der Parapsychologie (Parapsychological Counseling Center of the Scientific Society for the Advancement of Parapsychology)

Director: Walter von Lucadou, D.Sc., Ph.D.

Hildastrasse 64
D 7800 Freiburg im Breisgau
Germany

Telephone: (0761) 77202

The Scientific Society for the Advancement of Para—

psychology, founded in 1981, is a private organization of Ger—man scientists from different backgrounds and some other per—sons. The goals of the Society are the advancement of serious parapsychological research and the dissemination of knowledge on parapsychology to the public. One of the Society's major activities is the organization of annual workshops. Since 1989, the Society has had a parapsychological counseling center, directed by Walter von Lucadou, formerly assistant at Mischo's university chair (1979—1985) and Visiting Research Fellow (1985—1987) at the Utrecht Parapsychology Laboratory. The Center offers information and counseling services, especially regarding "occult practices" among young people.

908 Spiritual Emergence Network (SEN)

Contact person: Jeneane Prevatt

Institute of Transpersonal Psychology
250 Oak Grove Avenue
Menlo Park, CA 94025

Telephone: (415) 327—2776

SEN was founded by Christina and Stanislav Grof in 1980 as the Spiritual Emergency Network. Its headquarters was at Esalen Institute until 1984, when it was transferred to the In—stitute of Transpersonal Psychology.

Now the Spiritual Emergence Network, SEN is a nonprofit international service that seeks to support people undergoing cer—tain types of psychospiritual crises. SEN operates a free informa—tion and referral service and works to inform the lay and profes—sional community about the forms, incidence, and treatment of transpersonal crises. SEN is supported solely from member—ships, donations, and grants.

SEN is the only international office to serve people under—going a phenomenon not yet recognized by the standard psychiatric and psychological establishment. This phenomenon is called "psychospiritual crisis," or more commonly, "spiritual emergence" or "spiritual emergency." These persons are usually undergoing altered states of consciousness, disorientation, strange bodily sensations, and other symptoms commonly associated with mental disorders. Some label their experiences as kundalini, psychic attack, possession, or other forms of psychic phenomena. Many are confused, isolated, frightened, and afraid they are psychotic.

SEN proposes that many such experiences are natural,

painful stages of development that have been clearly documented
within the world's great spiritual traditions. Given proper con—
text and support, a person can make it through them with mini-
mal damage and, having faced these "demons" in a supportive
context, may emerge with greater levels of maturity, creativity
and integration.

To assist such persons and to further the understanding of
these phenomena in the professional community, SEN integrates
a number of disciplines. Foremost among these are psychology,
psychiatry, comparative religion, anthropology, and parapsychol—
ogy.

Organizationally, SEN is two things——the central office in
California and a loose consortium of Coordinators, Helpers and
members worldwide. SEN's central office is located at the In—
stitute of Transpersonal Psychology (ITP) in Menlo Park,
California. It is here that its work is coordinated worldwide, in—
cluding publishing the SEN *Newsletter* and *Journal*, designing
and conducting workshops and training classes, sponsoring profes—
sional conferences, processing memberships and orders for the
books and articles sold as a public service, compiling statistical
data for research, and maintaining SEN's central information
and referral banks. Secondly, SEN is a loose consortium of
Regional Coordinators and Centers, of which there are 44: 23 in
the United States and 21 abroad. There is no legal connection
between the Regional Coordinators and the SEN central office.
However, new Coordinators are appointed through the central
office and receive as much back—up and support as is prac—
ticable with limited resources. A list of approved Coordinators is
available through the central office.

REGIONAL ORGANIZATIONS

909 Australian Institute of Parapsychological Research (AIPR)*

Contact Person: The Secretary

P.O. Box 445
Lane Cove
New South Wales 2066
Australia

Telephone: 02 660 7232

The AIPR is an incorporated nonprofit organization formed
in 1977 to promote scientific research into all aspects of para—
psychological and related phenomena. It serves as a resource

center; holds public meetings, seminars, and socials; and pub—
lishes a *Bulletin* [852] and a newsletter. The membership
categories are individual, family, and concessional (low income).
As of 1989 there were 130 members. The research objectives of
AIPR are to encourage "the study of objective (observational)
reality of the various reported occurrences of parapsychological
phenomena, including both spontaneous events and phenomena
observed under control." It also is concerned with "the study of
the subjective aspects of psychic phenomena, e.g., how the
phenomena are experienced . . . and how the observational
evidence can be distorted." The use of instruments to record
data wherever possible is encouraged so as to minimize reliance
on eyewitness testimony.

910 Parapsychology Research Group*

President: Saul–Paul Sirag, Ph.D.

3220 Sacramento Street
San Francisco, CA 94115

Telephone: (415) 528–0137

The Parapsychology Research Group (PRG) was formed in
1964 by Russell Targ, Charles Tart, David Hurt, and other in—
terested researchers to serve as a forum for technical discussions
of psi research.

There are several membership classes, three of them non—
voting: associate (dues $20); contributing (Dues $50); and life
($1,000 and more). Regular membership (dues $20) is open to
any person who is (a) professionally engaged in parapsychology
or a related field, and (b) is approved for regular membership
by a majority of the directors.

Monthly meetings are held for the general membership at
3118 Washington Street, San Francisco. The meetings generally
consist of a research presentation by an individual researcher fol—
lowed by a discussion. Smaller technical and business discus—
sions, open to regular members, are also held. A brief summary
of the topic of each meeting together with a biographical sketch
of the presenter, announcements of future meetings, current
research activities of PRG members, a list of current officers
and directors, as well as other PRG news is published in a
one–page monthly mailing. The PRG also sponsors some of the
research programs undertaken by various members and has a
Director of Research.

911   South African Society for Psychical Research

   Contact Person: G. Aubrey du Plessis

   P.O. Box 23154
   Joubert Park
   Johannesburg 2044
   Republic of South Africa

   Telephone: 011 660–3636 (Du Plessis)

   This organization was established in 1955 to investigate phenomena that are purportedly paranormal. There are approximately 120 members in the Johannesburg area and an additional 80 in Durban. Both groups hold monthly meetings in which public lectures are given. The South African Society also holds conferences on parapsychology, two of which have been international. It published the *Parapsychological Journal of South Africa* [860] from 1980 to 1986 and a newsletter entitled *Psi* from 1974 to 1986 (available only to members). From 1957 to 1970 it put out an occasional publication, *Parapsychologica* (9 issues). It also has a library housed at the University of Witwatersrand, Johannesburg. Its holdings were published in the *PJSA* in 1982, Vol. *2*(1), with supplements in later issues.

ORGANIZATIONS PRINCIPALLY INTERESTED IN SURVIVAL OF DEATH

912   College of Psychic Studies

   Contact Person: Brenda Marshall

   16 Queensberry Place
   London SW7 2EB
   England

   Telephone: 071 589 3292/3

   Organized in 1884 by the London Spiritualist Alliance, this British group became the College of Psychic Science in 1955. In 1970 it assumed its present name. This nonprofit organization is aimed at both experienced investigators and members of the general public who wish to experience, experiment, and discuss psi phenomena, with special attention to those phenomena that bear on the question of survival of death and postmortem com–

munication. The College has a 10,000—volume library on psychi—
cal research, spiritualism, and much wider aspects of the
hypothesis of survival; it organizes lectures; classes, courses, and
workshops; and it offers counseling. It publishes the journal
Light [837] three times a year. There are three classes of mem—
bership: associate, full, and fellow. In early 1989 there were ap—
proximately 2,000 members.

913 International Institute for the Study of Death (IISD)*

Director: Arthur S. Berger, J.D.

P.O. Box 8565
Pembroke Pines, FL 33084

Telephone: (305) 435—2730

The Institute was formed in 1985 to serve as a scholarly
and scientific affiliate of the Survival Research Foundation [914].
Its purpose "is to provide an international and interdisciplinary
forum in which leading representatives from many countries and
cultures and such diverse fields as medicine, psychology,
psychiatry, nursing social work, education, liberal arts,
philosophy, science, law, religion, and parapsychology can enjoy
an open and informed dialogue in the exploration of the central
questions raised by death, including questions of a philosophical,
spiritual, legal, practical, psychological, and parapsychological
nature. The exploration of these questions will be conducted in
the best scientific spirit and without advocating a particular
theory or doctrine."
The activities of the IISD include the participation of the
aforementioned representatives in the IISD and their contribu—
tion of new ideas, theoretical concepts, investigative techniques
and creative approaches. The IISD also organizes conferences
and symposia in which sections will be created to conduct spe—
cial lines of inquiry into various themes, including the dying
process; death—related problems (e.g., organ donation, bereave—
ment, legal matters, funeral arrangements, relationship of death
to the living process), and the possibility of an afterlife (whether
reincarnation, resurrection, immortality, or the survival of bodily
death for an indeterminate period of time of some conscious
component of the human being). It is intended that all sections
will interact to the maximum extent possible in order that the
several disciplines represented can relate to, learn from and
cross—fertilize one another so as to broaden understanding and
advance empirical research with respect to each line of inquiry.

The IISD also maintains a research division and publishes multilanguage periodicals, books, and proceedings from its con—ferences.   It also has established a branch in the United Kingdom.

Members of the IISD case from a wide range of disciplines and backgrounds. In general, they hold advanced degrees or use scientific methods, teach, or otherwise contribute theoretical or empirical knowledge to the subject areas explored by IISD. In 1989 there were 175 members.

914 Survival Research Foundation (SRF)*

President: Arthur S. Berger, J.D.

P.O. Box 8565
Pembroke Pines, FL 33024—0565

Telephone: (305) 435—2730

The Survival Research Foundation was incorporated in 1971 by Susy Smith.   It is an independent nonprofit organiza—tion devoted to research and educational activities concerning the question of human survival.

The current program of SRF involves its "Unrecorded In—formation" project, an experimental plan based on a new posthu—mous experiment designed by Arthur S. Berger in which sub—jects, following Berger's formula, convert a message into ciphers with one secret key word, send the ciphered message to the SRF, and plan to communicate the key word after death to decipher the message by SRF members and to obtain multiple replications of them.   In addition to research the Survival Research Foundation distributes literature and maintains a speaker's bureau to inform interested individuals and organiza—tions concerning tests for survival.   It also conducts lectures and workshops to disseminate information about parapsychology in general.

Arthur S. Berger serves as President and Chair of the Board of SRF.   The SRF Directors are Frank Tribbe, Susy Smith, Joyce Berger, and Daisy Zeitlin. Membership, which was approximately 100 in early 1989, is conferred when one par—ticipates in the Foundation's research. Everyone is qualified to participate. There are no dues, but voluntary donations are ap—preciated. Members benefit by being able to participate in the research to obtain evidence of their survival. Donors may also suggest research projects to be pursued.

RELIGIOUS AND HUMAN POTENTIAL ORGANIZATIONS WITH INTEREST IN PSI

915 The Academy of Religion and Psychical Research (ARPR)*

Executive Secretary: Boyce Batey

326 Tunxis Ave.
P.O. Box 614
Bloomfield, CT 06002

Telephone: (203) 242—4593

Founded in 1977 as an academic affiliate of Spiritual Fron—tiers Fellowship, the Academy has three purposes: (1) "To en—courage dialogue, exchange of ideas and cooperation between clergy and academics of religion and philosophy and scientists, researchers and academics of all scientific and humanistic dis—ciplines in the fields of psychical research and new disciplines as well as the historic sciences." (2) "To conduct an education program for these scholars, the Spiritual Frontiers Fellowship membership and the general public, blending data already avail—able in the area of their common interests with the interchange of views stimulated in these scholars, to the end that both the scientific and the religious communities may be better informed about such facts and their propriety, value and respectability." (3) "To work closely with, and offer good offices to, all reputable organizations having related interests in the fields of education, religion, science and psychical research." The Academy is governed by a 24—member board of trustees composed of people of academic stature holding a post—graduate degree or the equivalent in outstanding activity relevant to the interests of the Academy. There are several com—mittees: Fund—Raising, Program, Publications, and Research. There are three membership categories: patron, academic, and supporting. In 1989 the Academy reported a total of 300 mem—bers. The Academy holds an annual convention and publishes the quarterly *Journal of Religion and Psychical Research* [836] and an annual conference *Proceedings* [838]. It makes available audio cassette tapes of papers given at its annual conventions. Membership benefits include the Academy's publications, borrow—ing books from the Spiritual Frontiers Fellowship [926] library by mail, and a 20% discount when attending Academy—sponsored programs. The Academy holds an annual essay con—test on a prescribed topic, with the winning paper featured at the annual conference.

916 [327] **Alister Hardy Research Centre (AHRC)***

Director: David Hay, Ph.D.

29—31 George Street
Oxford OX1 2BR
England

Telephone: 01 (0865) 243006

American branch:
Alister Hardy Research Center (USA) Inc.

President: Dr. John Hickey

302 N. Oak Park Avenue, Apartment 3
Oak Park, IL 60302

Telephone: (312) 386—4031

The Center is a charitable organization founded in 1969 by Professor Sir Alister Hardy as the Religious Experience Research Unit. Shortly before his death in 1985, Sir Alister was awarded the Templeton Prize for Programs in Religion in recognition of the Center's work. After his death the present name of the Center was adopted in honor of its founder.

The Center was formed to try to answer such questions such as why do people have spiritual, transcendental or religious experiences? When, and how often, are they reported? Of what do they consist? How do they affect people's lives? The Center has collected over 5,000 firsthand accounts of these experiences. It conducts surveys and analyzes the results, usually publishing them in book form.

Membership benefits include free newsletters, invitations to lectures, seminars, and use of the Center's library, and opportunities to contribute to the Center's research activities. The Centre publishes a newsletter entitled *Numinis*, which consists of three—four brief articles one or two book reviews, and news of the Centre and its regional groups. In early 1989 the membership was 600.

917 Archaeus Project

Director: Dennis Stillings

Contact Person: Gail Duke

2402 University Avenue
St. Paul, MN 55114

Telephone: (612) 641–0171 (Stillings)
(612) 781–5012 (Duke)

Founded in 1977 to study current and historical processes and techniques for the detection and measurement of bioener-getic fields——electric, magnetic, and psychic——and to apply the results to the realization of human potential and to the diag-nosis, treatment, and prevention of disease. The Project is primarily interested in theoretical aspects of these phenomena with an eye to their potential useful application, especially in the health fields. The term Archaeus was first used by Paracel-sus to refer to the soul—like or psychoid ordering principle of life. It has traditionally embodied such ideas as "energy bodies," synchronicity, paranormal phenomena, and the relationship of consciousness to psychic and somatic conditions.

The Project holds meetings about nine times per year for which it invites speakers throughout the U.S. to present their work. It organizes a major conference every three years. It pub-lishes the annual *Archaeus* [842] and the quarterly *Artifex* [843]. It provides information service networking and library facilities to members. There are four classes of membership based on a graded scale of benefits/dues: member I, member II, student, and associate. There also are subscribers. Both members and subscribers receive the Project's publications. As of March 1989 there were 135 members.

918 Association for Research and Enlightenment (ARE)

President: Charles Thomas Cayce, Ph.D.

67th Street at Atlantic Avenue
Virginia Beach, VA 23451

Telephone: (804) 428–3588

This organization, which was founded in 1931, seeks to give physical, mental, and spiritual assistance through the inves-tigation and analysis of the 14,256 "readings" that form the psychic legacy of Edgar Cayce (1877–1945). The Edgar Cayce Foundation has custody of the readings and conducts a con-tinuous program of indexing, extracting, microfilming, and other—

wise treating the files (which are open to the public). ARE spon—
sors lectures, symposia, and workshops on psychic research,
prayer, and meditation; a summer camp; and over 1500 interna—
tional study groups. It also maintains a 50,000—volume library
on metaphysics, psychic phenomena, and related subjects, and a
bookstore (including mail—order sales). It has its own research
program directed by Douglas Richards, and it maintains contact
with a medical research program at the ARE Clinic in Phoenix,
AZ and the Harold J. Reilly School of Massotherapy. Research
is conducted on aspects of parapsychology, dreams, altered
states of consciousness, and transpersonal theories of human
growth and development. Research projects focus on subjects
testing hypotheses in home study format. It also has initiated
Atlantic University [893]. Publications consist of a bimonthly
magazine, *Venture Inward*, books, booklets, and studies based on
or related to the readings. It holds an annual convention in Vir—
ginia Beach. In early 1989 there were 90,000 members and a
staff of 150.

919 Center for Applied Intuition

Director: William H. Kautz, Sc.D.

2046 Clement Street
San Francisco, CA 94121

Telephone: (415) 221—1280

The Center is devoted to the practical application of intui—
tively derived information in the sciences. It is working to es—
tablish effective bridges between the inner, undeveloped creative
faculties of humans and those aspects of science and the social
sciences that are oriented toward the solution of this planet's
most serious problems. Toward this end The Center for Ap—
plied Intuition (CAI) uses a problem—solving method called
"Intuitive Consensus," a variant of majority vote analysis of the
statements of several psychics or intuitives. This method is
being offered to directors of scientific research programs,
laboratory managers, and research scientists in industry, govern—
ment, and nonprofit organizations. It works best with stubborn
scientific and technical problems whose solutions requires new
information, new perspectives, or new ideas. Because the valida—
tion of intuitively derived information is a key issue, the
problems selected for early application of the methods should be
those for which the *testing* of proposed solutions can be carried
out readily.

Other activities of CAI are a training program for assisting scientists and other professionals to enhance and expand their own intuitive abilities, the development of a large computer database of intuitively derived information in selected fields, and some of the more unusual, high—impact hypotheses that have already arisen from inquiries in physics, chemistry, and geophysics. The Center formerly published *Applied Psi* [853], and now publishes the magazine *Intuition: The Journal of Intuitive Discovery.*

920 Churches' Fellowship for Psychical and Spiritual Studies

The Priory
44 High Street
New Romney, Kent TN28 8BZ
England

Telephone: (0679) 66937

Reginald M. Lester and a group of clergy and laypersons interested in psychical research and Christianity founded the Churches' Fellowship for Psychical and Spiritual Studies in 1953. A registered charitable organization, its aim is "to interpret in the light of continuous research the Gifts of the Spirit as set out by St. Paul." It tries to help all aspirants towards the spiritual life with advice about meditation, prayer, and the psychology of the growth of the whole person. It also aims at bringing about a greater understanding of the varied kinds of spiritual healing and a more effective and intelligent practice of these ministers within the churches.

In order to serve its ends the Fellowship has divided England into 10 regions (the first including London, Home Islands, and Overseas). Groups of the Fellowship are found within each Region so that members in that area may meet for fellowship and study. Conferences and retreats are held in the various Regions, and the Fellowship publishes two periodicals, the *Quarterly Review of the Churches' Fellowship for Psychical and Spiritual Studies* and the *Christian Parapsychologist* [834]. Study material and booklets are published occasionally. A library is maintained at the London office for use of all members and books may be borrowed by mail.

The Fellowship is funded by its members. There are two classes of membership: members and associates. Members must be practicing members of a Christian Church and can vote and serve as officers of the Fellowship. Associates may speak at any general meeting of the Fellowship but may not hold office or vote.

921   Esalen Institute Study of Exceptional Functioning

Director: Michael Murphy
Contact person: Margaret Livingston, Director of Research

230 Forbes Ave.
San Rafael, CA 94901

Telephone: (415) 459–7270

The Project began in 1976 as The Transformation Project with a grant from Laurance Rockefeller. It was conceived in a novel written by Michael Murphy called *Jacob Atabet*, and since its beginning the archive has collected almost 10,000 articles on exceptional physical functioning. The ideas behind the project propose that further human development will include the power to radically alter the body through various psychophysical disciplines and liberated powers of the mind. Thus, the purpose of the project is to compile and analyze scientific and anecdotal evidence for bodily transformation from a wide variety of activities, including religious practice, sport, hypnosis, biofeedback, parapsychology, spiritual healing, and imagery.

The Project is housed in a small office outside of San Francisco, which includes its archive of over 10,000 articles (80% of them scientific). The Project's activities are limited to the collection of information and the publication of a monograph series, the first of which was published in 1988 by the Institute of Noetic Sciences [923]. It is entitled *The Physical and Psychological Effects of Meditation: A Review of Contemporary Research with a Comprehensive Bibliography 1931–1988*. Currently Michael Murphy is working on an ambitious compendium of information and theory entitled *Future of the Body* (to be published by Tarcher in 1991).

922   Foundation for Mind-Being Research

President: William C. Gough, Ph.D.

442 Knoll Drive
Los Altos, CA 94022

Telephone: (415) 941–7462

This nonprofit organization was established in 1980 to as—
sist in the evolution of consciousness studies and to advance
this new field as a science. The major project of the Founda—
tion is to facilitate the development of an integrated/interdis—
ciplinary model(s) of consciousness by encouraging a rather
unique group of creative researchers and scientists. Competent
individuals are sought who (1) are open to a multi—disciplinary
approach to knowledge; (2) recognize that they are an integral
part of any experiment; and (3) know and are prepared to ex—
tend their own level of consciousness. To accomplish its objec—
tives the Foundation holds monthly meetings of its members.
These alternate between lectures on cutting edge topics and in—
teractive discussions among members on a specific focus.
Copies of the final report of the first meeting, entitled *The
Status of Research on the Physics of Consciousness——Working
Models and Experiments* can be obtained for $3.85 from the
Foundation. (It emphasized remote perception and psychokinesis
and it formed part of the "Survey of Science and Technology
Issues, Present and Future" by the Committee on Science and
Technology, U.S. House of Representatives.) The Foundation
publishes a monthly Newsletter/Meeting Announcement plus
reports on specific topipcs. Each year the Foundation selects a
theme that serves as the focus of its lectures and other ac—
tivities. The 1989 topic was "Information, Energy and Meaning."
At the start of 1989 there were approximately 75 members.

923 Institute of Noetic Sciences (IONS)

President: Willis W. Harman, Ph.D.
Director of Communications: Barbara McNeill, Ph.D.
Vice—President for Research: Brendan O'Regan, Ph.D.

475 Gate Five Road
Suite 300
Sausalito, CA 94965

Telephone: (415) 331—5650

The Institute of Noetic Sciences (IONS) was founded in
1973 by Apollo 14 astronaut Edgar D. Mitchell, Ph.D. It is an
independent, tax—exempt, nonprofit public foundation dedicated
to the study and development of human consciousness. The
Institute's name derives from the Greek word "nous," meaning
variously, knowledge, consciousness, or intelligence. The In—
stitute has as part of its mission to explore and study all ways
of knowing in the human experience. The Institute promotes

the view that a fundamental change is occurring on the planet which includes the development of new scientific approaches that expand or replace the traditional, reductionist models of scien-tific research. Therefore, the Institute conducts and funds research on many aspects of human consciousness that employ new models of research.

IONS is funded by a combination of membership contribu-tions, voluntary donations, and research grants. Membership is open and offered at various levels, depending on the size of the contribution. Student and senior citizen memberships are also available. As of early 1989 there are over 20,000 members.

Principal research conducted by the Institute explores the link between the mind and the body as it relates to the inner mechanisms of the healing response. Psychoneuroimmunology is one important area of research, and a large bibliography on spontaneous remission has been assembled. Other areas include exceptional human conditions and abilities, such as the autistic-savant syndrome. Specific projects have been funded to study causality, cancer and imagery, biofeedback, near−death ex-periences, and multiple personality disorder.

Equally as important in its program as its research, the Institute is involved with various global peace activities and educational programs. Each year, the Institute sponsors and conducts the Temple Award for Creative Altruism, given to in-dividuals and groups who demonstrate outstanding altruism and service in their communities. Nominations are accepted from the public each spring. The Institute is involved in disseminat-ing information about global mind change and shifting paradigms of consciousness through its publications and books. The Institute's publications include *Noetic Sciences Review* [848] (quarterly), a *Catalog* (biannually with supplements), and oc-casional special reports. These are currently available to mem-bers only. The Institute was instrumental in publishing *The Home Planet*, and several other books. Catalog prices include a 15% discount to members. The Institute's library now com-prises over 3,000 volumes.

924 Spiritual Frontiers Fellowship International (SFFI)

President: Paul B. Fenske

P.O. Box 7868
Philadelphia, PA 19101

Telephone: (215) 222−8459

This inter—faith group of ministers and laypersons was formed in 1956 "to sponsor, explore and interpret the growing interest in psychic phenomena and mystical experience within and outside the church, wherever these experiences relate to ef— fective prayer, spiritual healing and survival of consciousness." Activities involve some research programs, Prayer—Healing Minis— teries, study courses, seminars, conferences, and tours. It main— tains a library of some 13,000 volumes on parapsychology, religion, and metaphysics. It holds six regional summer retreats per year plus an annual national conference. It publishes a quar— terly journal, *Spiritual Frontiers* [839], and a monthly newsletter. Current committees are Children's Work, Church Relations, Col— leges and Seminaries, Endowment, Leadership Training and Group Organization, Research, Retreats and Seminars, and Study Groups. As of early 1989 there were 5,000 members. SFFI has one affiliate: The Academy of Religion and Psychical Research.

In 1989 SFFI moved into its new Philadelphia head— quarters. The new quarters will not only house the office facilities and the lending and research libraries, but it contains kitchen facilities, meeting rooms, and some space for small group study and retreat activities. Possible new services include a small media program production capability, a university without walls program, and a desktop publishing operation.

925 Unitarian Universalist Psi Symposium (UUPS)

President: Rev. Peter Lanzillotta

Unitarian Universalist Church
101 Forest Avenue
Swampscott, MA 01907

Telephone: (617) 595—8836

This organization was formed in 1969 by a group of Unitarian Universalist ministers, including Robert Slater and Richard Fewkes, under the preliminary title of "Unitarian Universalists Interested in ESP." It is devoted to the study and discussion of psi phenomena and to relating the results and implications of such studies to a liberal religious and philosophi— cal perspective. The symposium seeks to promote interest in this field throughout the denomination, the Unitarian Univer— salist Association, and beyond. In 1970 it was formally af— filiated with the Unitarian Universalist Association. It was then the only parapsychology interest organization in the U.S. af—

filiated with a religious denomination. A statement of purpose revised in 1989 states: "The purpose of the UU Psi Symposium is to provide an organization for Unitarian Universalists and others interested in the study of metaphysics, holistic health, meditation, psi, spiritual development and related subjects."

Currently UUPS has 12 affiliated chapters and a membership that varies between 600 and 650. The UUPS publishes an annual journal and quarterly newsletters. It holds regional conferences in the spring and fall, and a weeklong summer conference at Ferry Beach in Saco, ME. It also holds monthly chapter meetings, lectures and workshops at the Unitarian Universalist General Assembly, and provides a Program Packet for chapters, churches, and others available for a fee.

IV
Information on Parapsychology in General Sources of Information*

QUICK INDEX TO SECTIONS IN THIS CHAPTER

*This chapter is based on a PSI Center pamphlet entitled "Using the Library to Find Out About Parapsychology." The only editorial changes that have been made are those that would integrate it with the rest of this book. Because it is aimed at grade school students through specialist adults, the language used is as simple as possible. Although the material listed here does list basic information on parapsychology in nonparapsychological sources, unfortunately the material available is too general or of too poor a quality to be recommended without reservation, but there are exceptions.

Both as a reference librarian in a mid–sized public library where questions on parapsychology frequently arise and as the director of a specialized parapsychology information center that receives many requests from students for information on parapsychology for term papers and other purposes, I have tried to find ways of answering the needs of the general public and students in particular. At the Parapsychology Sources of Information Center we have compiled specialized bibliographies on aspects of parapsychology, including one geared to teachers and students, a guide to information on carrying out experiments, a leaflet on careers in parapsychology, and an education packet for students. This entire book, with the exception of this chapter, is on specialized parapsychological sources of information. But in this chapter I will describe the *general* information sources likely to be available in public, school, and university libraries that could answer the needs of students and inquisitive laypersons. Throughout the chapter I will deal primarily with sources of information on parapsychology in general, as opposed to specific aspects of the subject.

The types of sources I checked were general and specialized encyclopedias, general interest magazines, nonparapsychological journals, chapters, psychology textbooks, and general reference works. What I looked for were reasonably sound introductions to the field of parapsychology as a whole. I also looked for paired articles or chapters presenting pro and con views of parapsychology.

In a sense this chapter is both a replacement for and an expansion of the chapter on encyclopedias in *Sources*. Encyclopedias are covered here, but not quite as intensively. Instead they are viewed as one possible source of information on parapsychology that is generally available.

The best choices in each category are indicated by a letter in parentheses following a citation. Sometimes a document will be suited for more than one class of readership. The symbols used are G (for grade school to young adult), H (for high school and general adults), and U (for undergraduates and serious adults). The first place to start when using a library* is with a reference librarian, who is trained to get you off on the right path better than you could do on your own. However, even librarians cannot be expected to be able to locate sources of parapsychological information in any depth––hence

---

*General guidance for using the library to locate information on parapsychology is the subject of two articles by White (1965, 1976). Note: Citations such as the preceding in which an author's name is followed by a date or dates are to items in the reference list at the end of this chapter.

this chapter, which is aimed at both librarians and library users who want to find out what a particular library has to offer on parapsychology.

THE CARD CATALOG

The first place to start is with the card catalog (which in some libraries will now be in a computer). It is very important to look under the proper subject headings because the information you need may be available but not under the subject heading you might think to look under. For example, if you look under PARAPSYCHOLOGY, you will almost always find that that heading is only used as a *see* reference to PSYCHICAL RESEARCH. You therefore must look under PSYCHICAL RESEARCH to find a list of books on parapsychology. If you are interested in mediums or mediumship or survival of death, information will most likely by found under the headings of SPIRITUALISM, GHOSTS, APPARITIONS, or MEDIUMS. For precognition use PROPHECY. For unorthodox healing use MENTAL HEALING and FAITH CURE.

CLASSIFICATION SYSTEMS

Dewey Decimal

Another way to look for books on parapsychology is to bypass the card catalog and go directly to the shelves. Most libraries that use the Dewey Decimal classification system would shelve books on parapsychology in 133.107, 133.8, or 133.9, although some may simply use 133. If you go to the library and can't remember the number for parapsychology, look under PSYCHICAL RESEARCH and check the upper lefthand corner of the cards with that heading to see what number is most often used.

Library of Congress

The Library of Congress classification, which combines letters and numbers, is used primarily in college and university and large public libraries. The most relevant classifications are BF1001−1389 (psychical research, psychology of the unconscious) and BF1068−1389 (sleep, dreaming, hypnotism, and spiritualism).

ENCYCLOPEDIAS

A basic reference tool for information on many subjects is the general encyclopedia. Most encyclopedia editors have asked either parapsychologists or acknowledged critics to write the material on parapsychology. To get a balanced picture, read articles representing both sides. In the descriptions of encyclopedia entries on parapsychological topics that follow, the authors are identified as parapsychologists or critics. Some of the articles in encyclopedias described in *Sources* are still of value but will not be repeated here.* Of special note are *Catholic Encyclopedia for School and Home* [285], *Chamber's Encyclopedia* [286], *New Catholic Encyclopedia* [297], *The New Outline of Modern Knowledge* [299], *Encyclopedia of Philosophy* [304], *Encyclopedia of Psychology* [305], *Encyclopedia Science Supplement for 1966* [308], *International Encyclopedia of the Social Sciences* [310], and *McGraw–Hill Encyclopedia of Science and Technology* [311]. The summary of the chapter on encyclopedias in *Sources* (pp. 189–191) is still worth reading.

General Encyclopedias

General encyclopedias are aimed at general audiences and they attempt to cover the whole range of human knowledge. Sometimes they are geared to a particular age range. The general encyclopedias checked are listed here alphabetically by title. The age range for each one is indicated.

The general encyclopedias reviewed in *Sources* that are still of value are *Catholic Encyclopedia for School and Home* [285], *Chamber's Encyclopedia* [286], *New Catholic Encyclopedia* [297], and *The New Outline of Modern Knowledge* [299].

927 *Academic American Encyclopedia.* Princeton, NJ: Arete Publishing Co., 1981. 21 vols. (G) (H)

For students in junior high school, high school, or college and the inquisitive adult. Note: Also published in North America as the *Lexicon Universal Encyclopedia* and abroad as the Grolier. There is a one and a half–page article on PARA–PSYCHOLOGY written by a prominent critic, Ray Hyman. There is a creditable bibliography of general books on the subject.

---

*Exceptions are newer editions of some of the titles if significant changes have been made.

928 *Collier's Encyclopedia With Bibliography and Index.* New
York: Macmillan Educational Company, 1984. 24 vols.
Aimed at adults, and college and advanced high school
students. (H) (U)
Contains a four—and—a—half—page article on PARA—
PSYCHOLOGY by psychologist/parapsychologist John Beloff.
Good bibliography.

929 *Collier's 1976 Year Book Covering the Year 1975.* New
York: Macmillan Educational Company, 1975. [Same
readership as *Collier's Encyclopedia*] (H)
Contains a six—page article on parapsychology by psy—
chologist/parapsychologist Stanley Krippner. He provides a brief
history and an overview of research with a balanced view of
criticisms. The dream—ESP experiments that Krippner pioneered
with Montague Ullman are emphasized. There is a brief glossary
of terms and a list of nine books.

930 *Encyclopedia Americana.* International edition. Danbury,
CT: Grolier Incorporated, 1988. 30 vols. For adults and
college and advanced high school students. (H) (U)
There is a one—and—a—half—page article on PARAPSY—
CHOLOGY, one—and—a—half—page article on EXTRASEN—
SORY PERCEPTION, and a one—page article on
PSYCHOKINESIS by psychologist/parapsychologist Gertrude
Schmeidler. Three—four good references are cited after each.

931 *Great Soviet Encyclopedia.* New York: Macmillan, 1978.
For adults and serious students.
Contains an article under PARAPSYCHOLOGY by two
Soviet psychologists, V.P. Zinchenko and A.N. Leont'ev. The
article deals first with parapsychology in general, then European
parapsychology, and finally with Soviet parapsychology. The
tone is moderately skeptical. There are seven references, of
which three are by Americans, two by Russians, and two by
Europeans.

932 *Merit Students Encyclopedia.* New York: Macmillan Educa—
tional Co., 1981. 20 vols. For 5th—grade students
through high school. (G) (H)
Contains a two—page article on PARAPSYCHOLOGY by
psychologist/parapsychologist Robert L. Morris. References are
only to two journals.

933 *New Book of Knowledge.* Danbury, CT: Grolier Incor—
porated, 1986. 21 vols. Preschool children through the

age when they are ready for an adult encyclopedia.
Contains a two–page article on EXTRASENSORY PER–
CEPTION that is slightly critical in tone. It was reviewed by
Ernst Z. Rothkopf of Bell Telephone Laboratories, but no
author is named. No literature citations are given.

934 *New Encyclopaedia Britannica.* 15th ed. Chicago: En–
cyclopaedia Britannica, 1986. 32 vols. For educated
adults and serious students.
The new format of the *Britannica* has not served para–
psychologists well. A third of a page is devoted to "parapsy–
chological phenomenon[*sic*]," and it is unsigned and very out–
dated.

935 *World Book Encyclopedia.* Chicago: World Book, Inc.,
1988 (c1987). 22 vols. Students in grade school, junior
high, and high school. (H)
This is unchanged from the 1972 edition reviewed in
Sources [**301**] except that the paragraph on TELEPATHY is
written by critic James E. Alcock.

Specialized Encyclopedias

Specialized encyclopedias are aimed at readers interested in
a specific subject area. Although the entries are written by
specialists, they are usually aimed at general adult readers. In
the case of parapsychology, the information provided in the spe–
cialized encyclopedias, even though they purposely emphasize a
specific subject angle, is usually of a better quality (and quan–
tity) than that in many of the entries found in the general en–
cyclopedias.
The specialized encyclopedias reviewed in *Sources* that are
still of value are *Encyclopedia of Philosophy* [**304**], *Encyclopedia
of Psychology* [**305**], *Encyclopedia Science Supplement for 1966*
[**308**], *International Encyclopedia of the Social Sciences* [**310**], and
McGraw–Hill Encyclopedia of Science and Technology [**311**].

936 Adelman, George (Ed.). *Encyclopedia of NeuroScience.*
Boston: Borkhauser, 1987. 2 vols. Aimed at non–
specialist adults. (H) (U)
There is a four–page article under PSYCHIC
PHENOMENA by Robert G. Jahn. This is one of the best
sources for current research (the author's own research at Prin–
ceton Anomalies Research Laboratory is emphasized). There are
five references to further reading.

937 Corsini, Raymond J., and Ozaki, Bonnie D. (Eds.). *En–*

cyclopedia of Psychology. New York: John Wiley, 1984. 4 vols. Aimed at the "average intelligent layman." (H) (U)

Overview of parapsychology by J.B. Rhine's successor at FRNM, K.R. Rao. He presents key terms, history, types of research, scoring patterns, theories, and criticisms.

938 Deutch, Albert (Ed.). *Encyclopedia of Mental Health.* New York: Franklin Watts, 1963. 6 vols. For the general reader. (H)

Psychiatrist/parapsychologist Montague Ullman in an article "Parapsychology and the Occult" provides a half page each to describing eight key parapsychology terms. He also describes the prominent people in the field, organizations, the relationship of psi to mental health and of parapsychology to psychotherapy. He closes with a discussion of possible applications and the significance of psi research.

939 Eliade, Mircea (Ed.). *The Encyclopedia of Religion.* New York: Macmillan, 1987. 16 vols. For general students.

There is a four—page article entitled "Modern Perspectives" on miracles by Morton Kelsey. Half of the article is on "Scientific Study of Paranormal Experience." It contains an outline of parapsychological types and concludes that "the most recent studies seem to suggest that the issue of miracle or paranormal experience remains an open question in the Western World" (Vol. 9, p. 551). There are two citations to works on parapsychology.

940 Goldenson, Robert M. *The Encyclopedia of Human Behavior: Psychology, Psychiatry, and Mental Health.* Garden City, NY: Doubleday, 1970. 2 vols. Aimed at general adults.

There are three articles, presumably by Goldenson: overviews of extrasensory perception, psychical research (parapsychology), and telepathy in psychotherapy. All are quite thorough and objective.

941 Harré, Rom, and Lamb, Roger (Eds.). *The Encyclopedic Dictionary of Psychology.* Cambridge, MA: MIT Press, 1983. 1 vol. Aimed at laypersons, students, and social scientists. (H)

There is a two—page article by parapsychologist/psychologist Charles T. Tart under PARAPSYCHOLOGY. It provides a good history as well as an overview of contemporary ESP and PK research. There is a six—item bibliography.

942 Kaplan, Harold I., Freedman, Alfred M., and Sadock, Benjamin J. (Eds.). *Comprehensive Textbook of Psychiatry/III* (3rd ed.). Baltimore, MD: Williams & Wilkins, 1980. 3 vols. Aimed at professionals. (U)
There is an excellent article under PARAPSYCHOLOGY by psychiatrist/parapsychologist Montague Ullman. It begins with a general overview of the evidence, a history, and a survey of major directions. The second part is on parapsychology and psychiatry with the emphasis on psychoanalysis and parapsychology. There are sections on presumably telepathic dreams. The article closes with a section on implications. There is a bibliography of two pages (four columns).

943 Kaplan, Harold I., and Sadock, Benjamin J. (Eds.). *Comprehensive Textbook of Psychiatry* (4th ed.). Baltimore, MD: Williams & Wilkins, 1985. 1 vol. Aimed at specialists.
The only reference to parapsychology consists of a half page in a two-page article on Gardner Murphy. The subsection is a summary of Murphy's views on parapsychology.

944 *McGraw-Hill Encyclopedia of Science & Technology* (5th ed.). New York: McGraw-Hill, 1982. 20 vols.
The latest edition of this standard work does not even mention parapsychology or any other parapsychological term! If available, see the 4th ed.

945 *New Book of Popular Science.* Danbury, CT: Grolier Incorporated, 1988. 6 vols. Upper elementary grade students through high school.
Contains an unsigned paragraph on parapsychology under the general heading of PSYCHOLOGY.

946 *The Raintree Illustrated Science Encyclopedia.* Milwaukee, WI: Raintree Publishers, 1983. 20 vols. This science encyclopedia is aimed at elementary school and junior high school users.
There is what amounts to a one-page article on EXTRASENSORY PERCEPTION on pp. 587–588 of Vol. 7. Written by two named writers on the editorial staff of the encyclopedia, it provides a fair overview including definitions, history, scientific status, and potential importance of parapsychology.

947 *Science and Technology Illustrated: The World Around Us.* Chicago, IL: Encyclopaedia Britannica, 1984. 28 vols. For older students and adults.

Contains an unsigned article on PARAPSYCHOLOGY in Vol. 19, pp. 2378–2379 that emphasizes historical aspects rather than current parapsychology. It touches on the need for theory and the scientific status of the field. Although this encyclopedia emphasizes instrumentation, none of the apparatus developed by parapsychologists is described. Fortune–telling devices are mentioned in a picture caption.

948 Wolman, Benjamin B. (Ed.). *International Encyclopedia of Psychology, Psychoanalysis, & Neurology.* New York: Van Nostrand Reinhold, 1977. 12 vols. Aimed at educated adults and specialists. (H) (U)

There is a two–and–a–half–page article on general parapsychology by psychologist/parapsychologist Gertrude R. Schmeidler that summarizes all aspects. There is also a four–page article by psychiatrist/parapsychologist Montague Ullman on PARAPSYCHOLOGY AND PSYCHIATRY. The first half is an historical overview and the second is a review of the psychiatric implications of psi. There are two citations for the Schmeidler article and eight after the Ullman piece.

PERIODICALS

Magazines

The National Magazine Index was searched online back to 1959 through January 1989 for general articles on parapsychology using the descriptor PSYCHICAL RESEARCH. (Although hundreds of articles were located, most of them were on specific aspects of parapsychology––such as criminal investigation and psi or out–of–body experiences––or on fringe topics not truly parapsychological, such as UFOs, and astrology.) Considering the time span involved, the yield was amazingly low. Because students are not likely to have access to the National Magazine Index online, I am listing the articles located that I consider provided fairly general coverage of parapsychology. None of them is very good. Probably the best ones are those by Frumkes and by Wade. Hoover's coverage is fair.

Donovan, Jennifer Boeth. Psychic powers. Fact? Fantasy? Fraud? *Woman's Day*, 1987 (Nov 24), *51*(1), 102–104, 220–221. (G)

Popular article emphasizing psychics. Includes a questionnaire.

Frumkes, Lewis Burke. Psychic phenomena: Miracle or myth?

Harper's Bazaar, 1980 (May), *113*(3222), 126—127, 42—44.
(G)
    Basic nontechnical introduction to the fundamentals of
parapsychology, including criticisms and a do—it—yourself
ESP test.

Hoover, Eleanor Links. Science's stepchild. *Human Behavior*,
    1978 (Jun), *7*(6), 14—15. (H)
    Based on talks with Norma Bowles and Fran Hynds
about their book, *Psi Search* [508], a well—received introduc—
tion to parapsychology.

Huyghe, Patrick. Exploring the unexplained.   *Science Digest*,
    1983 (Dec), *91*(12), 68—71, 127.
    About the 1983 annual meeting of the Society for
Scientific Exploration. Included is a description of psi
research at Princeton, SRI International, and Syracuse
University.

Marks, David F. Explaining the paranormal. *World Press
    Review*, 1986 (May), *33*(5), 53—54.
    Excerpt from the British science weekly *Nature*.
Marks, a critic, discusses what disposes people to believe in
the paranormal.

Myers, David G. ESP and the paranormal: Supernatural or su—
    perfraud? *Christianity Today*, 1983 (Jul 15), *27*(11), 14—17.
    Critical review of parapsychology by a psychologist
who explains everything away as illusion.

Randi, James. Parapsychology: A doubtful premise. *The
    Humanist*, 1984 (Nov/Dec), *44*(6), 23—27, 42.
    Predictably critical article from the "Amazing Randi."

Wade, Nicholas. Psychical research: The incredible in search of
    credibility. *Science*, 1973 (Jul 13), *181*(4095), 138—143. (H)
    (U)
    A fair overview emphasizing scientific acceptance of
parapsychology.

Wolkomir, Richard. Do we really have a sixth sense?   *McCall's*,
    1983 (Nov), *111*(2), V—10. (G)
    Positive piece. The author interviewed parapsychologist
Loyd Auerbach.

    Because the yield was so low, I turned to the section of
Parapsychology Abstracts International [828] that covers articles

on parapsychology in general interest magazines. Although *PAI* did not begin publishing until 1983, it covers earlier as well as current articles. I located the following additional titles, includ—ing some slated for publication in future issues of *PAI*. A num—ber of the articles listed are by persons active in para—psychology: John Beloff, R. McConnell, G. Murphy, J.B. Rhine, D.S. Rogo, J.H. Rush, I. Stevenson, M. Ullman (and S. Krippner), and A. Vaughan. The results of Greeley's survey should be general interest. The articles by O'Neil and Policoff are good introductions.

Beloff, John. Coming to terms with parapsychology. *Encounter*, 1980 (Jan), *54*, 86–91. (H) (U)
Parapsychologist/psychologist offers a few pointers and guidelines for those who are still undecided but who wish to take a position on parapsychology.

Beloff, John. Why parapsychology is still on trial. *Human Na—ture*, 1978 (Dec), *1*(12), 68–74. (H) (U)
Review of the evidence for psi phenomena and assess—ment of the current position of the field.

Cherry, Lawrence. Physicists explain ESP. *Science Digest*, 1980 (Sep/Oct), *88* (Special Edition), 84–88, 116. (H)
In the course of reviewing some ideas borrowed from physics that might help explain psi, Cherry cites several research findings and describes the basics of psi research.

Cohen, Daniel. ESP: Science or delusion? *Nation*, 1966 (May 9), *202*(19), 550–553. (G)
Science writer describes current research and deflates some psychic claims popularized by the press.

Dean justifies psychic research. *Science News*, 1979 (Nov 24), *116*, 358–359.
Describes the parapsychology research of Dean Robert G. Jahn of the Princeton Engineering Anomalies Research Laboratory.

Gliedman, John. Mind and matter. *Science Digest*, 1983 (Mar), *91*(3), 68–72, 115.
Overview of David Bohm's implicate order theory that could "someday confirm the reality of ESP–like phenomena and retrieve from oblivion the long–forgotten past——since past, present and future coexist outside of time in a holographic universe" (p. 68).

Gray, Francine du Plessix. Parapsychology and beyond. *New York Times Magazine*, 1974 (Aug 11), 13+(6). (H)
Journalist and author reviews the current scene in parapsychology and discusses some of the theories offered to explain psi.

Greeley, Andrew. Mysticism goes mainstream. *American Health*, 1987 (Jan/Feb), *6*(1), 47—49. (G) (H) (U)
Report of a survey indicates that 42% of American adults feel they have contacted someone who has died and 67% report having had an ESP experience.

Holden, Constance. Parapsychology update. *Science*, 1983 (Dec 2), 997.
On congressional interest in parapsychology.

McConnell, Robert A. ESP—fact or fancy? *Scientific Monthly*, 1949 (Aug), *69*(2), 121—125. See 1949 (Oct), *69*, 280 and 1950 (Feb), *70*, 138—140 for discussion. (H)
Physicist/parapsychologist reviews the evidence, which is contested in a discussion with physicist Warren Weaver.

Miller, Laurence. Weird psi—ence: Getting a grip on ESP, psychokinesis and out—of—body experiences. *Health*, 1986 (May), *18*, 78—79+(5).
Reporter discusses the evidence for ESP and psychological and physiological correlates associated with it.

Murphy, Gardner. Parapsychology: New neighbor or unwelcome guest? *Psychology Today*, 1968 (May), *1*(12), 52—55+(5). (H)
Psychologist/parapsychologist reviews the status of parapsychology and emphasizes psychological variables. He lists five of the major needs of the field.

O'Neil, Paul. Science ponders the ESP mystery. *Smithsonian*, 1973 (Jul), *4*(4), 76—81. (G) (H)
Good elementary introduction to parapsychology. Some tests for ESP, PK, and precognition are described.

Policoff, Stephen Phillip. The spooky gift of ESP. *Cosmopolitan*, 1982 (Jun), *192*(6), 250—253, 267. (G) (H)
This journalistic account provides balanced overview of parapsychology.

Rhine, J.B., and Hall, Mary Harrington. A conversation with J.B. Rhine and Mary Harrington Hall, on psychology and parapsychology. *Psychology Today*, 1969 (Mar), *2*(10), 20—25,

68. (H)
This interview with J.B. Rhine, the founder of modern parapsychology, is a history of his and his wife's involve—ment in the field and a discussion of what they hoped to accomplish.

Rogo, D. Scott. The case for parapsychology. *The Humanist*, 1977 (Nov/Dec), *37*(5), 40–45. (H)
Popular writer presents the case for parapsychology and takes the critics to task for not keeping up.

Rush, Joseph H. The rediscovery of mind. *The Humanist*, 1960 (May/Jun), *22*(3), 143–153.
Physicist/parapsychologist holds that psi phenomena challenge our current conception of the objective world.

Stevenson, Ian. The uncomfortable facts about extrasensory per—ception. *Harper's*, 1959 (Jul), *219*, 19–25. (H)
A psychiatrist/parapsychologist reviews the evidence for psi—emphasizing interpersonal relations and psi. He dis—cusses spontaneous psi experiences as well as experiments and responds to the criticism of George R. Price.

This is what you thought about . . . extrasensory perception. *Glamour*, 1981 (Feb), *79*, 37.
Presents the results of a survey on ESP and other parapsychological experiences that *Glamour* conducted in the December 1980 issue.

Ullman, Montague, and Krippner, Stanley. ESP in the night. *Psychology Today*, 1970 (Jun), *4*(1), 46–50+(6).
A psychiatrist and a psychologist, both of whom are also parapsychologists, describe research on telepathy and dreams.

Vaughan, Alan. Toward a technology of psi. *Analog Science Fiction/Science Fact*, 1984 (Nov), 58–68. (H)
Summary of recent evidence for psi by a psychic/parapsychologist.

Weintraub, Pamela. Test your psychic abilities. *Omni*, 1987 (Oct), *10*, 19–20, 150.
Describes current psi research, primarily remote view—ing at SRI International. Includes a precognitive remote viewing test run by Russell Targ.

Wydler, Joseph. Psychic powers in animals? *Science Digest*,

1979 (May), *85*, 173—177.
Describes the various forms in which a number of animal species have exhibited behavior suggestive of psi.

Articles on Parapsychology in Nonparapsychological Journals

Parapsychology interfaces with many disciplines: anthropology, archaeology, biology, education, engineering, mathematics, philosophy, physics, psychology, religion, statistics, and more. Each one of these subjects has its own specialized reference works——indexes, directories, dictionaries, handbooks, etc. Sometimes one can find information about parapsychology by using these specialized tools——but except for journal indexes and encyclopedias, it is not recommended. The general sources are in most cases more useful.

An important exception is the nonparapsychological journals. Probably psychology is the discipline that most often carries articles on parapsychology in its journals. The primary index to journals in psychology, *Psychological Abstracts* (and its online counterpart *PsychINFO*), is not very useful unless one is prepared to spend hours or days seeking the information. This is because its index uses only the broadest terms to cover articles that are for the most part on narrow aspects of parapsychology. There is no way to tell which are general and which are not. Online *PsychINFO* lists over 1,000 articles under parapsychology but it would take hours and a lot of money to print them all out. *Social Sciences Index* is easier and quicker to use, but the range of journals is limited.

To provide some idea of the type of articles available, I again checked *Parapsychology Abstracts International*, using the section on nonparapsychological journals.* The articles written by one or more persons associated with parapsychology have a (P) following the citation. Those that are written by critics are followed by a (C). Those with no letter are not associated with either position.

Note that there is a wide range in publication dates represented here. Even though some of these journal articles are dated, some provide better introductions to parapsychology than some of the encyclopedic articles. For the most recent coverage see the articles by Hyman, Jahn, and Morris.

Blatt, S.L. ESP: Teaching "scientific method" by counter example. *American Journal of Physics*, 1975 (Dec), *43*, 1079—1082. (C)

---

*Also included are some articles slated for future issues of *PAI*.

Brier, Bob, and Giles, J. Philosophy, psychical research and parapsychology: A survey. *Southern Journal of Philosophy*, 1975 (Fal), *13*(3), 393—405. (P) (U)

Dilley, Frank B. Mind—brain interaction and psi. *Southern Journal of Philosophy*, 1988, *26*(4), 469—480. (P)

Ellison, A.J. Recent developments in psychical research. *Electronics and Power*, 1979 (Jan), *25*, 57—60. (P)

Flew, Antony. Parapsychology: Science or pseudo—science? *Pacific Philosophical Quarterly*, 1980 (Jan/Apr), *61*, 100—114. (C)

Gibbs-Smith, Charles H. Parapsychology in the modern world. *Social Service Quarterly*, 1976 (Jun), *49*, 139—140.

Grattan-Guinness, Ivor. Extra—sensory perception and its methodological pitfalls. *Methodology and Science*, 1979, *12*(1), 17—32. (P)

Hay, David, and Morisy, Ann. Reports of ecstatic, paranormal, or religious experience in Great Britain and the United States——A comparison of trends. *Journal for the Scientific Study of Religion*, 1978 (Sep), *17*(3), 255—268.

Heaney, John J. Some implications of parapsychology for theology. *Theological Studies*, 1979 (Sep), *40*, 474—494.

Helfer, Laurence S. Beyond the limits of sense perception. *Pastoral Psychology*, 1959 (Nov), *10*(98), 31—34, 39—40.

Hyman, Ray. Parapsychological research: A tutorial review and critical appraisal. *Proceedings of the Institute of Electrical and Electronic Engineers*, 1986 (Jun), *74*(6), 823—849. (C) (U)

Jahn, Robert G. The persistent paradox of psychic phenomena: An engineering perspective. *Proceedings of the Institute of Electrical and Electronic Engineers*, 1982 (Feb), *70*(2), 136—170. (P) (U)

Keil, H.H.J. Parapsychology——Searching for substance beyond the shadows. *Australian Psychologist*, 1980 (Jul), *15*(2), 145—168. (P)

King, Cecil H. Extra–sensory perception. *Proceedings of the Royal Institution of Great Britain*, 1970 (Apr), *43*, 85–102.

Mangan, G.L. How legitimate are the claims for ESP? *Australian Journal of Psychology*, 1959 (Jun), *2*(1), 121–125. (P)

McConnell, R.A. ESP and credibility in science. *American Psychologist*, 1969, *24*, 531–538. (P)

Morris, R.L. Parapsychology, paradox and exploration of the impossible. *Speculations in Science and Technology*, 1987, *10*, 303–309. (P)

Moss, Samuel, and Butler, Donald C. The scientific credibility of ESP. *Perceptual & Motor Skills*, 1978, *46*, 1063–1079. (C)

Murphy, Gardner. Trends in the study of extrasensory percep–tion. *American Psychologist*, 1958 (Feb), *13*(2), 69–76. (P)

Rao, K. Ramakrishna. On "The scientific credibility of ESP." *Perceptual & Motor Skills* 1979, *49*, 415–429. (P)

Stauffer, L. A look at ESP and philosophy. *Southwest Philosophi–cal Studies*, 1977 (Apr), *2*, 84–89.

Tobacyk, Jerome, Miller, Mark J., and Jones, Glenda. Paranor–mal beliefs of high school students. *Psychological Reports*, 1984 (Aug), *55*, 255–261.

Special Issues of Nonparapsychological Journals and Magazines

Occasionally a journal identified with a subject other than parapsychology will devote an entire issue to parapsychology. Three of these special issues were described in *Sources: Correc–tive Psychiatry and Journal of Social Therapy* [**343**], *International Journal of Neuropsychiatry* [**344**], and *Psychoanalytic Review* [**345**]. Since then the following special issues have appeared and are listed in alphabetical order.

Many of the articles in the special issues described here are fairly superficial and do not add significantly to our knowledge of parapsychology, although there are important ex–ceptions. However, the fact that they were written at all is wor–thy of note in itself––and in such a wide range of subject specialty journals! Of special note is the *Behavioral and Brain Sciences* issue, which is a monumental achievement. Also impor–tant are the articles in *Inquiry*, the *Journal of Communication*, the *Sociological Review Monograph*, and the *Transactions of the New York Academy of Sciences*.

949 *Behavioral and Brain Sciences* (U)
This is an interdisciplinary journal that provides a special service called "Open Peer Commentary" to researchers in any area of psychology, neuroscience, behavioral biology, or cognitive science who wish to solicit, from fellow specialists within and across these *BBS* disciplines, multiple responses to a particularly significant and controversial piece of work.

The lead article in the December 1987 issue was "The Anomaly Called Psi: Recent Research and Criticism," by para-psychologists K.R. Rao and J. Palmer. It was answered by a critique entitled "Parapsychology: Science of the Anomalous or Search for a Soul?" by critical psychologist J.E. Alcock. The Open Peer Commentary consisted of comments from 47 recognized scholars, including parapsychologists, other scientists, and critics. Then comes a response from Palmer and Rao and one from Alcock. There is a total of 18 columns of references for all of the contributions.

950 *Counseling and Values*
This is the official journal of the National Catholic Guidance Conference. In its February 1975 issue it carried a series of articles on "New Frontiers." Two of the articles were on parapsychology: "Scientism and Parapsychology," by J.J. Lang; and "Using ESP: Science Fiction or Applied Science," by Gertrude R. Schmeidler.

951 *The Humanist*
This magazine is the official organ of the American Humanist Association, which typically takes a negative stance towards parapsychology. The July/August 1976 issue carried eight articles on "Antiscience and Pseudoscience." Those of relevance to parapsychology were "ESP: Lying, Cheating or Absolute Idiocy," by D. Rawlins; "Uri Geller and the Chimera," by B. Richardson; an interview with James Randi; "The Scientific Attitude vs. Antiscience and Pseudoscience," and "Subjective Thinking," by M. Zimmerman.

952 *Impact of Science on Society*
Published by the United Nations, this magazine is specialized in the sense that it is mainly concerned with the effects of technology on society. Each issue is usually on a specific theme. The October–December 1974 issue was entitled "The Parasciences." The contributors represented many countries with the emphasis on Soviet physiological/theoretical approaches.

The first article is by Arthur Koestler and is adapted from a chapter in his *The Roots of Coincidence*. The other titles are "Psychotronics: The State of the Art," by Zdenek Rejdak; "Psychotronics and Psychology," by Michael Cernousek; "Psi, a New Dimension in the Sciences," by Charles Musés; "Biogravitation and Psychotronics," by Aleksandr Dubrov; "The Mechanism by which Psi—ability Manifests Itself," by Hiroshi Motoyama; "Induction of Psychotronic Effects in Altered States of Consciousness," by Stanley Krippner; "Does Para—optical Perception Exist?," by Yvonne Duplessis; and "Brazilian Research on Paraphenomenal Events," by Jarbas Marinho.

953 *Inquiry* (U)
Subtitled "An Interdisciplinary Journal of Philosophy," Vol. 30, 1987 was a special issue on philosophical issues of parapsychology. Included were "Parapsychology and the Mind—Body Problem," by J. Beloff; "How Parapsychology Could Become a Science," by P.M. Churchland; "Parapsychology and the Demarcation Problem," by R.L. Harris; "On the Philosophy of Psi," by M. Dybvig; and "Psi and Our Picture of the World," by S.E. Braude.

954 *Journal of Communication* (U)
The organ of the International Communications Associations, this journal published a special symposium of 11 papers on parapsychology in its Winter 1975 issue. Included were "On Psychical Research," by Gardner Murphy; "Error Some Place!," by Charles Honorton; "Building Experimental Models," by Robert L. Morris; "Psi Phenomena and Normal Communication," by Phillip W. Keezer; "Personal Differences in the Effective Use of ESP," by Gertrude R. Schmeidler; "Psi—conducive States," by William G. Braud; "Response Patterns in Extrasensory Performance," by Rex G. Stanford; "The Role of Imagery," by Montague Ullman; "Dreams and Other Altered States of Consciousness," by Stanley Krippner; "The Cuna Indians of Panama," by Robert L. Van de Castle; and "Yes, But Is It Really Communication?," by Larry Gross. Every author except for the last is or was associated with parapsychology.

955 *Journal of the American Society of Psychsomatic Dentistry and Medicine*
The October, 1968 issue (Vol. 15, No. 4) was devoted to parapsychology. Stanley Krippner contributed a guest editorial followed by "Telepathy and Pseudotelekinesis in Psychotherapy," by Berthold E. Schwarz; "Parapsychology," by D. Dillahunt; and "Experiments in Telepathy," by Stanley Krippner.

956 *Journal of the Mississippi State Medical Association*
The October 1965 issue (Vol. 6, No. 10) had four articles on parapsychology: "The Credibility of Parapsychology," by James L. Royals; "The Field of Parapsychology," by J.B. Rhine; "Parapsychology and Medicine," by J.B. Rhine, and "Spontaneous Psi Experiences," by L.E. Rhine.

957 *New Humanist*
This is the official journal of the Rational Press Association, a British group. The November 1975 issue carried a symposium on "Science and the Paranormal" that had been given at the 1975 annual convention of the Association. It consisted of "Laws, Miracles and Repeatability," by A. Flew; "A Scientist Looks at the Paranormal," by J. Taylor; "Research into Paranormal Phenomena," by C.E.M. Hansel; "A Non—Magical View of Psychical Research," by T.H. Hall; "Responsibilities and Parapsychology," by E.J. Dingwall; and "A Realistic View," by D. Bergler. These articles are skeptical in tone but the approach is more rational than that of the articles in its American counterpart, *The Humanist.*

958 *New York*
A special year—end/beginning number (Dec. 27, 1976/Jan. 3, 1977) was devoted to "mysteries of your body." Most of the contributions were on new types of body therapy and awareness, but three were on parapsychology: "Psychic Power: The Next Superweapon?," by William Stuckey; "The Psychic Purges: Giving People What They Want," by Perry Garfinkel; and "The Out—of—Body Trip: What a Way to Go!," by Joan Kron.

959 *OP/The Osteopathic Physician*
The April 1974 (Vol. 41) issue was devoted to parapsychology. It ran a guest editorial by Stanley Krippner and ten papers authored primarily by parapsychologists. "Parapsychology: Out of the Ivory Tower," by G.R. Schmeidler; "Out—of—Body Experiences and Autoscopy," by J. Mitchell; "Consciousness Localized in Space Outside the Body," by J. Palmer; "Instrumentation, Hypnosis, and ESP," by E.D. Chapman; "The 'Spirit Hypothesis' and Survival After Death," by E. Servadio; "Still Healing in the Shadow of Legal Regulations?," by Z. Rejdak; "Telepathic Training of a Blind Wood—Carver," by Y. Duplessis; "The Relationship Between Medicine and Parapsychology," by A. Angoff; "The Telepathic Dream: Lab Support for Clinical Findings," by M. Ullman; and "Signal Increasing in ESP," by C. Honorton.

960 *Pastoral Psychology*
Guest editor W.H. Clark organized a special issue of *Pas—toral Psychology* for September 1970. He wrote an editorial en—titled "Parapsychology and Religion: The Case for the Paranormal." The other contributions were "Parapsychology and Biblical Religion," by R.E. Grace; "Phantasms of the Living and the Dead," by G. Murphy; "Healing by the Laying on of Hands: Review of Experiments and Implications," by B. Grad; "Religious Implications of Paranormal Events Occurring During Chemically—Induced 'Psychedelic' Experience," by S. Krippner and R. Davidson; "Informal Methods of Research in Psychic Phenomena for Religious Believers," by K. Osis; and "Beyond the Intellectual," by B.R. Hoffmann.

961 *The Roche Report: Frontiers of Psychiatry*
The May 15, 1974 issue (Vol. 4, No. 10) was devoted to ESP. Of its ten pages, seven were devoted to two articles en—titled "ESP: Psychiatry's Ultimate Frontier?" and "ESP in the Psychiatrist's Office." Although neither one is signed, both ap—pear to be based——at least in part——on interviews with Jan Ehrenwald, Charles Honorton, and Montague Ullman.

962 *Sociological Review Monograph*
In 1979 (Vol. 27), this British journal had three articles on the sociology of science and parapsychology: "The Construc—tion of the Paranormal," by H.M. Collins; "Experimental Para—psychology as a Respected Science," by P.D. Allison; "Between Sceptism and Credulity: A Scientific Study of Victorian Scien—tific Attitudes to Modern Spiritualism," by Jon Palfreman; and "Investigation of Frames of Meaning in Science," by H.M. Col—lins.

963 *Transactions of the New York Academy of Sciences*
The December 1967 issue (Series II, Vol. 30) had three papers by well—known parapsycholgists: "Introductory Aspects of Modern Parapsychological Research," by G. Murphy; "Personality Interactions in Current Research," by G. Schmeid—ler; and "Rapid—eye Movement Dream—Monitoring Techniques in Current ESP Research," by M. Ullman.

964 *Twentieth Century*
This is a British magazine that dealt with contemporary life but is no longer published. Issue 1046, 1971, was devoted to "An Inquiry into ESP." Several British parapsychologists contributed. The authors and the titles of their articles are as follows: "The Forbidden Science," by Rosalind Heywood; "ESP and the Skeptics," by C.W.K. Mundle; "The Mind and the

Body," by Reneé Haynes; and "The Future for Psychical Research," by John Cutten.

GENERAL REFERENCE WORKS

There are hundreds of reference books but certain types more than others can be expected to have information on parap—sychology: bibliographies, directories, and indexes. A few of the major works in each group are described below.

Bibliographies

965 *Books in Print*
Perhaps the most important national bibliography is *Books in Print* (R.R. Bowker), an annual publication with a midyear supplement. This multi—volume work has author, title, and sub—ject indexes to all the books published in the United States cur—rently available for sale from publishers and in bookstores. The major subject headings for parapsychology are PSYCHICAL RESEARCH, EXTRASENSORY PERCEPTION, and for mediumistic studies, SPIRITUALISM. (There are many other more specific headings. Check the *see also* references under PSYCHICAL RESEARCH.)

966 *Forthcoming Books*
Bowker also publishes *Forthcoming Books* six times a year. It contains author, title, and subject listings of books that are in press and due out in the coming five months and books pub—lished since the last edition of *Books in Print* or its supplement. (When a book becomes listed in *Books in Print* it is removed from *Forthcoming Books*.) The subject headings for parapsychol—ogy in *Forthcoming Books* are the same.

967 *Cumulative Book Index (CBI)*
Another important annual publication is known as *CBI* (H.W. Wilson). It contains author, title, and subject listings of all books published in the English language worldwide during a given year. (Sometimes a book published in one year does not get into *CBI* until the following year.) The subject headings for parapsychology are the same as in *Books in Print*.

968 *American Book Publishing Record*
Known as *BPR*, this monthly paperback listing (with a hardcover annual cumulation) is an index to the listings in the *Weekly Record* published by R.R. Bowker. The *Record* lists cur—rently published books arranged by Dewey number. The main

listing of *BPR* is also by Dewey number, with author, title, and subject indexes. The Dewey numbers to check are the same as those one searches on library shelves: 133—133.9.

Indexes

969 *Reader's Guide to Periodical Literature* (G) (H)
    One of the most important general indexes for persons searching for information on parapsychology is *Reader's Guide to Periodical Literature* (H.W. Wilson). There is an annual hardcover volume and 11 paper supplements during the year. *Reader's Guide* covers a broad range of 200+ general interest magazines. Some examples are *Science, Science News, Scientific American, Smithsonian, Esquire, American Health, New Age, New Realities, Omni*, etc. Each magazine article is arranged by author, title, and subject. The important subject headings are PSYCHICAL RESEARCH, EXTRASENSORY PERCEPTION, PSYCHOKINESIS, MEDIUMSHIP, and SPIRITUALISM.

970 *Info TrackII and General Periodicals Index on InfoTrac* (H)
    These are CD Rom units that cover general interest magazines (400 and 800, respectively), and are produced by In—formation Access. *Info Track* can be searched electronically and most people find it fun to use. The major headings are PSYCHICAL RESEARCH, EXTRASENSORY PERCEPTION, and PSYCHOKINESIS, as well as many more specific headings and a wide range of subheadings.

971 *Book Review Digest*
    Published by H.W. Wilson, this annual volume with paper—back supplements throughout the year provides excerpts from and citations to reviews of current nonfiction and fiction in the English language in selected periodical in the humanities, social sciences, and general science from the U.S., Canada, and Great Britain. The main arrangement is by author. A brief summary of the book is also given, as is the number of words in the review. There are title and subject indexes (look under PSYCHI—CAL RESEARCH). Selection is limited, in the case of nonfiction books, to those with a minimum of two reviews, and four for fiction.

972 *Book Review Index*
    This index to book reviews is published by Gale Research and is issued in an annual hardcover volume with six paper supplements throughout the year. It covers many more journals and magazines than *Book Review Digest* and it lists a review of a book even if it is the only one. Publication of *Book Review*

Index is very prompt. The reviews are listed under the author's name, and there is a title index.

Directories

973 *Directories in Print*
If you need to contact persons or organizations in para-psychology, you need to use directories. The first directory to use is a "directory of directories:" *Directories in Print* (Gale), an annual publication with supplements that lists directories that are guides to specific subject areas. The directories in-cluded are listed alphabetically by title in 16 broad subject areas. The likeliest for parapsychology are "Science, Engineer-ing, and Computer Science"; "Information Sciences, Social Sciences, and Humanities"; and "Religious, Ethnic, and Frater-nal Affairs." There also is a supplementary subject index (look under "Paranormal Movements, Publications, etc." "Psychic and Occult Movements, Publications, etc.," and "Spiritual Move-ments, Publications, etc.") There is a title index.

974 *Encyclopedia of Associations*
This is the most important general directory and it will be found in most libraries. Published by Gale, the work consists of several volumes: 1. National Organizations of the United States; 2. Geographic and Executive Indexes; 3. New Associations and Projects, which serves as a supplement to Vol. 1 and describes newly formed and newly discovered associations not listed in the main volume; and 4. International Organizations (i.e., nonprofit membership organizations). Paragraph—length entries describe the primary facts on each organization. Parapsychology or-ganizations are listed under parapsychology in the section en-titled "Scientific, Engineering, and Technical Organizations." Some of the organizations you will find listed are not on what parapsychologists consider to be parapsychology, nor are all the major parapsychological organizations listed, but it is a good place to start.*

975 *Ulrich's International Periodicals Directory*
There are several directories of periodicals but Ulrich's is the best known. It is published annually by R.R. Bowker. It consists of information on each periodical (with address,

---

*In late 1989 the PSI Center plans to publish three direc—tories: *International Directory of Parapsychology Journals*, *Inter-national Directory of Parapsychological Organizations*, and *Inter-national Directory of Parapsychology Research Centers*. Also see Chapters II and III of this book.

coded information on contents, indication of where it is indexed,
etc.) under its major subject heading, with a title index.
Magazines and journals on parapsychology are listed under
"Parapsychology and Occultism."

976 *Directory of Special Libraries and Information Centers*
     Published annually by Gale, the directory lists information
centers and libraries in the U.S. and Canada specializing in
parapsychology alphabetically by name.   There is a subject in—
dex that uses the term PSYCHICAL RESEARCH.

977 *Research Centers Directory*
     This annual published by Gale contains three paragraphs
describing individual research centers arranged by broad subject.
Major parapsychological research centers are listed in the section
on Behavioral and Social Sciences.   The subjects covered in the
three paragraphs are "Organizational Notes," Research Activities
and Fields," and Publications and Services."   There are subject
and organization indexes.   The subject index is broken down fur—
ther by states.

PSYCHOLOGY BOOKS

Textbooks

     Another possible source of information is psychology
textbooks.   D. Scott Rogo (1980) reviewed 10 undergraduate
psychology books and found that the best (Hilgard and
Atkinson's *Introduction to Psychology*, 4th ed., 1967) was disap—
pointing.
     Child (1985) checked five books by psychologists (not
textbooks——with the possible exception of Neher [**355**]) to see if
scientific standards of judgment were used in reviewing the
Maimonides experiments on dreams and ESP.   He found that
"the experiments have received little or no mention . . . . [or] .
. . . they have been so severely distorted as to give an entirely
erroneous impression of how they were conducted" (p. 1219).
     I checked 13 introductory psychology textbooks on the
shelves of the East Meadow Public Library where I work.   Those
that deserve some mention are listed below alphabetically. (The
first item is an exception.*)

---

     *This book was not yet published when this chapter was
written.   This write—up is based upon a manuscript of the sec—
tion on psi kindly sent to me by Dr. Bem.

978 Atkinson, Rita L., Atkinson, Richard C., Smith, E.E., and Bem, Daryl J. *Introduction to Psychology* (10th ed.). San Diego, CA: Harcourt Brace Jovanovich, 1990. (H) (U)

The textbook originally written by Hilgard with various coauthors has long been considered the best or one of the best available. The 7th edition is described below. The 10th edition is described here. Daryl Bem, a Cornell University psychologist/magician, wrote the section on "Psi Phenomena," which appears in Chap. 6, "States of Consciousness." It provides a good up–to–date critical review of the experimental evidence, the debate over the evidence, anecdotal evidence, and includes a discussion of skepticism about psi. There is a 23–item bibliography.

979 Hilgard, Ernest R. *Psychology in America: A Historical Survey.* New York: Harcourt Brace Jovanovich, 1987.

There is a five–page subsection of Chapter 21 entitled "Parapsychology as an Alternative Psychology of Mental Activity." It begins with some definitions and a historical subsection on "spiritism" (spiritualism and psychical research). The longest subsection is on parapsychology (three pages), which concentrates on the work conducted at Stanford, Harvard, and Duke. Considerable space is devoted to Uri Geller. Hilgard considers "whether or not there is any substance in the claims that would lead to possible alternative paradigms for psychology" (p. 795). He gives the last word to physicist John Wheeler who repudiated a link between quantum physics and parapsychology (although he is a prominent advocate of this being a "participatory universe" based on quantum mechanics).

980 Hilgard, Ernest R., Atkinson, Richard C., and Atkinson Rita L. *Introduction to Psychology* (7th ed.). New York: Harcourt Brace Jovanovich, 1979.

Parapsychology is covered in a four–page article on EX–TRASENSORY PERCEPTION. Primarily it is a summary of criticisms, but the authors state: "These criticisms are not decisive. It is desirable to keep an open mind about issues that permit empirical demonstration, as some ESP phenomena do. At the same time, it should be clear that the reservations of the majority of psychologists are based on more than stubborn prejudice" (p. 153). Also, in fairness it should be pointed out that the experiments cited are those of Rhine and Soal in the 1940s and 1950s. None of the research of the 1970s using computers is touched on, but Wolman's *Handbook* [688] is cited at least. The parapsychology sections in the 8th and 9th editions of this work are substantially the same. However, see the write—up of the 10th edition by Atkinson, Atkinson, Smith, and Bem on page 24.

981 Roback, A.A. *Present—Day Psychology.* New York: Philosophical Library, 1955.

Of historical interest is that this textbook has a chapter on parapsychology by J.B. Rhine. Given the fact that the dis—cussion of parapsychology in most psychology textbooks is limited to research conducted in the 1950s or earlier, Rhine's chapter may still have considerable value. For a certainty its coverage is more complete than the cursory notice given to the subject in most psychology textbooks——if it is given any atten—tion at all.

982 Sanford, Fillmore H., and Wrightman, Lawrence S., Jr. *Psychology: A Scientific Study of Man* (3rd ed.). Pacific Grove, CA: Brooks/Cole, 1970.

There is a subsection on ESP (pp. 327—329) in the chap—ter on "Attention and Perception." It is very outmoded. Zener cards are described as a typical experiment and the old trans—mission theory or radio—wave model is the only one mentioned. They close with a discussion of psychologists and psi, concluding that the nature of ESP is "still shrouded in mystery and con—troversy" (p. 329). (This text is typical of the others I looked at that had anything at all on parapsychology. Usually the coverage was two—three pages, was listed under ESP in a chap—ter on perception, cited early ESP research and then what critics have said (usually C.E.M. Hansel), and concluded that ESP is nonproven and/or a waste of time for psychologists.)

983 Zimbardo, Philip G., in consultation with Floyd L. Ruch. *Psychology and Life* (9th ed.). Glenview, IL: Scott—Foresman, 1975.

There is a three—page article on EXTRASENSORY PER—CEPTION that emphasizes the potential revolutionary implica—tions of ESP and criticisms of the research, including a section on "criteria for acceptable evidence." One of the demands in—sisted upon (taken from Hansel) is that "the research design must make it *impossible* to attribute the results" to subject or experimenter fraud.

Reminder: the book on anomalistic psychology by Zusne and Jones [659] reviewed in Chapter I is a specialized psychol—ogy text that contains a significant amount of material on parapsychology.

State—of—the—Art Surveys

The first item below is the most recent in the entire bibli—
ography, and it is also one of the most sophisticated. It is
highly recommended.

984 Coleman, Andrew, and Beaumont, Graham (Eds). *Psychol—
ogy Surveys 7.* East Leicester, England: British
Psychological Society, 1989. (U)
There is a chapter on parapsychology by Robert L. Morris,
psychologist/parapsychologist and holder of the Koestler Chair
of Parapsychology at the University of Edinburgh in this British
series of reviews of psychological topics, which is analogous to
the American *Annual Review of Psychology.* Morris treats
parapsychology as an aspect of anomalistic psychology and
presents a communication model of psi. He posits six steps in
the psi communication process in each of which psychology can
provide insight as to where people may be misled in regard to
resemblance in "an apparent communication or transfer of in—
fluence between organism and environment" (p. 154). Once one
is convinced that there is a strong resemblance between two
sets of events, the next stage is to assess "the likelihood that
such a resemblance can be readily explained and does not in
fact constitute support for some new means of influence transfer
in operation" (p. 158). There are brief sections describing the
methodology of parapsychology and experimental findings in
parapsychology. There are 78 references.

985 Rosenzweig, Mark R., and Porter, Lyman W. (Eds.). *An—
nual Review of Psychology,* Vol. 32. Palo Alto, CA: An—
nual Reviews, 1981.
It is noteworthy that in this volume of the American
series the *Annual Review of Psychology* there is a chapter by
Leona E. Tyler entitled "More Stately Mansions—— Psychology
Extends Its Boundaries." Under the subheading "Two Doubtful
Extensions," parapsychology is included along with humanistic
and religious psychology. The former is briefly discussed on pp.
11—12.

CHAPTERS IN BOOKS ON SUBJECTS
OTHER THAN PARAPSYCHOLOGY

Sometimes a good article on parapsychology will appear as
a chapter in an anthology devoted to a psychological, philosophi—
cal, or other topic. They may be useful as information sources
for the general reader, but such chapters are difficult to locate

because they usually do not turn up in the card catalog. To lo—
cate the chapters listed below I checked the published bibliogra—
phies of some key parapsychologists (Gardner Murphy, J.G.
Pratt, and J.B. Rhine) and I also checked the items listed un—
der "Chapters in Books" in *Parapsychology Abstracts Interna—
tional.* The following bibliography of chapters on parapsychology
is listed alphabetically by author. All are fairly general except
for the two by Schmeidler and Woodruff in the volume dedi—
cated to Gardner Murphy. Their chapters are experimental
reports.

Two of the chapters listed here are guides to the litera—
ture. McCarthy's chapter provides an overview of the field
aimed at teachers of parapsychology. It also presents a prospec—
tus on parapsychology in the classroom and closes with an over—
view of resource materials and a bibliography. McInnis's chap—
ter describes research guides, substantive information sources,
bibliographic sources, literature reviews, and bibliographies.
Two are autobiographical chapters in the series entitled *The
Psychologists* edited by T.S. Krawiec: they are by Gardner Mur—
phy and J.B. Rhine.

The bulk of these chapters have been written by persons
associated with parapsychology and for the most part their
quality is high. However, because they are so outdated, none of
them has been recommended. Bear in mind, though, that the
information in most of the general encyclopedias and psychology
texts is also outdated. For basic information, these chapters
are recommended over the general encyclopedias and the psychol—
ogy books, with the exception of the chapter by Morris in the
volume edited by Coleman and Beaumont, and the section on
psi by Bem in Atkinson et al.

Beloff, John. The paranormal. Chap. 7 in his *The Existence of
    Mind* (pp. 212–253). New York: Citadel Press, 1964 (c1962).
    (U)

Beloff, John. Parapsychology. Chap. 8 in his *Psychological
    Sciences* (pp. 282–320). New York: Harper & Row, 1973.
    (U)

Child, Irvin L. Extrasensory perception. In his *Humanistic
    Psychology and the Research Tradition: Their Several Virtues*
    (pp. 109–130). New York: Wiley, 1973. (H) (U)

Comfort, A. The rank outsider: Is parapsychology real, and does
    it matter? Chap. 52 in his *Reality and Empathy: Physics,
    Mind, and Science in the 21st Century* (pp. 210–222). Al—
    bany, NY: State University of New York Press, 1987.

Eisenbud, J. Research in precognition. In Stanley R. Dean (Ed.), *Psychiatry and Mysticism* (pp. 101–110). New York: Harper, 1960.

Estabrooks, George H., and Gross, Nancy E. The paranormal and extrasensory perception. Chap. 7 in their *The Future of the Human Mind* (pp. 177–209). New York: Dutton, 1961.

Eysenck, Hans J. Telepathy and clairvoyance. Chap. 3 in his *Sense and Nonsense in Psychology* (pp. 106–141). Baltimore: Penguin Books, 1958.

Flew, A.G.N. Broad and supernormal precognition. Chap. 12 in P.A. Schilpp (Ed.), *The Philosophy of C.D. Broad* (pp. 41–435). New York: Tudor, 1959.

Haynes, R. Wrestling Jacob: Koestler and the paranormal. In H. Harris (Ed.), *Astride the Two Cultures: Arthur Koestler at 70* (pp. 175–186). New York: Random House, 1976.

Isaacson, Walter. ESP. In his *Pro and Con* (pp. 102–108). New York: Putnam's, 1983.

Koestler, Arthur. Extra–sensory perception. In his *Insight and Outlook* (pp. 190–194). New York: MacMillan, 1949.

Krippner, Stanley, and George, Leonard. Psi phenomena as re–lated to altered states of consciousness. Chap. 11 in Ben–jamin B. Wolman and Montague Ullman (Eds.), *Handbook of Altered States of Consciousness* (pp. 332–364). New York: Van Nostrand Reinhold, 1986. (U)

LeShan, Lawrence, and Margenau, Henry. The Domains of the parapsychologist. In their *Einstein's Space and Van Gogh's Sky* (pp. 205–223). New York: MacMillan, 1982.

McCarthy, Donald. Parapsychology. In Michael R. Abraham and Fred W. Fox (Eds.), *Science, Education, Society: A Guide to Interaction and Influence* (1979 AETS Yearbook) (pp. 103–128). Columbus, OH: ERIC Clearinghouse for Science, Mathematics, and Environmental Science, 1978.

McCurdy, Harold G. At the borders of science. Chap. 14 in his *The Personal World* (pp. 464–512). New York: Harcourt, Brace and World, 1961.

McInnis, Raymond G. Parapsychology. In his *Research Guide for Psychology* (pp. 471–476). Westport, CT: Greenwood Press, 1982.

Murphy, Gardner. Discussion. In H. Feifel (Ed.), *The Meaning of Death* (pp. 317–340). New York: McGraw–Hill, 1959.

Murphy, Gardner. Parapsychology. Chap. 6 in N.L. Farberow (Ed.), *Taboo Topics* (pp. 56–63). New York: Atherton, 1963.

Murphy, Gardner. Scientific approaches to the study of survival. Chap. 11 in Austin Kutscher (Ed.), *Death and Bereavement* (pp. 139–145). Springfield, IL: Charles C Thomas, 1969.

Murphy, Gardner. There is more beyond. In T.S. Krawiec (Ed.), *The Psychologists* (Vol. II, pp. 321–344). New York: Oxford University Press, 1974.

Price, H.H. Paranormal cognition, symbolism, and inspiration. Chap. 2 in his *Essays in the Philosophy of Religion* (pp. 21–36). Oxford: Clarendon Press, 1972.

Price, H.H. Some objections to behaviorism. Chap. 7 in Sidney Hook (Ed.), *Dimensions of Mind* (pp. 78–84). New York: University of New York Press, 1960.

Rhine, J.B. The bearing of parapsychology on human potentiality. In Herbert A. Otto (Ed.), *Human Potentialities* (p. 46–63). St. Louis, MO: Warren H. Green, 1968.

Rhine, J.B. The development of parapsychology. *Frontiers of the Mind.* Garden City, NY: Doubleday, 1979.

Rhine, J.B. ESP: What can we make of it? In Granada Television (Ed.), *Guildhall Lectures* (pp. 11–40). Manchester, England: MacGibbon and Kee, 1965.

Rhine, J.B. Extrasensory perception and hypnosis. In L.M. LeCron (Ed.), *Experimental Hypnosis* (pp. 359–368). New York: MacMillan, 1952.

Rhine, J.B. From miracle to experiment. In Alson J. Smith (Ed.), *Religion and the New Psychology* (pp. 13–22). Garden City, NY: Doubleday, 1951. (G)

Rhine, J.B. Man's nonphysical nature. In J.W. Cullen (Ed.), *Legacies in the Study of Behavior.* Springfield, IL: Charles C.

Thomas, 1974.

Rhine, J.B. On parapsychology and the nature of man. In Sidney Hook (Ed.), *Dimensions of Mind* (pp. 71–77). New York: New York University Press, 1960.

Rhine, J.B. Parapsychology and man. In E. Laszlo and J.B. Wilbur (Eds.), *Human Values and the Mind of Man*. New York: Gordon, 1971.

Rhine, J.B. Parapsychology and the study of altruism. Chap. 6 in P.A. Sorokin (Ed.), *Explorations in Altruism and Behavior* (pp. 165–180). Boston: Beacon Press, 1950.

Rhine, J.B. The parapsychology of religion: A new branch of inquiry. In *The Centrality of Sciences and Absolute Values: Proceedings of the Fourth International Conference on the Unity of the Sciences* (pp. 1–14). Tarrytown, NY: International Cultural Foundation, 1975.

Rhine, J.B. The science of parapsychology. In G. Pat Powers and Wade Baskin (Eds.), *New Outlooks in Psychology* (pp. 38–63). New York: Philosophical Library, 1968.

Rhine, J.B. Security versus deception in parapsychology. In Stanley R. Dean (Ed.), *Psychiatry and Mysticism* (pp. 111–131). Chicago: Nelson-Hall, 1975.

Rhine, J.B., and Rhine, L.E. A search for the nature of mind. In T.S. Krawiec (Ed.), *The Psychologists* (Vol. 3, pp. 181–205). Brandon, VT: Clinical Psychology Publishing Co., 1978. (U)

Roll, W.G. ESP and America's occult upsurge. In Stanley R. Dean (Ed.), *Psychiatry and Mysticism* (pp. 47–65). New York: Harper, 1960. (H)

Rucker, R. Time travel and telepathy. Chap. 10 in his *The Fourth Dimension: Toward a Geometry of Higher Reality* (pp. 165–191). Boston: Houghton Mifflin, 1984.

Schmeidler, Gertrude R. Changing field relations of an ESP experiment. In John G. Peatman and Eugene L. Hartley (Eds.), *Festschrift for Gardner Murphy* (pp. 94–105). New York: Harper, 1960.

Schmeidler, Gertrude R. The "fitting-in" of parapsychological experiments. In I.J. Good (Ed.), *The Scientist Speculates* (pp. 157–158). New York, Basic Books, 1962.

Scriven, Michael. The frontiers of psychology: Psychoanalysis and parapsychology. In Robert G. Colodny (Ed.), *Frontiers of Science and Philosophy, Vol. I* (pp. 79–129). Pittsburgh, PA: University of Pittsburgh Press, 1962.

Servadio, Emilio. Parapsychology and the paranormal. In his *Psychology Today* (pp. 265–293). New York: Garrett/Helix, 1965.

Talbot, M. Why is science afraid of the paranormal? Chap. 9 in his *Beyond the Quantum* (pp. 217–224). New York: MacMillan, 1986.

Targ, R., and Hurt, D.B. Use of an automatic stimulus generator to teach extrasensory perception. In Thomas B. Roberts (Ed.), *Four Psychologies Applied to Education* (pp. 482–486). Cambridge, MA: Schenkman, 1975.

Tart, Charles T. The physical universe, the spiritual universe, and the paranormal. In his *Transpersonal Psychologies* (pp. 115–151). New York: Harper & Row, 1975.

Tauber, Edwards S., and Green, Maurice R. Extrasensory perception. In their *Prelogical Experience* (pp. 106–112). New York: Basic Books, 1959.

Ullman, M. Vigilance, dreaming and the paranormal. In C. Musés and A.M. Young (Eds.), *Consciousness and Reality* (pp. 35–56). New York: Outerbridge & Lazard, 1972.

Woodruff, Joseph L. The effect on ESP scoring of an unexpected qualitative change in ESP material. In John G. Peatman and Eugene L. Hartley (Eds.), *Festschrift for Gardner Murphy* (pp. 106–116). New York: Harper, 1960.

PRO AND CON VIEWS

Students, especially those involved in debates, and persons trying to come to grips with parapsychology are best served by information sources that present both sides of the subject. Several of the items that have been cited in this chapter are paired with other items. Together they provide insight into both sides of the question. These contributions are pointed out below.

Among the magazine articles, the *Humanist*, a skeptical magazine, devoted their May/June 1977 issue to the topic of "The Psychics Debunked." In response to criticisms that they had been unfair to parapsychology by not distinguishing it from the occult and paranormal, they then invited D. Scott Rogo to present the case for parapsychology. Two critics, Martin Gardner and Ray Hyman, responded.

McConnell's "ESP––Fact or Fancy" presented the case for parapsychology in *Scientific Monthly* (Aug 1949) but a critical scientist, Warren Weaver (Oct 1949) took issue with McConnell. A second exchange between the two followed (Feb 1950).

Among the nonparapsychological journals there are two articles published in the *Proceedings of the Institute of Electrical and Electronic Engineers* that balance each other very nicely. In 1982 engineer Robert G. Jahn wrote an invited paper entitled "The Persistent Paradox of Psychic Phenomena," which presented as a conservative view of the field. In 1986, critical psychologist Ray Hyman contributed a "Tutorial Review and Critical Appraisal" of parapsychological research.

Moss and Butler, in "The Scientific Credibility of ESP," explain why academic psychologists are generally skeptical about the evidence for ESP (*Perceptual & Motor Skills*, 1978, *46*, 1063–1079). Parapsychologist R.A. McConnell, in *Perceptual & Motor Skills*, 1978, *47*, 875–878, points out eight statements about his writings that Moss and Butler misquoted with a response from Moss and Butler on p. 992 of the same issue. Parapsychologist K.R. Rao then attempts to refute Moss and Butler in *Perceptual & Motor Skills*, 1979, *49*, 415–429, with a response from Moss and Butler in the same journal for 1980, *50*, 502. The entire exchange was reprinted by J. Rubenstein and B.D. Slife in their *Taking Sides: Clashing Views on Controversial Psychological Issues* (3rd ed.). Similar treatment is provided by W. Isaacson in a chapter on ESP in his book, *Pro and Con*.

Finally, in 1989 *Fate* magazine ran two pro and con articles in its January issue. First came critic David F. Marks' "The Case Against the Paranormal" (*Fate*, 1989, *42*(1), 86–92) followed by parapsychology writer D. Scott Rogo's "The Case Against the Case Against the Paranormal" (*Fate*, 1989, *42*(1), 93–100).

RESEARCH PROJECTS IN PARAPSYCHOLOGY

An important need of students is ideas for research projects. To come up with any kind of worthwhile subject and to know how to carry out a demonstration or experiment of value almost necessitates that one be well—versed in the litera—ture of parapsychology, not just the popular literature. However, for whatever they are worth, I was able to locate some suggestions for experiments in the nonparapsychological literature and they are listed below.

Guerin, S.M. Experiments in ESP. *Science Activities*, 1979 (Sum), *16*, 31—38.

Nathanson, J.A. ESP——Factor or fiction? *Science Experimenter: A Science & Mechanics Annual*, 1964, No. 709, 16.

Schmidt, H. Randomness and the mind: Looking for psychic ef—fects in games of chance. *Creative Computing*, 1987, *9*(8), 180—186.

Spector, I.H. Build an ESP testing machine. *Popular Electronics*, 1974 (Jul), *6*, 27—32.

Try a little ESP with us. *Psychology Today*, 1969 (Mar), *2*(10), 26—28. (G)
   With the assistance of J.B. Rhine, a few experiments in parapsychology that one can do oneself are described.

Waite, M. ESP machine. In his *Sight, Sound and Sensation* (pp. 82—106). Indianapolis, IN: Howard W. Sams, 1964.
   For those interested in conducting surveys, suggestions for questions can be found from the questionnaire in the December 1980 issue of *Glamour* and the article by Greeley in *American Health* (Jan/Feb 1987).

Wichmann, Brian, and Hill, David. Building a random—number generator. *Byte*, 1987 (Mar), *9*(6), 127—128.

SUMMARY

   Taken as a whole, the quantity and quality of general sources available on parapsychology is disappointing. The most important sources are those that are not likely to be found in any but the larger libraries. The best material is provided by parapsychologists who have written for general sources. Al—though sufficient information is available to write a superficial term paper on parapsychology, it is generally not possible to carry out a high quality science experiment or to write an up—

to—date review of parapsychology,* or to prepare a term paper in any depth on any specific aspect of parapsychology (with the possible exception of dreams and ESP).

REFERENCES

Child, I.L. (1985). Psychology and anomalous observations: The question of ESP in dreams. *American Psychologist*, *40*, 1219–1230.

Rogo, D.S. (1980). Parapsychology and psychology textbooks. *Research Letter*, No. 10, 9–18.

White, R.A. (1976). The role of the library in education for parapsychology. In B. Shapin & L. Coly (Eds.), *Education in Parapsychology: Proceedings of an International Conference Held in San Francisco, California, August 14–16, 1975* (pp. 273–299). New York: Parapsychology Foundation.

White, R.A. (1965). The library and psychical research. Part I: Making the most of the library. *Journal of the American Society for Psychical Research, 59*, 266–308.

---

*Those fortunate enough to have access to the special issue of *Behavioral and Brain Sciences*, and the chapters by Morris and Bem emphasized in the section on psychology books

V

U.S. Government Interest in Parapsychology*

QUICK INDEX TO SECTIONS IN THIS CHAPTER

There continue to be rumors about U.S. and Soviet govern—
ment research in parapsychology, but both governments main—
tain an official silence on the subject. I am not in a position
to know whether one or both governments are supporting clan—
destine research or if either country has had, or currently has,
operational programs in which some application of parapsy—
chology is being attempted. Some who claim to have had some
official involvement in these areas say that both governments
are playing an active role. However, that is not our concern in
this chapter. Here I attempt to document any *overt* signs of
U.S. government interest in parapsychology.

I found that the United States government has evidenced
some interest in psi phenomena since the 1960s, as indicated by
the reports listed in Section 4, whereas the popular press did
not pick up on this interest until the 1980s (see Section 5).

---

*I would like to thank the following persons for their assis—
tance with this chapter: Martin Ebon, C.B. Scott Jones,
Theodore Rockwell, and Marcello Truzzi.

This interest has been documented in various publications and is shown by some government grants for psi research, a number of government reports on psi, and government—sponsored trans—lations of foreign language reports on aspects of parapsychology. The arrangement of the material in this section reflects these expressions of government interest and is in seven parts. The first is an abstract of "Parapsychology and the CIA," a paper given by David Clemens (5: 1982) at the 1981 annual conven—tion of the Parapsychological Association. Clemens reports on his examination of over 900 pages of CIA documents relating to parapsychology released to him under the Freedom of Informa—tion Act. The second section consists of excerpts from an up—date of that paper prepared by Clemens especially for this book. Boldface text in the latter indicates insertions by Dr. C.B. Scott Jones, who is special assistant to Senator Claiborne Pell of Rhode Island, and whose job it is to serve as liaison between serious parapsychologists and the government community in Washington, DC. The third is an assessment of the role of the U.S. government and parapsychology prepared for this volume by Dr. Jones (he stresses a little—discussed form of government interest in psi——the use of psychics in law enforcement. I have appended a bibliography of materials on criminal investigation and psi to this article). The fourth section is a bibliographic list—ing of reports on parapsychology sponsored and/or published by the government. The fifth is a bibliography of books and ar—ticles about U.S. government interest in parapsychology. Some of these citations also discuss Soviet interest in psi research, but that is not the subject of this chapter (for a comprehensive bib—liography of Soviet, East European, and Chinese psi research, see White and Vilenskaya [680]). The sixth part is a list of projects related to parapsychology funded wholly or in part by government grants. The last section is my own summary of some important events of the last 25 years that suggest a grow—ing rapprochement between the U.S. government and working parapsychologists

If the authors in this section or I make a reference to any books, articles, or other reports they will be cited in one of three ways. If the citation is to a book in Chapter 1, only the book number in boldface will be given, plus page references if relevant. If the item cited is listed in the bibliographies of Sec—tions 4 or 5 of this chapter, it will be indicated by giving in parentheses the section number by a boldface 4 or 5 followed by the first author's last name or the first few words of the title if there is no author, and the publication date, if there is more than one item listed by the author, and the pagination, if relevant. If an item is not listed elsewhere in the book, the cita—tion will be given in a footnote.

PARAPSYCHOLOGY AND THE CIA

David Clemens

Many parapsychologists have suspected for some time that various U.S. military and intelligence agencies have been in— volved in, or at least interested in, parapsychological research. Tart's survey on negative uses and government interest in psi (**5**: Tart, 1979) indicated that five out of 13 research laboratories polled had been approached by agents of the U.S. government acting in an official capacity in order to obtain in— formation on parapsychology. At least one well—known researcher was contacted by a CIA agent in the mid—seventies seeking to offer funds for the purpose of supporting covert research. DARPA (Defense Advanced Research Projects Agency) officials have been known to have visited at least one psi research facility.

Over 900 pages of CIA documents relating to parapsy— chology were released to the author under the Freedom of Infor— mation Act. These documents show that the CIA was involved in parapsychology over a 15—year period, from 1961 to 1976.

In 1961, a year—long study funded by the CIA commenced on an "Experimental Analysis of Extrasensory Perception." An internal CIA memo discussing the project stated that "any posi— tive results along these lines would have obvious utility for the Agency." The results of this study were not released.

A large portion of the documents consisted of translations of "essentially open" Soviet parapsychological literature. It seems that over a period of several years the CIA was trying to monitor and assess interest and progress of parapsychology in the USSR.

In 1975, the AiResearch Manufacturing Company, a defense contractor, completed a proposal entitled "Biophysical Aspects of Parapsychology" for the CIA (**5**: Wortz et al., 1975). This document was released with only the odd—numbered pages photocopied. A request for the even—numbered pages was ig— nored. In 1976, the final report of the project was completed for the CIA (**5**: Wortz et al., 1976). The report, "Novel Biophysical Information Transfer Mechanisms," was an analysis and assess— ment of Soviet work in the field of parapsychology.* The areas of Soviet research analyzed in the report were the application of statistical theories to psi mechanisms, the electros—

---

*A chapter based on these reports entitled "An Investiga— tion of Soviet Psychical Research" by the same authors was pub— lished in C.T. Tart et al. ([**505**], pp. 235—260)——R.A.W.

tatics of PK, remote physiological sensors, sensitivity of human subjects to magnetic fields, and biophysical information carrier mechanisms. In addition, the report speculates on the nature of possible covert laboratories and makes recommendations for future U.S. research and development. The research team makes the following conclusion: "It is our considered opinion that it is worthwhile for the United States government to initiate and support systematic research in this area." A CIA memo evaluating the final AiResearch report praised the quality of the report and stated that "the reports contain material which will substantially contribute to the preparation of a STIR." Much of the memo was blackened out.

The fact that the CIA was involved in parapsychology over a 15—year period would seem to indicate that investigation of the area was and is more than a passing fancy. Furthermore, the method and pattern the CIA used to release documents as well as the documents it did not release and portions of documents blackened out suggests that the CIA's findings in the area are sufficiently sensitive to prevent the release of much of the information. These same factors seem to indicate that the CIA is continuing its efforts in this area. When it is considered that, at least on a theoretical basis, psi processes could be extremely formidable weapons for a national military or intelligence agency to possess, it would seem that the CIA is making a logical decision if it is continuing its research.

U.S. GOVERNMENT INVOLVEMENT IN PARAPSYCHOLOGY: AN UPDATE*

David Clemens

Government involvement in parapsychology continues to be a topic of concern and interest to researchers and laypersons alike. Since the author's paper in 1981 detailing CIA involvement (4), interest and coverage has increased significantly. Yet despite three books on the subject (Ebon [370], McRae [375], and White [820]), the area has remained an enigma, with no clear picture emerging. Moreover, the media and some lay—authors have sensationalized the topic to the level of the absurd. A Jack Anderson column talked of a "hyperspatial nuclear howitzer" that could send incoming missiles into a "time warp" where they would explode harmlessly and "maybe kill a few dinosaurs."

---

*Material in boldface was inserted by Dr. C.B. Scott Jones.

When attempting to undertake an assessment of U.S. government involvement in parapsychology, certain questions are raised. Is the government currently involved in the area? If so, what is the nature and extent of the involvement? Are there multiple agencies involved, and if so, is it a coordinated effort? Is there one particular agency that administrates the research program? Is there anyone in the government or work—ing for them in a consulting capacity who is capable of under—standing the subject, separating the "wheat from the chaff," developing promising areas into operational programs, and managing the entire area effectively and competently?

The government's own behavior on the subject has been unpredictable and inconsistent. They have released certain docu—ments to some parties while not to others. Their official policy has been that they have no involvement in the area while their own documents prove otherwise. In February, 1982, the Pen—tagon purportedly announced that they had unsuccessfully used a Department of Defense psychic to try to locate kidnapped Army General James Dozier. Why they would admit this and open themselves up to possible ridicule is a mystery. Common sense suggests that a good portion of these incidents are simply government and bureaucratic incompetence, while a much smaller portion could be designated **either** disinformation or "water—testing."

Perhaps the most significant indication of current interest and involvement is the inclusion of parapsychology documents in the Defense Technical Information Center (DTIC) listings. The DTIC is an inter—agency library consisting of documents relat—ing to defense and national security. Various agencies and defense contractors have access to documents on the list on a need—to—know basis. The list itself is now classified and there is a fairly elaborate document control system to regulate access. Some documents on the list are so highly classified that the clas—sifications themselves are classified. The DTIC document list contains several titles on parapsychology.

What is the nature and extent of the involvement? As mentioned earlier, the CIA alone has been involved in parapsy—chology since the early fifties, or almost since its inception. Other agencies that have had documented or reported involve—ment in the area are the Defense Department, the Defense Ad—vanced Research Projects Agency (DARPA), the Army, Navy, and Air Force, The Defense Intelligence Agency (DIA), National Security Agency (NSA), and others. The National Technical In—formation Service (NTIS), a nonclassified government database, has many titles under the classification of parapsychology. Perhaps most interesting are numerous translations of Soviet works made by the Foreign Technology Division at Wright—

Patterson Air Force Base. It may be logical that Wright–Patterson was used, as it has Russian translators on staff, but the Foreign Technology Division is perhaps one of the most secret U.S. facilities in existence. For 35 years it has reputedly been the site of research so secret it is not spoken of or even alluded to.

Additionally, the NTIS published two Master's theses done at the Naval Postgraduate School in Monterey, California on parapsychology topics. One is titled *Questionnaire Results on the Prospects for Soviet Development of Parapsychology for Military Purposes* (4: Bray).

Is there a coordinated effort, and if so, is one agency administrating it? It appears that there is no coordination or single administration. While there **may be** sharing of information, documents, and research results, it appears that for the most part **activities** are initiated by individual agencies. There is little doubt that the level of activity and involvement in any given agency rises and falls in relation to the **interests**, attitudes, and beliefs of personnel as they come and go. **This may well describe the majority of parapsychological activities in government: when an individual who is interested in trying to apply parapsychology to an area in which he has responsibility, he does so in a very confidential way. The results of the efforts probably never will be known.**

Declassified documents indicate a naïveté about the subject. Even research conducted at Stanford Research Institute International in the 1960s and 1970s, which has been regarded by some as the most professional and respected work that the government has supported or **shown** interest in, has not been immune to criticism regarding experimental designs and controls.

In conclusion, it is the author's assessment that the U.S. government has been involved in **various levels of** parapsychology **activity** for at least 25 years and continues to be involved. **Some in government appear to be** convinced of the potential importance of psi phenomena and **definitely** are concerned with Soviet involvement. **Interest** within the government is fairly widespread, including the Defense Department, several branches of the military, and various intelligence agencies. Perhaps the most credible work is being conducted within some of the military branches. The government does not appear to have central administration, and they seem to lack competent personnel and capable management. With the right personnel and a relatively small commitment of funds, the government might be able to yield promising results in experimental, operational programs. For example, a remote viewing program using various formats and redundancy of participants for improved results might yield impressive results in both present and future

time frames. **Or several of the proposals for future research
made in two relevant government sponsored surveys: the U.S.
Army Report for the National Research Council [462] and the
report of the Office of Technology Assessment (4:** *Report of a
Workshop* **. . .) could be followed up.**

ASSESSMENT OF RECENT EVIDENCE OF GOVERNMENT INTEREST IN PARAPSYCHOLOGY

C.B. (Scott) Jones

Given the assessment that there is no coordinated effort of
government psi activity, and the likelihood that over the years a
number of small application efforts have been tried by in—
dividuals without their agency or department being fully
knowledgeable, the question of whether the government has
capable management of its psi activity does not arise. What is
important is how each of the "unofficial" psi application at—
tempts has been managed. It is speculated that this has been a
mixed bag. The potential for a series of "new starts" without
an experience base being established and passed on is very high.
It is unlikely that successes and failures have been critically
analyzed, or that the science of parapsychology has been or ever
will be advanced by this process. Resources probably have been
wasted on these wildcat operations, at least to the point that a
better investment would have been in a formal, sanctioned un—
dertaking with qualified consultants and with independent assess—
ment of the protocols and outcomes.

Most articles and books that address assumed government
activity in parapsychology center on national security issues with
the assumption that either intelligence activity to obtain informa—
tion or some form of remote behavior modification are the most
likely applications. In fact, it appears that there is a much
more active area in terms of number of personnel and level of
effort. The federal government is a heavy player here, but prob—
ably less so than state and local governments. The area is law
enforcement.

There is no known formal measurement of the extent of
such activity, but informal measurement, i.e., news and other
popular accounts* and interviews with psi sensitives who are
willing to talk generally and confidentially about their activities

-----

*A bibliography of books, chapters, journal articles, and
newspaper accounts selected from a section on criminal investiga—
tion and psi originally published in R.A. White [378], is
presented as an appendix to this section.

with various law enforcement agencies, suggests a great deal of applied parapsychology has been going on in this area for years.

It is typical that the official response from federal, state, and local law enforcement agencies is that they do not use psychics. It appears, however, that there is considerable flexibility on this issue as to what an individual agent or officer may do. Are these efforts cost effective? The answer must be positive because most psychics claim that they offer their serv—ices without charge. In many cases they even refuse to take expense money. Has the use of psychics in law enforcement situations been successful? There can be no general answer. Each particular case would have to be evaluated. There are abundant claims from psychics that their contribution to a case was critical. As expected, the law enforcement agency that has a formal policy of denial concerning the use of psychics rarely confirms the fact that a psychic not only was used but was helpful. Moreover, legally the testimony of psychics cannot be used directly. Psychics can provide law enforcement officials with clues, but they must find hard evidence to substantiate the psychic's impressions for use in court. As far as the court proceedings are concerned, the evidence must be located by nor—mal means, not psychic ones.

An argument can be made that it is better to have this sub—rosa applied psi activity than no activity at all. However, the stronger argument is that this is a most unprofessional relationship, and regardless of the incidence of successful applica—tion, it could be improved if it were formally sanctioned and if a modest amount of training on the subject were made avail—able to the law enforcement community. Additionally, some rigor would need to be introduced into the system as to how to judge the skill of the psi sensitive and how to evaluate the infor—mation received.* Cooperation or at least consultation with a bona fide parapsychologist would be an advance. Such coopera—tive effort would not only be helpful to law enforcement agents but the situation would appear to provide an important oppor—tunity for the research community to obtain data that

---

*In line with Dr. Jones's emphasis on the application of psi in criminal investigation, in 1982 *Psychic Criminology* [371] was published, notable because it was *not* written by or about psychics, but by two respectable criminal investigators: Whitney S. Hibbard, a forensic hypnotist and instructor and a licensed private investigator, and Raymond W. Worring, who has similar qualifications. Both were associated with the Great Falls/Cascade County Crime Attack Team. The subtitle of the book is "An Operations Manual for Using Psychics in Criminal Investigations."

may be helpful in approaching the many unanswered questions about the phenomena and its mechanisms.

There can be no doubt about interest in psi phenomena by individual members of governmental bodies, particularly on its potential applications in a wide variety of areas. A baseline for this statement was the private conference on applications held outside of Washington, DC in late 1983. Forty—seven government personnel attended, representing 18 different govern—ment agencies, offices, bureaus, and departments. If there was something new to say at the present time concerning specific applications then discussed by the dozen senior parapsychological researchers who made presentations at that symposium, it would be easy to draw several hundred interested government scientists and managers from dozens of different government organizations.

The Congress has recently played a constructive role in advancing the quality of research into parapsychology. The report of a one—day workshop conducted by the Office of Tech—nology Assessment (OTA) on problems of experimental para—psychology was released in February 1989. The nine par—ticipants of this workshop, unlike the composition of the com—mittee of the National Research Council that produced the nega—tive chapter on parapsychology in the National Academy of Sciences' controversial report, *Enhancing Human Performance* [462], were both critics of and critical researchers in the field. The OTA report is an important unbiased critical assessment of problems of research in parapsychology. It should be widely cir—culated.

APPENDIX

BIBLIOGRAPHY ON LAW ENFORCEMENT AND PSI

Most of the titles listed here are popular accounts of cases in which psychics or psychic experiences have helped to solve crimes. Primarily they are first—person accounts by the psychics themselves or are second— or third—hand accounts. Some of the better documents are by Allison and Jacobson; Cox and Pfeiffer; Nolan with Vilenskaya; Tabori; Tanous; and Trubo. One of the most rehashed cases is that of the Dutch psychic Gerard Croiset who was investigated by W.H.C. Tenhaeff. Hoebens has criticized this work. For a criticism of journalistic accounts in general, see Ejvegaard and Johnson. Other critical accounts listed are any of the pieces by Hoebens, and any items cited from the *Skeptical Inquirer* and the *Zetetic Scholar*. Also see the book reviews listed.

For serious literature by police officials see Gordon and Tobias; Hibbard and Worring; Marshall; and Reiser et al. (2 items). For the approach of serious parapsychologists, see Ej—vegaard and Johnson; Osis; Roll; and Weiner. Interesting discus—sions on legal questions and psi are authored by Reichbart; Rodr; and Schall (2 items). There is a master's thesis listed (Farabee), but I have not seen it.

Allison, D., and Jacobson, S. *Dorothy Allison: A Psychic Story.* New York: Jove Publications, 1980.

Archer, F. *Crime and the Psychic World.* New York: William Morrow, 1969.

Bird, C. Fruitful searches. *New Realities*, 1982, *4*(5), 56—62.

Blank, J.P. The woman who sees through psychic eyes. *Reader's Digest*, 1978 (Dec), *113*, 107—112. [On Dorothy Allison]

Bloom, M.T. Is it Judge Crater's body? *Harper's*, 1959 (Nov), *219*, 41—47. [On Gerard Croiset]

Boeth, R., and Agrest, S. "Visions" of the dead: Psychic D. Allison's ability to locate missing bodies. *Newsweek*, 1978 (Apr 17), *91*, 52.

Browning, N.L. *Peter Hurkos: I Have Many Lives.* Garden City, NY: Doubleday, 1976.

Browning, N.L. *The Psychic World of Peter Hurkos.* Garden City, NY: Doubleday, 1970.

Browning, N.L., and Stone, C.W. Sensitives I have known. In their *The Other Side of the Mind* (pp. 90—103). New York: Paperback Library, 1967 (c1964). [On Gerard Croiset and Peter Hurkos]

Christopher, M. Peter Hurkos——Psychic sleuth. In his *Mediums, Mystics and the Occult* (pp. 66—76). New York: Thomas Y. Crowell, 1975.

The clairvoyant and the policeman. *Tomorrow*, 1952—53, *1*(2), 117—118. [On Bertha Harris]

Cohen, D. *ESP: The Search Beyond the Senses* (pp. 123—132). New York: Harcourt Brace Jovanovich, 1973. [On Gerard Croiset and Peter Hurkos]

Cortesi, L. Psychic tracks a killer. *Fate*, 1987 (Jul), *40*(7), 68—

75. [On Ann Fisher]

Courtroom psychics. *Omni,* 1987 (Jul), *9*(10), 12.

Cox, R.V., and Peiffer, K.L., Jr. *Missing Person: The True Story of a Police Case Resolved by the Clairvoyant Powers of Dorothy Allison.* Harrisburg, PA: Stackpole Books, 1979.

Crenshaw, J. Court admits psychic evidence. *Fate,* 1979 (Nov), *32,* 46–54.

DeLouise, J. *Psychic Mission.* Chicago, IL: Regnery, 1971. [Autobiography]

Doran, B. Extrasensory detection. *TWA Ambassador,* 1979 (Aug), *12*(8), 108+.

Druffel, A., with Marcotte, A. (1983). *The Psychic and the Detective.* San Diego, CA: ACS Publications, 1983. [On Armand Marcotte]

Dykshoorn, M.B., as told to R. H. Felton. *My Passport Says Clairvoyant.* New York: Hawthorn Books, 1974. [Autobiography]

Ejvegaard, R., and Johnson, M. Murderous ESP: A case of story fabrication. *European Journal of Parapsychology,* 1981, *4*(1), 81–98.

Farabee, C.R., Jr. *Contemporary Psychic Use by Police in America.* Unpublished master's thesis. Department of Criminology. California State University, 1981. (Described in *Applied Psi Newsletter,* 1983, *2*(3/4), 12–14.)

Fodor, N. The Mona Tinsley murder. *Journal of the American Society for Psychical Research,* 1938 (Feb), *32*(2), 63–64. [On Estelle Roberts]

Frank, G. *The Boston Strangler* (pp. 87–120). New York: New American Library, 1966. [On Peter Hurkos]

Frazier, K. CSICOP cautions against use of police psychics. *Skeptical Inquirer,* 1981 (Sum), *5*(4), 6–7.

Friedman, R. (Ed.). *Murder Most Eerie: Homicide and the Paranormal.* Virginia Beach, VA: Donning Co., 1982. (Best of Fate Set) [Anthology]

Fuller, C. Peter Hurkos and the Jackson family murders. *Fate*, 1960 (Oct), *13*, 49–54.

Fuller, C. Psychics and police. *Fate*, 1979 (Apr), *32*, 8–14. [On unidentified female psychic]

Gaddis, V. Psychic detective. *Fate*, 1953 (Sep), *6*, 51–55. [On Arthur P. Roberts]

Godsell, H.P. From the files of the mounted police: How mentalist solved a murder. *Fate*, 1959 (Jan), *12*, 36–43. [On "Professor" Gladstone]

Gordon, T.J., and Tobias, J.J. Managing the psychic in criminal investigations. *Police Chief*, 1979 (May), 58–59.

Guarino, R. The police and psychics. *Psychic*, 1975 (May/Jun), *6*(2), 9, 14–15. [On Peter Hurkos and some unnamed psychics]

Haaxma, E. Psychic phenomena related to law. In W.G. Roll, R.L. Morris, and J.D. Morris (Eds.), *Research in Parapsychology 1972* (pp. 36–37). Metuchen, NJ: Scarecrow Press, 1973.

Harper, R. The psychic and missing persons. I. *Psi–Scene*, 1981, *2*(11), 1–2, 6.

Harper, R. The psychic and missing persons. II. *Psi–Scene*, 1981, *2*(12), 1, 3, 6.

Harrison, S., and Franklin, L. *The Psychic Search.* Portland, ME: Guy Gannett, 1981. [Autobiography]

Hibbard, W.S., and Worring, R.W. *Psychic Criminology: An Operations Manual for Using Psychics in Criminal Investigation.* Springfield, IL: Charles C Thomas, 1982.

Hoebens, P.H. Croiset and Professor Tenhaeff: Discrepancies in claims of clairvoyance. *Skeptical Inquirer*, 1981–82 (Win), *6*(2), 32–40.

Hoebens, P.H. Croiset: Double Dutch? *The Unexplained; Mysteries of Mind, Space, and Time*, 1983, No. 132, 2630–2633.

Hoebens, P.H. Gerard Croiset: Investigations of the Mozart of "psychic sleuths"––Part I. *Skeptical Inquirer*, 1981 (Fal), 6(1), 17–28.

Hoebens, P.H. The mystery men from Holland, I: Peter Hurkos' Dutch cases. *Zetetic Scholar*, 1981 (Jul), No. 8, 11–17.

Hoebens, P.H. The mystery men from Holland, II: The strange case of Gerard Croiset. *Zetetic Scholar*, 1982 (Mar), No. 9, 21–32.

Hoebens, P.H. The mystery men from Holland, III: The man whose passport says clairvoyant. *Zetetic Scholar*, 1982, No. 10, 7–16.

Hoebens, P.H. *The Unexplained: Mysteries of Mind, Space, and Time*, 1983, No. 138, 2754–2757.

Hoebens, P.H., with Truzzi, M. Reflections on psychic sleuths. In P. Kurtz (Ed.), *A Skeptic's Handbook of Parapsychology* (pp. 631–643). Buffalo, NY: Prometheus Books, 1985.

Holt, D., and Simons, P.E. A tip from beyond? *Newsweek*, 1978 (Sep 18), *92*, 35–36. [On Teresa Basa murder]. *See also* O'Brien and Baumann.

Hunt for the strangler. *Newsweek*, 1964 (Feb 24), *63*, 31. [On Peter Hurkos]

Hurkos, P. *Psychic*. Indianapolis: Bobbs, Merrill, 1961. [Autobiography]

Hynd, A. Mind–reader who trapped Alberta's mad murderer. *Fate*, 1958 (Mar), *11*, 49–54. [On Maximilian Langsner]

Interview: Irene F. Hughes. *Psychic*, 1971, *3*(3), 4–7, 32–35.

Interview: Peter Hurkos. *Psychic*, 1970, *1*(5), 5–7, 32–37.

Jones, M. Psychics and police mix it up in L.A. *Fate*, 1979 (Oct), *33*(10), 64–71. [On Dorothy Allison, Roberta Jacobson, Kebrina Kinkade, M. Kathlyn Rhea, and Lotte von Strahl]

Jordan, P. Psychic's search for a missing child. *Fate*, 1977 (Aug), *30*(8), 60–65. [Autobiographical]

King, H.S. Sheriff who was not Wyatt Earp. *Fate*, 1961 (Jan),

14, 48–51. [On psychic lawman John Horton Slaughter]

Kitaev, N., and Ermakov, N. Usage of parapsychology in criminal investigation. *International Journal of Paraphysics*, 1978, *12*(5/6), 111–116. (Originally published in Russian in *Ural Review*, 1977, No. 1.) [On Stefan Ossowicki and various unnamed psychics and mediums]

Kobler, J. Man with the x–ray mind. *True*, 1986 (Jun), 36–37, 73–76. [On Peter Hurkos]

Kramer, G., Pratt, J.G., and St. Paul, U.V. Two direction experiments with homing pigeons and their bearing on the problem of goal orientation. Durham, NC: Duke University, September 30, 1957. 12p. Nonr–1181(03).

Kyro, L. Bevy Jaegers, psychic sleuth. *Fate*, 1979 (Mar), *32*(3), 39–43.

Law enforcement looks at psychic evidence to determine usefulness. *Brain/Mind Bulletin*, 1979 (Mar 5), *4*(8), 1–2.

Love, M. The woman who solves crimes. *Tomorrow*, 1959 (Spr), *7*(2), 9–18. [On Florence Sternfels]

MacGregor, R., and Janeshutz, T. The psychic vs. the killer. *Fate*, 1984 (Jul), *37*(7), 34–41. [On Renie Wiley]

Mandel, P. Now a seer helps stalk the Boston Strangler. *Life*, 1964 (Mar 6), *56*, 49–50. [On Peter Hurkos]

Marion, F. In my mind's eye. *Fate*, 1952 (Jan), *5*, 193–212. [Autobiographical]

Marshall, E. Police science and psychics. *Science*, 1980 (Nov 28), *210*, 994–995.

McGraw, W. Peter Hurkos and the Boston Strangler. *Fate*, 1967 (May), *20*, 48–58.

McMurran, K. Some days, say police, this New Jersey psychic can indeed see forever. *People*, 1979 (Jul 16), *12*, 95–101. [On Dorothy Allison]

Meerloo, J.M. Psychic detectives. Review of *Beschouwingen over het Gebruik van Paragnosten* [Notes on the Use of Telepathy for Police and Other Practical Purposes], by W.H.C.

Tenhaeff. *Tomorrow*, 1957, *5*(3), 66—67.

Messing, W. Wolf Messing about himself. (Excerpt from Tatyana Lungina, *Wolf Messing——Mystery of the Twentieth Century.* Ann Arbor, MI: Hermitage, 1982.) *Psi Research*, 1982 (Dec), *1*(4), 85—91.

Miles, E. The body in the well. *Fate*, 1974 (Nov), *27*(11), 70. [On Vera McNichol]

Murray, H.A., and Wheeler, D.R. A note on the possible clair—voyance of dreams. *Journal of Psychology*, 1937, *3*, 309—313. [Failure of attempt to solicit dreams about the Lindbergh kidnapping]

Myers, E. L. The use of psychics to aid police investigators in solving crimes. *Applied Psi Newsletter*, 1982 (Jul/Aug), *1*(3), 1—2.

Nolan, R., interviewed by Vilenskaya, L. My view of psi in criminology. *Psi Research*, 1984 (Sep/Oct), *3*(3/4), 118—130.

O'Brien, J., and Baumann, E. Accused of murder by a voice from the grave. *Ebony*, 1978 (Jun), *33*, 56—58+. [Voice came through stranger who was not a psychic. On Teresa Basa murder]

Osis, K. The application of ESP to criminal investigations, locat—ing missing persons, and cases of airplane disasters. In C.B. (Scott) Jones (Ed.), *Proceedings of a Symposium on Applica—tions of Anomalous Phenomena* (pp. 241—279). Santa Bar—bara, CA: Kaman Tempo, 1984.

Pettinella, D.M. Croiset locates a suicide. *Fate*, 1977 (Apr), *30*, 43—44.

Police are calling on psychics for aid. *New York Times*, 1978 (Nov 26), 61.

Pollack, J.H. Croiset aids FBI in civil rights murders. *Fate*, 1966 (Feb), *19*(2), 74—80.

Pollack, J. H. *Croiset the Clairvoyant.* Garden City, NY: Doubleday, 1964.

Pollack, J.H. Florence Sternfels, the amazing mind reader who solves crimes. *Parade*, 1964 (Apr 26).

Posinsky, S.H. The case of John Tarmon: Telepathy and the law. *Psychiatric Quarterly*, 1961, *35*(1), 165—166. [On Peter Hurkos]

Psi in criminal investigation (Some attempts in Poland and the USSR). *Psi Research*, 1982 (Dec), *1*(4), 82—84. [On Klimuszko and unnamed female psychic]

Psychic detectives. *Alpha*, 1979 (Mar/Apr), No. 1, 3. [On Nella Jones]

Ralston, J. Can psychics see what detectives cannot? *McCall's*, 1983 (Feb), *110*(5), 72—73. [On Dorothy Allison; John Catchings]

Randi, J. Allison and the Atlanta murders: A follow—up. *Skeptical Inquirer*, 1983—84 (Win), *7*(2), 7. [On Dorothy Allison]

Randi, J. Atlanta child murderer: Psychics' failed visions. *Skeptical Inquirer*, 1982 (Fal), *7*(1), 12—13. [On George Chang, Joseph DeLouise, Beverley Jaegers, John Monti, and George Murphy]

Reichbart, R. Western law and parapsychology. *Parapsychology Review*, 1981 (Mar/Apr), *12*(2), 9—11.

Reiser, M., and Klyver, N. A comparison of psychics, detectives, and students in the investigation of major crimes. In M. Reiser (Ed.), *Police Psychology: Collected Papers* (pp. 260—267). Los Angeles, CA: LEHI, 1982.

Reiser, M., Ludwig, L., Sate, S., and Wagner, C. An evaluation of the use of psychics in the investigation of major crimes. *Journal of Police Science and Administration*, 1979, *7*(1), 18—25.

Robbins, S. *Ahead of Myself: Confessions of a Professional Psychic.* Englewood Cliffs, NJ: Prentice—Hall, 1980. [Autobiography]

Roberts, E. Murder and suicide. In her *Fifty Years a Medium* (pp. 76—91). New York: Avon Books, 1969. [Autobiography]

Rodr, F. Psychotronics and the law. *Proceedings of the Second International Congress on Psychotronic Research*, 1975, 250—252.

Rogo, D.S. Psychic Sherlocks. *Human Behavior*, 1979 (May), *8*, 48. [About Reiser's survey]

Roll, W. G. Public safety agencies: Should we become involved? [abstract]. In W.G. Roll, R.L. Morris, and R.A. White (Eds.), *Research in Parapsychology 1981* (pp. 19–20). Metuchen, NJ: Scarecrow Press, 1982.

Roll, W.G., and Grimson, R.C. A majority–vote study of im—pressions relating to a criminal investigation. In R.A. White and R.S. Broughton (Eds.), *Research in Parapsychology 1983* (pp. 72–77). Metuchen, NJ: Scarecrow Press, 1985.

Rudley, S. *Psychic Detectives.* New York: Watts, 1979. [On Gerard Croiset, Irene Hughes, Peter Hurkos, and Beverley Jaegers]

Schall, S.L. Legal issues related to the use of psi. *Archaeus*, 1985 (Sum), *3*, 47–52.

Schall, S.L., and Kautz, W.H. Legal issues related to psi applica—tion in law enforcement. *Applied Psi*, 1984/1985 (Win), *3*, 7–12.

Stearn, J. *The Door to the Future.* New York: MacFadden—Bartell Books, 1964. (Ch. 12: The Crime Busters, & Ch. 13: The Psychic Machine) [On Peter Hurkos]

Steiger, B. Clairvoyance, cops, and dowsing rods. *ESP: Your Sixth Sense* (pp. 13–14, 67–78). New York: Award Books, 1966. [On Edgar Cayce, Gerard Croiset, and Mrs Runnels]

Steiger, B. Irene Hughes and the Canadian kidnappings. *Fate*, 1971 (Jul), *24*(7), 36–44.

Steiger, B. *Irene Hughes on Psychic Safari.* New York: Warner Paperback Library, 1972.

Steiger, B. *The Psychic Feats of Olof Jonsson* (pp. 11–19; 129–133). New York: Popular Library, 1971.

Steiger, B. Psychic Olof Jonsson solves mass murder. *Fate*, 1972 (Dec), *25*, 77–85.

Stemman, R. Croiset: The psychic detective. In *The Un—explained: Mysteries of Mind, Space, and Time* (vol. 4, pp.

488−489). New York: Marshall Cavendish, 1983.

Stephens, I. Linked precognitive dreams of a murder. *Journal of the Society for Psychical Research*, 1960 (Sep), *40*(705), 334−342.

Stokes, D., with Dearsley, L. *Voices in My Ear−−The Autobiography of a Medium* (pp. 140−147). London: Futura Publications, 1980.

Tabori, P. *Crime and the Occult.* New York: Taplinger, 1974.

Tanagras, A. Case: The detection of a theft through information given by a medium. *Journal of the Society for Psychical Research*, 1939, *31*, 81−83.

Tanagras, A. The case of the stolen jewels: A further note. *Journal of the Society for Psychical Research*, 1939, *31*, 95−96.

Tandler, M. Noreen Renier: Psychic detective. *Frontiers of Science*, 1980 (Nov− Dec), 25−28.

Tanous, A., with Ardman, H. Helping the police psychically. In their *Beyond Coincidence* (pp. 61−88). Garden City, NY: Doubleday, 1976. [Autobiography]

Tenhaeff, W.H.C. The Croiset experiments: 2. Aid to the police. *Tomorrow*, 1953 (Aut), *2*(1), 10−18.

Tenhaeff, W.H.C. The employment of paragnosts for police purposes. *Proceedings of the Parapsychological Institute of the State University of Utrecht*, 1960 (Dec), No. 1, 15−31. [On Gerard Croiset]

Tenhaeff, W.H.C. Police use of sensitives. *Light*, 1962 (Spr), *82*, 14−23. [On Gerard Croiset]

Tenhaeff, W.H.C. Practical uses of ESP. *Light*, 1962 (Sum), *82*, 16−20. [On Mr. W.B. and Gerard Croiset]

Tenhaeff, W.H.C. Psychoscopic experiments on behalf of the police. In *Proceedings of the First International Conference of Parapsychological Studies* (pp. 107−109). New York: Parapsychology Foundation, 1955. [On Gerard Croiset]

Tenhaeff, W.H.C. Success and error in sensitives. *Light*, 1962 (Aut), *82*, 18−30. [On "Mr. Alpha" (Croiset)]

Trubo, R. Psychics and the police. *Psychic*, 1975 (May/Jun), *6*(2), 8, 10—12. [On Gerard Croiset, Shirley Harrison, Irene Hughes, Peter Hurkos]

Truzzi, M. Editorial. *Zetetic Scholar*, 1981, No. 8, 3—4. [On his Psychic sleuths project]

Two newspaper stories on psychic detection. *Applied Psi Newsletter*, 1982 (Mar/ Apr), *1*(1), 3. [On unnamed psychics]

The use of psychics by local police departments. *Parapsychology Information Network Newsletter*, 1984 (Mar), *1*(9), 1—2, 4.

Vaughan, A. Police: How to use psychics. *Psychic*, 1975 (May/Jun), *6*(2), 22—23. [Psychic offers advice to police]

Walther, G. Crime detection by clairvoyance. *Tomorrow*, 1952, *1*(1), 30—37. (Originally published in German) [On Mrs. Günther—Geffers and other unnamed persons]

Weiner, D.H. The use of psi in law enforcement. (Abstract of paper presented at the 1980 SERPA Conference symposium on the use of psi.) *Journal of Parapsychology*, 1980 (Mar), *44*(1), 71—72.

Weird clue in the Crater mystery. *Life*, 1959 (Nov 16), *47*, 42—44. [On Gerard Croiset]

West, D.J. The identity of "Jack the Ripper": An examination of an alleged psychic solution. *Journal of the Society for Psychical Research*, 1949 (Jul/Aug), *35*(653), 76—81. [On Robert J. Lees]

Wharton, B. 17—year old murder reenacted. *Fate*, 1961 (Jan), *14*, 67—73.

Wilcox, T. *Mysterious Detectives: Psychics.* Milwaukee, WI: Raintree, 1977.

Wilson, C. The art of psychic detection. In his *Psychic Detectives: The Story of Psychometry and Paranormal Crime Detection* (pp. 195—226). San Francisco, CA: Mercury House, 1986. [The rest of this book is not on psychic detectives, despite the title.]

Winsey, V.R. Unbidden messages: Review of *My Passport Says*

Clairvoyant, by M.B. Dykshoorn. *Saturday Review*, 1975 (Jun 14), *2*, 37—38.

Wintereich, J.T. Time and telepathy. Review of Jack H. Pol— lack, *Croiset the Clairvoyant*. *Saturday Review*, 1965 (Sep 11), *48*, 44—45.

Wolkimer, R., and Wolkomir, J. Clairvoyant crime busters. *McCall's*, 1987 (Oct), *115*(1), 162—164. [On Dorothy Allison, Peter James Brown, Bertie Catchings, and John Catchings]

Woodward, K.L. The strange visions of Dorothy Allison. *McCall's*, 1978 (Sep), *105*(12), 28—38.

Wright, G.H. Book review of J.H. Pollack, *Croiset the Clair— voyant*. Garden City, NY: Doubleday, 1964. *Spiritual Fron— tiers*, 1964 (Jun), *9*, 107.

Yeterian, D. *Casebook of a Psychic Detective*. New York: Stein and Day, 1982. [Autobiography]

Zorab, G. A case of clairvoyance? *Journal of the Society for Psychical Research*, 1956 (Jun), *38*(688), 244—248. [On Gerard Croiset]

Zorab, G. Review of Jack Harrison Pollack's *Croiset the Clair— voyant*. *Journal of the Society for Psychical Research*, 1965 (Dec), *43*(726), 209—212. Discussion: *Journal of the Society for Psychical Research*, 1966 (Jun), *43*(728), 331—332.

BIBLIOGRAPHY OF U.S. GOVERNMENT REPORTS ON PARAPSYCHOLOGY

Adamenko, V. *Close to the Theme*. Wright—Patterson AFB, OH: Foreign Technology Division, March, 1981. 5p. (Originally published in Russian.) (FTD—ID(RS)T—2032— 80).

Adams, Ronald L., and Williams, R.A. *Biological Effects of Electromagnetic Radiation (Radiowaves and Microwaves)— — Eurasian Communist Countries (U)*. Washington, DC: Defense Intelligence Agency, 1976. (DST—1810S—074—76)

Anisimov, V. *Autogravitation——Bordering on the Fantastic*. Wright—Patterson AFB, OH: Foreign Technology Division, August, 1974. 9p. (Originally published in Russian.) (FTD—

HT−23−0003−75).

Anisimov, V. *On the Edge of the Fantastic.* Wright−Patterson AFB, OH: Foreign Technology Division, April, 1977. 12p. (Originally published in Russian.) (FTD−ID(RS)T−0306−77).

Arkadyev, V. *Electromagnetic Hypothesis of the Transmission of Mental Suggestion.* Wright−Patterson AFB, OH: Foreign Technology Division, September 15, 1977. 19p. (Originally published in Russian.) (FTD−ID(RS)T−1445−77).

Arseniev, K. *Phenomena Become Explainable.* Wright−Patterson AFB, OH: Foreign Technology Division, July 1981. 15p. (Originally published in Russian.) (FTD−ID(RS)T−0573−81).

Bateman, J.B., and Hewitson, V.S. *European Scientific Notes. (Vol. 30, No. 11).* London, Eng: Office of Naval Research, November, 1976. 52p. (ESN−30−11).

Bearden, T.E. *Field, Formon, Superspace, and Inceptive Cyborg: A Paraphysical Theory of Noncausal Phenomena.* Washington, DC: Army Medical Intelligence and Information Agency, December 1974. 52p. (MIIA−1−74).

Beaumont, R.A. C^{nth}?: On the strategic potential of ESP. *Signal: Journal of the Armed Forces Communications and Electronics Association,* 1982 (Jan), *36*(5), 39−45.

Bleikher, V.M. *Parapsychology: Science or Superstition?* Springfield, VA: National Technical Information Service, December, 1973. 78p. (Originally published in Russian.) (JPRS−60883).

Bongard, M., and Smirnov, M. *A Telepathic Experiment: Necessary Requirements.* Wright−Patterson AFB, OH: Foreign Technology Division, June 21, 1977. 29p. (Originally published in Russian.) (FTD−ID(RS)T−0324−77).

Bray, J.D. *Questionnaire Results on the Prospects for Soviet Development of Parapsychology for Military or Political Purposes* (Master's thesis). Monterey, CA: Naval Postgraduate School, December, 1979. 70p.

Bul, P.I. *Hypnosis and Suggestion.* Arlington, VA: Joint Publications Research Service, August 12, 1975. 88p.

Cazzamalli, F. *The Radiating Brain.* Wright—Patterson AFB, OH: Foreign Technology Division, March, 1966. 252p. (Originally published in Italian.) (FTD—TT—65—759).

Committee on Science and Technology. U.S. House of Representatives, 97th Congress. Research on the physics consciousness (parapsychology). In *Survey of Science and Technology Issues Present and Future* (pp. 59—60), John Fuqua, Chair. Washington, DC: Government Printing Office, 1981.

Dagle, E.F., Hill, M.D., & Mott-Smith, J. *Testing for Extrasensory Perception with a Machine.* Hanscom Field, MA: Air Force Cambridge Research Labs, May 1963. 31p. (AD—414 809).

Diaconis, P., and Graham, R. *The Analysis of Sequential Experiments with Feedback to Subjects.* Stanford, CA: Stanford University, Department of Computer Science, November, 1979. 50p. (STAN—CS—79—775).

Dodge, C.H. *Research into "Psi" Phenomena: Current Status and Trends of Congressional Concern.* Washington, DC: Science Policy Research Division, Congressional Research Service, Library of Congress, 1983. 35p. (83—511 SP).

Dodge, C.H. *The Human Mind: What Are Its Potentials and Limits, and How Can We Determine and Use Them?* Prepared for the Chautauqua on "Mind Research" at the request of Congressional Clearinghouse on the Future. Washington, DC: Library of Congress, Congressional Research Service, November 28, 1979. 4p.

Draganescu, Gh.E. *The Treatment of Psychotronic Experiments by Statistical Dynamics Methods.* Wright—Patterson AFB, OH: Foreign Technology Division, October, 1977. 5p. (FTD—ID(RS)T—1500—77).

Druckman, D., and Swets, J.A. *Enhancing Human Performance: Issues, Theories, and Techniques.* Washington, DC: National Academy Press, 1988.

Dumasova, J. *Parapsychology: Science Unknown.* Wright—Patterson AFB, Ohio: Foreign Technology Division, November 15, 1965. 13p. (FTD—TT—65—733).

Edelstam, C.G. *Long Range Navigation in Animals.* Springfield,

VA: National Technical Information Service, September 15, 1968. 39p. (AF—EOAR—81—65, AF—EOAR—62—108, AF—9777, 97701)

Faust, D., Gross, G.L., Kyler, H.J., and Pehek, J.O. *Investigations into the Reliability of Electrophotography.* Washington, DC: Defense Advanced Research Projects Agency, September 30, 1975. 83p. (AD—A018 806/0).

Foreign Science Bulletin (Vol. 4, No. 8, 1968). Washington, DC: Library of Congress, Aerospace Technology Division, August, 1968. 44p.

Givens, R.B., and Hudson, R.L. *A Low Cost Telemetry System.* Washington, DC: National Science Foundation, Aug 7, 1973. 5p. (NSF—73—57—GI—12) [Extrasensory perception and mental telepathy given as descriptors]

Graue, L.C. *Distant Goal Orientation in Birds.* Springfield, VA: National Technical Information Service, October 1, 1969. 7p. (NONR—4331[00], NR—301—537).

Graue, L.C. *Orientation of Homing Pigeons Exposed to the Electromagnetic Fields at Project Sanguine's Wisconsin Test Facility.* Washington, DC: Department of the Navy, Office of Naval Research, 1974. 23p. (NR101—633).

Graue, L.C., and Pratt, J.G. *Directional Differences in Pigeon Homing in Sacramento, California and Cedar Rapids, Iowa.* Springfield, VA: National Technical Information Service, June 25, 1959. 8p. (Microfilm only) (NONR—1181[03]).

Griffin, Donald R. *Airplane Observations of Homing Pigeons.* Springfield, VA: National Technical Information Service, September 3, 1952. 30p. (Microfilm only) (N6ONR26409).

Harlacher, W.M. *Bombs from Hyperspace.* Fort Detrick, MD: Army Medical Intelligence and Information Agency, April, 1979. 13p. (USAMIIA—K—9741).

Harman, W.W. *The Potential Use and Misuse of Consciousness Technologies.* Prepared for the National Science Foundation, Directorate for Scientific, Technological, and International Affairs, Division of Policy Research and Analyis, 1977. 19p. (Contract NSF/STP76—02573).

Houston, J. *Brain Research and the Human Potential.* Prepared for the Chautauqua on "Mind Research" at the request of Congressional Clearinghouse on the Future. Washington, DC: Library of Congress, Congressional Research Service, December 5, 1979. 3p.

International Congress on Psychotronic Research (2nd). (Selected Articles). Wright—Patterson AFB, OH: Foreign Technology Division, September, 1977. 141p. (Originally published in Russian, Czech, etc.) (FTD—ID(RS)T—1415—77).

International Congress on Psychotronic Research (3rd) Part II. Wright—Patterson AFB, OH: Foreign Technology Division, September, 1977. 50p. (Originally published in Russian, Czech, etc.) (FTD—ID(RS)T—1576—77).

Kantorovich, M. *Thoughts at a Distance.* Wright—Patterson AFB, OH: Foreign Technology Division, December, 1963. 5p. (FTD—TT—63—1035).

Karamfelov, I. *Is Parapsychology a Science?* Wright—Patterson AFB, OH: Foreign Technology Division, June 10, 1977. 15p. (Originally published in Russian.) (FTD—ID(RS)T—0892—77).

Kazhinsky, B.B. *Biological Radio Communications.* Springfield, VA: Clearinghouse for Federal Scientific and Technological Information, Apr. 1, 1963. 171p. (#AD 415: 676).

Kholodov, Y.A. *The Effect of Electromagnetic Fields on the Central Nervous System.* Springfield, VA: Clearinghouse for Federal Scientific and Technological Information, 1967. (Originally published in Russian.)

Kogan, I.M. *The Information Theory Aspect of Telepathy.* Santa Monica, CA: Rand Corp., July, 1969. 26p. (Originally published in Russian.) (P—4145).

Kogan, I.M. *Telepathy, Hypotheses, and Observations.* Washington, DC: Joint Publications Research Service, October 19, 1967. 7p. (Originally published in Russian.) (JPRS—43028).

Kolodny, Lev. *Wireless Telegraphy Number 2.* Wright—Patterson AFB, OH: Foreign Technology Division, March 24, 1977. 10p. (Originally published in Russian.) (FTD—ID(RS)I—0324—77).

Kozyrev, N.A. *Possibility of Experimental Study of the Properties of Time.* Arlington, VA: Joint Publications Research Service, Dept. of Commerce, 1968. 29p. (Originally published in Russian.) (JRPS—45238).

Kramer, G., Pratt, J.G., and St. Paul, U.v. Directional dif— ferences in pigeon homing. Springfield, VA: National Techni— cal Information Service, August 29, 1955. 2p. (Microfilm only) (NONR—1181[03]).

LaMothe, J.D. *Controlled Offensive Behavior USSR (Unclassified).* Washington DC: Defense Intelligence Agency, 1972. 176p. (# StCS0116972).

Maire, L.F., and LaMothe, J.D. *Soviet and Czechoslovakian Parapsychology Research (Unclassified).* Prepared by the U.S. Army Medical Intelligence and Information Agency, Office of the Surgeon General. Washington, DC: Defense Intelligence Agency, 1975. 71p. (# DST1810S38775).

Mitchell, E. *Implications of Mind Research.* Prepared for the Chautauqua on "Mind Research" at the request of Congres— sional Clearinghouse on the Future. Washington, DC: Library of Congress, Congressional Research Service, Decem— ber 19, 1979. 2p.

Nature Journal (Selected articles). [Translated by SCITRAN from *Ziran Zazhi*, 1980 (Apr), *3*(4)]. Wright—Patterson Air Force Base, OH: Foreign Technology Division, 1981. 74p. (Originally published in Chinese.) (FTD—ID(RS)T—1766—80).

Naumov, E. *The Riddle of "Psi" Phenomena.* Wright—Patterson AFB, OH: Foreign Technology Division, Air Force Systems Command, July 13, 1964. 2p. (Originally published in Russian.) (FTD—TT64 288; TT—64 71182).

Naumov, E.K., Vilenskaya, L.V., and Klychnikov, I.V. *Bibliogra— phies on Parapsychology (Psychoenergetics) and Related Sub— jects.* Washington, DC: Joint Publications Research Service, March 28, 1972. 101p. (Originally published in Russian.) (JPRS #55557).

Nonverbal Communication. 1964—December 1981 (Citations from the NTIS data base). Springfield, VA: National Technical In— formation Service, January, 1982. 149p.

Nonverbal Communication. 1964–February 1983 (Citations from the NTIS data base). Springfield, VA: National Technical In—formation Service, April, 1983. 154p.

Ostrander, S., and Schroeder, L. *Cultural and Historic Views of Mind Research in Other Countries.* Prepared for the Chautauqua on "Mind Research" at the request of Congres—sional Clearinghouse on the Future. Washington, DC: Library of Congress, Congressional Research Service, Decem—ber 5, 1979.

Palmer, J.A. *An Evaluative Report on the Current Status of Parapsychology.* Alexandria, VA: U.S. Army Research In—stitute for the Behavioral and Social Sciences, December, 1985. 254p. (DAJA 45–84–M–0405).

Paraphysics R & D Warsaw Pact. Washington, DC: Defense In—telligence Agency, March 30, 1978. 132p. (DST1810S20278).

A Person Who Reads Minds. Wright–Patterson AFB, OH: For—eign Technology Division, June 9, 1965. 2p. (Originally pub—lished in Russian.) (FTD–TT–65–155; TT–65–63919).

Pickenhain, L. *On the Classification of Psychic Phenomena as the Highest Plane of Interaction (With Notes on the Altered States of Consciousness Phenomenon).* Arlington, VA: U.S. Joint Publications Research Service, 1976. 8p. (Originally published in German.) [JPRS:66814]

Pinneo, L.R., Hall, D.J., and Wolf, D.E. *Feasibility Study for Design of a Biocybernetic Communication System.* Menlo Park, CA: Stanford Research Institute, March, 1973. 75p. (DAHC15–72–C–0167, ARPA Order–2034).

Pratt, J.G. *Research on Animal Orientation, with Emphasis Upon the Phenomenon of Homing in Pigeons.* Springfield, VA: National Technical Information Service, October, 1962. 17p. (NONR–1181[03])

Pratt, J.G. *Research on Animal Orientation, with Emphasis on the Phenomenon of Homing in Pigeons.* Springfield, VA: Na—tional Technical Information Service, January 26, 1954. 11p. (NONR–1181[03]).

Pratt, J.G. *Research on Animal Orientation, with Emphasis on the Phenomenon of Homing in Pigeons.* Springfield, VA: Na—tional Technical Information Service, February 7, 1955. 17p.

(NONR—1181[03]).

Pratt, J.G. *Testing for an ESP Factor in Pigeon Homing.*
Springfield, VA: National Technical Information Service, Sep—
tember 30, 1956. 15p. (NONR—1181[03]).

Pratt, J.G., and Thouless, R.H. *Homing Orientation in Pigeons
in Relation to Opportunity to Observe the Sun Before
Release.* Springfield, VA: National Technical Information
Service, April 3, 1954. 18p. (NONR—1181[03]).

Price, J.R., and Jureidini, P. *Witchcraft, Sorcery, Magic, and
Other Psychological Phenomena and Their Implications on
Military and Paramilitary Operations in the Congo.* American
University, Washington, DC: Special Operations Research
Office, August 8, 1964. 10p. (SORO—CINFAC—6—64).

*Proceedings of the Conference on Psychotronic Research (1st).
Held in Prague in 1973.* Arlington, VA: Joint Publications
Research Service, September 6, 1974. 2 vols. (Originally pub—
lished in Czech, Russian, etc.) (JRPS—L/5022—1, JPRS—
L/5022—2).

Psychotronic (Selected Articles). Wright—Patterson AFB, OH:
Foreign Technology Division, February, 1978. 41p.
(Originally published in German.) (FTD—ID(RS)T—2256—
77).

Pukhovskaya, S. *Seeing Through the Earth.* Wright—Patterson
AFB, OH: Foreign Technology Division, April, 1977. 9p.
(Originally published in Russian.) (FTD—ID(RS)T—0310—
77).

Pushkin, V. *"No"—to Parapsychology.* Wright—Patterson AFB,
OH: Foreign Technology Division, May, 1977. 20p.
(Originally published in Russian.) (FTD—ID(RS)T—0435—
77).

Pushkin, V., and Fetisov, V. *Intuition and Its Experimental
Study.* Washington, DC: Joint Publications Research Service,
April 1, 1969. 14p. (Originally published in Russian.)
(JPRS—47762).

Puthoff, H. *Remote Viewing.* Prepared for the Chautauqua on
"Mind Research" at the request of Congressional Clear—
inghouse on the Future. Washington, DC: Library of Con—
gress, Congressional Research Service, December 12, 1979.

2p.

Ranvaud, R. *Study of Homing Pigeons in Brazil: Some Preliminary Results.* Washington, DC: National Aeronautics and
Space Administration, October 1981. 16p.

**Ranvaud, R., Schmidt-Koenig, K., Kiepenheuer, J., and
Gasparotto, O.C.** *Initial Orientation of Homing Pigeons on
the Magnetic Equator, With and Without Sun Compass.*
Washington, DC: National Aeronautics and Space Administration, May 1984. 9p.

Raven, B.H., and Fishbein, M. *Acceptance of Punishment and
Change in Belief.* Washington, DC: National Technical Information Service, October 4, 1960. 26p. (AD–684 814)

Raven, B.H., Anthony, E., and Mansson, H.H. *Group Norms
and Dissonance Reduction in Belief, Behavior, and Judgment.*
Washington, DC: National Technical Information Service,
December 20, 1960. 24p. (AD–684 816).

Reeves, C.T. *Use of Kirlian Photography in Fatigue Assessment:
Final Report.* Army Materiel Command. Texarkana, Texas
Intern Training Center, December, 1975. 59p. (USAMC–
ITC–02–06–076–403).

Report of a Workshop on Experimental Parapsychology (Held on
September 30, 1988 in Washington, DC). Washington, DC:
Office of Technology Assessment, February, 1989. 23p.

Roux, A., Crussard, C., and Bouvaist, J. *Psychokinetic Experiments with Metal Test Samples.* June, 1979. 44p. (UCRL
Trans. 11522)

Ryzl, M. *Model of Parapsychological Communication.* Wright–
Patterson AFB, OH: Foreign Technology Division, July,
1965. 2p. (Originally published in Czechoslovakian) (FTD–
TT–65–366).

Schmidt, H. *Anomalous Prediction of Quantum Processes by
Some Human Subjects.* Springfield, VA: National Technical
Information Service, February 1969. 50p. (D1–82–0821).

Schmidt, H. *New Correlation Between a Human Subject and a
Quantum Mechanical Random Number Generator.* Sprinfield,
VA: National Technical Information Service, November 1967.
53p. (D1–82–0684).

Schmidt-Koenig, K. *Internal Clocks and Homing.* Springfield, VA: National Technical Information Service, December, 1960. 5p. (Microfilm only) (NONR−1181[03]).

Scientists Investigate Results of Telepathic Experiments. Washington, DC: Joint Publications Research Service, July 18, 1968. 15p. (Originally published in Russian.) (JPRS−45922).

Sergeyev, V. *Scientists "Examine" Tofik Dadashev.* Wright−Patterson AFB, OH: Foreign Technology Division, October 23, 1974. 11p. (Originally published in Russian.) (FTD−HT−23−0004−75).

Shevalev, A., and Filatova, V.P. *From Sensational Stir to Serious Research: Experiments Confirm, Yes, "Skin Sight" Exists.* Wright−Patterson AFB, OH: Foreign Technology Division, April 27, 1977. 10p. (Originally published in Russian.) (FTD−IJ(RS)I−0560−77).

Soviet Parapsychology: Annotated Bibliography (Surveys of Soviet−Bloc Scientific and Technical Literature). Washington, DC: Library of Congress, Aerospace Technology Division, July 23, 1964. 25p. (ATD U−64−77).

Spirkin, A. *Getting to Know Psycho−Biophysical Reality.* Wright−Patterson AFB, OH: Foreign Technology Division, September, 1980. 15p. (Originally published in Russian.) (FTD−ID(RS)T−1175−80).

Stanescu, S. *Telepathy in Relation with Science.* Wright−Patterson AFB, OH: Foreign Technology Division, March 24, 1977. 7p. (Originally published in Russian.) (FTD−ID(RS)I−0324−77).

Surveys of Soviet−Bloc Scientific and Technical Literature: Soviet Parapsychology, Annotated Bibliography. Washington, DC: Library of Congress, Aerospace Technology Division, July 23, 1964. 17p. (ATD U6477).

Tarasenko, F.P. *In Regard to I.M. Kogan's Article, "Is Telepathy Possible."* Wright−Patterson AFB, OH: Foreign Technology Division, August 17, 1977. 9p. (Originally published in Russian.) (FTD−ID(RS)T−1437−77).

Van, N. *An Analysis of the Relationship Among Meditation, Per−*

sonality Type and Control of Brain Wave Production. (Master's thesis). Monterey, CA: Naval Postgraduate School, September 1973. 56p. (AD−776 313/9). [Parapsychology listed as a descriptor.]

Varshavskii, K.M. *Hypnosuggestive Therapy (Treatment by Suggestion in Hypnosis).* Wright−Patterson AFB, OH: Foreign Technology Division, December 4, 1974. 203p. (Originally published in Russian.) (FTD−HC−23−2047−74).

Vasiliev, L.L. *Experimental Studies of Mental Suggestion.* Arlington, VA: Joint Publications Research Service, May 31, 1973. 198p. (Originally published in Russian.) (JPRS−59163).

Vasiliev, L.L. *Mysterious Phenomena of the Human Psyche.* Wright−Patterson AFB, OH: Foreign Technology Division, March 10, 1967. 192p. (Originally published in Russian.) (FTD−HT−66−336).

Volodin, B. *We Were Assembled to Investigate, Not See Mediumistic Phenomena.* Wright−Patterson AFB, OH: Foreign Technology Division, July 29, 1977. 45p. (Originally published in Russian) (FTD−ID(RS)T−1236−77).

Vondracek, V. *Valuation and Its Disturbances from the Standpoint of Psychiatry.* Wright−Patterson AFB, OH: Foreign Technology Division, February 17, 1966. 242p. (Originally published in Russian.) (FTD−TT−65−532).

Water Dowsing. Reston, VA: U.S. Department of the Interior, Geological Survey, 1977. 15p. (GPO Item No. 621).

Wortz, E.C., Eerkens, J.W., Bauer, A.J., and Blackwelder, R.F. *Biophysical Aspects of Parapsychology.* Torrance, CA: AiResearch Manufacturing Company of California, 1975. 36p. (Document No. 75−11096A).

Wortz, E.C., Bauer, A.J., Blackwelder, R.F., Eerkens, J.W., and Saur, A.J. *Novel Biophysical Information Transfer Mechanisms (NBIT). Final Report.* Torrance, CA: AiResearch Manufacturing Company of California, 1976. (Document No. 76−13197). 90p. Contract No. XG−4208(54−20)75S.

Zavala, A., Van Cott, H.P., Small, V.H., and Orr, D.B. *Research on Human Non−Visual Perception.* Silver Spring,

MD: American Institutes for Research, October, 1966. 54p. (AIR—E—70—10/66—FR).

Zavala, A., Van Cott, H.P., Small, V.H., and Orr, D.B. *Research on Human Non—Visual Perception* (Supplement to final report). Silver Spring, MD: American Institutes for Research, October, 1966. 38p. (AIR—E—70—10/66—FR—Suppl).

Zincenko, V., and Leontov, A.N. *Parapsychology——Fiction or Reality.* Charlottesville, VA: Army Foreign Science and Tech—nology Center, March, 1977. 21p. (Originally published in Russian.) (AST—1810I—115—76, FSTC—995—76).

PARTIAL BIBLIOGRAPHY ABOUT U.S. GOVERNMENT INTEREST IN PARAPSYCHOLOGY

Alexander, J.A. Enhancing human performance: A challenge to the report. *New Realities*, 1989 (Mar/Apr), *9*(4), 10—15, 52—53.

Alexander, J.A. The new mental battlefield: "Beam me up, Spock." *Military Review*, 1980 (Dec), *60*, 47—54.

Anderson, J. CIA toys with extrasensory weapons. *Washington Post*, 1981 (Mar 30), B13.

Anderson, J. Ouija—board group in Pentagon pushes for ESP weaponry. *Newsday*, 1981 (Feb 15), 68.

Anderson, J. Pentagon, CIA cooperating on psychic spying. *Washington Post*, 1984 (May 3), *107*, B15.

Anderson, J. Pentagon invades Buck Roger's turf. *Washington Post*, 1981 (Jun 8).

Anderson, J. Pentagon and CIA coordinate efforts on psychic spying. *Newsday*, 1984 (May 3), 94.

Anderson, J. Psychic studies might help U.S. explore Soviets. *Washington Post*, 1984 (Apr 23), *107*, B14.

Anderson, J. The race for "inner space." *San Francisco Chronicle*, 1984 (Apr 24).

Anderson, J. Secret psychic research. *San Francisco Chronicle*,

1984 (Apr 23).

Anderson, J. U.S. intelligence is racing the Soviets in psychic research. *Newsday*, 1984 (Apr 23), 44.

Anderson, J. "Voodoo gap" looms as latest weapons crisis. *Washington Post*, 1984 (Apr 24), *107*, C13.

Anderson, J. Yes, psychic warfare is part of the game. *Washington Post*, 1981 (Feb 5).

Anderson, J., and Van Atta, D. Sources claim ESP—ionage psychs out secrets. *Washington Post*, 1985 (Aug 12).

Anderson, J., and Van Atta, D. Soviets winning the "mind" race. *Washington Post*, 1985 (Jul 17), *108*, E25.

Badger, D. Report review: Research into psi phenomena by Christopher Dodge. *Applied Psi*, 1986 (Spr), *5*(1), 2, 12—13.

Bartlett, L.E. *Psi Trek* (pp. 277—283). New York: McGraw—Hill, 1981. [On the so—called psychic arms race.]

Broad, W.J. Pentagon is said to focus on ESP for wartime use. *New York Times*, 1984 (Jan 10), *133*, 17.

Broad, W.J. Pentagon reportedly spent millions on ESP. *San Francisco Chronicle*, 1984 (Jan 11), 16.

Caldwell, C. Beyond ESP. *New Times*, 1978 (Apr 3), *10*(7), 42—50.

Ebon, M. *Psychic Warfare: Threat or Illusion?* New York: McGraw—Hill, 1983.

An E.S.P. gap: Exploring psychic weapons. *Time*, 1984 (Jan 23), *123*(4), 17.

Fitzgerald, R. Pentagon's secret psychic task force. *Fate*, 1981 (Jul), *34*(7), 86—92.

Frazier, K. Improving human performance: What about para—psychology? *Skeptical Inquirer*, 1988 (Fal), *13*(1), 34—45.

Fuller, C. Psychic cold war. *Fate*, 1984 (May), *37*(5), 7—14.

Government report on psi. *Parapsychology Review*, 1982

(Mar/Apr), *13*(2), 18.

Holden, C. Parapsychology update. *Science,* 1983 (Dec 2), *222*, 997.

Honegger, B., and Mishlove, J. Security implications of applied psi: An historical summary. *Applied Psi Newsletter,* 1982 (Nov/Dec), *1*(5), 1–5.

Levine, A., with Fein, E., and Doder, D. The Communists' psychic edge. *U.S. News and World Report,* 1988 (Dec 5), *105*(22), 30.

Levine, A., and Fenyvesi, C., with Emerson, S. The twilight zone in Washington. *U.S. News and World Report,* 1988 (Dec 5), *105*(22), 24–26, 30.

McRae, R.M. *Mind Wars: The True Story of Government Research into the Military Potential of Psychic Weapons.* New York: St. Martin's, 1984.

National Research Council. Enhancing human performance: A summary of the findings. *New Realities,* 1989 (Mar/Apr), *9*(4), 10–11, 52–53. (Excerpts from NRC report)

Office for Technology Assessment sponsors parapsychology working group. *FRNM Bulletin,* 1988 (Aut), No. 39, 4.

Palmer, J.A., Honorton, C., and Utts, J. Reply to the National Research Council Study on Parapsychology. *Journal of the American Society for Psychical Research,* 1989 (Jan), *83*(1), 31–49.

Pell pushes for new look at psi proposals. *Skeptical Inquirer,* 1989 (Win), *13*(2), 126.

"Psi" R & D: Congress gets favorable report. *Science and Government Report* (Washington, DC), 1983 (Oct 1), *13*(16), 1–3.

Psychic war and the Pentagon. *Science Digest,* 1984 (May), *92*(5), 38.

Radin, D.I. Parapsychology bushwacked. *Fate,* 1989 (Feb), *42*(2), 36–43.

Reppert, B. US and the Soviets working on mind–control

weapons. *Philadelphia Inquirer*, 1983 (Nov 7), 5.

Rhine, J.B. Location of hidden objects by a man–dog team. *Journal of Parapsychology*, 1971 (Mar), *35*(1), 18–33.

Rossman, M. On some matters of concern to psychic research. In his *New Age Blues: On the Politics of Consciousness* (pp. 167–258). New York: Dutton, 1979.

Skeptical eye: Paranormal Pentagon. *Discover*, 1981 (Mar), *2*(3), 15.

Squires, S. Army research finds possible military use for sleep learning; parapsychology's validity doubted. *Washington Post*, 1988 (Apr 17), C3.

Squires, S. The Pentagon's twilight zone. *Washington Post*, 1988 (Apr 17), *111*, C3.

Starr, D., and McQuaid, C.P. PSI soldiers in the Kremlin: They labor in the bowels of government searching for ways to harness the mind. *Omni*, 1985 (Aug), *7*(11), 80, 82, 104–106.

Stuckey, W.K. Psi on Capitol Hill: Official circles. *Omni*, 1979 (Jul), *1*(10), 24, 142.

Stuckey, W.K. Psychic power: The next superweapon? *New York*, 1976 (Dec 27)/1977 (Jan 3), *10*(1), 47–50, 55.

Sturrock, P.A. Commentary on NRC report: *Enhancing Human Performance*. *Explorer*, 1988 (Apr), *4*(2), 3.

Targ, R., and Harary, K. U.S. and Soviet psi research. In their *The Mind Race* (pp. 1–108). New York: Villard Books, 1983.

Tart, C.T. A survey of expert opinion on potentially negative uses of psi, United States government interest in psi, and the level of research funding in the field. In W.G. Roll (Ed.), *Research in Parapsychology 1978* (pp. 54–55). Metuchen, NJ: Scarecrow Press, 1979.

Tart, C.T. A survey on negative uses, government interest and funding of psi. *Psi News: Bulletin of the Parapsychological Association*, 1978 (Oct), *1*(2), 2.

A Twilight zone defense? *Newsweek*, 1987 (May 11), 5.

Vilenskaya, L.V. Epilogue: Psi research in the Soviet Union: Are they ahead of us? In R. Targ and K. Harary, *The Mind Race* (pp. 247—260). New York: Villard Books, 1983.

Voss, S. Mind games: ESPionage and the arms race. *Washington Times*, 1984 (Aug 28), 1B, 4F—6F.

Weberman, A.J. Mind control: The story of Mankind Research Unlimited, Inc. *Covert Action*, 1980 (Jun), No. 9, 15—21.

Weeks, A.L. Soviet studies in hostile parapsychology put U.S. at loss in "psychic arms race." *New York City Tribune* (New York), 1987 (Aug 17), 1, 2.

Wilhelm, J.L. The politics of psi. In his *The Search for Super— man* (pp. 240—256). New York: Pocket Books, 1976.

Wilhelm, J.L. "Psychic" spying? *Outlook/Washington Post Sun— day Magazine*, 1977 (Aug 7), B1, B5.

GOVERNMENT GRANTS FOR RESEARCH ON PARAPSYCHOLOGY

It is difficult to be very definitive in listing government grants for psi research. Story has it that some parapsychological research has been government financed under the cover of dif— ferent names, such as biofeedback or hypnosis. Rumor also has it that some psi research has been funded through private foundations that served as a front for government funding. There is also the ever present reality that many of those who speak or write about government funding are not really in the know, while those who really know won't talk. What is presented here is provisional and incomplete. At least it sum— marizes much of what is *publicly* known.

The only published statistics available were reported by psychologist/parapsychologist Charles Tart in a survey sum— marized in *Research in Parapsychology 1978* (see preceding bibliography). Tart polled the directors of 14 active para— psychology research centers in the U.S. All but one replied. One of the questions asked was: "Have you or one of your laboratory staff been approached by agents or officials of the U.S. government, acting in an *official* capacity, in the last five years, in order to gather information on parapsychology or any government agency?" Of the 13 who responded, eight had never been approached, one had been approached once, and four had been approached "several times."

There follows a list of grants for psi research that are not classified. It makes no claim to be exhaustive, but it is all we could locate. Many of the titles listed are of reports of research, not necessarily the grant title (or subject). Similarly, the dates are often the date of a report on the research, not the date the funds were allocated. In many cases the subject of an article or report is narrower than the purpose for which government funds were granted. If a title is known to be that of a report rather than the subject of a grant, it will be as-terisked. I am indebted in large part to a survey of federally-funded research on exceptional functioning and optimum develop-ment conducted by the Institute of Noetic Sciences and the *PsiLine* database for the following information.

DATE UNKNOWN

Title:	**Conceptual Background of Eighteenth Century Medicine**
Principal Investigator:	Lester King
Research Facility:	Unknown
Granting Agency:	NIMH
Amount:	$13,250

Title:	**Kirlian Photography**
Principal Investigator:	William Eidson and David L. Faust
Research Facility:	Drexel University, Department of Physics and Atmospheric Science
Granting Agency:	Dept. Army
Amount:	$145,000

Title:	**Investigations of Out-of-Body Ex-periences and Unusual Dream States**
Principal Investigator:	Roy D. Salley
Research Facility:	Veterans Adm, Medical Center, Topeka, KS
Granting Agency:	Veterans Administration, R&D
Amount:	$9,552

Title:	**Spirit Possession and Stress in Psychosocial Development**
Principal Investigator:	L. Metcalf
Research Facility:	Unknown
Granting Agency:	ADAMHA
Amount:	$9,552

Title: **Studies of Techniques for the Enhancement of Human Performance***
Principal Investigator: National Academy of Sciences Executive Office
Research Facility: Unknown
Granting Agency: Army Research Institute for the Behavioral & Social Sciences
Amount: Unknown

1960

Title: **Group Norms and Dissonance Reduction in Belief, Behavior, and Judgement***
Principal Investigators: Bertram H. Raven, Edwin Anthony, and Helge H. Mansson
Research Facility: University of California (Los Angeles), Department of Psychology
Granting Agency: Office of Naval Research NONR– 233(54); NR–171–350
Amount: Unknown

Title: **Acceptance of Punishment and Change in Belief***
Principal Investigator: Bertram H. Raven, and Martin Fishbein
Research Facility: University of California (Los Angeles), Department of Psychology
Granting Agency: Office of Naval Research NONR– 233(54); NR–171–350
Amount: Unknown

1962

Title: **Research on Animal Orientation, with Emphasis upon the Phenomenon of Homing in Pigeons***
Principal Investigator: J.G. Pratt
Research Facility: Duke University, Parapsychology Laboratory
Granting Agency: Office of Naval Research NONR118103
Amount: Unknown

1963

Title: **Research With ESP Testing Machine***
Principal Investigator: Cambridge Research Laboratories

Research Facility:	Unknown
Granting Agency:	Air Force
Amount:	Unknown

1964

Title:	**Witchcraft, Sorcery, Magic, and Other Psychological Phenomena and Their Implication for Military Operations in the Congo**
Principal Investigator:	J. Price
Research Facility:	Unknown
Granting Agency:	Dept. Army
Amount:	Unknown

1971-72

Title:	**Comparison of Pre-Sleep & "Extrasensory" Stimuli in Dreams**
Principal Investigator:	Montague Ullman & Charles Honorton
Research Facility:	Maimonides Medical Center, Dream Laboratory
Granting Agency:	NIMH
Amount:	$52,000

1972

Title:	**Precognition Experiments Involving Learning**
Principal Investigator:	P. Phillips, Washington University
Research Facility:	Unknown
Granting Agency:	NIMH
Amount:	$7,500

1973

Title:	**Feedback in Computer Psi Tasks**
Principal Investigator:	W.F. Vitulli
Research Facility:	Unknown
Granting Agency:	U.S. Army Research Command
Amount:	Unknown

Title:	**Phenomenological Reality and "Post-Death Contact"** *
Principal Investigator:	R.A. Kalish & D.K. Reynolds
Research Facility:	UCLA, School of Public Health
Granting Agency:	NIMH, Center for Studies of Suicide

	Prevention (MH 20822)
Amount:	Unknown

Title:	**Technique to Enhance Extraordinary Human Perception**
Principal Investigator:	Russell Targ & Harold E. Puthoff
Research Facility:	Stanford Research Institute
Granting Agency:	NASA
Amount:	$80,000

1974

Title:	**Comparison of Pre-Sleep and Extrasensory Stimuli in Dreams**
Principal Investigator:	Montague Ullman
Research Facility:	Maimonides Medical Center, Dream Laboratory
Granting Agency:	NIMH
Amount:	$31,993

Title:	**History of Parapsychology**
Principal Investigator:	S. Mauskopf & M. McVaugh
Research Facility:	Duke University/University of North Carolina
Granting Agency:	NSF
Amount:	$40,000

Title:	**Kirlian Photography**
Principal Investigator:	D.E. Lord
Research Facility:	University of California (Livermore), Lawrence Livermore Laboratory
Granting Agency:	U.S. Army Research Command, AEC Contract W−7405−Eng−48
Amount:	Unknown

1975

Title:	**Investigations into the Reliability of Electrophotography--Phase III**
Principal Investigator:	David E. Faust
Research Facility:	Logical Technical Services Corp.
Granting Agency:	Advanced Research Projects Agency, Contract No. MDA 903−75−1028
Amount:	Unknown

| Title: | **Use of Kirlian Photography in Fatigue Assessment*** |

Principal Investigator:　　C. Thomas Reeves
Research Facility:　　Unknown
Granting Agency:　　Army Materiel Command
Amount:　　Unknown

1975-77

Title:　　**Mediums, Social Process & Change**
Principal Investigator:　　Edward Schieffelin
Research Facility:　　Unknown
Granting Agency:　　NIMH
Amount:　　$48,249

1976

Title:　　**Potential Uses and Misuses of Consciousness Technologies**
Principal Investigator:　　Willis W. Harman
Research Facility:　　Center for the Study of Social Policy, Stanford Research Institutes
Granting Agency:　　NSF STP76-02573
Amount:　　Unknown

1976-77

Title:　　**Physiological Correlates of PK (Psychokinesis) Performance**
Principal Investigator:　　Edward Kelly, Duke University
Research Facility:　　Unknown
Granting Agency:　　NIMH
Amount:　　Unknown

1977-78 and 1983-85

Title:　　**Translation of Weyer's "De Praestigis Demonum"**
Principal Investigator:　　George Mora
Research Facility:　　Unknown
Granting Agency:　　NIMH
Amount:　　$71,421

1978

Title:　　**Conversion, Testimonial & Healing**
Principal Investigator:　　James Peacock
Research Facility:　　Unknown
Granting Agency:　　NIMH

Amount: Unknown

1979

Title: The Analysis of Sequential Experi-
                               ments With Feedback to Subjects*
Principal Investigator: P. Diaconis & R. Grahman
Research Facility: Stanford University, Department of
                               Computer Science
Granting Agency: NSF-MCS77-23738 and NSF-
                               MCS77-16974
Amount: Unknown

1983

Title: Evaluating Performance in Continuous
                               Experiments With Feedback to Sub-
                               jects*
Principal Investigator: F. Samaniego & J. Utts
Research Facility: University of California (Davis), Divi-
                               sion of Statistics
Granting Agency: Air Force Office of Scientific Research
Amount: Unknown

Title: Immediate Feedback and Target-
                               Symbol Variation in Computer As-
                               sisted Psi Tests*
Principal Investigator: William F. Vitulli
Research Facility: University of South Alabama, Depart-
                               ment of Psychology
Granting Agency: USARC Grant No. 3-61194
Amount: Unknown

1987

Title: The Effects of Spiritual Healing
Principal Investigator: R.A. Burbank
Research Facility: Unknown
Granting Agency: Veteran's Administration
Amount: Unknown

1987-88

Title: The Representational Validity of Self-
                               Disclosure Coding
Principal Investigator: Kathryn Dindia
Research Facility: Unknown

Granting Agency: NIMH
Amount: $42,886

CHRONOLOGICAL RECORD OF EVENTS INDICATING
A RAPPROCHEMENT BETWEEN THE U.S. GOVERN-
MENT AND THE PARAPSYCHOLOGICAL COMMUNITY*

Rhea A.White

    The first   sign of serious government   interest in any
given topic is generally indicated by the preparation of overviews   and assessments, with bibliographies, of the past and current status of the subject under review.   The first report I
could locate was *Surveys of Soviet–Bloc Scientific and Technical
Literature. Soviet Parapsychology: Annotated Bibliography* (4:
1964),   compiled under the aegis of the Aerospace Technology
Division of the Library of Congress. It contains annotations of
35 Soviet documents held by the Library of Congress and a list
of 23 additional items known to exist but not owned. The documents reflect "Soviet research efforts and the state of Soviet
research in telepathy" (p. ii).
    In the 1970s, the U.S. government, under various auspices,
undertook a number of additional assessments, with interest centered on Soviet parapsychology and the possible military applications of psi.   John D. LaMothe prepared a report entitled
"Controlled Offensive Behavior––USSR" under the auspices of
the U.S. Army, Office of the Surgeon General, Medical Intelligence Office. Published by the Defense Intelligence Agency in
July 1972,   this report reviews the military applications of
various psychological, physiological, and parapsychological techniques developed in the Soviet Union. (The author says his
main source of information on the parapsychology material is a
popular article by Ostrander and Schroeder, the two authors of
the sensationalistic book, *Psychic Discoveries Behind the Iron
Curtain!* [Prentice–Hall, 1970].)
    Also in 1972, the Joint Publications Research Service trans-
lated and published *Bibliographies on Parapsychology
(Psychoenergetics) and Related Subjects––USSR*, which had been
compiled by three Soviet parapsychologists: E.K. Naumov, L.
Vilenskaya, and I.V. Klychnikov. It lists 1,077 items in several
categories. (In 1981 a revision of this bibliography was prepared
and published by the Parapsychological Association [668].) Louis
F. Maire III and John D. LaMothe prepared a report entitled

_____
    *All of the documents cited here are listed in Section 4
unless otherwise indicated.

Soviet and Czechoslovakian Parapsychology Research, which was published by the Defense Intelligence Agency in 1975. They ex—amined Soviet and Czechoslovakian parapsychological research since 1950 in several areas: telepathy, telepathic behavior modification, psychotronic generator research, PK, OBEs, remote viewing, and apports. (Some relevant research conducted at the Stanford Research Institute is also touched on.) In each case they point out the possible military applications of the research. Soviet personnel and facilities are listed in an appendix. They conclude that the Soviets are ahead of Western parapsy—chologists in developing instrumentation for learning how to con—trol and replicate paranormal phenomena and in discovering the mechanisms involved.

Under a government grant a report was prepared in 1975 by E.C. Wortz and others at AiResearch Manufacturing Com—pany of California entitled "Biophysical Aspects of Parapsychology." This report reviews "parapsychological processes . . . . under the general categories of psychophysiol—ogy, phenomenological parapsychology, physical constraints, and information theory and pattern recognition" (p. 1). Western as well as Soviet sources were taken into consideration.

The same team prepared a final report the following year entitled "Novel Biophysical Information Transfer Mechanisms (NBIT)." Their summary states: "The body of the report treats Soviet application of statistical theories, research done on electrostatics, the development of remote sensors, hypothesized carrier mechanisms, human sensitivity to magnetic fields, and performance training to improve NBIT" (p. 1—1).

A glance at the documents listed in Section 4 indicates that the U.S. government was actively monitoring Soviet para—psychological research throughout the 1970s, and that most of the reports dealt with military applications. The list of grants in Section 7 indicates that the government was also interested in parapsychology at home, if only sporadically. The first writ—ten record of this interest that I could locate consists of a few pages in a book by R.A. McConnell [**568**], who describes a proposed meeting of parapsychologists and government officials. According to McConnell:

> In early 1973, apparently as a result of curiosity about parapsychology among scattered individuals in several government agencies, an exploratory memorandum was prepared by a "plans and process analyst" of the National Institute of Mental Health proposing a one—week state—of—the—art con—ference on "psychic research."
> The memo was sent to more than a dozen

individuals, scattered over the U.S.A., none of whom belonged to the Parapsychological Associa—tion. Three months later the same memorandum, along with some informal responses to it, was sent for comment, under new (HEW) aegis——this time to 20 individuals (including myself) who were en—gaged in parapsychological research or otherwise closely associated with the field.

The memo described tentative plans for a con—ference of "very 'blue ribbon' scientists and profes—sionals across a wide range of disciplines to explore . . . and to recommend firmly developed . . . stan—dards that can be used to identify . . . the poten—tials for advancing [parapsychological] knowledge and its *use*." [italics added]

There were to be "two days spent in examin—ing the state of the art; another day devoted to identification of scientific and technical questions that need to be raised to identify the boundaries of the domain of inquiry; and finally a two— or three—day work session to develop specific state—ments of research that [would] be needed in order to establish psychic research and *development* as a legitimate field of scientific inquiry [italics added]. . . . . [This] final portion of the conference [would] address itself to such questions as: What are the criteria of credibility? . . . What kinds of data and findings will establish credibility? . . . *What type of research work is worth doing?* [italics added] . . . How should it be undertaken?" (pp. 293—294)

McConnell advised against the proposal on the ground that parapsychological research was still too immature to be handled constructively in a meeting such as that proposed by the NIMH. Presumably others concurred, for the meeting never took place.

In March 1978, the Defense Intelligence Agency issued another report entitled "Parapsychology R & D——Warsaw Pact." It presented historical reviews and a summary of the cur—rent status of parapsychological research in the U.S.S.R., Bul—garia, Czechoslovakia, and Poland.

The next parapsychologist to deal with government interest in psi was Charles Tart (5: 1979) of the University of California (Davis) in a report of a survey already summarized on pages 478—479.

In 1981 there apparently were a sufficient number of mem—bers of Congress——in addition to the well—known involvement of Senator Claiborne Pell (D—R.I.) and Congressman Charles

Rose (D—N.C.)——interested in psi research to request that the
Congressional Research Service prepare a report on the status of
psi research. The report, authored by Christopher Dodge (4:
1983), emphasized the potential application of psi in the areas
of education, the military, criminal investigation, and medicine if
psi effects could become repeatable.   Research in the United
States, Soviet Union, and the People's Republic of China was
described.   Dodge concluded   that to advance further, increased
recognition must be given to the inherently interdisciplinary na—
ture of parapsychology.   Although the report recognized the
problems involved in taking parapsychology seriously, it also
held out hope that those difficulties could be surmounted.

The bibliography in Section 5   reveals that popular and
journalistic works on the subject did not become common until
the decade of the 1980s.

An important step toward rapprochement was taken on
November 30—December 1, 1983 when an unusual symposium
organized by Scott Jones under the auspices of Kaman Tempo
was held in Leesburg, VA.   The papers given at the symposium
have already been described [see **373**].   The discussion here will
center not on what was said but on why it was held and its
potential impact on the government.   It was unusual because,
as David G. Foxwell, General Manager of Kaman Tempo,
points out: its objective was "to provide a venue where outside—
of—government researchers [in parapsychology] could present
government managers and scientists details of their research and
an assessment of the potential of application of their research."
One impetus for the symposium was to take a small step for—
ward toward the ultimate goal of initiating government support
for parapsychological research.   As Scott Jones [**373**] pointed
out: "As of this date we seem to be depending upon nongovern—
ment funding and the bold office or agency head who is willing
to invest in order to find out what the fruits of research mean
for supporting his area of responsibility. . . . The fact that the
conference was held is an unspoken consensus from that group
that the country can and should do better" (p. vi).   Another
aim of the conference was to consider under what disciplinary
umbrella or umbrellas parapsychology belongs.   Jones pointed
out the administrative and bureaucratic significance of this ques—
tion, especially in Washington, D.C.

In 1985, with the passage of the Gramm—Rudman—
Hollings Act and cutbacks by the Reagan Administration, the
trickle of government funding for psi research all but dried up.
And in 1988, with the publication of the notoriously critical
U.S. Army—sponsored report of the National Research Council
[**462**] on the benefits of various techniques for enhancing human
performance (including parapsychology), a blow was struck at

the very heart even of government interest in psi phenomena——
let alone funding for research! At a Washington, DC press con—
ference convened to announce the release of the NRC report,
the Committee Chair, John A. Swets, stated: "Perhaps our
strongest conclusions are in the area of parapsychology" (5:
quoted in Palmer, Honorton, & Utts, p. 31). The conclusion was
that the Committee found "no scientific justification from
research conducted over a period of 130 years for the existence
of parapsychological phenomena" (p. 22). However, the Com—
mittee did admit that it might be worthwhile to continue to
monitor some types of· research, and they cite some Soviet
research and the work currently being conducted at Princeton
University (Robert Jahn), Psychophysical Research Laboratories
(Charles Honorton), Mind Science Foundation (Helmut Schmidt),
and SRI International (Edwin May).

This was a grievous blow for parapsychology and for the
course of solidifying the relationship between psi researchers and
the government. It was particularly bitter because, although
the investigation had been conducted by a highly prestigious
governmental science body, the National Research Council of the
National Academy of Sciences, in fact the Committee chose as
its adjudicators two of parapsychology's best known critics,
James Alcock and Ray Hyman, whereas it ignored the reports
of two persons they had commissioned who were in favor of
parapsychology: John Palmer and Robert Rosenthal.

In an effort to set the record straight, three members of
the Parapsychological Association, the international professional
society of parapsychology, John Palmer, Charles Honorton, and
Jessica Utts, prepared on the Association's behalf a reply to the
National Research Council (5: Palmer, Honorton, & Utts, 1989)
in which they document some of the criticisms of the
Committee's report. In their concluding remarks they state:

> The Committee's primary conclusion regarding
> parapsychology is not merely unjustified by their
> report, it is directly contradicted by the
> Committee's admission that it can offer no
> plausible alternatives. This concession, coming as
> it does from a Committee whose principal
> evaluators of parapsychology were publicly com—
> mitted to a negative verdict at the outset of their
> investigation, actually constitutes a strong source of
> support for the conclusion that parapsychology has
> identified genuine scientific anomalies.
>
> We have documented numerous instances
> where, in lieu of plausible alternatives, the
> Committee's attempts to portray parapsychology as

"bad science" have been based upon erroneous or
incomplete descriptions of the research in question,
rhetorical enumeration of alleged "flaws" that by
its own admission frequently have no
demonstrable empirical consequences, selective
reporting of evidence favorable to its case, and the
selective omission of evidence not favorable to its
case.   Moreover, with respect to the Committee's
central mission for the U.S. Army, we have shown
that the Committee's prejudice against parapsy—
chology has led it to ignore research, the further
development of which could have important implica—
tions for our national security.

The scientific and defense communities are
entitled to a rigorous unbiased assessment of this
research area.   A strong prima facie case has been
made for the existence of psi anomalies, and mean—
ingful relationships between such events and
psychological variables have been reported in the
literature.   Further efforts and resources should be
expended toward the identification of underlying
mechanisms and the development of theoretical
models, either conventional or "paranormal," that
can provide adequate understanding. (5: Palmer et
al., 1989, pp. 43—44)

What may well be the most important of the efforts to
bring together government and the parapsychological community
is the most recent development, which is a report of a
workshop on experimental parapsychology held under the
auspices of the International Security and Commerce Program's
Office of Technology Assessment of the U.S. Congress.   The
workshop was held on September 30, 1980 and brought together
the two major authors of the critical NRC report, Professor
James Alcock of Glendon College, Toronto and Ray Hyman of
the University of Oregon, along with five persons associated
with psi research: Charles Honorton of the Psychophysical
Research Laboratories, Professor Robert G. Jahn, of Princeton
University Engineering Anomalies Research Laboratory, John Pal—
mer of the Institute of Parapsychology, Theodore Rockwell, and
Jessica Utts of the University of California (Davis).   Also
present were two persons very knowledgeable about parapsy—
chology but not identified with it: Professor Daryl Bem of Cor—
nell University and Professor Marcello Truzzi of Eastern
Michigan University.   The workshop was chaired by Alan Shaw
of the Office of Technology Assessment.   The report of the
workshop was published in February, 1989 (4: "Report of a

Workshop . . . ," 1989).

The impetus behind the workshop was the debate over the scientific status of parapsychology in general and the NRC report in particular. Known as the "OTA workshop," the Office of Technology Assessment (OTA) examined parapsychology at the request of the Technology Assessment Board, which serves as its oversight body and which had expressed an interest in the human potential aspect of psi research. Rather than con— tinuing the debate, OTA held the workshop to "illuminate it by identifying the main points of contention and the reasons for them" (p. 7). The report examines experimental para— psychological methodology, the issue of replicability, criteria for acceptable experiments, and data analysis. The meaningfulness of the experimental results was also considered as well as the question of what would constitute proof of psi, and the relation of parapsychology to the broader world of science. The report closes with an important discussion of future choices para— psychologists might make to bring "the field further inside the edifice of established science" (p. 23), and by implication, at— tract government backing.

It is my personal assessment that in principle the U.S. government has come as close to parapsychology as could reasonably be expected, given the scientific status of the research. The major error of the government agencies who have looked into parapsychology is that for the most part they did not consult mainline parapsychologists until the recent OTA workshop. Had they done so sooner, probably a significant amount of money could have been better spent. The other side of the coin is that parapsychologists might well take seriously some of the recommendations in the Dodge report, the NRC report, and especially the OTA report. In any case, with the OTA workshop the two sides met at last and a fruitful inter— change took place. It can be expected that this is but the begin— ning of productive cooperation that will be of benefit to both government and parapsychology and, ultimately, to all of us.

VI

New Views of Parapsychology and Parapsychological Phenomena*

QUICK INDEX TO SECTIONS IN THIS CHAPTER

I mentioned in the Introduction that various factors have been responsible for new views of parapsychology, especially in the last decade. If one is keenly interested in following the lead—ing edge of parapsychological thinking, one of the best sources of information is the series *Advances in Parapsychological Research**** edited by Stanley Krippner [**520-524**].*** The series of five volumes (with a sixth in press) contain chapters written by key persons who review the past findings and future prospects of specific areas of psi research

---

*I want to express my deep gratitude to Stanley Krippner, John Palmer, and Dean Radin for reading this chapter, as well as to Linda Henkel for laborious copy—editing.

**This invaluable series suffers from lack of support. Although not unreasonably priced by today's standards [$35.00], the publisher, McFarland, has not been able to break even, and the series may end with Volume 6, which is scheduled for publication in 1990. I strongly urge libraries especially, but also interested individuals, to buy Volume 6 at least, and preferably all the back volumes. Together they provide an incomparable source of information on all aspects of parapsychology, including both pro and con views, and cover research findings, methodology, and theory.

***Citations in brackets and boldface are to items described in earlier chapters. Citations in parentheses are to items in the reference list at the end of this chapter. Citations to *chapters* found in books in Chapter One will be listed—separately in the reference list.

and theory. Another valuable source is *Zetetic Scholar* [**850**], a
journal produced single–handedly by sociologist Marcello Truzzi
of Eastern Michigan University, who describes himself as a
"public doubter" (Truzzi, 1987a) of psi, but who through his
journal has provided a forum for debate between proponents
and critics of parapsychology. For example, in No. 5, 1979,
Truzzi published "A Dialogue on Statistical Problems in Psi
Research." In No. 6, July 1980, psychologist/parapsychologist
John Beloff described what he considered to be seven experi–
ments that provided good evidence for psi, followed by commen–
taries pro and con by 13 parapsychologists and critics, with a
reply by Beloff to his commentators. In No. 8, 1981, para–
psychologist Jeffrey Mishlove contributed an article entitled
"The Schism in Parapsychology," which was followed by the
remarks of 10 knowledgeable commentators and a response by
Mishlove.

In his own writings in this decade, Truzzi has also made
valuable theoretical contributions to parapsychology. I am
heavily indebted in writing this chapter to his introduction to
Volume 5 of *Advances in Parapsychological Research* [**524**]. In a
brief five–page introduction he anticipated and solidified several
of the main straws in the wind I describe here.

In the 1950s, when I spent four years as a Research Fel–
low at the Duke University Parapsychology Laboratory, the
general consensus of J.B. Rhine and the senior staff members of
the laboratory was ESP, PK, and probably precognition were
experimentally proven. Forty years have passed, and now the
consensus among many of the leaders of the field is that psi
has yet to be verified, and parapsychology has still not been ac–
cepted by the elite of the scientific community, although some
inroads in achieving this goal have been made.

Does this mean that all of the experimental and other in–
vestigations conducted thus far have been for nought—–that
parapsychology has not advanced at all? I would reply with an
emphatic *no*. Much has been learned. Many possibilities have
been suggested that have not yet been fully followed up.
Moreover, I think the analogy of a spiral is appropriate here:
Once we take a further turn of the spiral, much of what parap–
sychologists have learned, viewed from that higher angle, will
become meaningful and usable, though maybe not for the pur–
pose it was originally intended.

Before I joined the staff of the Parapsychology Laboratory,
I had steeped myself in the writings of the mystic/science writer
Gerald Heard. In reference to the (then) "new physics" of
relativity and quantum mechanics, he once told me something
that seemed very exciting and meaningful. He suggested that in
science as in life we were learning that, as he phrased it, "we

do not have laws; we have hypotheses." I found that statement liberating then, and I still do.

If we cannot expect to have laws, then that holds not only for parapsychology but for anthropology, psychology,* biology, physics, for everything! *Anything* is possible, although some events are more probable than others, and parapsychological events are among the least probable. As John Palmer pointed out in his 1979 presidential address to the Parapsychological Association, quantum theory proposes that "the universe, at least at the microscopic level, is probabilistic . . . . The laws of quantum physics are statistical laws, in contrast to the deterministic laws of Newtonian physics" (Palmer, 1980, p. 190). Palmer goes on to observe that parapsychologists have tended to apply the standards of nonprobabilistic science to parapsychology. (Rhine did this when he contended——as he did for 50 years——that psi is nonphysical. The physics he referred to was that of Newton, not quantum physics.) Palmer argued that parapsychology should make the most of being a probabilistic science, and he offers suggestions as to how this might work.

Parapsychology today is not a collection of laws but consists of a body of data and many hypotheses——some of them very exciting. In losing what it thought it had, I think it stands to gain more than parapsychologists previously imagined. In this chapter I discuss five areas that currently appear to be changing the nature of parapsychology and the ways in which the phenomena it studies are viewed. The impetus for the first five has come in part from outside parapsychology. The areas are:

1. The rise and development of "anomalies research."
2. Recent views of the nature of science, particularly in regard to the demarcation between science and pseudoscience.
3. The influence of electronic data processing.
4. The development of constructive criticism both from out-side and from within the parapsychological community.
5. Changes taking place within parapsychology.

This essay is intended to serve both as an overview of recent changes in parapsychology and as a guide to further sources of information. To accomplish both ends, I will not only

---

*There are signs of breaking loose from the machine model in psychology as well. In the lead chapter of a recent *Annual Review of Psychology*, Leona E. Tyler (1981) assesses the current psychology scene and describes several nonmechanistic models that have been proposed for psychology, including probabilistic functionalism. She also devotes two pages (11–12) to parapsychology as a possible (she uses the word "doubtful") extension of psychology's purview.

describe the changes, but I will also describe how and where to locate additional information. The references cited are given in the reference list at the end of the chapter unless they refer to a book, periodical, or organization described in previous chapters, in which case the item number is given in boldface within brackets.

ANOMALIES RESEARCH

Even as it is difficult to find an explanation for anomalies, it is hard to find a clearcut definition of the term "anomaly." Marcello Truzzi (1987b) comes to the rescue with an excellent discussion of the nature of anomalies. He points out that it is a multidimensional term, and he cites several ways of viewing anomalies. What they reduce to, however, is that an anomaly "is a fact in search of an explanation" (Truzzi, 1987b, p. 13).

Attempts to study anomalies, or anomalous phenomena, have existed ever since the sciences themselves became sufficiently advanced to be in the position to set boundaries as to what can exist (according to their theories and previous empirical findings) and what cannot. The residual of alleged phenomena the accepted sciences cannot explain are called anomalies, and they are either thought not to exist at all or are considered too dubious to be worthy of consideration. Probably hundreds of anomalous phenomena could be cited. A few examples are Bigfoot, UFOs, stigmata, the Loch Ness and other monsters, showers of frogs and fishes, cattle mutilations, vampires, etc. None of these are considered parapsychological (primarily because there is no indication that they are linked with a human observer who is outside sensory contact with the phenomena), but it is entirely conceivable that when parapsychologists learn more about the characteristics of psi, in principle it may serve as a helpful construct in explaining the above anomalies.

However, many of the phenomena studied by parapsychology (some would say all of them) can be classified as anomalous simply because they run counter to the basic limiting principles of what *Newtonian* physics deems possible. But, as Truzzi puts it: "One can fully endorse the legitimacy of parapsychology without endorsing the legitimacy of psi" (Truzzi, 1987a, p. 5). However, it has been argued (Akers, 1984; Morris, 1987; Truzzi, 1985) that at this stage, parapsychology is a protoscience rather than an established science.

There have long been many associations devoted to the study of various anomalistic phenomena. What is new in the past decade is that persons interested in various anomalies are banding together to study anomalies in general or certain classes of anomalies. This is evidenced by the fact that in Chapters 2 and 3 there are separate subsections of journals and organiza- tions dealing with groups of anomalies. This has led to the sharing of findings and the development of generalizations con- cerning anomalies. These organizations include many members who are established scientists. The most important group is the Society for Scientific Exploration [**905**]. A member of this group, chemist Henry H. Bauer, recently published an important review article (Bauer, 1988) in which he surveyed the *com- monalities* in controversies about alleged anomalies. The abstract of his paper is in itself very informative and indicates both the problems involved in studying anomalies and the poten- tial value in doing so. It is as follows:

There are a number of features that seem to be common to controversies about claimed anomalies. Foremost perhaps is the very fact of controversy. Typically, the anomaly runs counter to the expecta- tions of established orthodoxy, and there is often a populist tone to the argument. Questions concern- ing the demarcation of science from pseudoscience and of epistemology in general are typically raised. It becomes important to distinguish between the pros and cons of a particular claim and what is said by the disputants; an examination of the ways in which belief and disbelief are distributed among various groups can be useful in clarifying the issues. It is also vital that one distinguish between the occurrence and the reporting of events. As with interdisciplinary work, it is problematic to es- tablish what parts of existing knowledge might be relevant; and anomalies bring to attention large and sometimes unsuspected areas of ignorance. There are pitfalls in assuming that anomalies with superficial similarities have any functional or neces- sary relation to one another. The manner in which anomalies are perceived is clearly influenced by contemporary science and by contemporary socie- tal beliefs. For many reasons that go far beyond the possible reality of any given anomalous claim, then, the study of anomalies can be interesting and enlightening. (Bauer, 1988, p. 1)

The Society for Scientific Exploration (Sturrock, 1987) is dedicated "to using the highest standards of scientific inquiry to study the ideas and concepts that challenge our current paradigms" (Howard, 1987, p. i). Several parapsychologists are active members of this group, and papers on parapsychology comprise a good proportion of those given at its annual conven— tion. About a quarter of the articles published thus far in its *Journal of Scientific Exploration* have been on parapsychology.

An older group, founded in 1975, the Committee for the Scientific Investigation of Claims of the Paranormal (CSICOP) [902], consists primarily of science writers, philosophers, and a disproportionate number of magicians. Although the word "scientific" appears in its name, this is mainly a group of avowed skeptics. Hansen (1988) has recently pointed out that CSICOP is really an advocacy group devoted to debunking that is rapidly growing. A total of 26,000 subscribe to its journal, the *Skeptical Inquirer* [849]. Morris (1982b) has evaluated "the objectivity and effectiveness of the strategies . . . employed by CSICOP members and those who readily affiliate with them" (Morris, 1982b, pp. 257—258). Although CSICOP cannot be counted on to play a positive role in fostering cooperation be— tween parapsychology and its critics, at least it plays a major role as the "devil's advocate." This is an important ingredient.

An offshoot of this increased interest in applying the prin— ciples of scientific inquiry to anomalistic phenomena is the rise of a new discipline called "anomalistic psychology" (Zusne, 1982). Zusne and Jones [659] have prepared a textbook for use in the increasing number of courses offered on the subject in American universities. Anomalistic psychology, according to them, is

> the scientific approach to all those psychological phenomena that do not fit the current scientific world view by the criteria of most psychologists, as well as paranormal phenomena of other kinds that at least in part can be explained in terms of known psychological principles. (p. vii)

Until recently such subjects were only touched on briefly, if at all. Anomalistic psychologists attempt to treat them sys— tematically. They consider the entire range of occult beliefs, which is much broader than parapsychology, but parapsychology is included. Basically, they search for "complex physiological and psychological processes that influence behavior" [659], popularly termed paranormal.

Unfortunately, in practice Zusne and Jones often lapse into the stance of those who have a will to disbelieve. In a lengthy review, J. Rush (1983) points out their shortcomings. In addi— tion to many fallacies, omissions, and outright errors, they reveal a definite materialistic bias that apparently blinds them to those aspects of parapsychology that cannot be explained away. Zusne and Jones do not even deal with the ESP research of the last 20 years, such as the important work with computers and the Ganzfeld experiments. In other words, they deal almost exclusively with what *can* be explained away, and they shy away from the research results that cannot be dis— missed easily. Their coverage of the literature is also biased. For example, they deal with psychologist E. Girden's (1962a, 1962b) critical review of the evidence for PK (updated in Gir— den & Girden, 1985), but they do not mention G. Murphy's (1962) rebuttal. Thus, the potential value of anomalistic psychology to parapsychology is not realized by Zusne and Jones, although they do a creditable job of providing normal psychological explanations for many occult beliefs.

I. Child (1987), in a summation of the approach of Zusne and Jones, points out that Zusne, in particular, seems to be of the opinion that parapsychology is not a science. He says that Zusne (1985) expresses "the view that since parapsychologists study consistent relations between events without having an ac— ceptable account of the causal relation between them, they are engaging in magical thinking. His account of the thought processes of parapsychologists seems . . . a personal fantasy" (Child, 1987, p. 221).

There are some other works that could be classed as anomalistic psychology that make positive contributions to ex— tending the province of psychology into areas previously con— sidered to be occult or paranormal (e.g., see Leahey & Leahey [649], Neher [355], and Reed [813]). Within parapsychology the most notable example is Susan Blackmore [400, 597], who calls for, astonishingly, a parapsychology without psi. Also important are Robert Morris's attempts to delineate what parapsychology is *not* (Morris, 1986b) and to thoroughly review normal psychological explanations for what superficially appears to be psi (Morris, 1989).

The potential importance of anomalistic psychology to parapsychology is that it provides a meeting ground for para— psychologists and psychologists. (For a thorough discussion of this topic see Truzzi, 1985.) Anomalistic psychology attempts to delineate and to extend the boundaries of the normal into ter— ritory once thought to be paranormal. Critics of parapsychology also attempt to show how so—called normal means can be used to explain paranormal phenomena, but they put the emphasis on explaining them away and tend to dwell on deception, whether deluded or deliberate. Idealistically, anomalistic psychologists are more constructive in their approach and seek to preserve the phenomena and understand the dynamics in— volved. Although anomalistic psychology may reduce the number of phenomena considered to be genuinely psi—mediated, it also could be said to be working to purify parapsychology——to aid in the process of distilling that which is truly paranormal. Anomalistic psychologists would probably say that their aim is to abolish any residue. If they succeed, it would be the end of parapsychology. But to the extent that they fail to find normal explanations for what appear to be psi—mediated behavior, in effect they would be strengthening the case for psi.

THE NATURE OF SCIENCE

There are also active discussions underway concerning the nature of science and the scientific method. A central theme is the question of whether science is a body of facts or a method of investigation that can be applied to any subject. If the former, then parapsychology has amassed a large amount of data, and if the latter, it could be argued that its legitimacy has already been established. There is much current discussion of the demarcation problem, or where a line can be drawn between science and so—called pseudoscience. At Virginia Polytechnic Institute there is a Center for the Study of Science in Society that organizes and supports research and graduate education in the field of science and technology. Its special aim is to advance the understanding of the nature of science and technology and to study the ways in which they interact with the cultures and societies of which they are a part. In 1983 the Society held a workshop on "The Demarcation Between Science and Pseudo—science." In summarizing the workshop in the introduction to its proceedings, Rachel Laudan noted:

> As the workshop progressed some shared perspectives developed during the discussions. Most participants agreed that scientists and others frequently resort to demarcation criteria. It was also generally acknowledged . . . that these had to be taken into account in assessing any particular assertions of the scientificity or non—scientificity of a particular theory or activity. There was less consensus about whether demarcation criteria adopted for non—epistemic reasons reflected any underlying cognitive differences between science and pseudo-science or non—science. (Laudan, 1983, p. 6)

In part the need for such discussions has developed in order to defend the investigation of anomalous phenomena such as UFOs, Bigfoot, and parapsychological phenomena from the as—

saults of those who are active disbelievers rather than scientists. The will to disbelieve prevents the skeptic from viewing such problems dispassionately just as much as the will to believe can lead proponents to self—delusion. Disbelief can lead to a form of debunking that is fueled by the same passion that led to the witchcraft trials of an earlier day. This attitude and associated activities have been criticized not only by proponents of the investigation of various anomalies but by detached observers who are concerned that the truth is more important than bias, pro or con. The word "scientism" has been applied to the attempts of believers and disbelievers who ape the scientific method in order to mold the facts to fit their particular prejudices.   Certain members of CSICOP have been guilty of scientism.   Both Theodore Rockwell, either alone (Rockwell, 1979) or with his sons (Rockwell, Rockwell, & Rockwell, 1978a, 1978b), and Brian Inglis (1986) have written persuasively against scientism, especially as regards attacks on parapsychology. Lang (1975) shows how scientism also exists within parapsychology and more broadly, psychology, as does Eugene Taylor (1987).

Chemist Henry Bauer (1987) has presented some enlightening observations in an article entitled "What Do We Mean by 'Scientific?'"   He devotes considerable space to pointing out what science is *not*. It once was (but can no longer be) characterized by the certainty of its knowledge or by its theories. He writes: "Facts turn out to be slippery rather than graspable, and they change significantly as theories change——as our ways of looking at things change" (p. 119).   Nor can it be defined in terms of method.   The most important determining factor seems to be acceptance by the scientific community.   Bauer points out:

> Individuals who do scientific work can employ any style they choose, and success or failure are not necessarily determined by that style . . . . But the work that is done does not really become Science until it is acceptable to the scientific community. . . . And, of course, the scientific community agrees that, in the testing of candidates for entry into science, it is proper to employ logic, to demand clear evidence, to avoid unexplained contradiction, and to respect the validity of existing knowledge, methods, and theories. A criticism commonly made of Science is that it resists strikingly new claims. That, of course, is to be expected under the scheme just described.   Candidates for entry must be judged against what has already been incorporated; the more strikingly the claimed

novelty contradicts existing scientific knowledge, the
more overwhelmingly must the novelty be sup—
ported by the evidence if it is to be communally
accepted. (Bauer, 1987, p. 125)

Truzzi (1978) notes that extreme stances of critic and
proponent alike can result in a stalemate: "On the one hand we
may have the claimant offering evidence that is insubstantial for
the critic, and on the other hand we may have a critic giving
insubstantial indication of what it would take to force the critic
to accept the evidence" (Truzzi, 1978, p. 12). Truzzi attempts
to alleviate this situation by a discussion and classification of
extraordinary explanations in the context of extraordinary claims
(Truzzi, 1978).

Also in an attempt at clarification of the issues involved in
scientific acceptance, Bauer (1987) notes that established scien—
tists deal with three components: method, knowledge, and
theory. Work likely to succeed "is carried on by accepting or—
thodoxy in two of those and . . . novelty [is sought] in the
third" (p. 126). He says that the problem with anomalous
phenomena is that they are not grounded in any of the three.

I submit that a good case can be made that in para—
psychology, at least, we have both method and knowledge. It
is necessary now to concentrate on theory in order to gain
entrance to the halls of Science. (For a relevant discussion see
Churchland, 1987.) Only when the findings of parapsychology
make good sense theoretically will the data be acceptable as
part of science at large. Although only a few of the theories
put forth to explain psi have received exposure in the popular
literature, in fact, many have been proposed within the field.
Stokes (1987) recently surveyed these theories, and concluded:

As can be seen, despite the occasional contentions
of some skeptics, parapsychologists are active
theorizers. One of the most exciting developments
in the past two decades has been the growing
realization that psi phenomena need not be in con—
flict with established laws of science. A most excit—
ing development in the past five years has been
the experimental confirmations of the principle of
nonlocality in quantum mechanics and the realiza—
tion of the importance of that principle for a
theory of psi phenomena. At present, theorizing in
parapsychology is held back by the lack of a reli—
able data base and repeatable psi effects upon
which a theory of psi might be constructed and
refined. Thus many of the theories discussed in

this chapter represent mere presentations of
"theoretical environments" in which more testable
theories might be constructed. Those theories that
are already highly testable in principle, such as
Schmidt's, are still awaiting empirical confirmation.
(p. 189)

Few systematic attempts have been made to emphasize
and capitalize on potential novel approaches that could be
spawned by parapsychology itself. Such approaches might have
far—reaching implications for the other sciences as well, which
could speed up the legitimization of psi. Three areas of poten—
tial large—scale influence outside parapsychology are the role of
the experimenter/investigator in science (e.g., Schmidt, 1974b;
White, 1984a, 1984b), psi—determined human—machine interac—
tion (see Morgan, 1986; Morris, 1986a), and——what may be
most important——attempts to use psi itself to investigate parap—
sychological problems. If this approach should prove to be
worthwhile, then it might make a major contribution to science
itself (see Kautz, 1982; White, 1980, 1987).

Although it is essential in science not to let one's precon—
ceptions determine beforehand what can or cannot be true,
science——physics especially——has demonstrated repeatedly in
this century that not only can the observer not be eliminated
from the investigation but that the role of the observer may be
a necessary component of the phenomena being studied. Capra
(1975), Dossey [796], and Zukav (1979), to name only three of
many, have argued persuasively for this view. This question, in—
cluding the possible role of parapsychology, was discussed at a
roundtable that was organized by physicist Elizabeth Rauscher
at the 1986 convention of the Parapsychological Association en—
titled "Do We Discover and/or Create Reality?"

Within parapsychology itself, the issue of the experimenter
effect has become one of paramount importance. There is
general agreement that such an effect exists, and there are
several review articles on the subject (Kennedy & Taddonio,
1976; Thouless, 1976; White, 1976a, 1976b, 1977).
Some parapsychologists, led by Rex Stanford (1981a), hope that
the experimenter's influence can be minimized and ultimately
controlled for. Palmer, on the other hand, insists: "We must
stop treating it as artifact and start treating it as essence if
progress is to be made" (Palmer, 1985, p. 3). Proponents of the
observational theories (Broughton, 1979; Millar, 1979) hold that
without the experimenter there wouldn't be any statistically sig—
nificant data to study!

INFLUENCE OF ELECTRONIC DATA PROCESSING AND OTHER TECHNOLOGICAL INNOVATIONS

The advent of the microcomputer and extensive incorpora—tion of its uses, especially in conjunction with random number generators (RNGs), has transformed and improved psi research in many ways. As will be seen in the next section on critics, experimental parapsychology has been subjected to many criticisms from within and from outside the field since J.B. Rhine published his first experimental monograph (Rhine, 1934). The computerized psi experiments have risen phoenix—like from the battleground of criticisms over earlier experiments, and an amazing number have been conducted under conditions that are much better controlled against sensory cues and rational in—ference than in the past.

Initially, computers were used in parapsychology primarily to analyze data gathered by traditional methods. Tart (1966) pioneered an early ESP test device called ESPATESTER as well as a random generator (Tart, 1967). Davis (1974) was one of the first to review ways in which computers could be used to advantage in the actual conduct of experiments. He also described some caveats involved in computerized psi research. Schmidt (1974a, 1975) followed with detailed descriptions of in—strumentation in parapsychology. For more recent reviews, see Broughton (1982) and McCarthy (1982).

Probably the easiest way to catch a glimpse of this work is in the pages of *Research in Parapsychology* (*RIP*) [**527-529, 533-540, 543-547**], which contains lengthy abstracts of the reports given at the annual convention of the Parapsychological Association. For serious students and scholars the information in *RIP* is too truncated to allow an in—depth survey (for that purpose one should consult the full papers printed in the *Proceedings* distributed at each convention and available from the Program Chair or——better yet——the refereed journals [**823, 825, 826, 827**], which publish the PA papers that survive the peer scrutiny involved in attaining journal publication). However, for an overview, *RIP* is valuable because of the range of experiments reported and the very fact that the abstracts, al—though lengthy, are still not long enough to clog the reader's mind with too many details. Some of the ways in which automation has methodologically improved and innovatively ex—tended the range of experimentation in parapsychology is evi—dent from the following titles taken from the three latest *RIP*s (1985—1987).

From *RIP 1985*: "A Preliminary Review of Performance Across Three Computer Psi Games," "Further Studies Using a Competitive Computer Game," "A Cooperation—Competition PK

Experiment with Computerized Horse Races," "Variance Effects in REG Series Score Distributions," "Psi Experiments with Ran—dom Number Generators: Meta—Analysis Part I," "Human PK Effort on Prerecorded Random Events Previously Observed by Goldfish," "Complexity Dependence in Precognitive Timing: An Experiment with Pseudorandom Number Sequences," "An Automated Psi Ganzfeld Testing System," "Data—Base Manage—ment for Anomalies Research," and "Psi Experiments with Ran—dom Number Generators: An Informational Model."

From *RIP 1986*: "Psi Effects Without Real—Time Feed—back Using a Psilab//Video Game Experiment," "Psi Effects Without Feedback: Will the Real Psi Source Please Stand Up?" "Operator—Related Anomalies in a Random Mechanical Cascade Experiment," "Random Event Generator PK in Relation to Task Instructions," "Ganzfeld Target Retrieval with an Automated Testing System: A Model for Initial Ganzfeld Success," "Successful Performance of a Complex Psi—Mediated Timing Task by Unselected Subjects," "Missing and Displace—ment in Two RNG Computer Games," "Testing the Intuitive Data Sorting Model with Pseudorandom Number Generators: A Proposed Method," "Recent Advances in the IDS Model," "Exploratory Test of the IDS Model," and "On Some Statistical Implications of the IDS Hypothesis."

From *RIP 1987*: "Precognition of Probable versus Actual Futures: Exploring Futures That Will Never Be," "Explorations of Some Theoretical Frameworks Using a New PK—Test Environment," "Psi Effects Without Real—Time Feedback," "Building a PK Trap: The Adaptive Trial Speed Method," "In Search of 'Psychic Signatures' in Random Data," and "Anomalous Human—Computer Interaction (AHCI): Towards an Understanding of What Constitutes an Anomaly."

What are all these computerized experiments about and how have they transformed psi research? To answer this ques—tion I will discuss the role of computers and psi in 11 key areas.

Home Experimentation. In the past it has been noted that sub—jects tend to do better if they participate in psi experiments in a friendly, familiar environment, ideally in their home surround—ings (Pratt, 1961). The laboratory experiment can try to emu—late a home environment, but it is no substitute. With the in—troduction of random generators and microcomputers, it has be—come possible for subjects to provide reliable data at a place and time of their own choosing. In earlier days subjects might be encouraged to work at home in order to become acquainted with psi test techniques, but if they scored significantly, they had to be brought into the laboratory in the hope that they could still score well under controlled conditions. This was changed in 1969 when Schmidt (1969a, 1969b, 1970) introduced a portable RNG that would revolutionize the nature of psi testing. Also useful for this purpose was an inexpensive portable electronic ESP tester proposed by B. Millar (1973).

Experimenter as Subject. Experimenters themselves are potentially the best subjects for several reasons: (1) their motivation is high, (2) they understand why controls are important, and (3) they are trained observers who are in the best position to record all pertinent data and to understand the nature of the psi process as it operates in any given experimental situation. In addition, by participating as subjects they are better equipping themselves to devise experiments for others and to instruct subjects in responding in psi experiments. In the past, successful experimenter subjects had to perform under observed laboratory conditions in order to lend credence to their solo work. A prime example is the Swedish engineer Haakon Forwald, a star PK subject and developer of the PK placement test (see Rhine, 1952), who served as the subject in many home experiments (for a complete bibliography see White, 1985a) but whose work could not be taken seriously until he was tested under observation at the Duke University Parapsychology Laboratory (Pratt & Forwald, 1958) and at the University of Pittsburgh (McConnell & Forwald, 1967a, McConnell & Forwald, 1967b, McConnell & Forwald, 1968). With improved automated techniques that greatly reduce the possibility of unintentional experimenter error and deception, experimenters can serve as their own subjects and can work in the relaxed and familiar atmosphere of home and office, yet provide reliable data that can be trusted to be free from major criticisms of self–delusion, recording errors, and sensory cues because a computer can maintain an independent record of both the targets and the responses, while the experimenter/subject has no knowledge of the results. Such self–experimentation has several advantages. For a recent example, see Radin, 1988.

Improvements in the Psi Testing Situation. Richard Broughton (1982) points out that automating an experiment enables the experimenter to concentrate on the subject and on the interpersonal relationship between subject and experimenter that is an important ingredient in experiments that succeed.

Use of a computer makes it much easier to include blind conditions in an experiment. Because the computer can be programmed to generate the order in which various conditions are presented, not even the experimenter need know the order until after the experiment is conducted. The computer also makes it possible to run simultaneous control conditions.

Computers make it possible to safely provide feedback to subjects at the time of the experiment, facilitating the ruling out of sensory cues and inference. This is important because of the possibilities of sensory cues and rational inference. Feed—back is psychologically very important to the subject and may be an important ingredient in his or her success. In experi—ments aimed at training subjects in how to develop psi ability, it is essential. The computer has made feedback instantaneous and free from many of the criticisms formerly leveled against it such as sensory cues.

In studies of the role of feedback itself, computers have enabled psi researchers to do what previously was impossible——create a condition in which no person ever knows what the in—dividual trial scores are/were. The targets are generated by the computer, the computer registers the subject's responses (for which no hard—copy record is made), and the computer tallies the results, providing hard copy for total scores only. Schmeid—ler (1964) capitalized on this use of the computer in a study of precognition with and without feedback. (However, in this test—ing situation, as in any other, it is not possible to rule out the experimenter as the psi source.)

Critics have long urged parapsychologists to employ control runs at the time of the experiment. Computers make it easy to comply. However, nothing in the world of experimental para—psychology is as simple as it appears. Berger (1988) reviews previous observations and presents new data that indicate that "silent data" generated in computerized psi tests "suggest that there may be no such thing as a 'control' condition in psi research in the traditional meaning of the term" (p. 3). The experiments of Berger (1988) and incidental observations of other experimenters suggest that so—called "control" data can be influenced by real—time subject—induced psi. Berger (1988) discusses the theoretical implications of this possibility.

Another computer application is the confirmation of earlier work with special subjects with subsequent computerized tests. A case in point is the British psychic Malcolm Bessent who par—ticipated in two precognitive dream experiments in the 1970s (Krippner, Ullman, & Honorton, 1971; Krippner, Honorton, & Ullman, 1972). Bessent's precognitive ability was confirmed in an RNG study conducted at the same time (Honorton, 1971). Recently, Honorton, with much more sophisticated controls, again reported significant results with Bessent (Honorton, 1987).

Broughton and Millar (1980) pioneered techniques that be—came known as the "Edinburgh split." They programmed a computer so that it could handle split analyses of the data of a single experiment. Thus, one part of the data of an experiment could serve as a pilot test from which predictions could be made, and the remaining part could serve as a confirmatory experiment. Both experimenter and subject remain blind as to which condition is which while the experiment is being run.

Split analyses become essential when trying to determine whether a psi—mediated experimenter effect is present or in testing various theories, most notably, the observational theories (Broughton, 1982; Broughton & Millar, 1980).

New Techniques for Processing Data. Computers have spawned methodologies in mathematics and statistics that could not have been applied to parapsychological data earlier except by those with access to a mainframe or minicomputer. For example, meta—analysis, a statistical method that makes it possible to analyze whole groups of experiments in a specific area in which the individual experiments are treated as single data points or observations of one overall experiment, is increasingly being applied in psychology and now in parapsychology. (For overviews see Palmer, 1986d, and Schechter, 1987. For examples, see Honorton's (1985) meta—analysis of the Ganzfeld research, Radin and Nelson's (1987) meta—analysis of replication in REG experiments, and Schechter's (1984) review of hypnosis/ESP experiments. For a critical evaluation of meta—analysis in psi research, see Akers, 1985.) Similarly, in the past one had to be well trained in statistics in order to carry out experiments with the sophisticated designs that could yield a correspondingly high amount of complex information, but now it is much easier to apply complicated statistics because of the availability of statistical and mathematical software for computers (Goldstein, 1989; Grafton & Permaloff, 1988; Jerrell, 1988; Permaloff & Grafton, 1988; Raskin, 1989; Simon, 1989). The presentation of results has also been greatly improved by graphics packages that make it possible to create professional looking graphs, charts, and tables (Kaplan, 1989) without having to pay professionals to do the work—−a great boon for parapsychologists, many of whom work with slim budgets or with no financing whatsoever.

Computers have also made it possible to conduct reanalyses of old data that would have been formidable undertakings if carried out by hand (for example, see Pratt's reanalyses of the data of high—scoring subjects Gloria Stewart and C.J. in Pratt, 1967a, 1967b; Pratt, Martin, & Stribic, 1974). Computers have also been used to reanalyze data to detect indications of fraud (e.g., see Markwick, 1978). Schouten (1979a, 1979b, 1982, 1983) has pioneered in computerized reanalyses of spontaneous cases originally analyzed by hand tabulation.

Computerized Games. Personal computers have made it possible to develop a number of computerized ESP and PK tests in the form of games that are much more fun to take part in than the older, boring (for some) card—guessing techniques. Both Charles Honorton and his colleagues at the Psychophysical Research Laboratories (e.g., see Berger, Schechter, & Honorton, 1986; Hansen, 1986; Schechter, 1987) and Richard Broughton and James Perlstrom of FRNM (Broughton & Perlstrom, 1985a, 1985b, 1986) have led in this area. The beauty of these psi games is not just that they help to create a psychological atmos—phere conducive to psi, but they (and any experiment using a computer) provide an objective and accurate record of targets and responses that is difficult for subjects to tamper with. PRL has also pioneered in developing video—game REGs (Berger, 1987).

Automation of Experiments. Many experiments using random event generators are fully automated, and Honorton has pioneered in automating the Ganzfeld experiments in order to minimize the possibility of experimenter influence (Berger & Honorton, 1985, 1986; Honorton & Schechter, 1987). This would not only advance the scientific status of psi experimentation, but it would represent a large step forward in creating an ex—perimental paradigm that theoretically anyone could repeat.

Schmidt (1977b) has pointed out that computers may also be used to automate testing situations that the subject favors but that are very difficult to evaluate. He uses the example of a randomly swinging pendulum as a PK target. Although a subject may favor the task of trying to influence the pendulum, physically it is very difficult to measure and evaluate the pendulum's normal movement. However, a computer can be programmed to make an instrument needle function just as a randomly fluctuating pendulum would, thus providing a test situation that is psychologically favored by the subject, but one in which the computer is able to evaluate results with accuracy.

Database Applications. The use of database software has also improved the conduct of psi experimentation, as illustrated by the research of the Princeton Engineering Anomalies Research Laboratory, described by Nelson and duPont (1985). They point out that microcomputers have especially benefited experimental research with their capacity to deal with large databases and to use sophisticated statistical techniques. An effective database management system can perform several functions such as "support for data acquisition and storage, security and per—manent archiving, and flexible access to both particular and global aspects of experimental results" (p. 1). McCarthy has proposed

> the long—term development of a data base that
> would contain information on the performance of a
> large number of subjects in an assortment of psi
> tasks, along with contemporaneous information on
> moods, personality traits and whatever else can be
> obtained in an inoffensive and unobtrusive manner
> as part of an overall interaction of subjects with
> the computer. Perhaps, ultimately, a number of
> profiles can be constructed from this data base
> that could be used in instructing the computer to
> administer certain types of tasks to subjects display—
> ing certain characteristics under certain cir—
> cumstances. If such an approach were even
> moderately successful, it could provide a powerful
> technique for psi—optimization as well as affording
> potential insight into the phenomena. (McCarthy,
> 1982, p. 91)

The use of text databases has made it possible for para—
psychologists to obtain many citations to the literature on any
given research topic through computerized searches. See the
entry on *PsiLine* [659a], a database devoted specifically to
parapsychological literature.

Graphics Applications. McCarthy points out that computer
graphics can be exploited "to develop a particularly effective set
of target materials or an especially attractive task" (McCarthy,
1982, p. 89). Use of color is of course recommended for these
purposes. McCarthy adds that the graphics capabilities of com—
puters

> can be used in developing graphics displays that
> are interesting and pleasant to watch; sometimes
> they are engrossing, even hypnotic. These can be
> used as rewards for successful performance in psi
> tasks, as well as in induction techniques and target
> materials. (McCarthy, 1982, p. 89)

Finally, McCarthy (1982, p. 89) points out that computer
graphics displays "can be augmented considerably by interfacing
the computer with video equipment," that is, full motion TV
video.

Modems. By using a modem, a device that connects computers
via telephone lines, parapsychologists can keep in touch directly

or via bulletin boards (see the Parapsychology Information Net—
work [**889**] in the chapter on organizations). McCarthy (1982)
describes how modems also can be employed in setting up
long—distance experiments, cooperation among parapsy—
chologists in joint experiments, and in promoting and allowing
"independent investigators to share information contained in a
common data base" (McCarthy, 1982, p. 88).

Apparatus for Use in Field Investigations. Tart (1965) has
described electronic devices that could be employed in the inves—
tigation of possible physical effects in hauntings and in polter—
geist cases as a means of determining whether the observed ef—
fects are due to normal or paranormal causes. Lenz, Kelly, and
Artley (1980) have described a computer—based facility for
monitoring physiological correlates of psi processes, especially
momentary relationships between brain areas.

Dunseath, Klein, and Kelly (1981) describe several devices
they have constructed to make physiological recordings in field
investigations of hauntings and poltergeist phenomena. Hearne
(1982) has developed a portable, battery—powered unit for
monitoring rate of respiration that could be adapted for testing
psi in lucid dreams. As McCarthy points out:

> Once a computer is interfaced with physiological
> monitoring devices, a whole range of new pos—
> sibilities suggest themselves. For example, the com—
> puter can be programmed so that psi trials are con—
> ducted only when specified physiological conditions
> are met; or comparisons can be made between per—
> formance on a psi task during periods when
> physiological conditions were met and when they
> were not met. This last approach can be used to
> test the effectiveness of potentially psi—conducive
> states, such as relaxation. (McCarthy, 1982, p. 84)

Bierman (1985) and Beloff (1985) recommend audio— and
video—taping sessions with psychics. Other examples could be
given, but these should be sufficient to illustrate some of the
ways in which parapsychologists are or could take advantage of
technological innovations to improve the quality of research. It
should also be pointed out that there are other areas of inter—
face that have not yet been touched. In a personal communica—
tion, engineer/parapsychologist Dean Radin cites a few: chaos
analysis, neural networks, biochemical assays, and exploratory
data analysis techniques.

Theoretical Changes. Although computers have played a very important methodological role in the conduct of experiments, their impact on theory has been equally important. Computers have not only served as research tools, but they have become the *object* of research: the so—called "target." And they have completely disrupted the old comfortable paradigm of ex—perimenter and subject such that we no longer can discern who is doing what in an experiment; so parapsychologists have started to call the person or persons to whom a psi effect can be attributed the *source*, because at this stage we cannot be any more exact than that. What was known as PK now has center stage, except it may not be what we thought. Some parapsychologists think PK is not mind over matter so much as "probability shifting" (e.g., Broughton, 1988; Schmidt, 1987). In Section 5 of this chapter I will try to clarify some of the major theoretical changes that have occurred or are occurring, many of them attributable to use of computers.

Limitations of Computers in Psi Testing. It should not be as—sumed that because the personal computer has revolutionized psi testing it has not created problems of its own! Davis and Akers (1974) have reviewed methods of randomization and com—puter tests of randomness. More recently, Radin (1985) reviewed computer—based alogorithmic pseudorandom number generators. Most recently, Nelson, Bradish, and Dobyns (1989) have described extensive qualification and calibration procedures used to insure that the results of the psi experiments conducted at the Princeton Engineering Anomalies Research Laboratory are free of artifacts, biases, or irregularities. Akers (1984) points out that even automated testing devices have limitations. Paraphrasing Davis (1974), he writes: "A subject who has access to the computer, knows the data format, and has sufficient programming knowledge, might subvert experimental precau—tions. Davis also noted that software 'bugs' might allow easy access to computer targets, in a system which was supposedly secure" (Akers, 1984, p. 124). (Of course, this is true of an experimenter working independently as well.) It is equally im—portant to use an adequate experimental design that insures that experimental equipment is not malfunctioning. Another safeguard is that of preserving a record of the target order used after all the calls have been recorded. Ideally, control trials should also be generated at the same time as the experimental trials. Akers (1984) notes that it is essential that REGs be tested for randomness "in the actual experimental environment, with all peripheral equipment attached" (p. 122). Radin (1985) has discussed the limitations of various types of pseudo—random number generators in psi research. Also, even in an automated experimental set—up such as was introduced by

Schmidt (1969a, 1969b), working alone introduces the possibility of human error. However, Schmidt (1980) has since outlined an experimental design that excludes such errors (including deliberate data fabrication) and has also carried out a successful experiment implementing this design (Schmidt, Morris, & Rudolph, 1986).

INTERCHANGE BETWEEN
CRITICS AND PARAPSYCHOLOGISTS

Many critics of parapsychology choose qualitative psi ex—periments (e.g., those using pictorial materials such as drawings as targets) to criticize (see Morris, 1982b). When they do criticize quantitative experiments they tend to select J.B. Rhine's (1934) first book or the spate of experiments that fol—lowed. Thus, they tend to ignore much of the best work in parapsychology that was done later. Until recently, probably the best criticisms of psi experiments have come from para—psychologists themselves, but in the last few years some critics have turned their attention to some of the better psi experi—ments. (For reviews of criticisms, see Akers, 1984; Child, 1987; Hövelmann, 1985; Hyman, 1985; Ransom, 1971; Stokes, 1985.)

Another positive development that is leading to change in the way parapsychologists conceptualize their data is the detente between critics and parapsychologists. Until the past decade, critics were for the most part debunkers fueled by a strong will to disbelieve and this is still the case with cities associated with CSICOP. (An exception was the period in the early 1940s when Gardner Murphy and Bernard Reiss were editors of the *Journal of Parapsychology* (*JP*) and J.B. Rhine was still actively teach—ing psychology. At Murphy's insistence the *JP* had a "Board of Review" headed by Saul Sells and composed of psychologists who provided constructive criticism of all articles submitted to the *JP* [Murphy & Reiss, 1939].) For the most part, however, the disbelievers remained in their own camp, as did the believers, who were the parapsychologists. This is not to say that parapsychologists did not have their own high standards of criticism or that they did not profit from outside criticism. They increasingly improved their methodology until a peak of criticism was reached with the landmark publication of an ar—ticle by a chemist, George R. Price (1955), in *Science*, the offi—cial organ of the American Association for the Advancement of Science. The first part of his long article was a review of the evidence for parapsychology. Each of the major criticisms of the preceding 70 years was taken up and refuted. The article to that point could have been written by a parapsychologist. Then

Price went on to say, with reference to David Hume's dictum on miracles, that the claims of parapsychology were so outland—ish that by the law of parsimony it was more likely that para—psychologists were cheating than that the data parapsychologists assembled were well founded.*

Accusations of experimenter fraud such as that offered by Price cannot be answered, although Johnson (1975) and Schmidt (1980) have described experimental designs that in Akers' opinion "would satisfy all but the most hardened skeptic" (Akers, 1984, p. 158). In any case, when it comes to ex—perimenter fraud, parapsychology should not be considered, as many critics have done, as a field where it is rampant (e.g., Hansel, 1966, 1980; Price, 1955). Broad and Wade (1982) con—cluded from a survey of science as a whole that "fraud is en—demic" (p. 224). The truth is that there is fraud in every science, including parapsychology.

In recent years, in large part due to the efforts of Mar—cello Truzzi on the perimeter of parapsychology, and such parapsychologists as Gerd Hövelmann, Stanley Krippner, Robert Morris, and John Palmer, a constructive interchange has developed between the two camps that, at the leading edge at least, has resulted in a consensus that can no longer be depicted as the product of just believers or disbelievers. This consensus is being worked out by persons who truly want to understand what is happening in psi experiments and who are concerned with applying the highest standards of criticism to methodological considerations and to the interpretation of results. Truzzi, especially, has fostered a brand of skepticism rooted in a view that "the goal of science is to *explain* rather than *explain away* phenomena. Skepticism should seek to be *con—structive*, to advance science, rather than purely destructive and perhaps thereby to block inquiry" (Truzzi, 1987b, p. 12).

Palmer (1986c) has examined criticisms of parapsychology made by those who wish to discredit psi research. He found seven premises and strategies that are commonly used. He dis—cusses these strategies and ways in which parapsychologists could react to them in a constructive manner. Moreover, he

---

*Seventeen years later, Price (1972) apologized for his blanket accusation.

Ironically, Price's article was an admission that para—psychologists had accomplished what Henry Sidgwick at the first general meeting of the Society for Psychical Research in 1882 urged psychical researchers to aim for: to reach the point where they could say: "We have done all that we can when the critic has nothing left to allege except that the investigator is in on the trick" (Sidgwick, 1882, p. 12).

points out that critics have been able to discredit parapsychol—
ogy to the scientific community because both the critics and
many parapsychologists as well "have implicitly agreed upon a
set of ground rules for the psi controversy that allows it to hap—
pen. Parapsychologists must share a good deal of the respon—
sibility . . . for creating these ground rules" (p. 263). He dis—
cusses ways in which the ground rules can and should be
changed, primarily "by stripping psi of its theoretical and
metaphysical connotations" (p. 267). This should promote more
constructive interaction among parapsychologists and their critics
and should also facilitate scientific acceptance of parapsychology.
Palmer thinks that the major benefit of his proposed new
ground rules is that the new framework would force "critics to
assume their fair share of the responsibility for providing clearly
articulated, empirically validated conventional explanations of psi
anomalies, rather than being content with merely explaining
them away" (p. 267).

Another step toward constructive interchange between
critics and parapsychologists is that two parapsychologists, Gerd
Hövelmann and Stanley Krippner, in an article entitled
"Charting the Future of Parapsychology," have recommended

> that parapsychologists carefully consider the argu—
> ments of their critics——whoever those critics may
> be——and seek active exchange and collaboration
> even with those with whom they appear to dis—
> agree the strongest. In addition, we suggest that
> both parapsychologists and skeptics work hard on
> eliminating the proponents/skeptics dichotomy . . .
> . What parapsychology needs in the long run is
> nonadvocacy as well as active collaboration and
> common scientific endeavors of all those who prefer
> to see a solution to the problems of the paranor—
> mal, whatever that solution may turn out to be.
> (Hövelmann & Krippner, 1986, p. 3)

Their paper is an expanded version of some recommenda—
tions aimed at lessening the distance between parapsychologists
and critics set forth by Hövelmann (1983). It is an interesting
exercise to read a response to the Hövelmann and Krippner
recommendations by three Dutch parapsychologists, Millar,
Jacobs, and Michels (1988), who play the devil's advocate, and
warn that if too many of parapsychology's already limited
resources are devoted to efforts to improve the image of para—
psychology among established scientists and the critics, it will
have a deterring effect on the time and funds devoted to actual
research.

One way to describe the change that is taking place is that there are some persons associated with parapsychology who are becoming more stringently critical and some critics who are offering constructive help. As examples of the former, Charles Akers (1984) reviewed 54 psi experiments cited in Wolman [688] and elsewhere as being of a type where the repeatability of results is among the best. He examined them systematically for procedural flaws: randomization failures, sensory leakage, subject cheating, recording errors, classification and scoring errors, statis—tical errors, and experimenter fraud. In all, he found that 85% of the experiments were flawed in one or more ways. He then examined the remaining experiments individually and found that none could be considered unblemished. He concluded that "the research methods are too weak to establish the existence of a paranormal phenomenon" (p. 161). A second example is engineer/parapsychologist George Hansen, who has made critical contributions to several aspects of parapsychology. The one I consider the most important to date is his recent review of sub—ject fraud and ways to prevent it (Hansen, in press).

Marcello Truzzi describes this trend in parapsychology as follows:

> Some psi researchers now present their data as atheoretic anomalies in need of explanation rather than as presumed instances of extrasensory percep—tion or psychokinesis. In fact, the term "psi" itself seems to be undergoing redefinition. Many para—psychologists merely treat psi as a negative "wastebasket" category where effects mysteriously take place while sensory channels are absent. Some seek to define psi as a positively defined construct within what they hope will be a progressive new research program. And still others now seem to be defining psi research as the study of *reports* of psi experiences, whether or not psi phenomena really exist. Parapsychology seems to be opening up to new explanatory approaches (e.g., parastatistical, paraphysical, parabiological, and even para-geophysical), thus becoming more interdisciplinary and becoming allied with the broader field of anomalistics (cf. Truzzi, 1983). (Truzzi, 1987a, pp. 5—6).

An example of the latter is that a number of prominent magicians took part in a roundtable discussion on magicians and parapsychology at the 1983 convention of the Parapsy—chological Association. At the close of the meeting, the Council

of the Parapsychological Association prepared a formal state—
ment urging parapsychologists to cooperate with magicians in
the design of psi experiments ("PA Statement on Magicians,"
1984). There is no question that magicians can play a very con—
structive role in some aspects of parapsychology, although to
date their role has mainly been that of explaining away psi
phenomena.

It should be noted, however, that magicians know nothing
about statistics or experimental design or other aspects of
methodology. Their expertise is primarily in ruling out decep—
tion. (For a brief history of magicians and parapsychology and
an overview of magic for parapsychologists, see Hansen, 1985.)
The best published discussion of what would be involved in
cooperation between magicians and parapsychologists is provided
by Hansen (1990). At the Parapsychological Association panel,
several of the magicians present indicated that they personally
thought that psi was a reality. And, as the organizer of the
panel, Marcello Truzzi summarized it: "All indicated their sup—
port for the legitimacy of parapsychological investigation and
expressed their desire to be of help to serious researchers"
(Truzzi, 1984, p. 121). There also have been interchanges be—
tween critics in which one to some extent defends parapsychol—
ogy. For example, Truzzi (1980) has criticized Paul Kurtz's
critical stance toward parapsychology.

But no real detente can be reached by parapsychologists
and their critics unless the critics actively cooperate. Akers
points out that

> there is a danger that critics will overstep their
> bounds, and engage in speculative criticism long
> after the major methodological   problems have
> been solved. Honorton (1975, 1979) believes that
> this is already the situation in parapsychology.
> However, the present survey suggests that there are
> methodological issues still to be solved, and that
> the critics have something positive to offer.
> Indeed, I feel that critics have a responsibility to
> do more than just criticize; they must offer con—
> structive suggestions, and work with parapsy—
> chologists in the design of new experiments, which
> exclude all counterhypotheses. (Akers, 1984, pp.
> 162—163)

One example of cooperation between parapsychologist and
critic is a remote—viewing experiment conducted by skeptical
psychologist Ray Hyman and sociologist/parapsychologist James
McClenon (McClenon & Hyman, 1987). Another cooperative

venture that was heralded as a "first" was a joint statement by Ray Hyman and Charles Honorton (1986) summarizing the status of the Ganzfeld experiments, one of the most promising areas of psi research.

In an important paper given at the 1985 annual meeting of the American Psychological Association, Truzzi lays the groundwork for a potential rapprochement between para—psychologists and their critics and notes that:

> There are strong signs today of a major breakthrough coming between psi proponents in parapsychology and their skeptical critics in anomalistic psychology. A number of leading parapsychologists now believe that protocols today exist for psi experiments where replication levels are substantial (some claim above 50% levels). These claims are further supported by recent meta—analyses. The main new psi research areas for which robustness are being claimed include: (1) remote viewing experiments, (2) ganzfeld experi-ments, and (3) micro—PK studies in which random number generator outputs seem affected. Leading proponents in these areas have expressed their con—victions to me that the time is now ripe for reasonable expectations for positive psi results from experiments to be conducted by skeptics. Towards that end, a proposal will soon be put forward by some psi proponents expressing their endorsement and support for those funding requests by psi critics who are willing to work cooperatively to replicate psi experiments. In other words, proponents are willing to help critics who properly want to prove them wrong. This proposal has al—ready been generally discussed informally and has met with enthusiasm by several leading parapsy—chologists and psi critics. The psi proponents ask only for prior agreement between them and their critics on (1) the experimental protocols, and (2) the data analyses to be conducted. The rules of the game must be agreed upon before play begins. Obviously, if the experimental results prove nega—tive to psi, this would be most damaging to those psi proponents who put forward this challenge. However, if the results come out positive, the proponents do *not* demand any direct endorsement of the existence of psi. Critics who replicate the parapsychologists' results are asked only to admit

(1) that a legitimate (non—spurious) anomaly was
found to be present in their experiments——it may
be a new source of error or artifact, some non—psi
variable which I earlier termed delta——and (2)
that this anomalous result is likely to be important
since it may be a contaminating artifact in other
psychological experiments where we thought it did
not exist. (Truzzi, 1985, pp. 8—9)

Some parapsychologists, such as R.A. McConnell and T.
Rockwell, feel it is wrong to use the term "anomaly" to charac—
terize the results of psi experiments.   As Rockwell puts it:
"From being a systematic attempt to explain certain widely ex—
perienced human phenomena, parapsychology would become
merely a matter of cataloging curiosities" (Rockwell, 1989, p.
201).   McConnell, in a letter of February 3, 1988 to Dr. Frank
Press, President of the National Academy of Sciences, protesting
the negative National Research Council report on parapsychol—
ogy, notes:

If ESP and PK were "mere statistical anomalies"
no attention would have been paid to Dr. Rhine
and no controversy would exist today.   The world
is full of trivial anomalies, statistical and otherwise,
which scientists and laymen alike have learned to
ignore.   To call parapsychological phenomena
"mere anomalies" is disingenuous at best. (p. 7)

Stanford (1982) takes a practical stance between these two
extremes. He writes:

Parapsychologists have sometimes failed to recog—
nize that in parapsychology the raw data are
anomalies and that the question is how to interpret
those anomalies.   Any interpretation which indi—
cates under what circumstances they should and
should not be observed is capable of scientific test.
Any interpretation which does not do this is in—
capable of scientific test.   (The latter, incidentally,
includes the claim that the anomalies are due to
"unknown sensory communication" unless that inter—
pretation can be more clearly specified.) (Stanford,
1982, p. 247)

Perhaps the most outstanding example of rapprochement is
that parapsychology is the featured topic of the December 1987
issue of *Behavioral and Brain Sciences*, whose subtitle is *An In—*

ternational Journal of Current Research and Theory with Open Peer Commentary. The case for parapsychology is presented in a lead article by K.R. Rao and J. Palmer (1987), and it is fol—lowed by an article by an acknowledged critic, James Alcock (1987a). The open peer commentary that follows contains responses by 49 parapsychologists, philosophers, psychologists, and physicists. Finally, the original authors, Palmer and Rao (1987) and Alcock (1987b), respond at length.

It should be pointed out that most of the criticism leveled at parapsychology from *within* the field is motivated by the desire to discover what *is.* Parapsychologists, for the most part, want to improve their methodology so that they will not be fooled by themselves or others. Much of the criticism coming from *outside* parapsychology, on the other hand, appears to be motivated by an active will to disbelieve that is the counterpart of the "will to believe" that these same critics accuse para—psychologists of being guilty of. Some suggestions for guiding principles aimed at critics were published by Stevenson and Roll in 1966, and they are still relevant today.

More recently, Palmer (1986e) discusses the basically *un—critical* attitude of the irrational skeptics. He proposes that they be called "conventional theorists" rather than skeptics, and proposes a new approach called "progressive skepticism" that could include both parapsychologists and critics who have faith in the ability of the scientific method to eventually provide the correct answers.

The will to disbelieve is reflected in many of the articles in the *Skeptical Inquirer* [849] and in the titles published by Prometheus Books (see the section on Criticisms). The most recent example of the will to disbelieve is the highly prestigious report of the National Research Council, which conducted research for the U.S. Army on various disciplines claiming to enhance human performance [462], including parapsychology. In the latter case, at least, the Committee's report was quite biased, as is evident from a reply to the National Research Council prepared by members of the Parapsychological Associa—tion (Palmer, Honorton, & Utts, 1989). Subsequently, some lead—ing parapsychologists and the critics who prepared the NRC report met at a workshop under the auspices of the Office of Technology Assessment (OTA) of the U.S. Congress. The out—come of this meeting was quite favorable to parapsychology. The report of the meeting, which was circulated by the OTA, ("Report of a workshop," 1989), was reprinted in the October 1989 issue of the *Journal of the American Society for Psychical Research.*

CHANGES WITHIN PARAPSYCHOLOGY

In the post—Rhinean era of parapsychology it is no longer believed necessary that to progress one must hold to the view that psi has been proven or that the basic building blocks of parapsychological phenomena can be neatly delineated as clair—voyance, telepathy, precognition, PK, etc. In fact, in its first position paper, which outlines the terms and methods of para—psychology, the Parapsychological Association defined parap—sychology simply as the study of "apparent anomalies of be—havior and experience which exist apart from currently known explanatory mechanisms which account for organism—environment and organism—organism information and influence flow" (Parapsychological Association, 1988, p. 353).

This view of psi is in part a reflection of new ways of con—ducting research. The Rhinean school concentrated largely on proof—oriented research; the aim was to establish that psi exists and to delineate its characteristics. This approach can be likened to trying to scoop up a handful of sand and somehow mold it into a durable shape with a specific function. Such at—tempts were only moderately successful, due to such difficulties as being able to repeat experiments, the possible contaminating factor of the experimenter effect, altered expectations of sub—jects, etc. Increasingly, parapsychologists have turned to process—oriented research, or the design of experiments aimed at exploring the variables that affect a specific phenomenon. Some 40 years of this type of research has revealed that we may not necessarily be dealing with specific aspects of psi that differ intrinsically from one another, but rather that what we call psi will behave in one way when we set up our experiments in a way that eliminates the possibility of anyone knowing the cor—rect target, and it will behave in another way in a test situa—tion involving an agent who does know the target, and in yet a third way when the target has not yet been determined. It could even be that it will behave in as many different ways as there are experimenters to conceive of them! In this view psi is not a handful of sand to be molded. Here one scoops up a handful of sand and then lets it slip through the fingers and fall to the ground in order to study any patterns that emerge, very much as in the process of sand painting.

Set and setting and the role of the observer are increas—ingly viewed as the primary determinants of psi research, not simply as background influences. Summing up the results of a century of proof—oriented experimentation in a recent textbook, John Palmer expressed this view:

It would seem that how much weight we attach to

any experiment as evidence for ESP comes down in the final analysis to the likelihood we assign to the various alternatives, based on our past experiences and world views. Perhaps the biggest mistake both sides have made in the ESP controversy is to imply that it can be settled exclusively by appeal to objective criteria, unless we mean by objectivity the collective subjectivity of the community of scientists. (1986a, p. 182)

Although it may be impossible to convince everyone by means of proof—oriented research that psi occurs, paradoxically, process—oriented research does not require that psi be established. Its existence need only be accepted as a "working hypothesis." Yet——again paradoxically——as Palmer (1986b) points out in the same textbook, "the demonstration of a significant and reliable relation between psi and some other variable is in itself evidence for the reality of psi" (p. 185).

This is because in the course of trying to experimentally investigate the psi process, significant evidence of a psi effect mounts up, and in investigating aspects of a particular problem area such as Ganzfeld experiments, replications multiply. A good example is a recent survey of binary RNG psi experiments published between 1969 and 1984 (Radin, May, & Thomson, 1986). They performed a meta—analysis of the 332 studies published and found that 71 of the experiments were significant at $p < .05$, two—tailed, which results in an impressive overall binomial probability of $p < 5.4 \times 10^{-43}$. Additional analyses showed that the effect was not due to a handful of the 28 principal investigators involved, nor can the "file—drawer problem" account for the results. Thus, this study provides strong prima facie evidence suggesting that psi is real.

Another major finding of parapsychology is the psi—mediated experimenter effect. Palmer (1986b) has called it "the most important challenge facing modern experimental parapsychology" (pp. 220—221). The existence of such an effect could undermine the results of all process—oriented research because the results of any experiment could be attributable to experimenter expectancies. However, it could also provide us with a key to open the way to more psi and a better understanding of the psi process itself. But to do so it becomes necessary to include the experimenter in the experimental equation and learn how to control for his or her influence on the results.

Another major change from the 1950s is that parapsychologists will now theorize about psi *before* they have data, which was not the case in Rhine's day. Parapsychologists have become more sophisticated and more scientific in that they are more aware that overt experiment is embedded in certain theoretical assumptions and that it is good science to be aware of those assumptions. (For a good discussion of this point, see Lang, 1975.) This, in turn, leads to experiments that deliberately test specific theories, such as the observational theories or the IDS hypothesis. The "thought experiment," of—ten employed by physicists, where one logically designs and carries out an experiment "in one's head" or "on paper," has also finally made its way (infrequently, still) into parapsychol—ogy. Not only do these approaches advance psi research as such, they also mark a step toward the legitimization of para—psychology as a science, since theory is a foremost need.

In the last decade a number of quite sophisticated theories have been proposed with which parapsychological research has yet to catch up. (For a review of theories in parapsychology see Stokes, 1987, which continues an earlier survey by Rao, 1978). Among the first were Stanford's (1974b, 1974c) PMIR model, followed by his conformance model (Stanford, 1977a, 1977b, 1978), Irwin's (1979) information processing model, Morris's (1980) communication model, the observational theories (Millar, 1978), the informational model of von Lucadou (1988), Schmidt's mathematical model (Schmidt, 1975), and Walker's quantum mechanical theory (Walker, 1976, 1979). Several others have also written about quantum mechanics and psi (a partial list includes Bierman, 1983; Bohm, 1986; Chari, 1956; Jahn & Dunne, 1986; Josephson, 1975; Lawden, 1980; Mattuck, 1982; Temkin, 1982; Villars, 1983; Whiteman, 1973). In 1974, the Parapsychology Foundation devoted an entire conference to quantum physics and psi [621].

These and other theories have spawned experiments designed to test them, but insufficient work has been done. Nonetheless, they have called for innovations in research and the interpretation of results, and they have generated much discus—sion. For example, physicist/parapsychologist Peter Phillips (1984) has challenged Walker's quantum mechanical theory and received a lengthy response from Walker (1987b). Walker has also responded to other criticisms of his theory (Walker, 1984) and has shown how his theory differs from that of Schmidt (Walker, 1977). Philosopher/parapsychologist Stephen Braude (1979) has taken on the observational theories, which led to an interchange between Millar (1988) and Braude (1988). Schmidt has proposed both teleological and quantum collapse models and has compared and contrasted them (Schmidt, 1984). Gardner (1985) is skeptical of any connection between quantum mechanics and psi, but apparently does not understand either.

One of the first experimental parapsychologists to examine the implicit theoretical base of parapsychology's first 90—odd years or so was Rex Stanford (1974a). Next (Stanford, 1977a), he pointed out that although parapsychology had often been criticized as not having a viable theory and although its primary proponent, J.B. Rhine, had eschewed theory, there had been a model guiding research in parapsychology right along, particularly in the work of J.B. and L.E. Rhine. Stanford called it the psychobiological model, and he tried to explicitly state its major tenets (Stanford, 1977a). Once stated, it be— came relatively easy to criticize the model as only being ap— plicable to a part of the body of observation in parapsychology concerning both spontaneous and experimental psi. Stanford notes:

> It may well be that the prevalence of the psychobiological paradigm, quite stringently adhered to by our small in—group of psychologically— and biologically—oriented psi researchers, has placed empirical and conceptual blinders upon us which have prevented our seeing the possibilities for con— ceptualizing our phenomena in alternative, more basic fashions and has caused us to ignore certain features of our own data which may call into ques— tion that very paradigm. (Stanford, 1977a, p. 7)

Stanford writes that he really began to question the psychobiological model when he reviewed the experimental PK literature for a chapter in Wolman's *Handbook of Parapsychology* (Stanford, 1977c). It seemed to him that the goal—oriented character of psi should become the cornerstone in building a new theory of psi (Stanford, 1977a, p. 11). In the remainder of this paper he introduced his theory of conformance behavior, which enabled parapsychologists to think about ESP and PK in entirely new ways. One of the most important components of the theory is that of causation. As Stanford put it: "Causation in this theoretical construction does not arise from something the subject does, but from the relationship between the subject's disposition and a REG" (Stanford, 1977a, p. 14). He closes with the hope that by studying the characteristics of REGs "we will for the first time discover physical boundary con— ditions for psi function and physical lawfulness in its operation" (p. 15), to the possibility of which the psychobiological model had blinded us.

Stanford gave a fuller presentation of the conformance be— havior model the following year (Stanford, 1978). He dismisses both the perceptual and cognitive models of psi, positing instead that "all psi processes (both ESP and PK) are goal—oriented rather than intrinsically based upon mechanical processing of in—

formation" (p. 197).   In the article following Stanford's, philosopher/parapsychologist Hoyt Edge presents "A Philosophical Justification for the Conformance Behavior Model" (Edge, 1978) by supporting a teleological rather than the tradition empiricist or cybernetic view of nature.

In a discussion of the model three years later, he points out that in the conformance model psi is viewed "as somehow organizing loose, disorganized or random processes such that their outcomes accord with the dispositions of someone or some organism which has an interest in or a concern about those outcomes" (Stanford, 1981b, p. 91).   He adds: "From the prospective of conformance, both ESP and PK events are really the same kind of event occurring in different circumstances" (Stanford, 1981, p. 91).

Palmer (1985) classifies the conformance behavior model as belonging to the "correspondence paradigm," which he discussed in his presidential address to the Parapsychological Association (Palmer, 1980).   However, he considers two assumptions of Stanford's (1974b, 1974c) earlier PMIR model to be the most important theoretical developments of the previous decade. These assumptions are "that psi can occur (a) without conscious effort on the part of the psi source to produce it and (b) without awareness of what needs to be accomplished to produce it" (Palmer, 1985, p. 3).   This opens up entirely new vistas, for as Palmer points out later:

> The most likely psi source in a given case . . . is not necessarily the person attempting to produce an effect, but rather the person who most wants or needs the effect.   Moreover, if we accept the independence of psi from space, then the physical proximity of the psi source to the target is inconsequential; what counts is *psychological* proximity. (p. 4) [*Italics* added]

For a review of experimental micro—PK data and a good discussion concerning the postulates of the conformance behavior model and the observational theories, which lead to quite different predictions, see Varvoglis (1986).

A major change has occurred in thinking about PK, or what used to be conceived of as mental influence on matter without physical contact.   In the early days of PK research, dice were used as the targets to be influenced.   Researchers observed in their experiments that the best results were obtained with moving rather than stationary targets.   As Schmidt (1983) points out: "What makes a rolling die different is the element of randomness that enters into the outcome.   And today it ap—

pears that PK might be able to act wherever truly random processes are in progress" (p. 180). Schmidt, who is a physicist, has pioneered in the design of psi tests that are grounded in the quantum mechanical view that "part of nature may be governed by pure chance" (p. 180). He developed psi testing machines ("Schmidt machines," now better known as ran—dom event generators, or REGs*) that were able to incorporate random processes such as radioactive decay. Providing a ran—dom source for targets had been important from the first psi experiments in order to be able to determine statistically what level of scoring could reasonably be attributed to chance, but it is one thing to say that for methodological reasons random ar—rangement of targets is needed and something far more revolu—tionary to say that the targets likeliest to be affected are those that exhibit random movement. The next step was to imple—ment RNGs with computers or to use the computer's own capacity for random number generation. Experimental subjects then tried to use PK to alter the random targets. Small but significant effects have been obtained in the presence of human operators that are not observed when a computer is left "to its own devices." Nonetheless, the specter of a possible psi—mediated experimenter effect cannot at present be ruled out.

In the old Rhinean view, physics and parapsychology were like oil and water. Psi, by definition, was viewed as "nonphysical." This is probably because to Rhine "physics" was synonymous with classical or Newtonian physics. But Hel—mut Schmidt, who was very familiar with quantum physics and the problems it posed as regards the ability of physicists to study micro—events objectively, saw that the *same* situation ob—tained in parapsychology. In a sense, quantum physics and parapsychology are dealing with the same problems, and Schmidt (1974b, 1978) has suggested that parapsychology may even help to clarify some of the problem situations in modern physics. Another physicist, Fritjof Capra (1979) has suggested that the place in which parapsychology now finds itself is analogous to the situation of physics in the early days of quan—tum theory, and he suggests that parapsychologists use the same method to attack their problems that early quantum physicists used for theirs, that is, a nonanalytic and feeling ap—proach rather than one of logic and analysis.

This is an enlightened and enlightening view of para—psychology vis—à—vis physics. The very existence of psi phenomena indicates that Newtonian physics does not hold for all situations. But current physics has revealed a world that does not conform to Newtonian mechanics any more than psi

---

*also random number generators, or RNGs.

does (Capra, 1975, 1982; Koestler, 1972; Zukav, 1979). In this world, parapsychology may have an important contribution to make. Walker (1975) holds that his quantum mechanical view of psi applies not only to psi phenomena but to consciousness itself. Schmidt (1987) takes the more limited view that psi phenomena may be a weak violation of causality "in that only the outcome of chance processes is effected, whereas the non—statistical laws of physics like the conservation laws for energy, momentum, symmetry, etc. are upheld" (p. 107). In a paper directed at physicists, Schmidt (1978) has presented a model of a noncausal world that has axiomatically well—specified properties that demonstrate that "noncausal systems can be dis—cussed in a logically consistent manner so that noncausality might well exist in the real world as a weak, but so far over—looked, effect" (p. 463).

To my knowledge Schmidt (1969a, 1969b, 1977a) was the first to refer to PK as a means of slightly altering quantum processes. Schmidt also introduced the concept of micro—PK for this psi effect. Richard Broughton, in his 1987 Presidential address to the Parapsychological Association, points out that micro—PK studies have shifted our view of PK from that of a "psychic lever that lifts tables or moves objects around——that is, a forcelike function——to a view of PK as the ability to shift probabilities of events, to bias probability distributions" (Broughton, 1988, pp. 191—192). Now the best guess is that psi may not be operating alongside of or outside of probability but may be a means of *altering probability itself!*

The criterion of what is required to establish that "psi exists" has changed from what it was a decade or more ago. Traditionally, it was thought that a conclusive experiment was required, but this has proven to be a will—o'—the—wisp. As John Palmer observes: "'Proof—oriented' experiments in para—psychology . . . are better regarded not as proving the existence of psi . . . but rather as defining the nature and plausibility of the normal interpretations necessary to explain away the anomalous results" (Palmer, 1985, p. 1). The new approach to establishing the reality of psi is that of "well—designed experi—ments which nonetheless are inconclusive by themselves" (Palmer, 1985, p. 1). The emphasis is being placed on pro—cess—oriented studies and on replication.

A new approach to the problem of the existence of psi as regards spontaneous psi phenomena is also emerging, its most forthright exponent being the Dutch psychologist/para—psychologist Sybo Schouten. He feels that justification for research on parapsychological phenomena does not depend on the existence of psi, and that trying to "prove" the existence of psi creates many of parapsychology's problems (Schouten, 1986).

He favors a "pragmatic approach," which he describes in the fol—
lowing paragraph:

> The pragmatic approach is based on the notion
> that all human experiences and activities are a
> legitimate subject for scientific research, including
> such human experiences as those found in spon—
> taneous cases. These experiences have at least one
> common property, namely, that the persons who
> reported these cases felt that these experiences were
> something special, that they had a paranormal
> character. It is a legitimate scientific question to
> ask why those persons, who must have had
> numerous "normal" experiences in life, singled out
> one or a few of these and attributed a paranormal
> character to them. It might be that they were jus—
> tified in their belief that some unknown paranormal
> process was responsible for these experiences, but
> other explanations are also possible. However, be—
> cause the contributors, themselves, considered these
> experiences as owing to some paranormal process,
> and considering the historical tradition of this belief
> and all the experimental evidence, we are justified
> in accepting this hypothesis as worthwhile to con—
> sider and to see how far it gets us. But basically,
> we study a certain class of human experiences, like
> hallucinations or emotions. From that point of
> view, parapsychology is just another psychological
> discipline. (Schouten, 1986, p. 35)

Greeley (1987) has recently reported that 67% of
Americans in a recent poll believed they had had at least one
spontaneous psi experience. Following Schouten's line of
thought, this gives parapsychologists a wide database with which
to work, regardless of whether or not there is anything paranor—
mal in the experiences themselves.

In a later paper, Palmer argues that "psi as a
'paranormal' process transcending generally accepted limiting
principles of nature" [the cornerstone of the Rhinean paradigm]
has not been established, but that "psi as anomaly *has* been es—
tablished" (Palmer, 1987, p. 111).

Another change that has occurred is in the views held con—
cerning the role of the experimenter. Fifty years ago the sub—
ject was considered all important, and experiments concentrated
on locating subjects who possessed a high degree of what was
thought to be native psi ability. Thirty years ago it was
pretty well recognized that the results of psi experiments were a

joint product of the experimenter and the subject. Some sub—
jects scored well with one experimenter and poorly with
another. Although recognized, the importance of the
experimenter—subject relationship has never been fully inves—
tigated. In the last decade the hypothesis has been raised that
the experimenter himself or herself is the real source of any
results they obtain from their subjects (Broughton, 1979; Millar,
1979). Moreover, there is a real possibility that the nature of
the results obtained appears to be related to the experimenter's
attitudes and beliefs even when the subjects have no normal
means of learning what those attitudes and beliefs are! Stan—
ford (1977a, 1985), especially, has urged parapsychologists to
improve their efforts to rule out or at least to control for the
experimenter effect.

Another change is that until the 1970s, studies of ESP far
outnumbered those of PK. ESP was considered better estab—
lished than PK and parapsychologists were surer of its
properties. (Some parapsychologists who were convinced of the
reality of ESP were still doubtful about PK.) With the intro—
duction of RNGs, PK has dominated experimental parapsychol—
ogy. Although RNGs can be used to test ESP as well as PK,
they are especially well adapted for PK research because usually
what the subject attempts to PK is the machine itself.
Schmidt (1976) and others (e.g., Broughton, 1977) have even
experimentally demonstrated that PK apparently can go back—
wards in time and affect target generation that occurred earlier!

Research with RNGs has shown that both conceptually
and experimentally precognition and PK seem to be closely re—
lated. When a parapsychologist tests for precognition (i.e., asks
the subject to predict what a target order not yet determined
will be), it is not yet possible to distinguish whether the subject
who succeeds in the task did so simply by predicting the future
random event or used PK to influence the RNG to produce tar—
gets that conform with the prediction. (For a discussion of
precognition and retroactive PK as regards experimental support
for true precognition, see Morris, 1982a.)

But that is not all. In an analysis of 425 RNG experi—
ments conducted over a 15—year period, May, Radin, Hubbard,
Humphrey, and Utts (1986) propose that the data support the
conclusion that an informational model called "Intuitive Data
Sorting" (IDS) rather than PK best accounts for the results. In
the IDS model, "individuals 'sort' locally deviant subsequences
from a random sequence into significant and nonsignificant bins"
(p. 120).

The status of the IDS model is in flux. The model itself
is still being worked out, and experiments are being designed to
test it. Preliminary discussions and results have been reported

in a roundtable at the 1986 Parapsychological Association con—vention (see *Research in Parapsychology 1986*, pp. 136—144) and in a session of three papers given at the 1987 convention (see *Research in Parapsychology 1987*, pp. 1—13). Walker (1987a) has argued against the IDS model, holding that it is not consis—tent with the data of RNG experiments.

Another example of employing computerized experimental design to ask questions that previously could not be explored empirically is a recent experiment by Radin (1988), who found that "in a computer—controlled precognition experiment . . . the author as subject showed significant evidence of precognition and tended to select more probable than actual futures" (p. 209). He speculates that just as according to the IDS hypothesis micro—PK cannot be distinguished from an informa—tional process like precognition, it may be that micro—PK, clair—voyance, and precognition are the *same* process. Such specula—tion is not new, but new technology and more sophisticated theories are enabling experimenters to fine—tune their questions and to approach them in a more sophisticated way than ever before.

Another major shift in emphasis worth mentioning could be viewed as an extension of interest in the experimenter effect. It consists of proposes made by parapsychologists that they not simply serve as their own experimental subjects but attempt to develop their own ESP (Ellison, 1982; White, 1984a) or related capacities such as unorthodox healing (LeShan, 1974), shamanism (Gruber, 1983), and mystical techniques (Servadio, 1983). E. Spinelli (1983) has suggested that the discrepancy in results he obtained with children and those of would—be replicators calls for what he calls "an 'alchemical' approach to the experimental situation in that the experimenter recognizes his/her lack of distinction from his/her experiment" (p. 25). He urges parapsychologists to "enter a period of observation and participation and forget about trying to prove anything for a while. As parapsychologists we should immerse ourselves in our research, expand it, make it part of our being and essence. We should allow it to encompass us, to take us where it wants, to lead us anywhere" (Spinelli, 1983, p. 26).

Another of the younger parapsychologists, Elmar Gruber (1983), finds it unfortunate that after a century of organized in—vestigation some parapsychologists find it necessary to "state, after all, that we do not really know if psi exists . . . . [They are] asking for strict and unambiguous scientific corroboration of something that most probably has been experienced by all parapsychologists" (p. 48). Gruber proposes that "the challenge is not to be worried about the correctness of one's perceptions and feelings . . . the sense of their objective reality, but rather to develop a sense of its inner 'correctness,' and its significance for understanding ourselves and nature" (p. 48).

But it is not only the young minds who question para—psychological methodology, although they predominate. E. Ser—vadio, who wrote his first paper on parapsychology in English in 1932, has recently taken the position that the future of parapsychology may be "a question of changing or dying" (Servadio, 1983, p. 146). He expresses the view that if para—psychology cannot succeed by following the tenets of the scien—tific tradition, then parapsychologists must "go back to the ap—proach of the Vedanta, of Yoga, of Zen, . . . of a Plotinus, of a Pico della Mirandola, of the neoplatonists in general, i.e., to a philosophy of nature that does not rule out imagination, self—realization, creativity, universality, and identity" (Servadio, 1983, p. 146).

Intermediate between these extreme positions are those who advocate that phenomenological approaches (Lang, 1975; Poynton, 1983; Schlitz, 1985; Walther, 1955; White, 1983) be applied in parapsychology to help us to understand what is hap—pening to the person who undergoes a psi experience as well as to the experimenter/investigator.

Finally, at the very frontier, there are those in para—psychology who propose that their discipline is bringing into question the very foundations of science itself. (For example, see Bierman, 1983; Schmidt, 1974b; Taylor, 1987; White, 1983, 1984a, 1984b, 1985b).

REFERENCES*

Akers, C. (1984).   Methodological criticisms of parapsychology. In S. Krippner (Ed.), *Advances in Parapsychological Research 4* (pp. 112—164). Jefferson, NC: McFarland.

Akers, C. (1985).   Can meta—analysis resolve the ESP con—troversy? In P. Kurtz (Ed.), *A Skeptic's Handbook of Parap—sychology* (pp. 611—627). Buffalo, NY: Prometheus Books.

Alcock, J.E. (1987a). Parapsychology: Science of the anomalous or search for the soul? *Behavioral and Brain Sciences, 10,*

---

*Because this chapter was originally written as a paper the style used in this reference list differs from that used elsewhere in the volume (which follows that of *Parapsychology Abstracts International*). The main differences are that an ampersand is used in citing authors/editors, issue numbers are omitted unless a periodical does not use continuous pagination, and the month is not named except in the case of technical reports.

553–565.

Alcock, J.E. (1987b). A to–do about dualism or a duel about data? *Brain and Behavioral Sciences, 10,* 627–634.

Bauer, H.H. (1987). What do we mean by "scientific?" *Journal of Scientific Exploration, 1,* 119–127.

Bauer, H.H. (1988). Commonalities in arguments over anomalies. *Journal of Scientific Exploration, 2,* 1–11.

Beloff, J. (1985). Research strategies for dealing with unstable phenomena. In B. Shapin & L. Coly (Eds.), *The Repeatability Problem in Parapsychology: Proceedings of an International Conference Held in San Antonio, Texas, October 28–29, 1983* (pp. 1–21). New York: Parapsychology Foundation.

Berger, R.E. (1987). Psi effects without feedback: Will the real psi source please stand up? [Abstract]. In D.H. Weiner & R.D. Nelson (Eds.), *Research in Parapsychology 1986* (pp. 9–12). Metuchen, NJ: Scarecrow Press.

Berger, R.E. (1988). Psi effects without real–time feedback. *Journal of Parapsychology, 52,* 1–27.

Berger, R.E., & Honorton, C. (1985). PSILAB//: A standardized psi–testing system [Abstract]. In R.A. White & J. Solfvin (Eds.), *Research in Parapsychology 1984* (pp. 68–71). Metuchen, NJ: Scarecrow Press.

Berger, R.E., & Honorton, C. (1986). An automated psi Ganzfeld testing system [Abstract]. In D.H. Weiner & D.I. Radin (Eds.), *Research in Parapsychology 1985* (pp. 85–88). Metuchen, NJ: Scarecrow Press.

Berger, R.E., Schechter, E.I., & Honorton, C. (1986). A preliminary review of performance across three computer psi games [Abstract]. In D.H. Weiner & D.I. Radin (Eds.), *Research in Parapsychology 1985* (pp. 1–3). Metuchen, NJ: Scarecrow Press.

Bierman, D.J. (1983). Towards a methodology to handle non–locality in space and time: The end of reductionism? In B. Shapin & L. Coly (Eds.), *Parapsychology's Second Century: Proceedings of an International Conference Held in London, England, August 13–14, 1982* (pp. 1–18). New York:

Parapsychology Foundation.

Bierman, D.J. (1985). Towards a better methodology in research with psychics. *European Journal of Parapsychology*, *5*, 403–404.

Bohm, D.J. (1986). A new theory of the relationship of mind and matter. *Journal of the American Society for Psychical Research*, *80*, 113–135.

Braude, S.E. (1979). The observational theories in parapsychology: A critique. *Journal of the American Society for Psychical Research*, *73*, 349–366.

Braude, S.E. (1988). Death by observation: A reply to Millar. *Journal of the American Society for Psychical Research*, *82*, 273–280.

Broad, W.J., & Wade, N. (1982). *Betrayers of Truth: Fraud and Deceit in the Halls of Science*. New York: Simon & Schuster.

Broughton, R.S. (1977). An exploratory study on psi–based subject and experimenter expectancy effects [Abstract]. In J.D. Morris, W.G. Roll, & R.L. Morris (Eds.), *Research in Parapsychology 1976* (pp. 173–177). Metuchen, NJ: Scarecrow Press.

Broughton, R.S. (1979). Repeatability and experimenter effect: Are subjects really necessary? *Parapsychology Review*, *10*(1), 11–14.

Broughton, R.S. (1982). Computer methodology: Total control with a human face. In B. Shapin & L. Coly (Eds.), *Parapsychology and the Experimental Method: Proceedings of an International Conference Held in New York, New York, November 14, 1981* (pp. 24–42). New York: Parapsychology Foundation.

Broughton, R.S. (1988). If you want to know how it works, first find out what it's for. In D.H. Weiner & R.L. Morris (Eds.), *Research in Parapsychology 1987* (pp. 187–202). Metuchen, NJ: Scarecrow Press.

Broughton, R.S., & Millar, B. (1980). Split analysis techniques for robust effects [Abstract]. In W.G. Roll (Ed.), *Research in Parapsychology 1979* (pp. 154–156). Metuchen, NJ:

Scarecrow Press.

Broughton, R.S., & Perlstrom, J.[R.]. (1985a). A competitive computer game in PK research: Some preliminary findings [Abstract]. In R.A. White & J. Solfvin (Eds.), *Research in Parapsychology 1984* (pp. 74–77). Metuchen, NJ: Scarecrow Press.

Broughton, R.S., & Perlstrom, J.R. (1985b). Results of a special subject in a computerized PK game [Abstract]. In R.A. White & J. Solfvin (Eds.), *Research in Parapsychology 1984* (pp. 78–81). Metuchen, NJ: Scarecrow Press.

Broughton, R.S., & Perlstrom, J.R. (1986). Further studies using a competitive computer PK game [Abstract]. In D.H. Weiner & D.I. Radin (Eds.), *Research in Parapsychology 1984* (pp. 4–5). Metuchen, NJ: Scarecrow Press.

Capra, F. (1975). *The Tao of Physics*. Berkeley, CA: Shambhala.

Capra, F. (1979). Can science explain psychic phenomena? *Re-Vision*, *2*(1), 52–58.

Capra, F. (1982). *The Turning Point: Science, Society, and the Rising Culture*. New York: Simon & Schuster.

Chari, C.T.K. (1956). Quantum physics and parapsychology. *Journal of Parapsychology*, *20*, 166–183.

Child, I.L. (1987). Criticism in experimental parapsychology, 1975–1985. In S. Krippner (Ed.), *Advances in Parapsychological Research 5* (pp. 190–224). Jefferson, NC: McFarland.

Churchland, P.M. (1987). How parapsychology could become a science. *Inquiry*, *30*, 227–239.

Davis, J.W. (1974). A developmental program for the computer–based extension of parapsychological research and methodology. *Journal of Parapsychology*, *38*, 69–84.

Davis, J.W., & Akers, C. (1974). Randomization and tests for randomness. *Journal of Parapsychology*, *38*, 393–407.

Dunseath, W.J.R., Klein, F.F., & Kelly, E.F. (1981). A portable facility for physiological recording in field settings. *Journal*

of the American Society for Psychical Research, *75*, 311–
320.

Edge, H.L. (1978). A philosophical justification for the confor–
mance behavior model. *Journal of the American Society for
Psychical Research, 72*, 215–231.

Ellison, A.J. (1982). Psychical research: After 100 years, what
do we really know? *Proceedings of the Society for Psychical
Research, 56*, 384–398.

Gardner, M. (1985). Parapsychology and quantum mechanics. In
P. Kurtz (Ed.), *A Skeptic's Handbook of Parapsychology* (pp.
585–598). Buffalo, NY: Prometheus Books.

Girden, E. (1962a).   A review of psychokinesis (PK).
Psychological Bulletin, 59, 353–388.

Girden, E. (1962b).   A postscript to "A review of psychokinesis
(PK)." *Psychological Bulletin, 59*, 529–531.

Girden, E., & Girden, E. (1985).   Psychokinesis: Fifty years af–
terward. In P. Kurtz (Ed.), *A Skeptic's Handbook of Parap–
sychology* (pp. 129–146).   Buffalo, NY: Prometheus Books.

Goldstein, R. (1989). Understanding numerical analysis: Working
with the odds. *PC Magazine, 8*(5), 94–99.

Grafton, C., & Permaloff, A. (1988). Microcomputer statistical
packages. *PS, 21*, 71–82.

Greeley, A. (1987). Mysticism goes mainstream. *American
Health, 6*(1), 47–49.

Gruber, E.R. (1983).   Outside and inside the paranormal.   In
B. Shapin & L. Coly (Eds.), *Parapsychology's Second Cen–
tury: Proceedings of an International Conference Held in Lon–
don, England, August 13–14, 1982* (pp. 39–54). New York:
Parapsychology Foundation.

Hansel, C.E.M. (1966).   *ESP: A Scientific Evaluation.*   New
York: Scribner's.

Hansel, C.E.M. (1980). *ESP and Parapsychology: A Critical Re–
Evaluation.* Buffalo, NY: Prometheus Books.

Hansen, G.[P.]. (1985). A brief overview of magic for para–

psychologists. *Parapsychology Review, 16*(2), 5—8.

Hansen, G.P. (1986). A cooperation—competition PK experiment with computerized horse races [Abstract]. In D.H. Weiner & D.I. Radin (Eds.), *Research in Parapsychology 1985* (pp. 5—9). Metuchen, NJ: Scarecrow Press.

Hansen, G.P. (1988). CSICOP and skepticism: An emerging social movement [Abstract]. In D.H. Weiner & R.L. Morris (Eds.), *Research in Parapsychology 1987* (pp. 157—161). Metuchen, NJ: Scarecrow Press.

Hansen, G.P. (1990). Deception by subjects in psi research. *Journal of the American Society for Psychical Research, 84*, 25—80.

Hearne, K.M.T. (1982). An automated technique for studying psi in home "lucid" dreams. *Journal of the Society for Psychical Research, 51*, 303—304.

Honorton, C. (1971). Automated forced—choice precognition tests with a "sensitive." *Journal of the American Society for Psychical Research, 65*, 476—481.

Honorton, C. (1975). Error some place! *Journal of Communication, 25*, 103—116.

Honorton, C. (1979). Methodological issues in free—response psi experiments. *Journal of the American Society for Psychical Research, 73*, 381—394.

Honorton, C. (1985). Meta—analysis of psi ganzfeld research: A response to Hyman. *Journal of Parapsychology, 49*, 51—91.

Honorton, C. (1987). Precognition and real—time ESP performance in a computer task with an exceptional subject. *Journal of Parapsychology, 51*, 291—320.

Honorton, C., & Schechter, E.I. (1987). Ganzfeld target retrieval with an automated testing system: A model for initial Ganzfeld success [Abstract]. In D.H. Weiner & R.D. Nelson (Eds.), *Research in Parapsychology 1986* (pp. 36—39). Metuchen, NJ: Scarecrow Press.

Hövelmann, G.H. (1983). Seven recommendations for the future practice of parapsychology. *Zetetic Scholar*, No. 11, 128—138.

Hövelmann, G.[H.]. (1985). Skeptical literature on parapsychol—
ogy: An annotated bibliography. In P. Kurtz (Ed.), *A
Skeptic's Handbook of Parapsychology* (pp. 449—490). Buf—
falo, NY: Prometheus Books.

Hövelmann, G.H., & Krippner, S. (1986). Charting the future of
parapsychology. *Parapsychology Review, 17*(6), 1—5.

Howard, R.A. (1987). Editorial. *Journal of Scientific Explora—
tion, 1*, i.

Hyman, R. (1985). A critical overview of parapsychology. In
P. Kurtz (Ed.), *A Skeptic's Handbook of Parapsychology* (pp.
3—96). Buffalo, NY: Prometheus Books.

Hyman, R., & Honorton, C. (1986). A joint communiqué: The
psi ganzfeld controversy. *Journal of Parapsychology, 50*,
351—364.

Inglis, B. (1986). *The Hidden Power*. London: Jonathan Cape.

Irwin, H.J. (1979). *Psi and the Mind: An Information Processing
Approach*. Metuchen, NJ: Scarecrow Press.

Jahn, R.G., & Dunne, B.J. (1986). On the quantum mechanics
of consciousness, with application to anomalous phenomena.
Foundations of Physics, 16, 721—772.

Jerrell, M.E. (1988). Computer programs to demonstrate
hypothesis—testing issues. *American Statistician, 42*, 80—81.

Johnson, M.U. (1975). Models of control and control of bias.
European Journal of Parapsychology, 1, 36—44.

Josephson, B. (1975). Possible connections between psychic
phenomena and quantum mechanics. *New Horizons, 1*,
224—226.

Kaplan, E. (1989). Scientific graphing software: Not just a
pretty picture. *PC Magazine, 8*(5), 259—286.

Kautz, W.H. (1982). Editorial: Fertilizing the scientific method.
Applied Psi Newsletter, 1(3), 3—4.

Kennedy, J.E., & Taddonio, J.L. (1976). Experimenter effects
in parapsychological research. *Journal of Parapsychology, 40*,
1—33.

Koestler, A. (1972). *The Roots of Coincidence.* New York: Random House.

Krippner, S., Honorton, C., & Ullman, M. (1972). A second precognitive dream study with Malcolm Bessent. *Journal of the American Society for Psychical Research, 66,* 269−279.

Krippner, S., Ullman, M., & Honorton, C. (1971). A precognitive dream study with a single subject. *Journal of the American Society for Psychical Research, 65,* 192−203.

Lang, J.J. (1975). Scientism and parapsychology. *Counseling and Values, 19*(2), 91−97.

Laudan, R. (1983). Introduction. In R. Laudan (Ed.), *The Demarcation Between Science and Pseudo−Science: Working Papers in Science & Technology, 2*(1), 3−6.

Lawden, D.F. (1980). Possible psychokinetic interactions in quantum theory. *Journal of the Society for Psychical Research, 50,* 399−407.

Lenz, J.E., Kelly, E.F., & Artley, J.L. (1980). A computer−based laboratory facility for the psychophysiological study of psi. *Journal of the American Society for Psychical Research, 74,* 149−170.

LeShan, L. (1974). *The Medium, the Mystic, and the Physicist: Toward a General Theory of the Paranormal.* New York: Viking.

Lucadou, W.v. (1988). The model of pragmatic information [Abstract]. In D.H. Weiner and R.L. Morris (Eds.), *Research in Parapsychology 1987* (pp. 18−22). Metuchen, NJ: Scarecrow Press.

Markwick, B. (1978). The Soal−Goldney experiments with Basil Shackleton: New evidence of data manipulation. *Proceedings of the Society for Psychical Research, 56,* 250−281.

Mattuck, R. (1982). A crude model of the mind−matter interaction using Bohm−Bub hidden variables. *Journal of the Society for Psychical Research, 51,* 238−245.

May, E.C., Radin, D.I., Hubbard, G.S., Humphrey, B.S., & Utts, J.M. (1986). Psi experiments with random number gen−

erators: An informational model [Abstract]. In D.H. Weiner
& D.I. Radin (Eds.), *Research in Parapsychology 1985* (pp.
119–120). Metuchen, NJ: Scarecrow Press.

McCarthy, D.J. (1982). The role of microcomputers in ex-
perimental parapsychology. In B. Shapin & L. Coly (Eds.),
*Parapsychology and the Experimental Method: Proceedings of
an International Conference Held in New York, New York,
November 14, 1981* (pp. 82–99). New York: Parapsychology
Foundation.

McClenon, J., & Hyman, R. (1987). A remote viewing experiment conducted by a skeptic and a believer. *Zetetic
Scholar*, Nos. 12/13, 21–33.

McConnell, R.A., & Forwald, H. (1967a). Psychokinetic placement: I. A reexamination of the Forwald–Durham experiment. *Journal of Parapsychology, 31*, 51–69.

McConnell, R.A., & Forwald, H. (1967b). Psychokinetic placement: II. A factorial study of successful and unsuccessful
series. *Journal of Parapsychology, 31*, 198–213.

McConnell, R.A., & Forwald, H. (1968). Psychokinetic placement: III. Cube–releasing devices. *Journal of Parapsychology,
32*, 9–38.

Millar, B. (1973). An inexpensive portable electronic ESP tester.
Journal of the Society for Psychical Research, 47, 90–95.

Millar, B. (1978). The observational theories: A primer. *European Journal of Parapsychology, 2*, 304–332.

Millar, B. (1979). The distribution of psi. *European Journal of
Parapsychology, 3*, 78–110.

Millar, B. (1988). Cutting the Braudian loop: In defense of the
observational theories. *Journal of the American Society for
Psychical Research, 82*, 253–271.

Millar, B., Jacobs, J.C., & Michels, J.A.G. (1988). Hövelmann
and Krippner's eleven commandments. *Parapsychology
Review, 19*(1), 6–9.

Morgan, K. (1987). Anomalous human–computer interaction
(A.H.C.I.): Towards an understanding of what constitutes an
anomaly (or how to make friends and influence computers).

Proceedings of Presented Papers: The 30th Annual Convention of the Parapsychological Association, 123–138.

Morris, R.L. (1980). Some comments on the assessment of parapsychological studies: A review of *The Psychology of the Psychic*, by David Marks and Richard Kammann. *Journal of the American Society for Psychical Research*, *74*, 425–443.

Morris, R.L. (1982a). Assessing experimental support for true precognition. *Journal of Parapsychology*, *46*, 321–336.

Morris, R.L. (1982b). Mainstream science, experts, and anomaly. *Journal of the American Society for Psychical Research*, *75*, 257–281.

Morris, R.L. (1986a). Psi and human factors: The role of psi in human–equipment interactions. In B. Shapin & L. Coly (Eds.), *Current Trends in Psi Research: Proceedings of an International Conference Held in New Orleans, Louisiana, August 13–14, 1984* (pp. 1–26). New York: Parapsychology Foundation.

Morris, R.L. (1986b). What psi is not: The necessity for experiments. In H.L. Edge, R.L. Morris, J. Palmer, & J.H. Rush, *Foundations of Parapsychology: Exploring the Boundaries of Human Capability* (pp. 70–110). Boston: Routledge & Kegan Paul.

Morris, R.L. (1987). Parapsychology and the demarcation problem. *Inquiry*, *30*, 241–251.

Morris, R.L. (1989). Parapsychology. In A. Coleman & G. Beaumont (Eds.), *Psychology Surveys 7* (pp. 153–168). East Leicester, England: British Psychological Society.

Murphy, G. (1962). Report on a paper by Edward Girden on psychokinesis. *Psychological Bulletin*, *59*, 520–528.

M[urphy], G., & R[eiss], B.F. (1939). Editorial. *Journal of Parapsychology*, *3*, 1–2.

Nelson, R.D., & duPont, P.S., V. (1985, June). *Data Base Management for Anomalies Research*. Princeton, NJ: Princeton Engineering Anomalies Research. (Technical Note PEAR 85002).

Nelson, R.D., Bradish, G.J., & Dobyns, Y.H. (1989, April).

Random Event Generator Qualification, Calibration, and Analysis. Princeton, NJ: Princeton Engineering Anomalies Research. (Technical Note PEAR 89001)

PA statement on magicians. (1984). *Parapsychology Review,* *15*(2), 16.

Palmer, J. (1980). Parapsychology as a probabilistic science: Facing the implications. In W.G. Roll (Ed.), *Research in Parapsychology 1979* (pp. 189–215). Metuchen, NJ: Scarecrow Press.

Palmer, J. (1985). Psi research in the 1980s. *Parapsychology Review,* *16*(2), 1–4.

Palmer, J. (1986a). The ESP controversy. In H.L. Edge, R.L. Morris, J. Palmer, & J.H. Rush, *Foundations of Parapsychology: Exploring the Boundaries of Human Capability* (pp. 161–183). Boston: Routledge & Kegan Paul.

Palmer, J. (1986b). ESP research findings: The process approach. In H.L. Edge, R.L. Morris, J. Palmer, & J.H. Rush, *Foundations of Parapsychology: Exploring the Boundaries of Human Capability* (pp. 184–222). Boston: Routledge & Kegan Paul.

Palmer, J. (1986c). Criticisms of parapsychology: Some common elements. In B. Shapin & L. Coly (Eds.), *Current Trends in Psi Research: Proceedings of an International Conference Held in New Orleans, Louisiana, August 13–14, 1984* (pp. 255–276). New York: Parapsychology Foundation.

Palmer, J. (1986d). Meta–analysis in process–oriented psi research. In R.A. White & J. Solfvin (Eds.), *Research in Parapsychology 1984* (pp. 113–114). Metuchen, NJ: Scarecrow Press.

Palmer, J. (1986e). Progressive skepticism: A critical approach to the psi controversy. *Journal of Parapsychology, 50,* 29–42.

Palmer, J. (1987). Have we established psi? *Journal of the American Society for Psychical Research, 81*(2), 111–123.

Palmer, J., & Rao, K.R. (1987). Where lies the bias? *Behavioral and Brain Sciences, 10,* 618–627.

Palmer, J.A., Honorton, C., & Utts, J. (1989). Reply to the National Research Council Study on parapsychology. *Journal of the American Society for Psychical Research, 83*, 31–49.

Parapsychological Association. (1988). Terms and methods in parapsychological research. *Journal of the American Society for Psychical Research, 82*, 353–357.

Permaloff, A., & Grafton, C. (1988). Top of the line: SPSS, SAS, and SYSTAT. *PS, 21*, 657–666.

Phillips, P.R. (1984). Measurement in quantum mechanics. *Journal of the Society for Psychical Research, 52*, 297–306.

Poynton, J.C. (1983). Nonevident psi and phenomenology. *Parapsychology Review, 14*(6), 9–10.

Pratt, J.G. (1961). On the question of control over ESP: The effect of the environment on psi performance. *Journal of the American Society for Psychical Research, 55*, 128–134.

Pratt, J.G. (1967a). Computer studies of the ESP process in card guessing: I. Displacement effects in Mrs. Gloria Stewart's records. *Journal of the American Society for Psychical Research, 61*, 25–46.

Pratt, J.G. (1967b). Computer studies of the ESP process in card guessing: II. Did memory habits limit Mrs. Stewart's ESP success? *Journal of the American Society for Psychical Research, 61*, 182–202.

Pratt, J.G., & Forwald, H. (1958). Confirmation of the PK placement effect. *Journal of Parapsychology, 22*, 1–19.

Pratt, J.G., Martin, D.R., & Stribic, F.P. (1974). Computer studies of the ESP process in card guessing: III. Displacement effects in the C.J. records from the Colorado series. *Journal of the American Society for Psychical Research, 68*, 357–384.

Price, G.R. (1955). Science and the supernatural. *Science, 122*, 359–367.

Price, G.R. (1972). Apology to Rhine and Soal. *Science, 175*, 359.

Radin, D.I. (1985).   Pseudorandom number generators in psi
research.   *Journal of Parapsychology, 49*, 303—328.

Radin, D.I. (1988).   Effects of a priori probability on psi percep—
tion: Does precognition predict actual or probable futures?
Journal of Parapsychology, 52, 187—212.

Radin, D.I., & Nelson, R.D. (1987, August). *Replication in Ran—
dom Event Generator Experiments: A Meta—analysis and
Quality Assessment.* Princeton, NJ: Human Information
Processing Group Technical Report 87001.

Radin, D.I., May, E.C., & Thomson, M.J. (1986). Psi experi—
ments with random number generators: Meta—analysis part
1 [Abstract]. In D.H. Weiner & D.I. Radin (Eds.), *Research
in Parapsychology 1985* (pp. 14—17). Metuchen, NJ:
Scarecrow Press.

Ransom, C. (1971).   Recent criticisms of parapsychology: A
review.   *Journal of the American Society for Psychical
Research, 65*, 289—307.   Reprinted with changes by the
author in R.A. White (Ed.), *Surveys in Parapsychology:
Reviews of the Literature with Updated Bibliographies* (pp.
401—423). Metuchen, NJ: Scarecrow Press, 1976.

Rao, K.R. (1978). Theories of psi. In S. Krippner (Ed.), *Ad—
vances in Parapsychological Research 2: Extrasensory Percep—
tion* (pp. 245—295). New York: Plenum.

Rao, K.R., & Palmer, J. (1987). The anomaly called psi: Recent
research and criticism. *Behavioral and Brain Sciences, 10*,
539—551.

Raskin, R. (1989). Statistical software for the PC: Testing for
significance. *PC Magazine, 8*(5), 103—255.

Report of a Workshop on Experimental Parapsychology. (1989,
February). Washington DC: Office of Technology Assess—
ment.

Rhine, J.B. (1934).   *Extra—Sensory Perception.*   Boston: Boston
Society for Psychic Research.

Rhine, J.B. (1952). The Forwald experiments with placement
PK. *Journal of Parapsychology, 16*, 273—283.

Rockwell, T. (1979). Pseudoscience? Or pseudocriticism? *Journal* of Parapsychology, *43*, 221–231.

Rockwell, T. (1989). Guest editorial: Psi is not a mere anomaly. *Journal of the American Society for Psychical Research, 83,* 201–203.

Rockwell, T., Rockwell, R., & Rockwell, W.T. (1978a). Irrational rationalists: A critique of *The Humanist's* crusade against parapsychology. *Journal of the American Society for Psychical Research, 72,* 23–34.

Rockwell, T., Rockwell, R., & Rockwell, W.T. (1978b). *The Humanist's* crusade against parapsychology: A discussion. Part Two. Context vs. meaning: A reply to Dr. Kurtz. *Journal of the American Society for Psychical Research, 72,* 357–364.

Rush, J.H. (1983). Review of *Anomalistic Psychology: A Study of Extraordinary Behavior and Experiences* by L. Zusne & W.H. Jones. *Journal of the American Society for Psychical Research, 77,* 171–180.

Schechter, E.I. (1984). Hypnotic induction vs. control conditions: Illustrating an approach to the evaluation of replicability in parapsychological data. *Journal of the American Society for Psychical Research, 78,* 1–27.

Schechter, E.I. (1987). Meta–analysis and parapsychology. *Parapsychology Review, 18*(2), 13–15.

Schlitz, M.J. (1985). The phenomenology of replication. In B. Shapin & L. Coly (Eds.), *The Repeatability Problem in Parapsychology: Proceedings of an International Conference Held in San Antonio, Texas, October 28–29, 1983* (pp. 73–97). New York: Parapsychology Foundation.

Schmeidler, G. (1964). An experiment on precognitive clairvoyance. Part I. The main results. *Journal of Parapsychology, 28,* 1–14.

Schmidt, H. (1969a). Clairvoyance tests with a machine. *Journal of Parapsychology, 33,* 300–306.

Schmidt, H. (1969b). Precognition of a quantum process. *Journal of Parapsychology, 33,* 99–108.

Schmidt, H. (1970). A PK test with electronic equipment. *Jour–*

nal of Parapsychology, *34,* 175–181.

Schmidt, H. (1974a). Instrumentation in the parapsychology laboratory. In J. Beloff (Ed.), *New Directions in Parapsychology* (pp. 13–37). Metuchen, NJ: Scarecrow Press.

Schmidt, H. (1974b). A new role of the experimenter in science suggested by parapsychology research. In A. Angoff & B. Shapin (Eds.), *Parapsychology and the Sciences: Proceedings of an International Conference Held in Amsterdam, The Netherlands, August 23–25, 1972* (pp. 266–279). New York: Parapsychology Foundation.

Schmidt, H. (1975). Toward a mathematical theory of psi. *Journal of the American Society for Psychical Research, 69,* 301–319.

Schmidt, H. (1976). PK effect on pre–recorded targets. *Journal of the American Society for Psychical Research, 70,* 267–291.

Schmidt, H. (1977a). Evidence for direct interaction between the human mind and external quantum processes. *Proceedings of the International Conference on Cybernetics and Society,* 534–536.

Schmidt, H. (1977b). The use of computers for channeling psi. *ASPR Newsletter, 3*(1), 2–3.

Schmidt, H. (1978). Can an effect precede its cause? A model of a noncausal world. *Foundations of Physics, 8,* 463–480.

Schmidt, H. (1980). A program for channeling psi data into the laboratory and onto the critic's desk [Abstract]. In W.G. Roll (Ed.), *Research in Parapsychology 1979* (pp. 66–69). Metuchen, NJ: Scarecrow Press.

Schmidt, H. (1983). Randomness and the mind: Looking for psychic effects in games of chance. *Creative Computing, 9*(8), 180–186.

Schmidt, H. (1984). Comparison of a teleological model with a quantum collapse model of psi. *Journal of Parapsychology, 48,* 261–276.

Schmidt, H. (1987). The strange properties of psychokinesis. *Journal of Scientific Exploration, 1,* 103–118.

Schmidt, H., Morris, R.L., & Rudolph, L. (1986). Channeling evidence for a PK effect to independent observers. *Journal of Parapsychology, 50*, 1−15.

Schouten, S.A. (1979a). Analysing spontaneous cases as reported in "Phantasms of the Living." *European Journal of Parapsychology, 2*, 408−455.

Schouten, S.A. (1979b). Analysis of spontaneous cases as reported in "Phantasms of the Living." *European Journal of Parapsychology, 4*, 9−49.

Schouten, S.A. (1982). Analysing spontaneous cases: A replication based on the Rhine collection. *European Journal of Parapsychology, 4*, 113−158.

Schouten, S.A. (1983). A different approach for analyzing spontaneous cases: With particular reference to the study of Louisa E. Rhine's case collection. *Journal of Parapsychology, 47*, 323−340.

Schouten, S.A. (1986). A different approach for analyzing spontaneous cases: With particular reference to the study of Louisa E. Rhine's case collection. In K.R. Rao (Ed.), *Case Studies in Parapsychology: Papers Presented in Honor of Dr. Louisa E. Rhine at a Conference Held on November 12, 1983 at Bryan University Center, Duke University, Durham, North Carolina* (pp. 31−45). Jefferson, NC: McFarland.

Servadio, E. (1983). Beyond parapsychology: The use and the meaning of psi phenomena. In B. Shapin & L. Coly (Eds.), *Parapsychology's Second Century: Proceedings of an International Conference Held in London, England, August 13−14, 1982* (pp. 137−152). New York: Parapsychology Foundation.

Sidgwick, H. (1882). [President's address, July 17, 1882.] *Proceedings of the Society for Psychical Research, 1*, 7−12.

Simon, B. (1989). Better tools for higher math: When numbers count. *PC Magazine, 8*(5), 289−310.

Spinelli, E. (1983). Paranormal cognition: Its summary and implications. In B. Shapin & L. Coly (Eds.), *Parapsychology's Second Century: Proceedings of an International Conference Held in London, England, August 13−14, 1982* (pp. 19−35). New York: Parapsychology Foundation.

Stanford, R.G. (1974a). Concept and psi. In W.G. Roll, R.L.
    Morris, & J.D. Morris (Eds.), *Research in Parapsychology
    1973* (pp. 137–162). Metuchen, NJ: Scarecrow Press.

Stanford, R.G. (1974b). An experimentally testable model for
    spontaneous psi events. I. Extrasensory events. *Journal of
    the American Society for Psychical Research, 68,* 34–57.

Stanford, R.G. (1974c). An experimentally testable model for
    spontaneous psi events. II. Psychokinetic events. *Journal of
    the American Society for Psychical Research, 68,* 321–356.

Stanford, R.G. (1977a). Are parapsychologists paradigmless in
    psiland? In B. Shapin & L. Coly (Eds.), *The Philosophy of
    Parapsychology: Proceedings of an International Conference
    Held in Copenhagen, Denmark, August 25–27, 1976* (pp.
    1–18). New York: Parapsychology Foundation.

Stanford, R.G. (1977b). Conceptual framework of contemporary
    psi research. In B.B. Wolman (Ed.), *Handbook of Parapsy–
    chology* (pp. 823–858). New York: Van Nostrand Reinhold.

Stanford, R.G. (1977c). Experimental psychokinesis: A review
    from diverse perspectives. In B.B. Wolman (Ed.), *Handbook
    of Parapsychology* (pp. 324–381). New York: Van Nostrand
    Reinhold.

Stanford, R.G. (1978). Toward reinterpreting psi events. *Journal
    of the American Society for Psychical Research, 72,* 197–
    214.

Stanford, R.G. (1981a). Are we shamans or scientists? *Journal
    of the American Society for Psychical Research, 75,* 61–70.

Stanford, R.G. (1981b). Cognitive constraints and ESP perfor–
    mance: On testing some implications of a model. In B.
    Shapin & L. Coly (Eds.), *Concepts and Theories of Para–
    psychology: Proceedings of an International Conference Held
    in New York, New York, December 6, 1980* (pp. 91–106).
    New York: Parapsychology Foundation.

Stanford, R.G. (1982). Is scientific parapsychology possible?
    Some thoughts on James E. Alcock's *Parapsychology:
    Science or Magic? Journal of Parapsychology, 46,* 231–271.

Stanford, R.G. (1985). Toward enhancement of inter–laboratory and inter—experimenter replicability in psi research. In B. Shapin & L. Coly (Eds.), *The Repeatability Problem in Parapsychology: Proceedings of an International Conference Held in San Antonio, Texas, October 28—29, 1983* (pp. 212—237). New York: Parapsychology Foundation.

Stevenson, I., & Roll, W.G. (1966). Criticism of parapsychology: An informal statement of some guiding principles. *Journal of the American Society for Psychical Research, 60,* 347—356.

Stokes, D.M. (1985). Parapsychology and its critics. In P. Kurtz (Ed.), *A Skeptic's Handbook of Parapsychology* (pp. 379—423). Buffalo, NY: Prometheus Books.

Stokes, D.M. (1987). Theoretical parapsychology. In S. Krippner (Ed.), *Advances in Parapsychological Research 5* (pp. 77—189). Jefferson, NC: McFarland.

Sturrock, P.A. (1987). A brief history of the Society for Scientific Exploration. *Journal of Scientific Exploration, 1,* 1—2.

Tart, C.T. (1965). Applications of instrumentation in the investigation of haunting and poltergeist cases. *Journal of the American Society for Psychical Research, 59,* 190—201.

Tart, C.T. (1966). ESPATESTER: An automatic testing device for parapsychological research. *Journal of the American Society for Psychical Research, 60,* 256—269.

Tart, C.T. (1967). Random output selector for the laboratory. *Psychophysiology, 3,* 430—434.

Taylor, E. (1987). Prospectus for a person—centered science: The unrealized potential of psychology and psychical research. *Journal of the American Society for Psychical Research, 81,* 313—331.

Temkin, A.Y. (1982). Parapsychology from the point of view of modern physics. *European Journal of Parapsychology, 4,* 257—270.

Thouless, R.H. (1976). The effect of the experimenter's attitude on experimental results in parapsychology. *Journal of the Society for Psychical Research, 48,* 261—266.

Truzzi, M. (1978). On the extraordinary: An attempt at clarifica—

tion. *Zetetic Scholar, 1,* 11—22.

Truzzi, M. (1980). A skeptical look at Paul Kurtz's analysis of the scientific status of parapsychology. *Journal of Parapsychology, 44,* 35—55.

Truzzi, M. (1984). The role of conjurors in psychical research and parapsychology. In R.A. White & R.S. Broughton (Eds.), *Research in Parapsychology 1983* (pp. 120—121). Metuchen, NJ: Scarecrow Press.

Truzzi, M. (1985). Anomalistic psychology and parapsychology: Conflict or detente? Paper presented at the annual meeting of the American Psychological Association, Los Angeles, CA.

Truzzi, M. (1987a). Introduction. In S. Krippner (Ed.), *Advances in Parapsychological Research 5* (pp. 4—8). Jefferson, NC: McFarland.

Truzzi, M. (1987b). Zetetic ruminations on skepticism and anomalies in science. *Zetetic Scholar,* Nos. 12/13, 7—20.

Tyler, L.E. (1981). More stately mansions——Psychology extends its boundaries. In M.R. Rosenzweig & L.W. Porter (Eds.), *Annual Review of Psychology* (pp. 1—20). Palo Alto, CA: Annual Reviews.

Varvoglis, M.P. (1986). Goal—directed— and observer—dependent PK: An evaluation of the conformance—behavior model and the observational theories. *Journal of the American Society for Psychical Research, 80,* 137—162.

Villars, C.N. (1983). Nonlocality and ESP. *Journal of the Society for Psychical Research, 52,* 189—193.

Walker, E.H. (1975). Foundations of paraphysical and parapsychological phenomena. In L. Oteri (Ed.), *Quantum Physics and Parapsychology: Proceedings of an International Conference Held in Geneva, Switzerland, August 26—27, 1974* (pp. 1—53). New York: Parapsychology Foundation.

Walker, E.H. (1976). Quantum mechanics/psi phenomena: The theory and suggestions for new experiments. *Journal of Research in Psi Phenomena, 1*(1), 38—52.

Walker, E.H. (1977). Comparison of some theoretical predictions of Schmidt's mathematical theory and Walker's quantum mechanical theory of psi. *Journal of Research in Psi Phenomena, 2*(1), 54–70.

Walker, E.H. (1979). The quantum theory of psi phenomena. *Psychoenergetic Systems, 3,* 259–299.

Walker, E.H. (1984). A review of criticisms of the quantum mechanical theory of psi phenomena. *Journal of Parapsychology, 48,* 277–332.

Walker, E.H. (1987a). A comparison of the intuitive data sort–ing and quantum mechanical observer theories. *Journal of Parapsychology, 51,* 217–227.

Walker, E.H. (1987b). Measurement in quantum physics revisited: A response to Phillips' criticism of the quantum mechanical theory of psi. *Journal of the American Society for Psychical Research, 81,* 333–369.

Walther, G. (1955). A plea for the introduction of Edmund Husserl's phenomenological methods into parapsychology. In *Proceedings of the First International Conference of Parapsychological Studies* (pp. 114–115). New York: Parapsychology Foundation

White, R.A. (1976a). The influence of persons other than the experimenter on the subject's scores in psi experiments. *Journal of the American Society for Psychical Research, 70,* 133–166.

White, R.A. (1976b). The limits of experimenter influence on psi test results: Can any be set? *Journal of the American Society for Psychical Research, 70,* 333–369.

White, R.A. (1977). The influence of experimenter motivation, attitudes, and methods of handling subjects on psi test results. In B.B. Wolman (Ed.), *Handbook of Parapsychology* (pp. 273–301). New York: Van Nostrand Reinhold.

White, R.A. (1980). On the genesis of research hypotheses in parapsychology. *Parapsychology Review, 11*(1), 6–9.

White, R.A. (1983). The future of parapsychology. *Journal of Religion and Psychical Research, 6,* 220–226.

White, R.A. (1984a). Gerald Heard's legacy to psychical research. *Holism in Religion and Psychical Research:*

Academy of Religion and Psychical Research Annual Con-ference Proceedings, 1982, 56—69.

White, R.A. (1984b). Parapsychology and the transcendent. *Christian Parapsychologist*, 5, 138—150.

White, R.A. (1985a). *Bibliography of Works By and About Haakon Forwald.* Dix Hills, NY: Parapsychology Sources of Information Center.

White, R.A. (1985b). The spontaneous, the imaginal, and psi: Foundations for a depth parapsychology. In R.A. White & J. Solfvin (Eds.), *Research in Parapsychology 1984* (pp. 166—190). Metuchen, NJ: Scarecrow Press.

White, R.A. (1987). Meaning, metanoia and psi. In B. Shapin & L. Coly (Eds.), *Parapsychology, Philosophy and Religious Con-cepts: Proceedings of an International Conference Held in Rome, Italy, August 23—24, 1985* (pp. 167—189). New York: Parapsychology Foundation.

Whiteman, J.H.M. (1973). Quantum theory and para-psychology. *Journal of the American Society for Psychical Research*, 67, 341—360.

Zukav, G. (1979). *The Dancing Wu Li Masters: An Overview of the New Physics.* New York: William Morrow.

Zusne, L. (1982). Contributions to the history of psychology: XXXII. On living with a spector: The story of anomalistic psychology. *Perceptual and Motor Skills*, 55, 683—694.

Zusne, L. (1985). Magical thinking and parapsychology. In P. Kurtz (Ed.), *A Skeptic's Handbook of Parapsychology* (pp. 685—700). Buffalo, NY: Prometheus Books.

VII
Chronological List of
Theses in Parapsychology

INTRODUCTION

This is a chronological and world—wide list of degrees con—ferred in part for a thesis or dissertation on parapsychology. Doctoral degrees are listed first, followed by a list of Master's degrees. Both lists begin with the earliest dates available. In the listing of doctorates no distinction is made between M.D. degrees, Ph.D.'s, Ed.D.'s, S.T.D.'s etc. Within the Master's de—grees, no distinction is made between M.S. and M.A. degrees. A third Miscellaneous Degrees section lists B.Litt, Dipl.The., and Honor's theses. In this chapter in *Sources* 49 degrees were listed, whereas here there are 328.

There are a total of 227 doctoral dissertations and 101 Master's theses listed, plus eight items in the Miscellaneous category. An analysis indicates that 1981 was the peak year for both dissertations (24) and theses (9). A geographic analysis shows the United States leading with 202 degrees, followed by England (19), Germany and Italy (13 each), India (11), Scotland (9), the Republic of South Africa (8), and the Netherlands (7). Some of the more active smaller countries were Sweden (6), Australia (5), the Philippines, and Switzerland (4). Regarding individual schools, City College of the City University of New York leads with 13 (5 Ph.D.s, 8 MAs). If all the University of California branches are lumped together, it shares second place with John F. Kennedy University (11 each). Next is the Univer—sity of Freiburg with 10; the University of Edinburgh and Saybrook Institute/Humanistic Psychology Institute with 9; then Duke University with 8; Andhra University and California State University (all branches) with 7; New York University, Union Graduate School, United States International University, and the University of Utrecht with 6 each; Boston University, California School of Professional Psychology (all branches), Catholic University of America, Harvard University, and the University of London with 5 each; Columbia University, Fuller Theological Seminary, Oxford University, University of Pennsyl—vania, University of the Philippines, and University of Surrey with 4 each; and Lund University, University of Cambridge,

University of Chicago, University of Chile, and University of Texas at Austin with 3 each.

DOCTORAL DISSERTATIONS

1893. Coste, Albert. Ph.D. **University of Montpellier.** *Psychic phenomena.* (In French)

1899. Goddard, Henry H. Ph.D. **Clark University.** *The effects of mind on body as evidenced by faith cures.*

1899. Thibault, E.J.A.J. M.D. **Faculté de Médecine et de Pharmacie de Bordeaux.** *The sensation of "Déjà vu."* (In French)

1902. Jung, Carl G. M.D. **University of Zurich.** *On the psychology of so—called occult phenomena.* (In German)

1905. Waddle, Charles W. Ph.D. **Clark University.** *Miracles of healing.*

1918. Liljencrants, Johann. Ph.D. **Catholic University of America.** *Spiritism and religion.*

1932. Lawton, George. Ph.D. **Columbia University.** *The drama of life after death: A study of the spiritualist religion.*

1932. Peerbolte, M.L. Ph.D. **Amsterdam University.** *On the problems of schizophrenia in its relation to a possible "wrong working" of the psi faculty.* (In Dutch)

1933. Bender, Hans. Ph.D. **University of Bonn.** *Psychic automatisms: Concerning the experimental psychology of the unconscious and extrasensory perception.* (In German)

1933. Tenhaeff, W.H.C. Ph.D. **University of Utrecht.** *Extrasensory perception and empathy.* (In Dutch)

1933. Thomas, John F. Ph.D. **Duke University.** *An evaluative study of the mental content of certain trance phenomena.*

1938. Hettinger, John. Ph.D. **University of London.** *The ultra—perceptive faculty.*

1941. Goldenson, Robert M. Ph.D. **Harvard University.** *The con—*

cept of intuition.

1941. Reuter, William L. Ed.D. **Temple University.** *An objective study in extrasensory perception.*

1941. Stuart, Charles E. Ph.D. **Duke University.** *An analysis to determine a test predictive of extrachance scoring in card– calling tests.*

1942. Björkhem, John. Ph.D. **Lund University.** *Hypnotic hal– lucinations.* (In Swedish)

1943. Bendit, Laurence J. M.D. **Cambridge University.** *Paranor– mal cognition: Its place in human psychology.*

1944. Kirk–Duncan, V.G. Ph.D. **Oxford University.** *A study of certain aspects of prima facie extrasensory perception.*

1945. Soal, Samuel G. D.Sc. **University of London.** Granted for the experiments described in *Modern Experiments in Telepathy.*

1946. Humphrey, Elizabeth M. Ph.D. **Duke University.** *Dis– crimination between high and low scoring subjects in ESP tests on the basis of the form quality of their response drawings.*

1948. Johnson, George A. Ph.D. **University of Chicago.** *From seeker to finder: A study of seventeenth century spiritualism.*

1950. Osis, Karlis. Ph.D. **University of Munich.** *Hypotheses of extrasensory perception.* (In German)

1951. Canavesio, Orlando. M.D. **National University of Córdoba.** *Electroencefalografía de los estados metapsíquicos.* (In Spanish)

1951. Gaya, John. Ph.D. **Catholic University of America.** *ESP phenomena and the regularities of human guessing: An ex– perimental study.*

1951. Sanso, John G. Ph.D. **Catholic University of America.** *ESP phenomena and the regularities of human guessing: An experimental study.*

1953. Cadoret, Remi J. M.D. **Yale University.** *The effects of* amytal and dexadrine on ESP performance.

1955. Bro, Harmon. Ph.D. **University of Chicago.** *The charisma of the seer.* [on Edgar Cayce]

1956. Binski, Sigurd. Ph.D. **University of Bonn.** *Experimental investigations in psychokinesis.* (In German)

1956. Flynn, Thomas J. Ph.D. **Fordham University.** *Empirical and metaphysical proofs for the immortality of the human soul.*

1956. Scriven, Michael. Ph.D. **Oxford University.** *Explanations of the supernatural.*

1958. La Roche, Robert. Ph.D. **Catholic University of America.** *Divination, with a supplement on superstition in Central Africa.*

1958. Marsh, Maurice. Ph.D. **Rhodes University.** *Linkage in extrasensory perception.*

1958. Strauch, Inge. Ph.D. **University of Freiburg.** *The problem of spiritual healing.* (In German)

1958. Sulkov, Martin. Ph.D. **New School for Social Research.** *An inquiry into certain proofs of the doctrine of personal immortality.*

1960. Torchiarolo, G.C. di. Ph.D. **University of Lecce.** *Psychology and parapsychology, telepathy.* (In Italian)

1961. Alberti, Giorgio. M.D. **University of Milan.** *Methodological criticism of the experimental parapsychological method.* (In Italian)

1961. Sannwald, Gerhard. Ph.D. **University of Freiburg.** *Relationship between "parapsychic" experiences and personality traits.* (In German)

1962. Agnelti, Alberto. Ph.D. **University of Rome.** *Statistical criticism of an ESP experiment.* (In Italian)

1962. Yaryan, Ruby. Ph.D. **University of London.** *The effect of various forms of conditioning upon liminal, subliminal and extrasensory perception.*

1963. Gross, Don H. Ph.D. **Harvard University.** *A Jungian analysis of New Testament exorcism.*

1963. Robbertse, J.H. M.D. **University of Utrecht.** *Applicability of the concept of telepathy in psychiatry.*

1965. Bhadra, B.H. Ph.D. **Sri Venkateswara University.** *The relationship of test scores to belief in ESP.*

1965. Urban, P. Ph.D. **Vienna University.** *Schopenhauer and the current state of affairs in parapsychology.* (In German)

1966. Moss, Thelma Shnee. Ph.D. **University of California, Los Angeles.** *A study of experimenter bias through subliminal perception, non−verbal communication, and ESP.*

1967. Nelson, Geoffrey K. Ph.D. **University of London.** *The spiritualist movement and the need for a redefinition of cult.*

1967. Zabotin, Vladimir. Ph.D. **Leningrad State University.** *Investigations of low−frequency electromagnetic fields occurring around living objects.* (In Russian)

1968. Bakker, L.F. M.D. **University of Utrecht.** *Kwakzalverij: en onbewegd hitoefen der Geneeokunst.* (In Dutch)

1968. Kanthamani, B.K. Ph.D. **Andhra University.** *An inquiry into the personality patterns of psi−hitters and psi−missers.*

1968. Kerr, Howard Hastings. Ph.D. **University of California, Los Angeles.** *Spiritualism in American literature, 1851−1886.*

1968. Leutin, V.P. Ph.D. **University of Leningrad.** *The study of condition−reflex changes in the EEG of a human being to a nonspecific stimulus.* (In Rus sian)

1968. Stein, Robert D. Ph.D. **Northwestern University.** *The impact of the psychical research movement on the literary theory and literary criticism of Frederic W.H. Myers.*

1969. Phillips, Preson Peek. Ph.D. **Bob Jones University.** *A Christian introduction to parapsychology.*

1970. Galbreath, Robert Carroll. Ph.D. **University of Michigan.** *Spiritual science in an age of materialism: Rudolf Steiner* and occultism.

1970. Raef, Yehia. Ph.D. **Cambridge University.** *Parapsychological phenomena in Islamic literature.*

1971. Nissen, K.D. Ph.D. **University of Freiburg.** *Attitudes of
German students of psychology toward parapsychological
and ideological problems.* (In German)

1971. Pienaar, D.C. Ph.D. **University of South Africa.** *Studies
in ESP: An investigation of distortion in ESP phenomena.*
(In Afrikaans)

1971. Wiesinger, Christofer. Ph.D. **University of Freiburg.** *In-
vesigations of extrasensory perception in the social field of
the classroom.* (In German)

1972. Halifax–Grof, Joan. Ph.D. **Union Graduate School.** *Indigenous healing systems and contemporary medicare:
Toward an integration of approaches.*

1972. Haraldsson, Erlendur. Ph.D. **University of Freiburg.**
*Vasomotor reactions as indicators of extrasensory percep-
tion.*

1972. Mindell, Arnold P. Ph.D. **Union Graduate School.**
*Synchronicity: An investigation of the unitary background
patterning synchronous phenomena.*

1972. Sailaja, P. Ph.D. **Andhra University.** *Experimental studies
of the differential effect in life setting.*

1972. Smith, Martha Stribling. Ph.D. **University of North
Carolina.** *A study of the realistic treatment of psychic
phenomena in selected fiction of William Dean Howells,
Hamlin Garland, Henry James, Frank Norris, and
Theodore Dreiser.*

1973. Bednarowski, Mary F. Ph.D. **University of Minnesota.**
*19th century American spiritualism: An attempt at scien-
tific religion.*

1973. Brown, Burton Gates. Ph.D. **Boston University.**
Spiritualism in nineteenth century America.

1973. Hammers, A.J. Ph.D. **University of Freiburg.** *Parapsychology as judged by Catholic theologians.* (In German)

1973. Kampman, Reima. Ph.D. **University of Oulu.** *Hypnotically induced multiple personality: An experimental study.*

1973. Lehman, N.B. Ph.D. **Ohio State University.** *The life of John Murray Spear: Spiritualism and reform in Antebellum America.*

1973. Long, J.K. Ph.D. **University of North Carolina.** *Jamaican medicine.*

1973. Moon, Marvin L. Ph.D. **Pennsylvania State University.** *Extrasensory perception and art experience.*

1973. Schouten, Sybo A. Ph.D. **University of Utrecht.** *The standard—method as instrument for parapsychological research.*

1974. Freeman, Bonita Ann. Ph.D. **University of Pennsylvania.** *The development of spiritualist mediums: Apprenticeship to a tradition.*

1974. Goodrich, Joyce. Ph.D. **Union Graduate School.** *Psychic healing——a pilot study.*

1974. Howard, Cary Elizabeth. Ph.D. **United States International University.** *Extrasensory perception training through altered states of consciousness: An pre—test and post—test study of teaching psychic sensitivy.*

1974. Locke, Ralph G. Ph.D. **University of Western Australia.** *Mainpulators of time: Symbolic orders in spirit medium—ship.*

1975. Asher, Maxine. Ph.D. **Walden University.** *Theories of in—tuitive perception applied to ancient anthropological in—quiry.*

1975. Carretto, Giovanni. Ph.D. **University of Lecce.** *Psychology and parapsychology.* (In Italian)

1975. Ceren, Sandra Levy. Ph.D. **United States International University.** *Personality and precognition in stock market speculation.*

1975. Davis, Ronald Julian. Ph.D. **University of North Carolina.** *Augustus Thomas and psychic phenomena.*

1975. Evans, Ted David. Ph.D. **Fuller Theological Seminary.** *A scientific evaluation of extrasensory perception: Telepathy and the effects of subjective emotional arousal on psychic phenomena.*

1975. Fair, Justin D. Ph.D. **United States International University.** *An investigation of the relationship between paranormal phenomena and altered states of consciousness.*

1975. Ironson, Dale. Ph.D. **Humanistic Psychology Institute [now Saybrook Institute].** *An investigation into the preconditions, characteristics and beliefs associated with out−of−body experiences.*

1975. Isaacs, Ernest. Ph.D. **University of Wisconsin.** *History. A history of nineteenth century American spiritualism as a religious and social movement.*

1975. Laria, Riccardo. Ph.D. **University of Cagliari.** *Healers and Article 3 of the penal code.* (In Italian)

1975. Michtom, Madeleine. Ph.D. **New York University.** *Becoming a medium: The role of trance in Puerto Rican spiritism as an avenue to mazeway resynthesis.*

1975. Rush, James Edward. Ph.D. **Hartford Seminary Foundation.** *Toward a general theory of healing.*

1975. Winston, Shirley. Ph.D. **Union Graduate School.** *Research in psychic healing: A multivariate experiment.*

1976. Banks, Hal N. S.T.D. **Geneva-St. Albans Theological College.** *An introduction to parapsychology: Theological, historical, and philosophical implications.*

1976. Casey, Thomas M. Ph.D. **University of Ottawa.** *Parapsychology and altered states of consciousness: Towards a point of integration with religious experience.*

1976. Condey, Andrew. Ph.D. **City College of the City University of New York.** *Non−verbal, non−proximic group and individual communication effects.*

1976. Cozzi, Giorgio. Ph.D. **University of Trent.** *Problems of parapsychology.* (In Italian)

1976. Donahoe, James Joseph. Ph.D. **California Institute of Asian Studies.** *Imagination and the paranormal.*

1976. Fitzgerald, Dale Kelley. Ph.D. **University of California, Berkeley.** *Spirit mediumship and seance performance among the Ga of Southern Ghana.*

1976. Fourie, David P. Ph.D. **University of South Africa, Pretoria.** *An investigation of the appearance of certain parapsychological phenomena in hypnosis.*

1976. Geissler, Winnifred Johnson. Ph.D. **Kansas State University.** *Psychic phenomena in Old English literature as an index to Anglo–Saxon attitudes toward the preternatural.*

1976. Grunzig, H.J. Ph.D. **University of Freiburg.** *Geometric–optical delusion and personality: An experimental investigation on the differential psychology of perception and on extrasensory perception.*

1976. Johannison, Karin. Ph.D. **Uppsala University.** *Magnetizing a point in space.* (In Swedish)

1976. Kobayashi, Ken. Ph.D. **Baylor University.** *A relationship between psi performance and EEG alpha activity of the cerebral hemispheres.*

1976. Kurzweg, A. Ph.D. **University of Berlin.** *The history of the Berlin Society for Experimental Psychology with special consideration of its starting situation and of the activity of Max Dessoir.* (In German)

1976. Montadon, Henry Eugene. Ph.D. **Columbia University.** *An experimental investigation of the Kirlian phenomena.*

1976. Tornatore, Nicholas. Ph.D. **Humanistic Psychology Institute [now Saybrook Institute].** *Employing the paranormal event in analysis as a psychotherapeutic tool.*

1976. Wagenfeld, Jeanne K. Ed.D. **Western Michigan University.** *A psychological investigation of the relationship of empathy, self awareness and telepathy in the counselor–client dyad.*

1976. Wiklund, Nils. Ph.D. **Lund University.** *Experiments in telepathy, clairvoyance and psychokinesis.*

1977. Aiazzi, Pierluigi. Ph.D. **University of Firenze.** *Yoga and psi phenomena.* (In Italian)

1977. Cassoli, Simonetta. M.D. **University of Bologna.** *Relation between parapsychology and neuro−psychiatry with regard to the phenomenon of so−called healers.* (In Italian)

1977. Goodman, Jeffrey. Ph.D. **California Coast University, Santa Ana.** *A psychic methodology for archaeological research.*

1977. Greene, Nancy Ellen. Ph.D. **California School of Professional Psychology, San Diego.** *Psychic abilities in the world of a psychological counselor.*

1977. Herold, Bruce Walter. Ph.D. **Pennsylvania State University.** *Precognition in selected dramas of Calderon de la Barca.*

1977. LeGassick, George Richard. Ph.D. **California School of Professional Psychology, Los Angeles.** *An exploration into the relationship between cerebral hemispheric functioning and extrasensory perception.*

1977. Lyall, James M. Ph.D. **University of Akron.** *The effects of relaxation and knowledge of results on ESP scoring.*

1977. Muller, L. Ph.D. **University of Freiburg.** *Parapsychology and trickery.* (In German)

1977. Noonan, Molly H. Ph.D. **University of Pennsylvania.** *Science and the psychical research movement.*

1977. Parker, Adrian. Ph.D. **University of Edinburgh.** *The experimenter−effect in parapsychology.*

1977. Rogers, Charles T.G. Ph.D. **Newport International University.** *The proposal of psi−flow as a generic term for the dynamics of paranormal phenomena in therapeutic relationships.*

1977. Rosin, U. Ph.D. **University of Freiburg.** *The paranormal and the occult as seen by Protestant parsons in the Federal Republic of Germany.* (In German)

1977. Shrager, Elaine Forman. Ph.D. **New York University.** *An experimental investigation of mother−child telepathy and re−*

lated personality variables.

1978. Broughton, Richard. Ph.D. **University of Edinburgh.** *Brain hemisphere differences in paranormal abilities with special references to the influence experimenter expectancies.*

1978. Chesser, Barbara E. Ph.D. **University of California, Los Angeles.** *Trafficking with spirits: A cross—cultural examination of the biosociocultural concomitants of spirit encounters.*

1978. Hearne, K.M.T. Ph.D. **Liverpool University.** *Lucid dreams: An electrophysiological and psychological study.*

1978. Millay, Jean. Ph.D. **Humanistic Psychology Institute** [now **Saybrook Institute**]. *The relationship between phase synchronization of brainwaves and success in attempts to communicate telepathy: A pilot study.*

1978. Stacy, J. Ph.D. **University of Oklahoma.** *Some philosophical implications of psi phenomena: Toward a reconcilliation of science and religion*

1978. Terry, James. Ph.D. **Humanistic Psychology Institute** [now **Saybrook Institute**]. *Research in parapsychology: A study of process.*

1978. Thurston, Mark. Ph.D. **Humanistic Psychology Institute** [now **Saybrook Institute**]. *An invesigation of behavior and personality correlates of psi incorporating a humanistic research approach.*

1978. Wedonoja, William A. Ph.D. **University of California, San Diego.** *Religion and adaption in rural Jamaica*

1979. Bernstein, Seth. M.D. **California School of Professional Psychology, Berkeley.** *An evaluation of psychic readers of human energy.*

1979. Brier, Robert M. Ph.D. **University of North Carolina.** *The problem of backward causation.*

1979. Heidt, Patricia R. Ph.D. **New York University.** *An investigation of the effects of therapeutic touch on anxiety of hospitalized patients*

1979. Jorgenson, Danny Lynn. Ph.D. **Ohio State University.** *Tarot divination in the Valley of the Sun: An existential sociology of the esoteric and the occult.*

1979. Kaplan, Pascal M. Ph.D. **Harvard University.** *Toward a theology of consciousness: A study in the logic and structure of esotericism.*

1979. McGarry, J. Ph.D. **Kent State University.** *Beliefs in paranormal phenomena: Their relationship to involvement in esoteric practices and locus of control.*

1979. Pasricha, Satwant K. Ph.D. **Bangalore University.** *An investigation into reported cases of persons who claim to have reincarnated.*

1979. Peduto, Angela. Ph.D. **University of Bologna.** *Mediumistic art and psychopathological painting.* (In Italian)

1979. Reeves, J.A. Ph.D. **Miami University, Ohio.** *Psychical phenomena in the United States in the nineteenth century.*

1979. Sargent, Carl. Ph.D. **Cambridge University.** *Intrasubject and extrasubject factors in psi test performance.*

1979. Selwa, Barbara I. Ed.D. **University of Georgia.** *An investigation of the relationship between figural creativity and clairvoyance in college students.*

1979. Van de Hooft, George A. M.D. **University of Leyden.** *The malapod dance——a transcultural—psychiatric study.* (In Dutch)

1979. Weissensteiner, H. Ph.D. **University of Salzburg.** *The history of the problems of parapsychology as viewed in the frame of scientific psychology.* (In German)

1979. Wind, J. Ph.D. **University of Salzburg.** *Development, results, and problems of parapsychology: Their meaning and their relationship to psychology.* (In German)

1980. Bedar, Bradford Bruce. Ed.D. **Boston University.** *An ESP—mediumship paradigm as a definition of mediumship.*

1980. Blackmore, Susan J. Ph.D. **University of Surrey.** *Extrasensory perception as a cognitive process.*

1980. Bonincini, Anna Maria. Ph.D. **University of Padova.** *Inves— tigation of the personality of so—called paranormal subjects.* (In Italian)

1980. Boucher, Faith Katherine. Ph.D. **University of California, Davis.** *The cadences of healing: Perceived benefit from treatment among the clientele of psychic healers.*

1980. Caire, Jill Bond. Ph.D. **Union for Experimenting Colleges and Universities.** *The transcendent visionary state as a discrete—altered state of consciousness and its impact on modern American women.* [Deals in part with altered states and psi.]

1980. Cerullo, John James. Ph.D. **University of Pennsylvania.** *The secularization of the soul: Psychical research in Britain, 1882–1920.*

1980. Francis, Lillie Mary. Ph.D. **University of North Dakota.** *"Night Journey to Meridian" and other stories.*

1980. Hagey, Rebecca Susan. Ph.D. **Case Western Reserve University.** *Healing entrepreneurship in the Philippines.*

1980. Ifkovits, G. Ph.D. **Vienna University.** *The influence of motivation in clairvoyance experiments.* (In German)

1980. LaBerge, S. Ph.D. **Stanford University.** *Lucid dreaming: An exploratory study of consciousness during sleep.*

1980. Lewis, Todd Vernon. Ph.D. **Louisiana State University.** *Charismatic communication and faith healers: A critical study of rhetorical behavior.*

1980. Martone, Kathryn Call. Ed.D. **University of Arkansas.** *Telepathy: An alternate form of communication between client and therapist.*

1980. Miller, Lloyd A. Ph.D. **California School of Professional Psychology, San Diego.** *Precognitive abilities among a general population.*

1980. Mishlove, Jeffery. Ph.D. **University of California, Berkeley.** *Psi—development systems: A disciplinary matrix for history, theory, evaluation, and design.*

1980. Roquemore, Gwendolyn Johnston. Ph.D. **Atlanta Univer-**

sity. *A study of the effects of training in empathy and meditation on the empathy and ESP scores of under-graduate subjects.*

1980. Schleip, H. Ph.D. **University of Freiburg.** *On the prag-matics of the laying-on-of-hands by healers.* (In German)

1980. Shargal, Susan K.Y. Ph.D. **City College of the City University of New York.** *ESP in children: Its relationship to age and personality.*

1980. Smith, Jane Gentry. Ph.D. **University of Texas, Austin.** *The mystery of the mind: A biography of William McDougall.* [Includes his contribution to parapsychology.]

1980. Van Dragt, Bryan. Ph.D. **Fuller Theological Seminary.** *Paranormal healing: A phenomenology of the healer's ex-perience. (Volumes I and II).*

1980. Young, Gloria Ann. Ph.D. **University of Texas, Austin.** *The new transcendentalism in post-industrial society: Life styles and the search for meaning in the new age.* [Focuses on two aspects: experimental parapsychology and Tibetan Buddhism.]

1981. Becker, Carl Bradley. Ph.D. **University of Hawaii.** *Sur-vival: Death and afterlife in Christianity, Buddhism, and modern science.*

1981. Dollar, Harold Ellis. D.Miss. **Fuller Theological Seminary.** *A cross-cultural theology of healing.*

1981. Espino Na Varrete, Conchita. Ph.D. **New York Univer-sity.** *Characteristics of hispanic patients reporting spiritualistic phenomena.*

1981. Faye, Jan. Ph.D. **University of Copenhagen.** *A natural philosophical essay on time and causality.* (In Danish) [On precognition.]

1981. Fey, Ann Quinn. Ed.D. **Columbia University Teachers College.** *A study of the relationship between choice of metaphor for time and precognitive ability.*

1981. Fried-Cassorla, Martha Jane. Ph.D. **Temple Univesity.** *An examination of the relationship between life purpose, death fear, religiosity and a belief in life after death.*

1981. Friedman, Harris Leonard. Ph.D. **Georgia State University.** *The construction and validation of a transpersonal mearsure of self—concept: The self—expansiveness level form.* [Relates scale to parapsychology.]

1981. Fulvia, Cariglia. Ph.D. **University of Urbino.** *The social context of parapsychology today.* (In Italian)

1981. Galanti, Geri—Ann. Ph.D. **University of California, Los Angeles.** *The psychic reader as shaman and psychotherapist: The interface between clients' and practitioners' belief systems.*

1981. Hirschlien, John Arthur. Ph.D. **University of Oklahoma.** *Construction of a life after death scale.*

1981. Jones, Funmilayo M. Ph.D. **Boston University.** *Strategies and techniques used in occasion maintenance: An examination of reader—client relationships in a tearoom.*

1981. Khilji, Anjum. Ph.D. **University of Punjab, Pakistan.** *New explorations in the realm of mind.*

1981. Lerner, Jeffrey Charles. Ph.D. **Columbia University.** *The public and private management of death in Britain, 1890—1930.* (Includes psychical research)

1981. McClenon, James M. Ph.D. **University of Maryland.** *Deviant science: The case of parapsychology.*

1981. Mitchell, Janet Lee. Ph.D. **City College of the City University of New York.** *Verbal and performance skills in relation to ESP.*

1981. Moeller, Charles Richard. Ph.D. **Ball State University.** *The effects of labeling on self—esteem, self—description, and behavior (Includes attitude toward ESP).*

1981. Neppe, Vernon Michael. Ph.D. **University of Witwatersrand.** *A study of déjà vu experience.*

1981. Neuberg, Matthew A. Ph.D. **Cornell University.** *Aeschylean universe: The ethical & metaphysical background of Aeschylus' Oresteia: responsibility, inevitability, precognition, persuasion, causality.*

1981. O'Sullivan, Michael Anthony. Ph.D. **University of Southern California.** *A harmony of worlds: Spiritualism and the quest for community in nineteenth–century America.*

1981. Rodriguez, Sylvia Belmont. Ph.D. **Stanford University.** *Ecstasy: Map and threshold, a cross–cultural study of dis– sociation.*

1981. Rose, Jonathan Ely. Ph.D. **University of Pennsylvania.** *The turn of the century: A study in the intellectual history of Britain 1895–1918.* (One chapter of six on psychical research)

1981. Stack O'Sullivan, Deborah Jean. Ph.D. **University of Connecticut.** *Personality correlates of near–death experiences.*

1981. Thalbourne, Michael Anthony. Ph.D. **University of Edinburgh.** *Some experiments on the paranormal cognition of drawings, with special reference to personality and at– titudinal variables.*

1981. Trausch, Clarence Phillip. Ed.D. **Northern Illinois University.** *Psi training through meditation and self–actualization as related to psi performance.*

1982. Anthony, James Stephen. Ph.D. **United States International University.** *Interrelationships among belief in psychic abilities, psychic experiences, and sensation seeking.*

1982. Krishna, Shanti. Ph.D. **Andhra University.** *Psi and the dif– ferential effect: A study of the bimodal response pattern in ESP tests.*

1982. Olson, Melodie Ann. Ph.D. **University of Texas, Austin.** *The relationship of the out–of–body experience and level of relaxation.*

1982. Quinn, Janet F. Ph.D. **New York University.** *An investiga– tion of the effects of therapeutic touch done without physi– cal contact on state anxiety of hospitalized cardiovascular patients.*

1982. Rao, P.V. Krishna. Ph.D. **Andhra University.** *Extrasen– sory (ESP) and subliminal perception (SP): A study of their relation.*

1982. Reinhardt, Douglas Edward. Ph.D. **University of North Carolina.** *Faith healing: Where science meets religion in the body of believers.*

1982. Shafer, Mark G. Ph.D. **University of California, Irvine.** *Self–actualization, mysticism and psychic experience.*

1982. Slater, George Richard. Ph.D. **Boston University Graduate School.** *Regression of cancer as an instance of Gotthard Booth's concept of Kairos.*

1982. Snel, F. Ph.D. **International University of Lugano, Switzerland.** *Biomedical ethics and the terminally ill: A link with unorthodox medicine.*

1982. Varvoglis, Mario P. Ph.D. **Adelphi University.** *Psychokinesis, intentionality, and the attentional object: Specificity and generality in mind matter.*

1982. Wittine, Bryan. Ph.D. **California Institute of Asian Studies.** *The crises and conflicts of spiritual awakening.* [Includes development of psi ability]

1983. Ballard, John Allen. Ph.D. **Purdue University.** *Affective assessment in a psi task.*

1983. Bowden, Robert Stephen. Ph.D. **Southern Baptist Theological Seminary.** *H.D. Lewis' defense of immortality: A response to the contemporary challenge of analytical philosophy.*

1983. Chappell, Paul Gale. Ph.D. **Drew University.** *The divine healing movement in America.*

1983. Dean, Eric Douglas Miller. Ph.D. **Saybrook Institute** [formerly **Humanistic Psychology Institute**]. *An examination of infra–red and ultra–violet techniques to test for changes in water follow–ing the laying–on of hands.*

1983. Gregory, Anita. Ph.D. **Polytechnic of North London.** *Problems in investigating psychokinesis in special subjects.*

1983. Houtkooper, Joop. Ph.D. **University of Utrecht.** *Observational theory: A research programme for paranormal phenomena.*

1983. Maher, Michaeleen Constance. Ph.D. **City College of the**

City University of New York. *Correlated hemispheric asym—
metry in the sensory and ESP processing of emotional and
nonemotional videotapes*

1983. McLaughlin, Steven Alexander. Ph.D. **Fuller Theological
Seminary**. *Near—death experiences and religion: A further
investigation.*

1983. Selley, April Rose. Ph.D. **Brown University**. *Behind the
veil: The 'voice from beyond' in American literature.*

1983. Spinelli, Ernest. Ph.D. **University of Surrey**. *Human
development and paranormal cognition.*

1984. Bohm, Evelyn Regina. Ed.D. **Columbia University
Teachers College**. *Nonverbal communication between in—
dividuals who share a close emotional bond: "ESP" com—
munication.*

1984. Helene, Nina. Ed.D. **Boston University**. *An exploratory
study of the near—death encounters of Christians.*

1984. Isaacs, Julian. Ph.D. **University of Aston, Birmingham**.
Some aspects of performance at a psychokinetic task.

1984. Pilkington, Rosemarie. Ph.D. **Saybrook Institute** [formerly
Humanistic Psychology Institute]. *The use of fantasy in a
children's ESP experiment.*

1984. Sudhakar, U. Vindhya. Ph.D. **Andhra University**. *Belief
and personality factors of participants: A study in an
ESP/Ganzfeld setting.*

1984. Valois, Winston Regis. Ph.D. **California School of Profes—
sional Psychology, Berkeley**. *The integrative aspects of pos—
session trance.*

1984. Winkelman, Michael. Ph.D. **University of California, Ir—
vine**. *A cross—cultural study of magico—religious prac—
tioners.*

1984. Zaleski, Carol Goldsmith. Ph.D. **Harvard University**.
*Otherworld journeys: A comparative study of medieval
Christian and contemporary accounts of near—death ex—
periences.*

1985. Goldenthal, Mark Howard. Ph.D. **Georgia State Univer—**

sity. *Childhood, development and personality functioning of the psychic.*

1985. Morote—La Torre, Gloria. Ph.D. **The Catholic University of America.** *Psychological and social adjustment as related to perceptions of cross cultural transition among young hispanic immigrants.*

1986. Delanoy, Deborah. Ph.D. **University of Edinburgh, Dept. of Psychology.** *The training of ESP in the Ganzfeld.*

1986. Harary, Keith. Ph.D. **Union Graduate School.** *A critical analysis of experimental and clinical approaches to apparent psychic experiences.*

1986. Milton, Julie. Ph.D. **University of Edinburgh.** *Psychology. Displacement effects, role of the agent, and mentation categories in relation to ESP performance.*

1986. Sanchez—Porter, Regina. Ph.D. **New York University.** *The relationship of empathy, diversity and telepathy in mother—daughter dyads.*

1986. Sondow, Nancy. Ph.D. **City College of the City University of New York.** *The relationship between hypnotizability, creativity, and psi in the ganzfeld.*

1987. Hardy, Christine. Ph.D. **University of Paris VII.** *Ethnology and parapsychology: A double perspective upon trance, altered states of consciousness, and paranormal phenomena.* (In French)

1987. Hess, David. Ph.D. **Cornell University.** *Anthropology. Spiritism and science in Brazil: An anthropological interpretation of religion and ideology.*

1987. Nelson, Peter L. Ph.D. **University of Queensland.** *The technology of the praeternatural: An operational analysis and empirical study of the psychophenomenology of mystical, visionary and remote perception experiences.*

1987. Roney—Dougal, Serena M. Ph.D. **University of Surrey.** *A comparison of psi and subliminal perception.*

1987. Supko, Joanne Dolores. Ph.D. **Saybrook Institute [formerly Humanistic Psychology Institute].** *The effect of the American funeral on the deceased and the bereaved.*

1988. Boerenkamp, Hendricus Godefricus. Ph.D. **University of Utrecht.** *A study of paranormal impressions of psychics.*

1988. McAleer, John P. Ph.D. **University of Ulster.** *A study of altered states of consciousness and psychic phenomena.*

1988. Meyer, Martin. Ph.D. **Saybrook Institute [formerly Humanistic Psychology Institute].** *The role of personality in meaningful coincidences.*

1989. Roll, William George. Ph.D. **Lund University.** *This world or that: An examination of parapsychological findings suggestive of the survival of human personality after death.*

1989. Smith, Susan M. Ph.D. **Union Graduate School.** *Psychics and mediums: The role of religion and belief in rationalizing paranormal phenomena.*

MASTER'S THESES

1931. Frick, Harvey L. M.A. **Duke University.** *Extrasensory cognition.*

1937. Kubis, Joseph F. M.S. **Fordham University.** *An experimental investigation of telepathic phenomena in twins.*

1937. Pegram, Margaret H. M.S. **Duke University.** *Some psychological relations of extrasensory perception.*

1937. Rouke, Fabian L. M.S. **Fordham University.** *An experimental investigation of telepathic phenomena in twins.*

1937. Smith, Burke M. M.S. **Duke University.** *The effect of benzedrine sulfate on extrasensory perception.*

1938. Crumbaugh, James C. M.A. **Southern Methodist University.** *A questionnaire designed to determine the attitudes of psychologists toward the field of extrasensory perception.*

1938. Rothera, Ralph E. M.A. **State Teachers College, Fitchburg.** *A study in extrasensory perception.*

1938. Taves, Ernest H. M.A. **Columbia University.** *An exploratory study of extrasensory perception.*

1939. Woodruff, Joseph L. M.A. **Duke University.** *Size of stimulus symbols in extrasensory perception.*

1940. Hutchinson, Lois. M.A. **Duke University.** *Variation of time intervals in pre—shuffling card—calling tests.*

1949. Eilbert, Leo. M.S. **City College of the City University of New York.** *A study of certain psychological factors in ESP performance.*

1951. Weinstein, Beatriz. M.A. **University of Chile.** *Parapsychology or extrasensory perception based on the Duke University investigations.* (In Spanish)

1952. Palmer, F. Claude. M.A. **University of London.** *Application of psychoanalytic techniques to ESP testing.*

1952. Powar, L.N. M.A. **Mysore University.** *Studies in psychokinesis.*

1952. Van de Castle, Robert L. M.A. **University of Missouri.** *Meaningfulness as a variable in ESP experiments.*

1955. Goba, Teresa C. M.A. **University of the Philippines, Manila.** *A study of drugless healing.*

1955. Gupta, Sita Ram. M.A. **Benares Hindu University.** *Precognitive elements in extrasenory perception.*

1955. Rao, K. Ramakrishna. M.A. **Andhra University.** *Paranormal cognition: An essay in survey of evidence and theories.*

1957. Gerber, Rebecca. M.S. **City College of the City University of New York.** *An investigation of relaxation and of acceptance of the experimental situation as related to ESP scores in maternity patients.*

1957. Isaacs, Ernest. M.A. **University of Wisconsin.** *A history of spiritualism: The beginnings 1845—1855.*

1958. Deguisne, Arnon. M.S. **Macmurray College.** *Two repetitions of the Anderson—White investigation of teacher—pupil attitudes and clairvoyance test results: High School Tests.*

1958. Goldstone, Gerald. M.S. **Macmurray College.** *Two repetitions of the Anderson—White investigation of teacher—pupil* attitudes and clairvoyance test results: Grade school tests.

1958. Waldron, Sherwood. M.S. **Harvard University.** *Clair-voyance of sheep versus goats when subjects' attitudes towards the experiment are manipulated.*

1959. Jach, Krzysztof. M.S. **Academic Polytechnic School, Warsaw.** *An Investigation of telepathy based on the hypothesis of electromagnetic wave transmission.* (In Polish)

1960. Perea, A. Cassigoli. M.A. **University of Chile.** *Concerning psi phenomena.* (In Spanish)

1961. Lanctot, John. M.S. **City College of the City University of New York.** Title unknown.

1962. Davis, Kenneth R. M.S. **University of Denver.** *The relationship of suggestibility to ESP scoring level.*

1964. Elguin, Gita H. M.S. **University of Chile.** *Psychokinesis in experimental tumerogenesis.*

1965. Moses, Helen. M.A. **City College of the City University of New York.** *A study of task-orientation and personality factors related to ESP performance.*

1966. Eastman, Margaret. M.A. **City College of the City University of New York.** *The relationship of ESP scores to knowledge of target location, and to birth order and family size.*

1968. Dagel, Lou. M.S. **Trinity University.** *A study utilizing reinforcement in a two-choice general ESP test situation.*

1970. Plug, C. M.A. **University of South Africa.** *A study of the psychological variables underlying the relationship between ESP and extraversion.*

1971. Eisenberg, Howard. M.S. **McGill University.** *Telepathic information transfer in humans of emotional data.*

1971. Johannesson, Goran. M.A. **Lund University.** *An attempt to control scoring direction by means of treatment of the subject.*

1971. Johnson, Rex V. M.S. **University of Washington.** *The influence of temperature and humidity on the low-frequency* capacitance and conductance across a philodendron leaf (a study of the Backster effect).

1971. Peterson, Clifton J. M.S. **Wisconsin State University.** *Relationships among attitudes, memory, and extrasensory perception.*

1972. Bopaiya, M. Sumathi. M.A. **Andhra University.** *Alpha rhythym, hypnosis, personality characteristics, and extrasensory perception*

1972. Brietstein, Howard. M.A. **City College of the City University of New York.** *Meditation, mood and ESP: A comparison of experienced and inexperienced mediators.*

1972. Gildmeister, Glen A. M.A. **Southern Illinois University.** *American spiritualism 1845–1855.*

1972. Wright, Linda Frances. M.A. **United States International University.** *A preliminary investigation into alpha traing and the relationship to intelligence, personality, ESP and subject response.*

1973. Allison, Paul D. M.A. **University of Wisconsin.** *Social aspects of scientific innovation: The case of parapsychology.*

1973. France, Gary A. M.S. **Illinois State University.** *Thought concordance in twins and siblings and associated personality variables.*

1973. Mason, Jeffery Adams. M.A. **United States International University.** *An experimental study of creativity and psi.*

1973. Warrick, Tom P. M.A. **University of Hawaii.** *Health implications of parapsychology and altered states of consciousness.*

1974. Artificio, Mariza V. M.A. **University of the Philippines, Manila.** *Union espiritista Christiana de Filipinas: A case study of faith healing.*

1974. Mackintosh, S. M.A. **University of Edinburgh.** *Skin resistance as a measure of ESP response to emotive stimuli.*

1974. Peterson, James. M.A. **University of California, Berkeley.** *Some profiles of non–ordinary perception of children.*

1974. Tobia, Cristina H. M.A. **University of the Philippines, Manila.** *Mass media exposure and the attitude of Loyola Heights housewives toward spiritual healing.*

1975. Ballard, John Allen. M.A. **University of Southern Mississippi.** *Unconscious perception of erotic, non–erotic, and neutral stimuli on a psi task.*

1975. Friedman, Ronald M. M.S. **Northeastern University.** *Intuitive decision making.*

1975. Gochoco, Victoria Angeles. M.A. **University of the Philippines, Manila.** *A study of the factors that effect the attitude of IMC students toward spiritual healing.*

1975. Gordon, Michael D. M.S. **Manchester University.** *The institutionalization of parapsychology: A study of innovation in science.*

1975. Greist, Walter. M.S. **University of Kentucky.** *Psi–speech communication system theory.*

1975. Hearne, K.M.T. M.S. **Hull University.** *Visual imagery and evoked responses.*

1975. Hebda, Hillard. M.A. **Governors State University.** *An inquiry into unorthodox healing: Psychic healing and psychic surgery.*

1976. Andrew, Edward Kennard. M.A. **Austin State University.** *Effects of an auditory tape presentation on alpha production in the cerebral hemispheres and possible effects on a parapsychological task.*

1976. Cameron, Constance C. M.A. **California State University, Fullerton.** *Archaeology and parapsychology.*

1976. Fishman, R.G. M.A. **State University of New York, Buffalo.** *Spiritualism, mediumship and faith healing: An adjunct to the medical profession.*

1976. Gambale, John. M.A. **City College of the City University of New York.** *Word frequency and associated strength in memory–ESP interaction: A failure to replicate.*

1976. O'Brien, Dennis P. M.A. **Connecticut College.** *Recall and* *recognition processes in a memory—ESP task.*

1976. Velissaris, Cynthia. M.A. **Governors State University.** *Experience—memory factors as indicators of psi rapport interaction.*

1976. Venturino, Michael R. M.A. **State University of New York, Geneseo.** *An investigation of the relationship between feedback—augmented EEG alpha activity and ESP performance.*

1976. Wollman, Neil Jay. M.A. **California State University, Fullerton.** *The effects of the experimenter in ESP research.*

1977. Collymore, James L. M.A. **Ball State University.** *The effects of visual and acoustic ganzfields in a forced choice ESP task: A study with auditory targets.*

1978. Bray, James David. M.A. **Naval Postgraduate School, Monterey, California.** *Questionnaire results on the prospects for Soviet development of parapsychology for military or political purposes.*

1978. Pekarcik, Karin Juua. M.A. **California State University, Fullerton.** *An ESP study analyzed in terms of the social psychology of the experient.*

1978. Peterson, D.M. M.A. **University of Edinburgh.** *Through the looking—glass: An investigation of the faculty of extrasensory detection of being stared at.*

1978. Rayet, Phillippe. M.A. **Bordeaux University.** *Mediums and healers in Britain.* (In French)

1979. Dunne, Brenda J. M.A. **University of Chicago.** *Precognitive remote perception.*

1979. Henegan, A. Hilary. M.Phil. **University of Edinburgh.** *ESP ability in young children.*

1979. Maher, Michaeleen Constance. M.A. **City College of the City University of New York.** *Cerebral lateralization effects in ESP processing.*

1979. Neppe, Vernon Michael. M.Med. **University of Witwatersrand.** *Investigation of the relationship between subjective paranormal experience and temporal lobe symptomatol—*

ogy.

1980. Biscop, P.D. M.A. 1980. **Simon Fraser University.** *By spirit possessed: A field exploration of some models for spirit possession among local spiritualist mediums.*

1980. Clancy, Patricia M. M.A. **University of Humanistic Studies, San Diego.** *An investigation of apparent paranormal sounds occurring on electronic recording equipment.*

1980. Knox, M. Crawford. M.S. **University of Surrey.** *The body—mind relationship and parapsychology.*

1980. Schweitzer, Susan Fredericka. M.S. **University of Nevada (Reno).** *The effects of therapeutic touch on short—term memory recall in the aging population: A pilot study.*

1980. Stapelberg, Monica—Maria. M.A. **University of Pretoria.** *Parapsychological aspects in the "Edda."* (In German)

1981. Alvarado, Carlos S. M.S. **John F. Kennedy University.** *ESP and out—of—the—body experiences: A spontaneous case survey.*

1981. Auerbach, Loyd N. M.S. **John F. Kennedy University.** *Fantasy role playing games as an ESP test strategy.*

1981. Birdsell, Pearl Griffin. M.A. **Holy Names College.** *An invesitgation into the opinions of a group of magicians regarding the relationship of the occult to modern magic.*

1981. Coppolino, Darlene Ann. M.S. **California State University, Long Beach.** *A pilot study on the interrelationships among ESP, creativity, age, sex, race, and birth order in a group of normal children.*

1981. Dempsey, Tom Gene. M.S. **California State University, Long Beach.** *The use of psychics by police as an investigative aid: An examination of current trends and potential applications of psi phenomena.*

1981. Farabee, Charles Roscoe, Jr. M.S. **California State University.** *Contemporary psychic use by police in America.*

1981. Honegger, Barbara S. M.S. **John F. Kennedy University.** *An experimental test of Harvey Irwin's structural—stratum hypothesis of psi—mediated long—term memory trace activa—*

tion using ambiguous figures.

1981. Irwin, Carol Plumlee. M.S. **John F. Kennedy University.** *The role of memory in free—response ESP studies: Is tar—get familiarity reflected in the scores.*

1981. Quider, Robert—Peter Francis. M.S. **John F. Kennedy University.** *Suggestology and parapsychology: The Lozanov method as a psi—conducive procedure.*

1981. Tchen, Yumin. M.A. **University of Edinburgh.** *Experiments in subliminal and extrasensory perception.*

1982. Giesler, Patric. M.S. **John F. Kennedy University.** *Multimethod study of psi and psi—related processes in the Umbanda ritual trance consultation.*

1982. Gonzalez—Scarano, Frances. M.S. **John F. Kennedy University.** *ESP and imagery in the sighted and the blind.*

1982. Kriesa, C. M.A. **Sonoma State University.** *The influence of distance on psychokinesis as measured by yeast metabolism.*

1982. MacDonald, Jeffery L. M.A. **Western Washington University, Bellingham.** *Transpersonal anthropology: History, theory, and method.*

1982. Myers, Susan Anne. M.S. **Saint Louis University.** *Personality characteristics as related to out—of—body experiences.*

1983. Hobson, Douglas. M.A. **Baylor University.** *A comparative study of near—death experiences and Christian eschatology.*

1983. Lang, Kathryn Holmes. M.S. **California State University, Sacramento.** *A system of training for use in transpersonal counseling.*

1983. Murray, Diane M. M.S. **John F. Kennedy University.** *A survey of reported psi and psi—related experiences in the Philippines among the Isnag of the Kalinga—Apayo province, & Sampaloc residents of Manila.*

1983. Ruiz, Joyce G. M.S. **John F. Kennedy University.** *Metabolic types and psi: A study of oxidizer type and ESP test scores.*

1983. Siegel, Cynthia E. M.A. **John F. Kennedy University.** *Experiences and changes in belief among PK party participants.*

1983. Taylor, G.G. M.Phil. **Reading University.** *Precognition: A philosophical and parapsychological enquiry.*

1983. Wallack, Joseph Michael. M.A. **West Georgia College.** *Testing for a psychokinetic effect on plants: Effect of "laying on of hands" on germinating corn seeds.*

1984. Munson, R. Jeffrey. M.S. **University of South Africa.** *Psychological perspectives on psi phenomena.*

1985. Davis, Lorraine. M.S. **John F. Kennedy University.** *How the unidentified flying object experience (UFOE) compares with the near—death experience (NDE) as a vehicle for the evolution of human consciousness.*

MISCELLANEOUS DEGREES

1959. Roll, W.G. B.Litt. **Oxford University.** *Theory and experiment in psychical research.*

1960. Keil, H.H.J. B.A. (Honors Thesis). **University of Tasmania.** *An investigation into the possibilities of advancing psychokinesis.*

1960? Ryzl, Milan. C.Sc. **Czechoslovakian Academy of Science.** *Theoretical basis for psi.* (In Czech)

1961. Green, Celia E. B.Litt. **Oxford University.** *An inquiry into some states of consciousness and their physiological foundations, with special reference to those in which ESP is reported to occur.*

1963. Rhally, Miltiade. Dipl.The. **C.G. Jung Institute.** *Analytic relationship and parapsychology.*

1976. Thalbourne, Michael Anthony. B.A. (Honors Thesis). **University of Adelaide.** *Closeness of relationship, and telepathy, personality and social intelligence.*

1981. Funkhouser, Arthur T. Dipl.The. **C.G. Jung Institute.** *Déjà vu: Déjà reve.*

1982. Williams, Linda. B.A. (Honors Thesis). **University of Adelaide.** *The feeling of being stared at: A parapsy—chological investigation.*

VIII
Glossary of Terms

The terms selected for definition below are either used in the main subject listings or appear in the titles of the books. Any term in those sources that it was thought might be un— familiar or unclear to the reader was included in this glossary. Also included are a number of terms commonly used in ex— perimental parapsychology. These terms may not appear in this book, but they are terms anyone interested in parapsychology should know. Note that terms followed by an asterisk are quoted by permission from the glossary in each volume of the *Journal of Parapsychology*. Most of them were borrowed or adapted from *A Glossary of Terms Used in Parapsychology* by Michael Thalbourne (London: Heinemann Ltd., 1982) [687]. Thalbourne's book is recommended highly for those who need an authoritative and comprehensive dictionary of terms used in parapsychology.

Absent sitting *see* **Proxy sitting.**

Agent.* In a test of GESP, the individual who looks at the in— formation constituting the target and who is said to "send" or "transmit" that information to a percipient; in a test of telepathy and in cases of spontaneous ESP, the individual about whose mental states information is acquired by a per— cipient. The term is sometimes used to refer to the subject in a test of PK.

Altered states of consciousness. Individual subjective distinction in which one clearly feels a qualitative shift in one's pattern of mental functioning that is different from one's usual state of consciousness.

Analytical psychology. The school of psychology founded by the Swiss psychologist, C.G. Jung.

Anomalies *see* **Anomalous phenomena.**

Anomalistic psychology. Scientific approach to those phenomena

580

of psychology that do not fit the current scientific world view (i.e., paradigm) by the criteria followed by the majority of psychologists as well as extraordinary phenomena that at least in part are explicable in terms of known psychological principles.

Anomalous phenomena. Any behavior, experience, or event that appears to violate any of the traditional basic limiting principles concerning reality.

Anpsi. General term for any form of psi exhibited by an animal. *See also* **Psi-trailing.**

Apparition. A visual appearance, usually not repeated, which suggests the presence of a deceased person or animal, or of a living person or animal outside the sensory range of the percipient. *See also* **Hallucination; Haunt or Haunting.**

Application of psi *see* **Applied psi.**

Applied psi. The deliberate use of psi for practical purposes. For example, *see* **Criminal investigation and psi** and **Military applications of psi.**

Astral projection *see* **Out-of-body experience.**

Automatic writing. Unconscious writing. *See also* **Ouija board.**

Backward PK *see* **Retroactive PK.**

Call.* (As noun), the overt response made by the percipient in guessing the target in a test of ESP; (as verb), to make a response.

Channel. Modern term for **Medium.**

Channeling. Ability to turn inward to connect intuitively with a universal source of insight and information beyond conscious awareness that is conveyed by means of writing or speaking to others, sometimes through the emergence of another personality or entity (i.e., **Control,** *which see*). Modern term for **Mediumship.**

Clairvoyance. Awareness of objects or objective events outside the sensory range of the percipient and of the consciousness of other persons. *See also* **Extra ocular vision** and **Remote viewing.**

Clairvoyant diagnosis *see* **Paranormal diagnosis.**

Closed deck.* A procedure for generating the target order for each run, not by independent random selection of successive targets, but by randomization of a fixed set of targets (e.g., a deck of 25 ESP cards containing exactly five of each of the standard symbols).

Collective hallucination. Hallucination experienced by two or more persons at the same time. *See also* **Reciprocal hallucination.**

Communicator. A personality who usually communicates through a medium and purports to be a deceased personality.

Confidence call.* A response the subject feels relatively certain is correct and indicates so before it is compared with its target.

Conformance behavior model. Theory of R.G. Stanford, who proposed that if a disposed organism (the subject) is contingently related to a random process (e.g., an REG in a PK experiment or the brain in an ESP experiment), in which one of the options offered by the REG serves a need of the organism, then the REG will be disposed toward that option.

Control. In mediumship, the "personality" who habitually relays messages from the **Communicator** (*which see*) to the **Sitter** (*which see*).

Criminal investigation and psi. The use of psi to obtain information useful in solving criminal investigations.

Cross correspondence. A series of independent communications through two or more mediums such that the complete message is not clear until the separate fragments are put together.

Cumberlandism *see* **Muscle-reading.**

Deathbed observations. Observations made by second parties of unusual experiences of dying persons such as visions, life review, out—of—body experiences, or hallucinations of persons not present or deceased.

Decline effect.* The tendency for high scores in a test of psi to decrease, either within a run, within a session, or over a longer period of time; may also be used in reference to the waining and disappearance of psi talent.

Déjà vu. An illusion of memory in which a person experiences an event as if he or she had experienced it before, when in reality he or she is experiencing it for the first time.

Development of psi *see* **Training psi.**

Diagnosis, paranormal *see* **Paranormal diagnosis.**

Differential effect. In psi experiments that employ contrasting conditions, the tendency of subjects to score above chance in one condition and below chance in the other.

Direct voice. In mediumship a voice not produced by the medium's vocal cords (sometimes) issuing from a trumpet that floats about the room or other such prop, but often without any. *See also* **Electronic voice phenomena.**

Displacement.* A form of ESP shown by a percipient who con-sistently obtains information about a target that is one or more removed, spatially or temporally, from the actual tar-get designated for that trial
Backward displacement: Displacement in which the target ex-trasensorially cognized precedes the intended target by one, two, or more steps (designated as −1, −2, etc.).
Forward displacement: Displacement in which the target ac-tually responded to occurs later than the intended target by one, two, or more steps (designated as +1, +2, etc.).

Divination. The use of various practices such as tea leaf read-ing, palmistry, scrying, *I Ching*, tarot cards, etc., in order to discern hidden knowledge.

Doctrinal compliance. Term introduced by psychiatrist J. Ehren-wald for the situation in psychotherapy wherein patients have a tendency to produce dreams, associations, and other material that validate the therapist's individual theories about therapy, sometimes by means of psi. This can also be noted in psi experiments where subjects tend to score in the direction desired by the experimenter even if they have no sensory means of discovering the experimenter's wishes. *See also* **Experimenter effect.**

Double. A replica of oneself, perceivable by oneself or alleged by others. *See also* **Out-of-body experience.**

Dowsing. A form of motor automatism in which a divining rod (forked twig or other device) is used to locate underground water, oil, or other concealed items by following the direction in which the rod turns in the user's hands.

Dream, lucid *see* **Lucid dream.**

Dual aspect target. A target with two features in which the subject can respond to both, e.g., in the case of playing cards, color and number.

Dual task. An ESP or PK task the purpose of which is to contrast the results under two conditions.

Electronic voice phenomena (EVP). Voices, allegedly from discarnate persons, impressed electronically on tape recordings made on standard apparatus and also sometimes audible on the "white noise" of certain radio bands.

ESP *see* **Extrasensory perception.**

ESP cards.* Special cards, introduced by J.B. Rhine, for use in tests of ESP; a standard pack contains 25 cards, each portraying one of five symbols, viz., circle, cross, square, star, and waves.

Experient. One who has a spontaneous psi experience.

Experimenter effect. Term for the fact that some experimenters in parapsychological experiments consistently obtain significant results, regardless of the hypothesis being tested, and others do not. Or, different experimenters can obtain opposite but significant results. Three main hypotheses have been proposed to account for these results: experimenter error, including artifacts, subject deception, and experimenter deception; psychological factors, in which one experimenter may be more skillful than another in creating the requisite psychological conditions under which subjects tend to be successful scorers; and experimenter psi, in which the experimenter's own psi is considered to be responsible for or influence the subject's results.

Extraocular vision. The ability to perceive veridical visual images without the use of the eyes or any external detection devices. *See also* **Clairvoyance.**

Extrasensory perception (ESP). Knowledge of or response to an external event or influence not obtained via known sensory channels.

Faith healing *see* **Diagnosis, paranormal; Unorthodox healing.**

Fantasy. Awareness of images produced by the imagination as opposed to images produced by sensory perception. Both dreams and daydreams are forms of fantasy.

Feedback. In psi tests notification of the subject in an experiment of the nature of the target to which he or she has responded. Feedback may be immediate (right after the subject has made his or her call/query) or delayed (at the end of a run, day, series, or other prespecified period of time).

Forced-choice test.* Any test of ESP in which the percipient is required to make a response that is limited to a range of possibilities known in advance.

Foreknowledge *see* **Precognition.**

Free material *see* **Free-response test.**

Free-response test.* Any test of ESP in which the range of possible targets is relatively unlimited and is unknown to the percipient, thus permitting a free response to whatever impressions come to mind.

Ganzfeld.* Term for a special type of environment (or the technique for producing it) consisting of homogenous, unpatterned sensory stimulation; an audiovisual ganzfeld may be accomplished by placing halved ping—pong balls over each eye of the subject, with diffused light (frequently re in hue) projected onto them from an external source, together with the playing of unstructured sounds (such as "pink noise") into the ears.

Geller effect. Phenomena associated with Uri Geller, specifically *metal bending* (*which see*) and the anomalous stopping or starting of watches or clocks.

General extrasensory perception (GESP).* A noncommittal technical term used to refer to instances of ESP in which the information paranormally acquired may have derived either from another person's mind (i.e., as telepathy), or from a physical event or state of affairs (i.e., as clairvoyance, or even from both sources.

Ghost. *See* **Apparition; Haunt; Poltergeist.**

Glossolalia. Speaking in an unknown or fabricated language ("speaking in tongues"), usually in a religious context. *See also* **Xenoglossy.**

Goal-oriented. * Term for the hypothesis that psi accomplishes a subject's or experimenter's objective as economically as possible, irrespective of the complexity of the physical system involved.

Hallucination. An experience with the qualities of sense perception (visual, tactile, olfactory, auditory, etc.) but that takes place in the absence of sensory stimulation. *See also* **Apparition, Collective hallucination, Reciprocal hallucination,** and **Veridical hallucination.**

Haunt or Haunting. The occurrence of apparently paranormal auditory and/or visual phenomena usually in connection with a specific locale and on a more or less regular basis. The phenomena are usually attributed to a deceased entity.

High-scoring subject. Person who consistently obtains statistically significant results in an ESP or PK experiment.

Immortality. An a priori assumption of the reality of everlasting existence. Contrast with **Survival research** (q.v.)

Intuition. Direct knowledge, insight, cognition, conviction without rational thought or inference.

Levitation. The raising of objects or persons, animals, or other organisms off the ground without the use of muscles or other normal means.

Lucid dream. A dream in which the dreamer is aware that he or she is dreaming.

Macro-PK. * Any psychokinetic effect that does not require statistical analysis for its demonstration; sometimes used to refer to PK that has as its target a system larger than quantum mechanical processes, including microorganisms, dice, as well as larger objects.

Magic. The use of rituals and other techniques to influence the dreams, thought processes, and actions of others at a distance.

Majority-vote technique (MV).* The so—called repeated or multiple guessing technique of testing for ESP. The symbol most frequently called by a subject (or a group of subjects) for a given target is used as the "majority—vote" response to that target on the theory that such a response is more likely to be correct than one obtained from a single call.

Materialization. A manifestation of physical mediumship in which supposed human or animal entities, or sometimes inanimate objects, become temporarily visible in various stages of solidity.

Mean chance expectation (MCE).* The average (or "mean") number of hits, or the most likely score to be expected in a test of psi on the null hypothesis that nothing apart from chance is involved in the production of the score.

Medium. A person who regularly receives communications purportedly coming from deceased persons or who produces paranormal physical effects sometimes attributed to discarnate agency.

Mediumship. The experiences and behavior involved in being a medium.

Meta-analysis. A statistical technique that is a form of exploratory data analysis that enables researchers to look for patterns across existing reports in a specific research area (e.g., Ganzfeld experiments, hypnosis ESP experiments, REG experiments, etc.).

Metal bending. Attempts to bend or otherwise alter the composition of metal not explicable in terms of known physical forces.

Micro-PK.* Any psychokinetic effect that requires statistical analysis for its demonstration. Sometimes used to refer to PK that has as its target a quantum mechanical system.

Multiple personality. Condition in which a person shifts from one personality to another (or others) that is often very different from the original personality. Each personality has no memory of the thoughts and actions of the others.

Muscle-reading. A type of "pseudo—telepathy" by means of which a person can locate a hidden person or object by means of subtle muscular cues provided by touching a person who knows the whereabouts of the person or object, even though the latter makes a conscious effort not to provide sensory clues. Also called Cumberlandism.

Mystical experience. Experience of the unity of all things and of contact with the divine or awareness of a benevolent non—physical power partly or entirely beyond and far greater than the individual self.

NAD. Sanskrit term for celestial or transcendental or paranormal music, e.g., the "music of the spheres."

NDE *see* **Near-death experience.**

Near-death experience (NDE). Subjective experience of a person physically close to death, that is, in a bodily state resulting from an extreme physiological catastrophe that would reasonably be expected to result in irreversible biological death in most instances, as in cardiac arrest, severe traumatic injury, or deep comatose situations.

Non-ordinary experience. Any experience which the subject perceives as different in quality from his or her ordinary conscious experience.

OBE *see* **Out-of-body experience.**

Object-reading *see* **Psychometry.**

Observational theory. Theory that the observer of psi test results influence those results in a fashion that itself involves psi.

Occult. Hidden or secret.

Occult medicine *see* **Unorthodox healing.**

Occultism. Belief in or the study of hidden powers, usually in connection with a metaphysical school or teaching, with a view to bringing them under control.

Open deck.* A procedure for generating a target order in which each successive target is chosen at random independently of all the others; thus, for example, in the case of a standard deck of ESP cards whose target order is "open deck," each type of symbol is not necessarily represented an equal num— ber of times.

Oracles. In ancient times a person through whom a deity was believed to speak or otherwise reveal hidden knowledge or the divine purpose.

Ouija board. The user touches with the fingertips a small board with a pointer that rests on a larger board marked with numbers and letters. By means of involuntary muscular movements the smaller board moves about, pointing at various letters or numbers to spell out a message.

Out-of-body experience (OBE). An experience in which one's center of consciousness seems to be in a spatial location separate from that of one's physical body.

Paranormal. A synonym for psychic or parapsychological; "para" means beyond what should occur if only the known laws of cause and effect were operating.

Paranormal diagnosis. Using clairvoyance or other means than those recognized by medical science to perceive diseased bodily organs or states for medical purposes.

Paranormal electrical effects. Spontaneous physical effects detected near psychic subjects.

Parapsychological phenomena. Phenomena characterized as being **paranormal** (*which see*). *See also* **Psi.**

Parapsychology. The study of behavior that cannot now be ex— plained or described in terms of known physical principles; modern term for psychical research.

Parascientific. Beyond the reach of the scientific method.

Percipient.* The individual who experiences or "receives" an ex— trasensory influence or impression; also, one who is tested for ESP ability.

Phantom *see* **Apparition.**

Phenomenology. The investigation of the subjective processes by means of which phenomena are presented. Also the study of occurrences or phenomena directly as they are experienced without superimposing an interpretation from outside.

Philosophy of science. Application of philosophical principles to the study of science as well as the study of the assumptions underlying science.

PK *see* **Psychokinesis.**

Poltergeist. Phenomena that involve the unexplained movement or breakage of objects, etc., and that often appear to be centered around the presence of an adolescent. They differ from haunts in that apparitions are rarely associated with them.

Position effect (PE).* The tendency of scores in a test of psi to vary systematically according to the location of the trial on the record sheet.

Possession. State of consciousness in which a person's normal personality is replaced by another. *See also* **Multiple personality.**

Precognition. Knowledge of a future event that could not have been learned or inferred by normal means.

Process-oriented.* Term for research whose main objective is to determine how the occurrence of psi is related to other factors and variables.

Proof-oriented.* Term for research whose main objective is to gain evidence for the existence of psi.

Proxy sitting. A sitting with a psychic or medium in which someone other than the person desiring information is present instead.

Pseudoscience. A body of assumptions, methods, and theories that are erroneously assumed to be scientific.

Psi. A general term for any or all phenomena that cannot be explained in terms of sensorimotor contact with the environment. Thus, psi includes both ESP and PK.

Psi-hitting.* The use of psi in such a way that the target at which the subject is aiming is "hit" (correctly responded to in a test of ESP, or influenced in a test of PK) more fre-

quently than would be expected if only chance were operat—
ing.

Psi-missing. In ESP or PK tests, the use of psi so that the tar—
get at which the subject aimed is missed significantly more
often than would be expected by chance.

Psi-trailing. The ability of a pet to "trail" its owner to a dis—
tant location where it has never been before and under cir—
cumstances in which normal means of tracking (e.g., by
scent) could not have been used.

Psilocybin. A hallucinogenic drug obtained from a fungus.

Psiology. Pop term for **Parapsychology**.

Psychic. As a noun: A person who has frequent psi experiences,
occasionally at will. As an adjective: Used as a synonym for
Psi.

Psychic criminology *see* **Criminal investigations and psi.**

Psychic healing *see* **Unorthodox healing.**

Psychic photography. The production of mental images on
photographic plates or film by apparently paranormal
means.

Psychic warfare *see* **Military applications of psi.**

Psychical research. Older term for **Parapsychology** (*which see*).

Psychodynamics. Psychological principles of behavior.

Psychokinesis (PK). Influence of objects or processes outside the
organism without the use of known physical energies or
forces.

Psychometry. The use of an object by a sensitive to obtain by
extrasensory means impressions relating to its history or per—
sons who have been in touch with it or associated with it.

Qualitative material *see* **Free-response material.**

Random event generator (REG). Electronic (usually) device that
employs an element capable of producing a random sequence
of outputs for use as targets in ESP and PK tests.

Reciprocal hallucination. Hallucination experienced in part or fully by two or more persons who are out of physical touch with each other. *See also* **Collective hallucination.**

Recitative xenoglossy *see* **Xenoglossy.**

Recurrent spontaneous psychokinesis (RSPK). Spontaneous physical effects (psychokinesis) that cannot be explained by any known physical energies and that recur over a period of time, usually in association with a particular person (poltergeist) or place (haunting).

REG *see* **Random event generator.**

Reincarnation. A form of survival in which the mind, or some aspect of it, is reborn in another body.

Remote action. Modern term for **Psychokinesis** (*which see*).

Remote viewing. The ability to view, by mental means, remote geographical or other targets. *See also* **Clairvoyance.**

Rescue circle. Term used by spiritualists in which sitter groups try to contact the dead for purposes of freeing them from an earthbound state.

Response bias.* The tendency to respond or behave in predictable, nonrandom ways.

Responsive xenoglossy *see* **Xenoglossy.**

Retro-PK *see* **Retroactive PK.**

Retroactive PK. PK that produces an effect backward in time.

Retrocognition. Knowledge of a past event that could not have been learned or inferred by normal means.

Retro-PK *see* **Retroactive PK.**

RNG (random number generator) *see* **Random event generator.**

RNG PK. A PK test in which a random event generator itself is the target system that a subject is asked to influence.

RSPK *see* **Recurrent spontaneous psychokinesis.**

Run.* A fixed group of successive trials in a test of psi.

Séance *see* **Sitting.**

Second sight *see* **Clairvoyance.**

Sensitive *see* **Psychic.**

Session. A (usually) predetermined number of trials carried out during a single test period. An experiment may consist of one or more sessions. In mediumship, the term used for session is **sitting** (q.v.).

Sheep-goat effect (SGE).* The relationship between one's acceptance of the possibility of ESP's occurrence under the given experimental conditions and the level of scoring actually achieved on that ESP test; specifically, the tendency for those who do not reject this possibility ("sheep") to score above chance and those who do reject it ("goats") to score below chance.

Sitter. A person who is present with a psychic or medium usually for the purpose of obtaining information for him- or herself or someone else. (In regard to the latter, *see* **Proxy sitting.**)

Sitter group. Group of persons who sit together regularly usually around a table for purposes of inducing PK phenomena. Modern term for **Table-tipping.**

Sitting. An interview with a psychic or medium. *See also* **Proxy sitting; Session.**

Sixth sense. Older term for ESP.

Sorcery. The practice of **Magic** (*which see*).

Speaking in tongues *see* **Glossolalia.**

Spirit photography. The projection of images, usually self-portraits, on film or photographic plates purportedly by the activity of deceased persons. *See also* **Psychic photography.**

Spiritism. French word for **Spiritualism.**

Spiritual healing. A form of **Unorthodox healing** (*which see*) that is based on religious faith.

Spiritualism. A religion with doctrines and practices based on the belief that survival of death is a reality and that com—munication between the living and the deceased occurs, usually by means of mediumship.

Spontaneous case *see* **Spontaneous psi experience.**

Spontaneous psi experience. An unanticipated experience of ESP or PK that occurs in the course of daily living.

Stacking effect. A spuriously high (or low) score in a test of ESP when two or more percipients make guesses in relation to the same sequence of targets; it is due to a fortuitous relationship occurring between the guessing biases of the per—cipients and the peculiarities of the target sequence.

State of consciousness *see* **Altered states of consciousness.**

Subject *see* **Percipient.**

Subjective paranormal experience. An experience that is as—sumed to be paranormal by the experient even though this conclusion has not been verified objectively.

Supernatural assault. Modern term for the incubus experience, or nightmare experience of someone, usually assumed to be an evil spirit, lying upon persons while they are asleep.

Superstition. A practice or belief that is based on ignorance or trust in chance or magic.

Survival research. The attempt to find empirical evidence that there is continued existence in disembodied form after bodily death. Contrast with **Immortality** (q.v.). *See also* **Reincarnation.**

Synchronicity. Term adopted by C.G. Jung for the simultaneous occurrence of a given subjective state and an external event which serves as a meaningful parallel to the subjective state, and vice versa, in the absence of any causal connection be—tween the two.

Table-tipping. A form of motor automatism in which several persons place their fingertips on a table top, causing it to move and rap out messages by means of a code. *See also*

Sitter group.

Target.* In a test of ESP, the object or event that the per—cipient attempts to identify through information paranor—mally acquired; in a test of PK, the physical system, or a prescribed outcome thereof, that the subject attempts to in—fluence or bring about.

Telekinesis. Older term for **Psychokinesis** (*which see*), still preferred in the USSR and Eastern Europe.

Telepathic impressions. Subjective imagery whose source is out—side the individual's organism.

Telepathy. Extrasensory awareness of another person's mental content or state.

Telesthesia *see* **Clairvoyance.**

Thought transference *see* **Telepathy.**

Thoughtography *see* **Psychic photography.**

Time displaced PK *see* **Retroactive PK.**

Token object. An object associated with a particular person or place that is held by the subject to provide a focus for and facilitate receiving impressions in psychometry tests.

Tracer effect. The appearance in the manifest content of a dream, hallucination, or other fantasy and distinctive fea—tures relating to a target event (Ehrenwald).

Training psi. The use of systematic programs to enable persons to improve their ability to use psi on demand.

Transpersonal. That which extends beyond the individual or per—sonal.

Trial.* An experimentally defined smallest unit of measurement in a test of psi: in a test of ESP, it is usually associated with the attempt to gain information paranormally about a single target; in a test of PK, it is usually defined in terms of the single events to be influenced.

Unorthodox healing. Healing by nonmedical means (e.g., by prayer, the "laying on of hands," etc.) and inexplicable in terms of present—day medical knowledge. *See also* **Paranormal diagnosis**.

Variance.* A statistic for the degree to which a group of scores are scattered or dispersed around their average; formally, it is the average of the squared deviations from the mean; in parapsychology, the term is often used somewhat idiosyncratically to refer to the variance around the theoretical mean of a group of scores (e.g., MCE) rather than around the actual, obtained mean.
Run—score variance: The variance around the mean of the scores obtained on individual runs.
Subject variance: The variance around the mean of the subject's total score.

Verbal material *see* **Free-response test**.

Veridical hallucination. A hallucination (*which see*) that cor—responds to events or circumstances not sensorially available or otherwise known to the percipient.

Water witching *see* **Dowsing**.

Xenoglossy. The act of speaking in a recognized foreign language not normally learned by the subject; in recitative xenoglossy the subject merely utters, as from rote memory, fragments of the language, whereas in responsive xenoglossy he or she converses more or less freely in it. *See also* **Glossolalia**.

Appendix 1
Books Containing Glossaries

The best glossary of parapsychological terms is that com—
piled by Thalbourne [687]. For older terms and those dealing
with primarily qualitative and field studies, Shepard [685, 686]
may be consulted. For more modern terms dealing with ex—
perimental parapsychology, the glossary appearing in each issue
of the *Journal of Parapsychology* [329] is a useful source.
For those who do not have access to the above tools or who
wish to supplement them, the books in Part I that contain glos—
saries are listed below in numerical order according to their
book numbers. For orientation as to contents, recency, and
length, the title, the original date of publication in paren—
theses, and the inclusive pagination of the glossary are
provided. The book numbers and page numbers are in boldface
type.

344 (1988) *Animal Magnetism, Early Hypnotism, and Psychical
Research, 1766–1925: An Annotated Bibliography.* **xxiii-
xxvii**

345 (1979) *Enigma: Psychology, the Paranormal and Self—
Transformation.* **167-170**

346 (1975) *Trance, Art, and Creativity.* **Inside front cover**

355 (1980) *The Psychology of Transcendence.* **341-350**

356 (1975) *States of Mind: ESP and Altered States of Con—
sciousness.* **173-176**

366 (1986) *The Logic of Intuitive Decision Making: A
Research—Based Approach for Top Management.* **147-148**

371 (1982) *Psychic Criminology: An Operations Manual for
Using Psychics in Criminal Investigations.* **97-99**

390 (1986) *The Golden Chalice: A Collection of Writings by the
Famous Soviet Parapsychologist and Healer.* **188-192**

394 (1988) *We Don't Die: George Anderson's Conversations with the Other Side.* **274-277**

413 (1980) *ESP: Your Psychic Powers and How to Test Them.* **53-54**

417 (1975) *ESP.* **58-59**

422 (1973) *ESP: The Search Beyond the Senses.* **175-178**

429 (1975) *On ESP.* **82-83**

434 (1977) *Second Sight: People Who Read the Future.* **151**

435 (1975) *PK: Mind Over Matter.* **163-167**

445 (1978) *The World of the Unknown: All About Ghosts.* **30-31**

453 (1979) *Is Your Child Psychic? A Guide for Creative Parents and Teachers.* **185-187**

454 (1988) *Parapsychology for Parents: A Bibliographic Guide.* **27-35** [same as glossary in this book]

462 (1988) *Enhancing Human Performance: Issues, Theories, and Techniques.* **252-261**

480 (1980) *The Truth About Uri Geller.* **233-234**

487 (1986) *Parapsychology: The Science of Psiology.* **261-273**

491 (1987) *Parapsychology for Teachers and Students: A Bibliographic Guide, 1987.* **27-35** [same as glossary in this book]

507 *New Directions in Parapsychology.* **xiv-xxvi**

508 (1978) *Psi Search.* **153-155**

512 (1978) *The Signet Handbook of Parapsychology.* **509-512**

513 (1977) *Inner Spaces: Parapsychological Explorations of the Mind.* **133-139**

516 (1982) *Psychical Research: A Guide to Its History, Principles and Practices.* **387-399**

526 (1974) *Psychic Exploration: A Challenge for Science.* **687-696**

527 (1975) *Research in Parapsychology 1974.* **229-231**

528 (1976) *Research in Parapsychology 1975.* **243-245**

529 (1977) *Research In Parapsychology 1976.* **263-265**

533 (1978) *Research in Parapsychology 1977.* **251-253**

534 (1979) *Research in Parapsychology 1978.* **185-186**

538 (1973) *Research in Parapsychology 1972.* **217-219**

539 (1974) *Research in Parapsychology 1973.* **219-221**

558 (1980) *The Elusive Science: Origins of Experimental Psychical Research.* **311**

563 (1982) *Chinese Ghosts and ESP: A Study of Paranormal Beliefs and Experiences.* **286-288**

566 (1977) *Extrasensory Ecology: Parapsychology and Anthropology.* **397-400**

579 (1987) *Channeling: Investigations on Receiving Information from Paranormal Sources.* **344-350**

580 (1988) *The Psychic Sourcebook: How to Choose and Use a Psychic.* **379-392**

581 (1979) *Mindsplit: The Psychology of Multiple Personality and the Dissociated Self.* **173-179**

650 (1980) *The Paranormal and the Normal: A Historical, Philosophical and Theoretical Perspective.* **220-228**

658 (1984) *The Family Unconscious: "An Invisible Bond."* **225-226**

630 (1982) *Psychokinesis: A Study of Paranormal Forces Through the Ages.* **234-238**

662 (1978 *Chicorel Index to Parapsychology and Occult Books.* **9-12**

665 (1980) *A Geo—Bibliography of Anomalies: Primary Access to Observations of UFOS, Ghosts, and Others Mysterious Phenomena, 1980.* xxv-xxxvii

667 (1986) *The Occult in the Western World: An Annotated Bibliography.* **191-200**

679 (1987) *Parapsychology: A Reading and Buying Guide to the Best Books in Print.* **90-97**   [same as glossary in this book]

680 (1989) *Parapsychology in the Soviet Union, Eastern Europe, and China: A Compendium of Information.*

688 (1977) *Handbook of Parapsychology.* **921-936**

696 (1974) *Born Twice: Total Recall of a Seventeenth—Century Life.* **185-189**

697 (1975) *Cases of the Reincarnation Type, Volume I. Ten Cases in India.* **361-367**

698 (1977) *Cases of the Reincarnation Type, Volume II. Ten Cases in Sri Lanka.* **363-367**

699 (1980) *Cases of the Reincarnation Type, Volume III. Twelve Cases in Lebanon and Turkey.* **375-377**

712 (1981) *The Way of Splendor: Jewish Mysticism and Modern Psychology.* **232-234**

727 (1981) *The Mask of Time: The Mystery Factor in Times—lips, Precognition and Hindsight.* **293-294**

729 (1975) *Apparitions.* **211-212**

737 (1983) *The Psychology of Déjà Vu: Have I Been Here Before?* **248-255**

751 (1978) *The Mediumship of the Tape Recorder.* **156-159**

761 (1971) *Life Without Death? On Parapsychology, Mysticism, and the Question of Survival.* **327-329**

773 (1975) *Life After Death?* **152-153**

797 (1985) *Spiritualist Healers in Mexico: Successes and*

Appendix 2
Books Containing Illustrations

Listed below in numerical order by Book Number are those titles in Section I that had 10 or more illustrations. In order to give the user some idea of the subject matter of the book the full title has also been listed followed by the number of illustrations in the book in boldface type.

347 *Beyond Biofeedback* **13**

348 *Realms of the Human Unconscious: Observations From LSD Research* **49**

353 *The Roots of Consciousness: Psychic Liberation Through History, Science and Experience* **287**

355 *The Psychology of Transcendence* **38**

357 *The Mind Possessed: A Physiology of Possession, Mysticism and Faith Healing* **31**

362 *Psychic Animals: A Fascinating Investigation of Paranormal Behavior* **14**

365 *Psychic Pets: The Secret World of Animals* **22**

368 *The Divining Hand: The 500–Year–Old Mystery of Dowsing* **206**

373 *Proceedings of a Symposium on Applications of Anomalous Phenomena* **49**

376 *The Secret Vaults of Time: Psychic Archaeology and the Quest for Man's Beginnings* **58**

383 *Double Vision* **60**

384 *Called to Heal: Releasing the Transforming Power of God* **27**

Appendix 3
Abbreviations

General Terms

ed.	edition	p. or pp.	page or pages
ed., eds.	editor, editors	Pt.	part
enl.	enlarged	rev.	revised
n.d.	no date	*Sources*	*Parapsychology:*
n.s.	new series		*Sources of*
no.	number		*Information*
		suppl.	supplement

Parapsychological Organizations

ASPR	American Society for Psychical Research
FRNM	Foundation for Research on the Nature of Man
PA	Parapsychological Association
PF	Parapsychology Foundation
PRF	Psychical Research Foundation
SPR	Society for Psychical Research

Parapsychological Periodicals

AIPRB	Australian Institute for Parapsychological Research Bulletin
AP	Applied Psi
ASPRN	ASPR Newsletter
ChrP	Christian Parapsychologist
EJP	European Journal of Parapsychology
IJParaphys	International Journal of Paraphysics
JASPR	Journal of the American Society for Psychical Research
JIndPsychol	Journal of Indian Psychology
JNDS	Journal of Near Death Studies
JP	Journal of Parapsychology
JRPP	Journal of Research in Psi Phenomena
JRPR	Journal of Religion and Psychical Research
JSPR	Journal of the Society for Psychical Research

609

P   Psychic
PAI   Parapsychology Abstracts International
PJSA   Parapsychological Journal of South Africa
PN   Parapsychology News
PR   Parapsychology Review
PS   Psychoenergetic Systems; also Psychoenergetics
PsiR   Psi Research
T   Theta
ZS   Zetetic Scholar

Nonparapsychological Periodicals

ABBookman   A B Bookman
Alpha   Alpha
AmAnthropol   American Anthropologist
AmDows   American Dowser
America   America
AmerRecGd   American Record Guide
AmEthnol   American Ethnologist
AmHistRev   American Historical Review
AmJPhysAnthropol   American Journal of Physical
                         Anthropology
AmJPsychiat   American Journal of Psychiatry
AmJPsychol   American Journal of Psychology
AmLit   American Literature
AmRefBksAnn   American Reference Books Annual
AmSpectator   American Spectator
AmTheos   American Theosophist
Analog   Analog
AngTheolRev   Anglican Theological Review
AntRev   Antioch Review
Archaeus   Archaeus
AREJ   ARE Journal
Artifex   Artifex
ASBYP   Appraisal: Science Books for Young People
AtlMon   Atlantic Monthly
AustJPhil   Australian Journal of Philosophy

BablBookworm   Babbling Bookworm
BCM   Book Collector's Market
BestSell   Best Sellers
BkRept   Book Report
BkWorld   Book World
Bks&Bkmn   Books and Bookmen
BksinCan   Books in Canada
Booklist   Booklist

Bookman	Bookman
Brain/MindBul	Brain/Mind Bulletin
BritBkNews	British Book News
BritJPsychol	British Journal of Psychology
CatholLibrWorld	Catholic Library World
CAY	Come—All—Ye
CenChildBksBul	Center for Children's Books. Bulletin
Choice	Choice
ChristCent	Christian Century
ChristSciMonit	Christian Science Monitor
ChristToday	Christianity Today
ChronHighEd	Chronicle of Higher Education
Churchman	Churchman
Coll&ResLibr	College & Research Libraries
ComGrd	Common Ground
Commonweal	Commonweal
ContempPsychol	Contemporary Psychology
ContempRev	Contemporary Review
ContempSociol	Contemporary Sociology
CrossCurr	Cross Currents
CurrRev	Curriculum Review
Dial	Dial
Discover	Discover
EarthSci	Earth Science
Economist	Economist
Encounter	Encounter
EngJ	English Journal
Esquire	Esquire
Essence	Essence
Explorer	Explorer
ExplorKnowl	Explorations in Knowledge
ExposT	Expository Times
FantRev	Fantasy Review
Fate	Fate
FronSci	Frontiers of Science
Futures (Engl)	Futures (England)
Futurist	Futurist
GiftedChildQ	Gifted Child Quarterly
Gregorianum	Gregorianum
GuardianWkly	Guardian Weekly
HastingCtrRep	Hastings Center Report

HeythropJ   Heythrop Journal
HiLo   High/Low Report
HistRevsNBks   History: Reviews of New Books
HornBk   Horn Book Magazine
HumBehav   Human Behavior
HumEv   Human Events
Humanist   Humanist

IdealisticStud   Idealistic Studies
ImagCogPers   Imagination, Cognition, and Personality
InstNoeticSciNewsl   Institute of Noetic Sciences Newsletter
Instructor   Instructor
Interpretation   Interpretation
IntJPsychosom   International Journal of Psychosomatics

JAmAcadRelig   Journal of the American Academy of
                        Religion
JAmFolklore   Journal of American Folklore
JAmHist   Journal of American History
JAnalPsychol   Journal of Analytical Psychology
JApplPsychol   Journal of Applied Psychology
JEcumStud   Journal of Ecumenical Studies
JMentImag   Journal of Mental Imagery
JNerv&MentDis   Journal of Nervous and Mental Disease
JPsychol&Theol   Journal of Psychology and Theology
JrBkshelf   Junior Bookshelf
JRead   Journal of Reading
JRelig   Journal of Religion
JSciExpl   Journal of Scientific Exploration
JSciStudRelig   Journal of the Scientific Study of Religion
JTranspersPsychol   Journal of Transpersonal Psychology

Kirkus   Kirkus
Kliatt   Kliatt: Kliatt Young Adult Paperback Book
                        Guide

Language   Language
LATimesBkRev   Los Angeles Times Book Review
LaughMan   Laughing Man
LibrJ   Library Journal
Light   Light
Listener   Listener
LitDig   Literary Digest
LondRevBks   London Review of Books
LucidLtr   Lucidity Letter
LutheranQuart   Lutheran Quarterly

Man	Man
Mankind	Mankind
Mind	Mind
ModChurchman	Modern Churchman
ModLangRev	Modern Language Review
Ms	Ms.
NatlObs	National Observer
NatlRev	National Review
Nature	Nature
NewAge	New Age
NewEngQ	New England Quarterly
NewHum	New Humanist
NewReal	New Realities
NewRepub	New Republic
NewRevBksRelig	New Review of Books in Religion
NewScientist	New Scientist
NewStatesman	New Statesman
Newsweek	Newsweek
New Yorker	New Yorker
NoeticNews	Noetic News
NoeticSciRev	Noetic Sciences Review
NYRevBks	New York Review of Books
NYTimes	New York Times
NYTimesBkRev	New York Times Book Review
NYTimesSatRev	New York Times Saturday Review
Observer	Observer
Omega	Omega
Outlook	Outlook
Parabola	Parabola
Personnel&GdJ	Personnel & Guidance Journal
PhilosRev	Philosophical Review
Phoenix	Phoenix
PopPhot	Popular Photography
PsychiatQ	Psychiatric Quarterly
Psychiatry	Psychiatry
PsychoanalRev	Psychoanalytic Review
PsycholPerspec	Psychological Perspectives
Psychoenergetics	Psychoenergetics
PsycholRec	Psychological Record
PsycholToday	Psychology Today
PublWkly	Publishers Weekly
Punch	Punch
Q&Q	Quill & Quire

Quadrant	Quadrant
QueensQ	Queens Quarterly
QueensQRev	Queens Quarterly Review
ReadTeacher	Reading Teacher
Ref&ResBkN	Reference and Research Book News
RefBkRev	Reference Book Review
Reflections	Reflections . . . The Wanderer Review of Literature, Culture, the Arts
ReformedRev	Reformed Review
RefServRev	Reference Services Review
ReligEduc	Religious Education
ReligStud	Religious Studies
ReligStudRev	Religious Studies Review
RevBks&Relig	Review of Books in Religion
ReVision	ReVision
RevReligious	Review for Religious
RevRevs	Review of Reviews
RevsAmHist	Reviews in American History
RevsAnthropol	Reviews in Anthropology
RQ	Reference Quarterly
SanFranChron	San Francisco Chronicle
SatEvePost	Saturday Evening Post
SatRev	Saturday Review
SchLibrJ	School Library Journal
Science	Science
SciAm	Scientific American
SciBooks&Films	Science Books and Films
SciDigest	Science Digest
SciFicChron	Science Fiction Chronicle
SciFic&FanBkRev	Science Fiction and Fantasy Book Review
SciFicRev	Science Fiction Review
SciTechBkN	SciTech Book News
ScotJTheol	Scottish Journal of Theology
SecondLook	Second Look
SewaneeRev	Sewanee Review
SkptInq	Skeptical Inquirer
Sky&Telescope	Sky and Telescope
SocForces	Social Forces
Sociology	Sociology
SociolRev	Sociological Review
SocSci&Med	Social Sciences and Medicine
SocSciQ	Social Science Quarterly
Spectator	Spectator
Specula	Specula
SpirFron	Spiritual Frontiers

StLukeJTheol   St. Luke Journal of Theology

Theology   Theology
TheologStud   Theological Studies
TheolToday   Theology Today
Time   Time
TimesEdSuppl   Times Educational Supplement
TimesLitSuppl   Times Literary Supplement
TopNews   Top of the News
Trav/Hol   Travel/Holiday
TribBks   Tribune Books (Chicago)

VaQRev   Virginia Quarterly Review
VentIn   Venture Inward
VictorianStud   Victorian Studies
VillVoice   Village Voice
VillVoiceLitSuppl   Village Voice Literary Supplement
VitalSigns   Vital Signs
VoiceYouthAdv   Voice of Youth Advocates

WestCoastRevBks   West Coast Review of Books
WilsonLibrBul   Wilson Library Bulletin

Zygon   Zygon

Appendix 4
Addresses of Publishers
Not Readily Available

American Society for Psychical Research
5 West 73rd Street
New York, NY 10023

A.R.E. Press
215 67th Street
Virginia Beach, VA 23451

Henry S. Dakin Co.
3220 Sacramento Street
San Francisco, CA 94115

Kaman Tempo
816 State Street
P.O. Drawer QQ
Santa Barbara, CA 93102

McConnell, R.A.
Order through local bookstore from:
"The Distributors"
702 South Michigan
South Bend, IN 46658

For single copies, if you are in a United Parcel Service delivery
area, order from:
University of Pittsburgh
Book Center
Pittsburgh, PA 15260
(Add $1.50 for shipping)

Mind Science Foundation
8301 Broadway, Suite 100
San Antonio, TX 78209

Parapsychology Foundation
228 East 71st Street
New York, NY   10021

Parapsychology Press
Box 6847, College Station
Durham, NC 27708

Parapsychology Sources of Information Center
2 Plane Tree Lane
Dix Hills, NY 11746

Princeton Engineering Anomalies Research Laboratory
C320 Engineering Quadrangle
Princeton University
Princeton, NJ   08540

Sourcebook Project
P.O. Box 107
Glen Arm, MD 21057

Washington Research Center
3101 Washington Street
San Francisco, CA   94115

Appendix 5
Have We Established Psi?[1]

One hundred years ago the American Society for Psychical Research was founded with its purpose being, as rephrased in the frontispiece of the Society's journal, "the investigation of ... all types of phenomena called parapsychological or paranormal."[2] At this centennial anniversary of the ASPR it is appropriate to look back and examine the fruits of this investigation. Parapsychology in the United States has grown dramatically since the ASPR was established as its first official organization, and countless other individuals and organizations have come to share in its quest. The annual convention of the Parapsychological Association, which represents workers in the field collectively, is therefore an appropriate forum for such an ASPR—sponsored examination to be presented.

If the purpose of the Society has been the investigation of phenomena "called parapsychological or paranormal," there is no doubt that this purpose has been achieved. Parapsychological inquiry has marched on, uninterrupted, for the 100 years of the Society's life. Although the inquiry has not enjoyed the financial or other tangible support enjoyed by other scientific fields and has yet to achieve more than token acceptance by the scientific community at large, there never has been a lack of courageous and independent scholars willing to make the sacrifices necessary to carry on the struggle. As long as the mysteries which the founders of the ASPR so daringly chose to confront remain mysteries, there will be at least a handful of such scholars willing to receive the torch from the preceding generation. Parapsychology may be defamed, ridiculed, perhaps even driven underground, but it will never die so long as a genuine curiosity about nature, a curiosity not satisfied by the unsubstantiated proclamations of either scientific orthodoxy or occult metaphysics, lives in the human spirit. Investigation *will* continue and the mysteries *will* be solved; the only questions are

---

[1]This is a revised version of a paper presented at the Twenty—Eighth Annual Convention of the Parapsychological Association, Tufts University, August 12—16, 1985. The paper was presented at a symposium commemorating the 100th anniversary of the American Society for Psychical Research [869], which was founded in 1895 by astronomer Simon Newcomb, philosopher William James and others.

[2]The Society also defined its domain of inquiry as including various "subconscious processes," such as automatic writing, "insofar as they may be related to paranormal processes."

when, and at what sacrifice.

Given that our mysteries remain mysteries, what has para—psychology accomplished in the past 100 years of investigating psychic phenomena? I address this question with some am—bivalence, because one part of me feels we expend far too much energy worrying about where we are and far too little energy worrying about where we are going. On the other hand, if kept in proper perspective I think the exercise is a useful one, primarily because a clear assessment of the present accomplish—ments of the field can be beneficial in helping to chart future directions. It is in this forward—looking spirit that I approach my task.

THE EXISTENCE OF PSI

If we substitute the term *psi* for "all types of phenomena called parapsychological or paranormal," then the Society's stated purpose becomes "the investigation of psi." Elsewhere (Palmer, 1986b), I have critically analyzed our use of the term *psi*, suggesting that it be employed in a "descriptive" sense to refer to our subject matter irrespective of its possible "paranormality," by which I mean the incompatibility of its ul—timate interpretation with conventional scientific theory. Such usage seems downright unnecessary in the present context, since we obviously cannot state whether or not a given event is paranormal until the investigation is underway. Thus, the Society's purpose only makes sense if we interpret the object of investigation as including *potentially* paranormal or *ostensibly* paranormal phenomena, that is, phenomena that *might* be paranormal. This usage corresponds closely to the definition of *psi* which I both proposed and advocated in the other paper (Palmer, 1986b): "A correspondence between the cognitive or psychophysiological activity of an organism and events in its ex—ternal environment that is anomalous with respect to generally accepted limiting principles of nature such as those articulated by C.D. Broad (1953/1969)."

Returning to the assessment of parapsychology's accomplish—ments, we may begin by asking whether the actual occurrence in nature of such correspondences has been established; in other words, have we established the existence of psi? Although my redefinition of *psi* has stripped this question of its heretofore momentous import, I nonetheless insist that the question is not a trivial one. After all, there are those who claim that para—psychology has no subject matter and/or that the correspon—dences in question are not even anomalous, let alone paranor—mal.

Despite such dissension, I think the answer to the question

of whether the existence of psi, as I have defined it, has been
established is clearly "yes." Moreover, in order to justify this af—
firmative answer we need not   appeal to our best controlled
laboratory experiments of recent decades. In fact, I would ven—
ture that in all likelihood the existence of psi had already been
established by the time the ASPR was founded, and in any
event it was surely established a mere one year later through
the monumental *Phantasms of the Living* (Gurney, Myers, &
Podmore, 1886/1970).

The cases in *Phantasms* are anomalous in part because
they have not yet been adequately explained by conventional
scientific theory. As I have noted elsewhere (Palmer, 1986a), ad
hoc explanations are not good enough. For scientists to pass off
such cases with all—too—common glib remarks like "it is ob—
viously just a coincidence" or "the person was obviously
deluded" makes a mockery of the scientific method. If the bur—
den of proof is on the claimant, then that burden must apply
whether the claim is being made on behalf of conventional
theory or paranormal theory. Until such claims are backed up
scientifically, our subject matter remains anomalous and the
mysteries remain mysteries.

An approach far more worthy of the label scientific is the
painstaking research on the *Phantasms* cases conducted by my
former colleague, Sybo Schouten, at the University of Utrecht.
Schouten (1979) coded the 562 usable cases from the *Phantasms*
collection with respect to 32 descriptive characteristics. Among
other things, Schouten was interested in determining whether
certain patterns in the collection could be explained by
reasonable conventional hypotheses. For example, he wondered
whether the preponderance of female percipients could be at—
tributed to females being more prone to report such cases. He
predicted that if this were true, the reporters should more likely
be female than male when the sample is restricted to those
cases where the reporter was not directly involved as percipient
or target person. It turned out that most of the reporters of
such cases were actually males, causing Schouten to reject his
hypothesis. There were several other instances in which predict—
ions based on the hypothesis of reporting artifacts or other con—
ventional hypotheses were refuted, leading Schouten to conclude
that he had been unable to find adequate conventional explana—
tions for these patterns in the data. Schouten (1981, 1982) later
found that many of these patterns also appeared in collections
from Germany and the United States, which suggested that they
could not be attributed to cultural idiosyncrasies.

Because Schouten based his conclusions on the systematic
evaluation of large samples of cases of diverse origins, there is
justification for generalizing his conclusions to similar cases not

included in his samples. Such inferences are not possible (or at least not legitimate) from the isolated case studies or "debunkings" so often cited by conventional theorists;[3] and for that reason, these debunkings have limited value as scientific evidence.

As Schouten takes pains to stress, his research does not prove the paranormal origins of the cases he analyzed. Other tests of his hypotheses may yet confirm conventional explanations of the internal patterns found in these samples or even of the one thing all the cases have in common——*ostensible* paranormality. The point is that no one has yet succeeded in doing this for any representative samples of spontaneous psi experiences. Until this is achieved, such experiences remain legitimate anomalies.

THE ESTABLISHMENT OF PARANORMALITY

The other common use of the term *psi* is to designate the paranormal principle or process hypothesized to account for the anomalous correspondences discussed in the previous section. In my opinion, the existence of psi in this sense has *not* been established, although not for the reasons customarily cited in defense of this conclusion. The customary argument states in its general form that psi has not been established because the research designed to establish it has failed to completely or adequately eliminate competing "normal" processes. In my judgment, this argument is a red herring. The real problem is that most of this research is incapable of demonstrating paranormality by the very logic of its conception. To bring home the point, even if *none* of the flaws or alleged flaws cited in Akers' (1984) thorough critical review of parapsychological research in the latest volume of *Advances in Parapsychological Research* occurred, that research would still not have established paranormality, or even have come close to establishing it.

The problem has its origin in the observation, which I believe was first made explicit by E.G. Boring (1966) and has since been echoed by many others (e.g., Schouten, 1979; Stanford, 1974), that *psi* (as a paranormal principle) is negatively defined.[4] This is certainly implicit in our research strategy,

---

[3]For reasons discussed elsewhere (Palmer, 1986b), I have chosen in most cases to use "conventional theorist" as a substitute for "skeptic."

[4]It is perhaps worth stressing that only in the explanatory mode is psi defined negatively. As a descriptive construct, psi is positively defined in the sense that the types of anomalies to which it refers can be clearly demarcated. For example, psi anomalies are restricted to certain kinds of cor-

which is to demonstrate psi not by confirming it but by eliminating its competition——a purely negative approach. It is this approach *per se* that I find bankrupt. To the extent that we are successful in eliminating known conventional explanations of our anomalies and nothing more, we have in fact maximized our *ignorance* about what they *are*. At the risk of sounding trite, ignorance is ignorance. We cannot say, on the one hand, that we know nothing about the causal mechanism behind our anomalies and, on the other hand, say that the causal mechanism is paranormal.

To put this another way, we can never be certain that we have conceived of all possible conventional explanations, let alone controlled for them, at a given point in time. All of us who do research in parapsychology know, for example, that each of two reviewers of a manuscript is likely to miss a "flaw" discovered by the other one. We also know that conventional explanations that are not discovered at the time a paper is published may emerge at a later date. Our experience with the research of S.G. Soal is a particularly poignant reminder of this fact. In other words, it is folly to think that because no adequate conventional explanation of an anomaly occurs to us at a particular point in time, no such explanation exists. This is especially true when we must depend on research reports in journals for our description of the conditions under which the anomaly occurred. An author may quite innocently leave out details of procedure which seem trivial at the time but that prove to be important in the final analysis. The strategy of trying to demonstrate paranormality by the process of elimination is based on the critical assumption that the population of competing explanations can be accurately defined. That assumption is untenable, whatever the apparent quality of the methodology.

It is significant that we are beginning to see this line of argument in the critiques of conventional theorists. Because of its simplicity, degree of automated control, and consistency of straightforward significant results, these theorists have found the research program of Helmut Schmidt to be a relatively difficult one with which to deal. Among those who have taken up the challenge is Ray Hyman (1981). Although Hyman suggests several conventional explanations of Schmidt's results (albeit implicitly, through the specification of what he considers to be procedural flaws), he also introduces another line of argument which implies that any definitive interpretation of Schmidt's results is premature until the research program is further developed and replicated. This line of argument, which I find

---

respondences between mental events and target events that mimic communication processes.

similar although not identical to the one I just outlined, is intro—duced independently of the traditional one, which suggests that it would be retained even if the "flaws" had not existed.

The implication of this, if I am reading Hyman correctly, is well worth pondering: Hyman would not have accepted Schmidt's results as evidence of paranormality even if the experiments had *not* been flawed by his (Hyman's) criteria. The moral is simple: We cannot get conventional scientists to accept the existence of psi (i.e., paranormality) simply by providing them with methodologically flawed experiments. Ultimately, they will retreat to the same *logical* argument I offered in the preceding paragraphs——and they will be right. The tragedy, for which some of our critics must share part of the responsibility, is that we have operated for so long under the myth that conventional scientists ever would or even could accept "flawless" demonstration experiments as proof of paranormality. Such acceptance would not follow even if it were agreed that the experiments were adequately replicable, because the replicability of an effect implies nothing about its cause. (However, increased replicability would doubtlessly make psi anomalies more *interesting* to conventional scientists.)

A POSITIVE RESEARCH STRATEGY

If we are ever to establish *paranormality*, it is necessary that we empirically confirm a theory or model built around a hypothetical paranormal process or principle.[5] The importance of theorizing in this connection has previously been stressed by Stanford (1974). It is not sufficient that a theory simply account for psi (in the descriptive sense) or even for the "effects," such as the sheep—goat effect, which are built around it. To be successful, a theory must predict *new* relationships, preferably but not necessarily involving psi, and these relationships must achieve enough independent replicability to be credible. Finally, they must be of such a nature that they would not be predicted by competing conventional theories. This requires that conventional theories be developed to the point that what they

---

[5]This requirement is isomorphic with the requirement I have proposed elsewhere (Palmer, 1986a) with respect to conventional theory; that is, explicit or implicit claims that psi anomalies are really conventional in nature cannot be sustained in the absence of empirical confirmation of specific conventional theories or models. If we don't know the explanation, we also don't know the nature of the explanation, that is, whether it is conventional or paranormal. Failure to abide by this principle is a major logical flaw in the approach of many of our critics to the psi controversy.

predict and do not predict is clear, lest they be used as "universal solvents" in the same way as "super—psi" has been used against the survival hypothesis (Palmer, 1975). Paranormal theorists must assure that conventional theorists fulfill their scientific responsibilities in this connection, just as conventional theorists have rightly insisted that paranormal hypotheses be falsifiable.

Implementation of this more theoretical approach requires shifting from a negative to a positive research strategy. This implies, in turn, a shift in priorities. Using the negative approach——which I submit predominates even in our so—called "process—oriented" research——the priority is elimination of "normal" processes (e.g., controls against sensory cues). In a positive approach, the priorities would be, first, to get the anomalies to occur reliably and, second, to explore their correlates or test hypotheses about their nature.

This shift of priorities has its most profound consequences in cases where the priorities conflict. Such a conflict would occur, for example, if one accepts the thesis suggested by Batcheldor (1984) and Reichbart (1978) that (genuine) paranormality is facilitated by lapse of controls. A researcher who both accepts this assumption and wishes to adopt a positive research approach is literally compelled to build lapses of control into his or her research design. Since normal processes are facilitated in such a design, the strategy would be to incorporate tests of the conventional hypotheses that would either confirm or refute them. In other words, the strategy would be to *detect* normal processes, not *eliminate* them. If predictions based on the conventional hypotheses failed and anomalous correspondences nonetheless were detected, correlates of these anomalies might well provide insights into the paranormal process involved that could contribute to paranormal theory, which might in turn be applied to allow the anomalies to occur in more tightly controlled contexts. Ideally, paranormal hypotheses would be tested directly in such experiments, and their confirmation in combination with the rejection of the conventional hypotheses would lend direct support to paranormal theory.

The idea of introducing lapses of control for any reason is likely to leave any red—blooded conservative parapsychologist with an uneasy feeling. It might help to note that research which allows sensory leakage, for example, is not considered flawed because of the sensory leakage per se, but because sensory leakage occurred in conjunction with the implicit or explicit claim that it had been eliminated. Whereas such a claim is required by the negative approach, it is not required by the positive approach. The latter approach very explicitly operates from the premise that the subject matter of parapsychology is the anomalous correspondences themselves and actively discourages any preconceptions about their paranormality or nonparanor—mality. As noted above, the best evidence for paranormality comes not by eliminating known normal processes but by con—firming hypotheses uniquely derived from paranormal theories or models. This is the kind of evidence, and the *only* kind of evidence, that can effectively silence a retort such as "he could have done it by sensory cues but you weren't smart enough to figure out how." Even though this "skeptical" retort is techni—cally correct (as it always will be, whatever the controls), it loses whatever scientific force it may have had whenever *positive* evidence exists for an alternative. Thus, the positive approach would be effective even if the strong form of Batcheldor's thesis were to prevail such that paranormality would occur reliably only in close conjunction with normal processes.

However, a positive approach does not mean "anything goes." The researcher will inevitably want to maintain control against some specific normal processes simply to reduce com—plexity. The rules of evidence remain the same as those in any experimental research, and proper principles of experimental design and procedure must be utilized so that the confirmation or refutation of hypotheses can withstand critical scrutiny. Whereas in a negative approach lapse of control is usually an oversight, in a positive approach it is a conscious decision based upon logical justification. In the absence of such justification, controls are maintained as always.

EXPERIMENTAL ACCOMPLISHMENTS

If psi (the anomaly) has already been established by our spontaneous cases and psi (the paranormal principle) has yet to be established at all, the question arises as to what, if anything, has been accomplished by the extensive amount of experimental research that has dominated parapsychology since the 1930s. I think significant accomplishments can be cited, but it is neces—sary to define them with some care so as not to conflict with the conclusions we have drawn up to now.

First of all, it is necessary to consider some respects in which our spontaneous cases and our experiments are isomor—phic. The key point to be made here is that both represent human testimonials to the witnessing of anomalous events. A lawyer observes an apparition and writes a description in his diary. A parapsychologist witnesses nonrandom output of an REG and writes it up in a research report. Each may or may not have other witnesses to the event who can share in the tes—timony. (In the case of the lawyer, he may share his diary and recollections with a researcher who then also shares in the testimony.) Finally, neither the lawyer nor the para—psychologist can guarantee that anyone who puts themselves in similar circumstances would observe a similar anomaly. Both must recognize that they are asking others to take their word for something out of the ordinary and, thus, must take steps to maximize their personal credibility.

Let me stress that my purpose is not to minimize the im—portance of the real differences that obviously exist between ex—periments and spontaneous cases. The point I am trying to make is that the fundamental role in the scientific process of most of our experiments has been the same as that of our spon—taneous case collections, namely to provide a documented source of relevant anomalies. Collectively, they provide the subject mat—ter, the raw material, of our field. They put before us what we need to explain. They do *not* explain or verify anything in them—selves.

With one qualification, this point also applies to those ex—periments and spontaneous case investigations that we call "process—oriented." In most cases, this research provides evidence for empirical generalizations consisting of relationships between psi scores or reports of psi and some psychological, physical, or demographic variable. Although such relationships serve as an important first step on the road to true explana—tion, in essence they are anomalies themselves that, along with psi, define what it is we need to explain. Only insofar as an experiment is designed to test a theory or model could it be removed from this category.

We are finally now ready to list the accomplishments of the experimental work in parapsychology. To a large degree, they coincide with the methodological advantages of laboratory research over field research; and it is now that the differences between the two to which I alluded earlier will be described.

First Class of Accomplishments

The first accomplishment of our experimental research that I wish to cite may, at first glance, seem too trivial to mention, but in fact it is quite important. I am referring to the fact that we have demonstrated that psi anomalies, whatever their cause, occur in laboratories. This fact is important because it reveals that we can apply experimental methods to the task of elucidat—ing the causes of psi that far exceed what we can do in the "real world." For example, it is a tremendous advantage that PK effects can be obtained by using a device as technologically elegant as a random event generator or that ESP can manifest in simple guessing tasks. Certainly, psi is not as tractable in the laboratory as we would like it to be, but the fact that it none—theless occurs there frequently and with at least rudimentary lawfulness is the single most important reason we have to believe that a scientifically adequate explanation will ultimately be found.

Second Class of Accomplishments

The second class of accomplishments, which is really an out—growth of the first, is that our experiments have allowed us to obtain better assessments of the likelihood and plausibility of various conventional explanations of the anomalies. As a general rule, the likelihood and plausibility of these hypotheses as ex—planations of the psi that occurs inside the laboratory is much less than that of the psi that occurs outside the laboratory. This is especially true in the case of the chance—coincidence hypothesis as applied to ESP. It is impossible to assess with any precision the likelihood of a spontaneous ESP experience being coincidental. Experimental procedures, on the other hand, provide means of establishing the "chance baseline" with great precision and enable us to determine exact p—values as to the likelihood of a particular outcome being attributable to random fluctuations from the norm. Despite complications related to mul—tiple statistical analyses and the like, there are numerous ex—amples of laboratory research demonstrating outcomes that clearly are of an extrachance character. The kinds of controls possible in experiments have also substantially reduced (whether or not they have entirely eliminated) mechanisms of gaining sen—sory access to information in ESP contexts or physical manipula—tion of events in PK contexts. Innocent errors in recording and reporting of the anomalous events have also been minimized. This is because, in contrast to most spontaneous cases, the ob—server in an experiment usually knows in advance where and when the anomalous event will occur (if it does occur) because he or she has set it up to allow this possibility. Again, none of this proves the paranormality of the anomalies, but it does make possible a better evaluation of some of the alternative pos—sibilities.

Third Class of Accomplishments

The third class of accomplishments of experimental research is the discovery of correlates of psi (or, more specifically of ESP) which show a modicum of statistical replicability. Among

individual difference variables, belief (Palmer, 1971) and extraver—
sion (Sargent, 1981) seem to provide the best case. Research
with the Defense Mechanism Test also seems promising
(Johnson & Haraldsson, 1984). Among manipulated variables,
there is good support for higher scoring rates under hypnosis
(Schechter, 1984), although the mechanism remains obscure and
could be distinct from hypnosis *per se* (Stanford, 1986). Other
altered states of consciousness induction procedures, such as the
Ganzfeld, also appear to be psi—facilitory (Honorton, 1977), but
the case here is weaker because only a handful of experiments
have provided any kind of baseline.

Perhaps the strongest, but at the same time, the most
provocative correlate of psi, is the so—called experimenter effect.
Although only a few studies have looked at the experimenter ef—
fect systematically (e.g., Taddonio, 1976; West & Fisk, 1953),
there seems to be widespread agreement that psi anomalies are
consistently more prevalent in some laboratories than in others.
Explanatory frameworks that particular constituencies find ap—
pealing include error or fraud by the experimenter, social
psychological factors, differences in subject populations, and
paranormal mediation by the experimenter. One of the more
sobering implications of the experimenter effect, whatever its ex—
planation, is that it may interact with the seemingly reliable cor—
relational patterns discussed in the preceding paragraph. A
recent series of experiments by Thalbourne exploring the
relationship of belief and extraversion to ESP (e.g., Thalbourne
& Jungkuntz, 1983) reinforces this possibility.

Fourth Class of Accomplishments

The fourth class of accomplishments consists of the growing
number of attempts to formulate coherent theories and models
that yield empirically testable predictions that are then actually
tested. I think the most successful of these attempts, both with
respect to the way it was done and the results it produced, is
Rex Stanford's (1977) psi—mediated instrumental response
(PMIR) model and research program. The model drew upon ex—
perimental and anecdotal data in parapsychology as well as
widely accepted principles from psychology, it was presented in
the form of clearly stated propositions, explicit predictions were
clearly derived from these propositions, and these predictions
were tested (and frequently confirmed) using appropriate ex—
perimental procedures.

The PMIR model dealt primarily with the "psychology" of
psi rather than with its "physics." Thus, it dealt only indirectly
with paranormality. More recent models, especially those in—

spired by quantum mechanics (e.g., the so—called observational
theories), promise to get closer to the heart of the matter. Al—
though some of these models have empirical implications and
have inspired research (e.g., see Houtkooper, 1983), they have
not yet been developed to the point where the intellectual gaps
are bridged as cleanly as in the PMIR model. Also, perhaps be—
cause of my training as a psychologist, I often find these models
mathematically elegant but conceptually rather barren. One ex—
ception is Harris Walker's (1975) model, especially its proposi—
tions concerning quantum tunnelling in the brain. What I find
appealing about this model, whatever its validity, is its attempt
to integrate physics with principles of psychology and physi—
ology in explaining psi. If this kind of theoretical approach can
be married to the structural elegance of Stanford's PMIR model
and associated research program, and if the marriage can yield
offspring in the form of novel predictions empirically confirmed,
we will be well on the road to establishing a strong scientific
case for the paranormality of psi——and fulfilling the legacy of
those who founded the ASPR 100 years ago.

REFERENCES

Akers, C. (1984). Methodological criticisms of parapsychology. In
S. Krippner (Ed.), *Advances in Parapsychological Research 4*
(pp. 112—164). Jefferson, NC: McFarland.

Batcheldor, K.J. (1984). Contributions to the theory of PK in—
duction from sitter—group work. *Journal of the American
Society for Psychical Research, 78,* 105—122.

Boring, E.G. (1966). Introduction: Paranormal phenomena. In
C.E.M. Hansel, *ESP: A Scientific Evaluation* (pp. xiii—xxi).
New York: Scribner's.

Broad, C.D. (1969). *Religion, Philosophy, and Psychical
Research.* New York: Humanities Press. (Original work pub—
lished 1953)

Gurney, E., Myers, F.W.H., & Podmore, F. (1970). *Phantasms
of the Living.* Gainesville, FL: Scholars' Facsimiles. (Original
work published 1886)

Honorton, C. (1977). Psi and internal attention states. In B.B.
Wolman (Ed.), *Handbook of Parapsychology* (pp. 435—472).
New York: Van Nostrand Reinhold.

Houtkooper, J.M. (1983). *Observational Theory: A Research Programme for Paranormal Phenomena.* Lisse, Netherlands: Swets & Zeitlinger.

Hyman, R. (1981). Further comments on Schmidt's PK experiments. *Skeptical Inquirer,* 5(3), 34–40.

Johnson, M., & Haraldsson, E. (1984). The Defense Mechanism Test as a predictor of ESP scores. *Journal of Parapsychology, 48,* 185–200.

Palmer, J. (1971). Scoring in ESP tests as a function of belief in ESP. Part 1. The sheep–goat effect. *Journal of the American Society for Psychical Research, 65,* 373–408.

Palmer, J. (1975). Some recent trends in survival research. *Parapsychology Review,* 6(3), 15–17.

Palmer, J. (1986a). Progressive skepticism: A critical approach to the psi controversy. *Journal of Parapsychology, 50,* 29–42.

Palmer, J. (1986b). Terminological poverty in parapsychology: Two examples [Summary]. In D.H. Weiner & D.J. Radin (Eds.), *Research in Parapsychology 1985* (pp. 138–141). Metuchen, NJ: Scarecrow Press.

Reichbart, R. (1978). Magic and psi: Some speculations on their relationship. *Journal of the American Society for Psychical Research, 72,* 153–175.

Sargent, C.L. (1981). Extraversion and performance in "extrasensory" perception tasks. *Personality and Individual Differences, 2,* 137–143.

Schechter, E.I. (1984). Hypnotic induction vs. control conditions: Illustrating an approach to the evaluation of replicability in parapsychological data. *Journal of the American Society for Psychical Research, 78,* 1–27.

Schouten, S.A. (1979). Analysis of spontaneous cases as reported in "Phantasms of the Living." *European Journal of Parapsychology, 2,* 408–454.

Schouten, S.A. (1981). Analysing spontaneous cases: A replication based on the Sannwald collection. *European Journal of Parapsychology, 4,* 9–48.

Schouten, S.A. (1982). Analysing spontaneous cases: A replica—
tion based on the Rhine collection. *European Journal of
Parapsychology, 4*, 113—158.

Stanford, R.G. (1974). Concept and psi. In W.G., Roll, R.L.
Morris & J.D. Morris (Eds.), *Research in Parapsychology
1973* (pp. 137—162). Metuchen, NJ: Scarecrow Press.

Stanford, R.G. (1977). Conceptual frameworks of contemporary
psi research. In B.B. Wolman (Ed.), *Handbook of Parap—
sychology* (pp. 823—858). New York: Van Nostrand Reinhold.

Stanford, R.G. (1986). Altered internal states and para—
psychological research: Retrospect and prospect [Summary].
In D.H. Weiner & D.J. Radin (Eds.), *Research in Parap—
sychology 1985* (pp. 128—131). Metuchen, NJ: Scarecrow
Press.

Taddonio, J.L. (1976). The relationship of experimenter expec—
tancy to performance on ESP tasks. *Journal of Parapsychol—
ogy, 40*, 107—114.

Thalbourne, M.A., & Jungkuntz, J.H. (1983). Extraverted sheep
vs. introverted goats: Experiments VII and VIII. *Journal of
Parapsychology, 47*, 49—51. (Abstract)

Walker, E.H. (1975). Foundations of paraphysical and para—
psychological phenomena. In L. Oteri (Ed.), *Quantum
Physics and Parapsychology* (pp. 1—53). New York: Para—
psychology Foundation.

West, D.J., & Fisk, G.W. (1953). A dual ESP experiment with
clock cards. *Journal of the Society for Psychical Research,
37*, 185—197.

*Institute for Parapsychology
Box 6847, College Station
Durham, North Carolina 27708*

Index of Names

This index includes authors or compilers, joint authors or compilers (A), editors (E), illustrators (I), translators (T), and writers of introductions, prefaces, forewords, and afterwords (P). When (S) follows a name it means that the person or the person's work is mentioned. Boldface item numbers are used for any names found in entries that have item numbers. Where no item numbers are used, page numbers are given. The item numbers are always listed first in numerical order. Next come page numbers, indicated by a single p (if there is only one page reference), or pp (if there are two or more page references), are listed in numerical order in regular type.

Johnson, George A., p553(A)
Johnson, M.U. *see* Johnson,
  Martin
Johnson, Martin, **868(E)**,
  881(S), pp454(A), 513(S),
  536(A), 628(S), 630(A)
Johnson, Rex V., p572(A)
Jolly, W.P., **403(A)**
Jones, C.B. (Scott), Jr.,
  373(E), **869(S)**, **900(S)**,
  pp488(S), 489(S)
Jones, Funmilayo M.,
  p565(A)
Jones, Glenda, p424(A)
Jones, M., p456(A)
Jones, Nella, **473(S)**
Jones, Warren H., **659(A)**
Jonsson, Olof, **479(S)**
Jordan, P., p456(A)
Jorgenson, Danny Lynn,
  p562(A)
Joseph of Copertino, **606(S)**
Josephson, B.D., **618(E)**,
  pp522(S), 536(A)
Jourdain, E.F., **723a(A)**
Jung, Carl G., **639(S)**,
  647(A)(S), p552(A)
Jungkuntz, J.H., p628(S),
  631(A)
Jureidini, P., p470(A)

Kalish, R.A., p481(A)
Kammann, Richard, **477(A)**
Kampman, Reima, p557(A)
Kamuda, Edward S., **805(P)**
Kanthamani, B.K., p555(A)
Kantor, Debra, **801a(J)**
Kantorovich, M., p467(A)
Kaplan, E., p536(A)
Kaplan, Harold I., **942(E)**,
  943(E)
Kaplan, Pascal M., p562(A)
Karamfelov, I., p467(A)
Kardec, Allan, **584(S)**
Kastenbaum, Robert, **762(E)**
Katz, Richard, **800(A)**
Kautz, William H., **374(A)**,

  856(E), **919(S)**, pp460(A),
  536(A)
Kazhinsky, B.B., p467(A)
Keane, Patrice, **869(S)**
Keene, M. Lamar, **475(A)**
Keezer, Phillip W., **954(S)**
Keil, H.H.J., **404(E)**,
  pp423(A), 578(A)
Keil, Jürgen *see* Keil, H.H.J.
Kelley, True, **429a(I)**
Kelly, Edward F., **350(A)**,
  pp483(A), 510(S), 533(A),
  537(A)
Kelsey, Morton T., **713(A)**,
  939(S)
Kennedy, J.E., p536(A)
Kenny, Michael G., **578(A)**
Kerr, Howard Hastings,
  p555(A)
Kettelkamp, Larry, **438(A)**,
  439(A), **440(A)**, **441(A)**
Khilji, Anjum, p565(A)
Kholodov, Y.A., p467(A)
Kiepenheuer, J., p471(A)
Kies, Cosette N., **667(A)**
King, Cecil H., p424(A)
King, H.S., p456(A)
King, Katie, **460(S)**
Kirk−Duncan, V.G., p553(A)
Kitaev, N., p457(A)
Klein, Aaron E., **442(A)**
Klein, F.F., pp510(S), 533(A)
Kleinman, Arthur, **797(P)**
Klimo, Jon, **579(A)**
Klychnikov, I.V., p468(A)
Klyver, N., p459(A)
Knight, David C., **443(A)**
Knox, M. Crawford, p576(A)
Kobayashi, Ken, p559(A)
Kobler, J., p457(A)
Koestler, Arthur, **477(S)**,
  496(A), **507(P)**, **895(S)**,
  952(S), pp437(A), 537(A)
Kogan, I.M., p467(A)
Kolodny, Lev, p467(A)
Komito, David, **896(S)**
Kozyrev, N.A., p468(A)

424(A), 438(A), 539(A)
Murphy, Michael, **921(S)**
Murray, Diane M., p577(A)
Murray, Gilbert, **874(S)**
Murray, H.A., p458(A)
Musés, Charles, **952(S)**,
  p440(E)
Myers, Arthur, **446(A)**,
  447(A)
Myers, David G., p418(A)
Myers, E.L., p458(A)
Myers, Frederic W.H.,
  354(A), p629(A)
  communicator, **747(S)**
Myers, Susan Anne, p577(A)

Nash, Carroll B., **487(A)**,
  488(A)
Nathanson, J.A., p442(A)
Naumov, E.K., **668(A)**,
  p468(A)
Neff, H. Richard, **388(A)**
Neher, Andrew, **355(A)**
Neihardt, John G., **631(S)**
Nelson, Geoffrey K., p555(A)
Nelson, Peter L., p569(A)
Nelson, Roger D., **544(E)**,
  882(S), pp507(S), 508(S),
  511(S), 539(A), 542(A)
Neppe, Vernon Michael,
  737(A), **860(E)**, pp565(A),
  575(A)
Nester, Marian L., **669(A)**,
  670(A)
Neuberg, Matthew A.,
  p565(A)
Newcomb, Simon, **869(S)**
Nicholson, Shirley, **569(A)**
Nissen, K.D., p556(A)
Nolan, R., p458(A)
Nolen, William, **802(A)**
Noonan, Molly H., p560(A)
Norman, Wayne, **873(S)**
Northage, Isa, **473(S)**
Noyes, R., Jr., **596(S)**

O'Brien, Dennis P., p574(A)

O'Brien, J., p458(A)
O'Keefe, Arthur S.T., **669(A)**,
  670(A)
O'Keeffe, Eleanor, **874(S)**
Olson, Melodie Ann, p566(A)
O'Neil, Paul, p420(A)
Oppenheim, Janet, **560(A)**
O'Regan, Brendan, **923(S)**
Orr, D.B., pp473(A), 474(A)
Osis, Karlis, **396(S)**, **397(S)**,
  408(S), **596(S)**, **706(P)**,
  750(S), **769(A)**, **869(S)**,
  960(S), pp458(A), 553(A)
Ossowiecki, S., **376(S)**
Ostrander, Sheila, **369(A)**,
  p469(A)
O'Sullivan, Michael Anthony,
  p566(A)
Oteri, Laura, **621(E)**
Otto, Herbert A., p438(E)
Owen, A.R.G., **530(A)**,
  859(E), **904(S)**
Owen, Iris M., **628(A)**
Ozaki, Bonnie D., **937(E)**

Pachita, doña, **801(S)**
Padfield, Suzanne, **618(S)**
Palfreman, Jon, **962(S)**
Palladino, Eusapia, **606(S)**,
  633(S), **811(S)**
Palmer, F. Claude, p571(A)
Palmer, John, **485(A)**, **949(S)**,
  959(S), pp469(A), 476(A),
  489(S), 491(S), 494(S),
  507(S), 513(S), 519(S),
  520–521(S), 524(S), 526(S),
  527(S), 540(A), 541(A),
  542(A), 619(S), 628(S),
  630(A)
Panati, Charles, **531(A)**,
  629(E)
Parise, Felicia, **633(S)**, **634(S)**
Parker, Adrian, **356(A)**,
  p560(A)
Pasricha, Satwant K.,
  p562(A)
Patanjali, **779(S)**

Index of Titles

This index includes titles of all books and journals cited (titles of journal articles are not listed). Titles that have been assigned item numbers are listed first, in numerical order and in boldface type. Also included are titles listed in reference lists and in the text for which no item number has been assigned. In the latter case, page numbers are cited. Page numbers are always listed after item numbers. They are in numerical order in regular type and are preceded by an initial p if only one page number is cited or pp if two or more page numbers are cited. Also listed are titles that are subject entries. These are titles of books or journals that are cited in the text. They may or may not be main entries. In any case, if a title is cited in an entry other than one devoted exclusively to that title, the citation to the subsidiary reference is indicated by an (S). An (S) follows each of these, regardless of whether item numbers or page references are cited.

All in the Mind: Reincarnation, Hypnotic Regression, Stigmata, Multiple Personality, and Other Little–Understood Powers of the Mind, **703**
Altered States of Consciousness and Psi: An Historical Survey and Research Prospectus, **350**
Alternate Realities: The Search for the Full Human Being, **352**
Alternate States of Consciousness: Unself, Otherself, and Super–self, **345a**
The Amazing Uri Geller, **624**
American Book Publishing Record, **968**
American Book Publishing Record Cumulative 1876–1949: An American National Bibliography, **660**
American Book Publishing Record Cumulative 1950–1977: An American National Bibliography, **661**
American Health, p420
Amityville Horror, **723(S)**
  criticisms, **473(S)**
Anabiosis: The Journal of Near–Death Studies *see* Journal of Near–Death Studies
Analog Science Fiction/Science Fact, p421
Animal Magnetism, Early Hypnotism, and Psychical Research, 1766–1925: An Annotated Bibliography, **344**
Annual Review of Psychology, **985**, p494
Anomalistic Psychology: A Study of Extraordinary Phenomena of Behavior and Experience, **659**
Anomaly, **899(S)**
Anthropology of Consciousness Newsletter, **852a**
Apparitions, **729**
Appearances of the Dead: A Cultural History of Ghosts, **726a**
The Application of Learning Theory to ESP Performance, **786**
Applied Psi, **856**
Applied Psi Newsletter *see* Applied Psi
Archaeus, **842**
Archaeus Project Newsletter *see* Artifex
Aristocracy of the Dead: New Findings in Postmortem Survival, **748**
Arthur C. Clarke's World of Strange Powers, **515**
Artifex, **843**
The Ashby Guidebook for Study of the Paranormal, **804**
ASPR Newsletter, **853**
ASSAP News, **899(S)**
The Astral Journey, **599**
At the Hour of Death, **769**
Awakening Your Psychic Powers, **814**

The Basic Experiments in Parapsychology, **498**
Behavioral and Brain Sciences, **949**, p518

Index of Subjects

This index consists of two types of subject entries: regular subjects, such as remote viewing and clairvoyance, and proper name subjects, such as named experiments, names of places, and names of organizations. [See also the Name Index for personal names for entries followed by an (S).] If a publication that has an item number is cited, the item number in boldface is used. If there is no item number, page numbers are cited in regular typeface preceded by an initial p or pp. After each entry item numbers are cited first, then page numbers.